SELECTED TOPICS IN ALGEBRAIC GEOMETRY

BY

VIRGIL SNYDER, AMOS H. BLACK, ARTHUR B. COBLE,
LEAMAN A. DYE, ARNOLD EMCH, SOLOMON LEFSCHETZ,
F. R. SHARPE, AND CHARLES H. SISAM

TWO VOLUMES IN ONE

CHELSEA PUBLISHING COMPANY
BRONX, NEW YORK

SECOND EDITION

THE PRESENT WORK IS A REPRINT IN ONE VOLUME, WITH THE CORRECTION OF ERRATA, OF TWO WORKS PUBLISHED UNDER THE TITLE, SELECTED TOPICS IN ALGEBRAIC GEOMETRY: BULLETIN OF THE NATIONAL RESEARCH COUNCIL, NUMBER 63 (WASHINGTON, 1928); BULLETIN OF THE NATIONAL RESEARCH COUNCIL, NUMBER 96 (WASHINGTON, 1934). THE ORIGINAL LIST OF BOOKS HAS BEEN SUPPLEMENTED BY INCLUSION OF LATER EDITIONS AND PRINTINGS, AND COLLECTED WORKS. THIS EDITION IS PUBLISHED AT NEW YORK, N. Y., AND IS PRINTED ON ACID-FREE PAPER FOR GREATER DURABILITY

LIBRARY OF CONGRESS CATALOG CARD NUMBER 78-113149

INTERNATIONAL STANDARD BOOK NUMBER 0-8284-0189-6

PRINTED IN THE UNITED STATES OF AMERICA

SELECTED TOPICS IN ALGEBRAIC GEOMETRY

Report of the Committee on Rational Transformations[1]

VIRGIL SNYDER, ARTHUR B. COBLE, ARNOLD EMCH,
SOLOMON LEFSCHETZ, F. R. SHARPE,
CHARLES H. SISAM

---

[1] The members of this committee of the Division of Physical Sciences are: Virgil Snyder, *Chairman,* Professor of Mathematics, Cornell University, Ithaca, New York; Arthur B. Coble, Johns Hopkins University, Baltimore, Maryland; Arnold Emch, University of Illinois, Urbana, Illinois; Solomon Lefschetz, Princeton University, Princeton, New Jersey; F. R. Sharpe, Cornell University, Ithaca, New York; C. H. Sisam, Colorado College, Colorado Springs, Colorado.

PREFACE TO VOLUME I

The purpose of this report is to give a brief, connected, and comprehensive survey of the literature on certain topics in algebraic geometry. It is designed primarily as an aid to investigators in this field, but it is hoped that it may be of service to a wider circle.

As the central theme of the report, the committee has chosen the Rational Transformation. It has been understood, however, that the authors were to feel free to depart from the theme whenever they deemed it advisable, either because a certain topic had been insufficiently covered in existing similar reports, or because they wished to present a rapid résumé of some question vital for other parts of the report.

Available reports of encyclopedic nature cover Rational Transformations only incidentally and therefore quite inadequately at best. This fact has had much to do with the choice of general subject by the committee. Although the committee has endeavored to cover the subject with a fair degree of completeness, the particular topics have been selected with due reference to the fields which the authors felt themselves most competent to discuss. It is the hope of the committee that other phases of the subject may be covered in the future.

The materials for the report have been collected, with few exceptions, directly from their original sources, then checked with current reviews. In this way, mistakes and omissions have been reduced to a minimum.

As to the arrangement of the contents of the report itself, each chapter is to be regarded as a unit. References from one chapter to another are made, but any chapter can be read almost independently of the others. Each author assumes full responsibility for the content of his chapter, but each chapter was read and criticized by all. In order to make the report as readable as possible, each chapter has a separate bibliography at its end, with articles arranged in an obvious order and with an assigned serial number. Citations in the text are made by name of author followed by serial number. Titles of papers were regretfully omitted, as their inclusion would have unduly increased the bulk of the report.

Acknowledgment is gratefully made to Colorado College, to the University of Illinois, the University of Kansas, Princeton University, and Cornell University for clerical assistance and traveling expenses. Moreover, Sharpe and Snyder were relieved from teaching for a considerable time through a grant from the Heckscher Foundation for the Advancement of Research at Cornell.

THE COMMITTEE.

PREFACE TO VOLUME II

Shortly after the appearance of Bulletin 63, Selected Topics in Algebraic Geometry, in 1928, the wish was generally expressed that provision should be made for a supplementary report on the same general lines, to cover the decade following the period included in the Report. In consequence of this demand, the Committee was continued. The admirable work of Berzolari in the Encyklopädie der mathematischen Wissenschaften: Algebraische Transformationen und Korrespondenzen has so nearly covered the transcendental theory and the mapping of algebraic surfaces that they have not been given a place in the present report. On the other hand, the mapping of curves and of systems of curves has been done in greater detail and wider inclusion of sources than was possible in the work of Berzolari. This was done mostly through the competent cooperation of Dr. L. A. Dye. Similarly, properties of surfaces, not always coming under the caption of rational transformations, have been collected and arranged by Dr. Amos H. Black. The Committee wishes to thank both of these men; by their contributions the value of the report has been materially increased.

While the other chapters have been largely the work of one member, it was with the assistance and support of the others; circumstances have been such as to prevent a more intensive cooperation.

THE COMMITTEE.

CONTENTS OF VOLUME I

CHAPTER I

QUADRATIC CREMONA TRANSFORMATIONS

ARNOLD EMCH

CHAPTER II

ANALYSIS OF SINGULARITIES OF PLANE ALGEBRAIC CURVES

ARNOLD EMCH

CHAPTER III

LINEAR SYSTEMS OF PLANE CURVES

F. R. Sharpe

CHAPTER IV

PLANAR CREMONA TRANSFORMATIONS

Arthur B. Coble

CHAPTER V

MULTIPLE CORRESPONDENCES BETWEEN TWO PLANES

Virgil Snyder

CHAPTER VI

INVOLUTIONS ON RATIONAL CURVES

VIRGIL SNYDER

CHAPTER VII

CORRESPONDENCES ON NON-RATIONAL CURVES

VIRGIL SNYDER

CHAPTER VIII

CREMONA TRANSFORMATIONS IN SPACE AND HYPERSPACE

ARTHUR B. COBLE

CHAPTER IX

(1, 2) CORRESPONDENCES BETWEEN S'_r AND S_r, $r>2$

VIRGIL SNYDER

CHAPTER X

REDUCTION OF SINGULARITIES OF SPACE CURVES AND SURFACES

ARNOLD EMCH

CHAPTER XI

MULTIPLE CORRESPONDENCE IN SPACE AND HYPERSPACE

VIRGIL SNYDER

CHAPTER XII

THE MAPPING OF A RATIONAL SURFACE ON A PLANE

F. R. SHARPE

CHAPTER XIII

THE MAPPING OF A RATIONAL CONGRUENCE ON A PLANE

F. R. SHARPE

CHAPTER XIV

INVOLUTIONS ON IRRATIONAL SURFACES

CHARLES H. SISAM

CHAPTER XV

TRANSCENDENTAL THEORY

S. LEFSCHETZ

CHAPTER XVI

SINGULAR CORRESPONDENCES BETWEEN ALGEBRAIC CURVES

S. LEFSCHETZ

PAGE

CHAPTER XVII

HYPERELLIPTIC SURFACES AND ABELIAN VARIETIES

S. LEFSCHETZ

CONTENTS OF VOLUME II

CHAPTER I

CURVES AND RULED SURFACES

VIRGIL SNYDER

CHAPTER II

THE MAPPING OF SYSTEMS OF CURVES

LEAMAN A. DYE

CHAPTER III

SYSTEMS OF LINES IN S_n; IRREGULAR SURFACES

VIRGIL SNYDER

CHAPTER IV

CREMONA TRANSFORMATIONS

VIRGIL SNYDER

CHAPTER V

MULTIPLE CORRESPONDENCES

VIRGIL SNYDER

CHAPTER VI

SURFACES AND VARIETIES

AMOS H. BLACK

LIST OF PERIODICALS; ABBREVIATIONS (I)

Acta Math.: Acta Mathematica.
Am. Ac. Mem.: American Academy of Arts and Sciences. Boston. Memoirs.
Am. Ac. Proc.: ——————— Proceedings.
Am. J.: American Journal of Mathematics.
Am. Math. Month.: American Mathematical Monthly.
Am. M. S. Bull.: American Mathematical Society Bulletin.
Am. M. S. Trans.: ——————— Transactions.
Amst. K. Ak. Wet. Verh.: Koninklÿke Akademie van Wetenschappen te Amsterdam Verhandelingen.
Amst. K. Ak. Wet. Versl.: ——————— Verslag.
Analyst: Analyst (1874-83).
Ann. Ac. Polyt. Porto: Annæs Scientificos da Academia Polytechnica do Porto.
Ann. di Mat.: Annali di Matematica Pura ed Applicata.
Ann. Éc. Norm.: Annales Scientifiques de l'École Normale Supérieure. Paris.
Ann. of Math.: Annals of Mathematics.
Arch. Math. Phys.: Archiv der Mathematik und Physik.
Arch. Néerl.: Archives Néerlandaises des Sciences Exactes et Naturelles.
Arxius: Arxius de l'Institut de Ciencies de Barcelona.
Assoc. Fr. Sc.: Association Française pour l'Avancement des Sciences.
Astron. Ab.: Astronomische Abhandlungen.
Berlin Ab.: Berlin. Preussische Akademie der Wissenschaften, Abhandlungen.
Berlin Monatsber.: ——————— Monatsberichte (to 1881).
Berlin Sitzungsber.: ——————— Sitzungsberichte (since 1881).
Bern. Mitt.: Berner Mitteilungen der Naturf. Gesellschaft.
Bibl. Math.: Bibliotheca Mathematica (to 1915).
Boll. Boncomp.: Boncompagni's Bollettino di Bibliografia e di Storia delle Scienze Matematiche e Fisiche (1868-87).
Boll. di Mat.: Bollettino di Matematica (Conti) (to 1911).
Boll. d. U. M. It.: Bollettino della Unione Matematica Italiana.
Boll. Loria: Bollettino di Bibliografia e Storia delle Scienze Matematiche (Loria).
Boll. Soc. It. Mat.: Bollettino della Mathesis Società Italiana di Matematica.
Bologna Mem.: Bologna. Reale Accademia delle Scienze dell' Istituto Memorie.
Bologna Rend.: ——————— Rendiconti delle Sessioni.
Bord. Mém.: Mémoires de la Société des Sciences de Bordeaux.
Bord. Proc.: Process verbaux des Séances de la Société ——————— de Bordeaux.
Brit. Ass. Rep.: British Association for the Advancement of Science. Report of meetings.
Brux. Ann. Soc. Sc.: Annales de la Société Scientifique de Bruxelles.
Brux. Bull.: Bruxelles. Académie Royale de Belgique. Bulletins.
Brux. Mém.: ——————— Mémoires. 8 vo (to 1904).
Brux. Mém. Cour.: ——————— Mémoires Couronnés et autres Mémoires. 8 vo.

Brux. Mém. Sav. Etr.: ———————— Mémoires Couronnés et Mémoires des Savants Etrangers. 4 to.
Brux. Mém. Sc.: ———————— Mémoires. Classe des Sciences. 8 vo (since 1904).
Bull. Soc. Sc. Hain.: Bulletin de la Société Scientifique de Hainaut. Belgium.
Bull. Sci. M.: Bulletin des Sciences Mathématiques. (Darboux.)
Calc. Bull.: Calcutta Mathematical Society Bulletin.
Cal. Ac. Proc.: Proceedings of the Academy of Science of California.
Cal. Pub.: Publications of the University of California.
Camb. Dub. Math. J.: Cambridge and Dublin Mathematical Journal (1846-54).
Camb. Math. J.: Cambridge Mathematical Journal (1837-45).
Camb. Ph. Soc. Proc.: Cambridge Philosophical Society. Proceedings.
Camb. Ph. Soc. Trans.: ———————— Transactions.
Časopis: Časopis pro Pěstováni Mathematiky a Fisiky.
Cat. Acc. Gio. Atti.: Atti della Accademia Gioenia di Scienze Naturali in Catania.
Cat. Acc. Gio. Boll.: Bollettino delle Sedute della Accademia ————————
Chris. Forhl.: Christiania Videnskabs Selskabets Forhandlinger.
Colo. Stud.: University of Colorado Studies.
C. R.: Comptes Rendus. Académie des Sciences de l'Institut de France. Paris.
Crac. Bull.: Bulletin International de l'Académie des Sciences de Cracovie.
Deutsche Math.-Ver.: Deutsche Mathematiker-Vereinigung. Jahresbericht.
Deutsche Math.-Ver. Ergb.: ———————— Jahresbericht Ergänzungsbände.
Dublin R. S. Proc.: Dublin. Royal Dublin Society. Scientific Proceedings.
Dublin R. S. Trans.: ———————— Scientific Transactions.
Dublin Proc.: Dublin. Royal Irish Academy. Proceedings.
Dublin Trans.: ———————— Transactions.
Edinb. M. S. Proc.: Edinburgh Mathematical Society. Proceedings.
Edinb. R. S. Proc.: Edinburgh Royal Society. Proceedings.
Edinb. R. S. Trans.: ———————— Transactions.
Educ. Math.: L'Education Mathématique (1898-1914).
Encycl.: Encyclopédie des Sciences Mathématiques.
Encykl.: Encyklopädie der Mathematischen Wissenschaften.
Enseign. Math.: L'Enseignment Mathématique.
Erl. Ber.: Sitzungsberichte der Physischen Medizinischen Sozietät. Erlangen.
Esp. Soc. M.: Revista de la Sociedad Matematica Española.
Form. Math.: Formulaire de Mathématique (1895-1908).
Fortschritte: Jahrbuch über die Fortschritte der Mathematik.
Gaz. Mat. Buk.: Gazeta Mat. Bukarest.
Gent. D.: Gentleman's Diary or Mathematical Repository (1741-1840).
Gent. Math. Comp.: Gentleman's Mathematical Companion (1798-1827).
Gergonne's Ann.: Gergonne's Annales de Mathématiques Pures et Appliquées (1810-32).
Giorn. di Mat.: Giornale di Matematiche.
Gött. Ab.: Göttingen. Königliche Gesellschaft der Wissenschaften. Mathematisch-Physikalische Klasse. Abhandlungen.
Gött. Nachr.: ———————— Nachrichten.
Gött. M. V. Ber.: Göttingen. Mathematischer Verein. Berichte.
Hamb. M. G. Mitt.: Hamburg. Mathematische Gesellschaft. Mitteilungen.
Il Pit.: Il Pitagora.
Ind. M. S.: Indian Mathematical Society. Journal.

Interméd.: L'Intermédiaire des Mathématiciens.
Jassy Ann.: Annales de l'Université de Jassy.
J. de Math.: Journal de Mathématiques Pures et Appliquées. Liouville.
J. de M. E.: Journal de Mathématiques Elémentaires.
J. Éc. Polyt.: Journal de l'École Polytechnique. Paris.
J. f. Math.: Journal für die Reine und Angewandte Mathematik.
J. Hopk. Circ.: Johns Hopkins University Circulars.
Kop. Dan. Vid. Sels. Skr.: Kongelige Danske Videnskabernes Selskabet Skrifter. Kopenhagen.
Ladies' D.: Ladies' Diary (1704-1840).
Lady's and Gents. D.: Lady's and Gentleman's Diary. (Volumes numbered in continuation of Ladies' Diary) (1841-71).
Leipz. Ab.: Leipzig. Königliche Sächsische Gesellschaft der Wissenschaften. Mathematisch-Physikalische Klasse. Abhandlungen.
Leipz. Ber.: ——————— Berichte.
Liège S. R. S. Mém.: Liège. Société Royale des Sciences. Mémoires.
Lomb. Ist. Mem.: Reale Istituto Lombardo di Scienze e Lettere. Milano. Memorie.
Lomb. Ist. Rend.: ——————— Rendiconti.
London M. S. Proc.: London Mathematical Society. Proceedings.
London M. S. J.: London Mathematical Society. Journal.
Lond. R. S. Proc.: Royal Society of London. Proceedings.
Lond. R. S. Trans.: ——————— Philosophical Transactions.
Lond. R. S. Year Book.: ——————— Year Book.
Mat. p. a.: Le Matematiche, pure ed applicate.
The Math.: The Mathematician (1843-50).
Math. Ann.: Mathematische Annalen.
Math. Gaz.: Mathematical Gazette.
Math. Mag.: Mathematical Magazine (1882-1910).
Math. Misc.: Mathematical Miscellany (1836-39).
Math. Mon.: Mathematical Monthly (1859-61).
Math. Nat. Ber. Ung.: Mathematische und Naturwissenschaftliche Berichte aus Ungarn.
Math. Qu.: Mathematical Questions and Solutions from the Educational Times.
Math. Rep.: Mathematical Repository (to 1835).
The Math. T.: The Mathematics Teacher.
Math. Zschr.: Mathematische Zeitschrift. Berlin.
Mathesis: Mathesis. Belgium.
Mem. Acc. Acireale: Memorie dell' Accademia di Scienze, Lettere ed Arti degli Zelanti.
Mess. of Math.: Messenger of Mathematics.
Modena R. Acc. Mem.: Memorie della Regia Accademia di Scienze, Lettere et Arti in Modèna.
Monatsh. f. Math.: Monatshefte für Mathematik und Physik.
Münch. Ab.: München Königliche Bayerische Akademie der Wissenschaften. Mathematisch-Physikalische Klasse. Abhandlungen.
Münch. Ber.: ——————— Sitzungsberichte.
Nap. Acc. Pont. Atti.: Accademia Pontaniana di Napoli. Atti.
Nap. R. A. Atti: Napoli. Reale Accademia delle Scienze Fisiche e Matematiche. Atti.

Nap. R. A. Rend.: ———————— Rendiconti.
Nieuw. Arch.: Nieuw Archief voor Wiskunde Uitgegeven door het Wiskundig Genootschap.
Note e Mem.: Note e Memorie di Matematica (Catania).
Nouv. Ann.: Nouvelles Annales de Mathématiques.
Nouv. Corr. Math.: Nouvelle Correspondance Mathématique (1874-80).
Nova Acta: Nova Acta Academiæ Cæsareæ Leopoldinæ Carolinæ (Halle).
N. Tijds. f. Wisk.: Nieuw. Tijdschrift voor Wiskunde.
N. Y. M. S. Bull.: New York Mathematical Society Bulletin (to 1904).
Oslo Selsk. For.: Oslo Videnskabs-Selskabets. Forhandlinger.
Oslo Selsk. Skr.: ———————— Skrifter.
Pad. N. Sag.: Padova Nuovi Saggi (to 1883).
Pal. Circ. Mat.: Palermo. Circolo Matematico. Rendiconti.
Par. Ac. Mém.: Académie des Sciences de l'Institut de France. Paris Mémoires.
Par. Ac. Mém. Sav.: ———————— Mémoires Présentés par Divers Savants.
Penn. Pub.: University of Pennsylvania. Publications: Mathematics.
Per. di Mat.: Periodico di Matematiche.
Per. di Mat. Supp.: ———————— Supplemento.
Pet. Ac. Sci. Bull.: Petrograd. Académie des Sciences. Bulletin Scientifique.
Pet. Ac. Sci. Mém.: ———————— Mémoires.
Phil. Mag.: Philosophical Magazine.
Poggendorff's Ann.: Poggendorff's Annalen.
Prac. mat. fiz.: Prace matematyczne fizyczne, Warsaw.
Prag. Ab.: Prager Abhandlungen der Kgl. Böhmischen Gesellschaft der Wissenschaften.
Prag. Ber.: ———————— Sitzungsberichte.
Prag. Cesk. Ak. Bull.: (Bulletin of the Franz Joseph Academy at Prague).
P. Rep.: Pascal's Repertorium der höheren Mathematik. 2d (German) edition.
El. Prog. Mat.: El Progreso Matematico (to 1900).
Quart. J.: Quarterly Journal of Pure and Applied Mathematics.
Ras. di mat.: Rassegna di matematica e fisica, Rome.
R. de Ci. Madr. Rev.: Revista de la Real Academia de Ciencias Exactes. Madrid.
Rec. Math.: Receuil Mathématique, Moscow.
Rev. Mat. Hisp. Am.: Revista Matematica Hispano-Americana. Madrid.
Rev. de Math.: Revue de Mathématiques (1891-1906).
Rev. de Math. Spec.: Revue de Mathématiques Speciales (1890-1914).
Rev. Sem.: Revue Semestrielle des Publications Mathématiques.
Riv. di Mat.: Rivista di Matematica, Torino.
Riv. F. M.: Rivista di Fisica Matematica e Scienze Naturali (1900-12).
R. L. Ven. Ann.: Annali delle Scienze del Regno Lombardo-Veneto. Padua, Venice (1831-45).
Rom. Pont. Acc. Atti: Roma. Pontificia Accademia dei Nuovi Lincei. Atti.
Rom. Pont. Acc. Mem.: ———————— Memorie.
Rom. Acc. L. Mem.: Roma. Reale Accademia dei Lincei. Memorie.
Rom. Acc. L. Rend.: ————————Rendiconti.
Rom. Acc. L. Trans.: ———————— Transunti.
Rom. Soc. It. Sc. Mem.: Roma. Società Italiana delle Scienze, Memorie di Matematica e di Fisica.
Roum. Ac. Bull.: (Bulletin de l'Academie Roumaine).
Schl. Ges. f. v. Kultur.: Schlesische Gesellschaft für vaterländische Kultur.

Soc. Ital. Prog. Sci.: Società Italiana per i Progressi delle Scienze.
Soc. M. F. Bull.: Société Mathématique de France. Bulletin. Paris.
Soc. M. F. C. R.: ——————— Comptes Rendus des Séances.
Soc. Phil.: Société Philomathique. Bulletin. Paris (to 1912).
Sph. Oe.: Sphinx-Oedipe. Nancy.
Stockh. Ark.: Arkiv. för Matematik K. Svenska Vetenskaps-Akademien i Stockholm.
Stuttg. Math. Nat. Mitt.: Stuttgart. Mathematisch-Naturwissenschaftlicher Verein in Würtemberg. Mathematisch-Naturwissenschaftliche Mitteilungen.
Sven. Vet. Ak. Bih.: Kongliga Svenska Vetenskaps-Akademie Handlingar. Bihang (from 1872). Stockholm.
Tidskr. f. Math.: Nyt Tidsskrift for Mathematik. Kopenhagen.
Tôhoku M. J.: Tôhoku Mathematical Journal.
Tôhoku Sc. R.: Tôhoku Imperial University. Science Reports.
Torino Atti.: Torino. Reale Accademia delle Scienze. Atti.
Torino Mem.: ——————— Memorie.
Toulouse Ann.: Toulouse. Faculté des Sciences de l'Université. Annales.
Toulouse Bull.: Toulouse. Académie des Sciences, Inscriptions et Belle-Lettres. Bulletin (1897-1900).
Toulouse Mém.: ——————— Mémoires.
Ven. Ist. Atti: Venezia. Reale Istituto Veneto di Scienze, Lettere ed Arti. Atti.
Ven. Ist. Mem.: ——————— Memorie.
Wash. Nat. Ac. Sc. Proc.: Washington. National Academy of Sciences. Proceedings.
Wash. Pub.: Publications of the University of Washington.
Wektor: Wektor. Warsaw (Polish).
Wiad. Mat.: Wiadomosci Matematyczne. Warsaw.
Wien. Ber.: Wien. Kaiserliche Akademie der Wissenschaften. Mathematisch-Naturwissenschaftliche Klasse. Sitzungsberichte.
Wisk. Tijds.: Wiskundig Tijdschrift. Harlem-Rotterdam.
Zschr. Math. Phys.: Zeitschrift für Matematik und Physik (Schlömilch).
Zschr. f. Math. Unt.: Zeitschrift für Mathematischen und Naturwissenschaftlichen Unterricht.
Z. Vierteljschr.: Zürich. Naturforschende Gesellschaft. Vierteljahrsschrift.

LIST OF PERIODICALS; ABBREVIATIONS (II)

Abh. Hamb. Sem.: Abhandlungen aus dem mathematischen Seminar der Hamburgischen Universität.

Acta Math.: Acta Mathematica.

Acta Soc. Helv.: Actes de la Société Helvetique des sciences naturelles.

Am. J.: American Journal of Mathematics.

Am. Math. Month.: American Mathematical Monthly.

Am. M. S. Bull.: American Mathematical Society, Bulletin.

Am. M. S. Trans.: ——————, Transactions.

Amst. K. Ak. Wet. Verh.: Koninklijke Akademie van Wetenschappen te Amsterdam, Verhandelingen.

Amst. K. Ak. Wet. Versl.: ——————, Verslag.

Amst. K. Ak. Wet. Proc.: ——————, Proceedings.

Anais de Soc. Porto: Anais de la Faculte des sciences, Porto.

Ann. di Mat.: Annali di Matematica Pura ed Applicata.

Ann. di Pisa: Annali della R. Scuola normale superiore di Pisa.

Ann. Éc. Norm.: Annales scientifiques de l'École Normale Supérieure, Paris.

Ann. of Math.: Annals of Mathematics.

Ann. Soc. Pol.: Annales de la Société Polonaise de mathématique.

Arch. Néerl.: Archives Néerlandaises des sciences exactes et naturelles.

Berlin M. G. Math.: Sitzungsberichte der Berliner mathematischen Gesellschaft.

Berlin Monatsber.: Berlin. Preussische Akademie der Wissenschaften. Monatsberichte (to 1881).

Berlin Sitzungsber. ——————. Sitzungsberichte (since 1881).

Bol. Congr.: Atti del Congresso internazionale dei Matematici, Bologna, 1928.

Boll. d. U. M. It.: Bollettino della Unione Matematica Italiana.

Brux. Bull.: Bruxelles. Académie Royale de Belgique. Bulletin.

Buch. Bull.: Bulletin de mathématique et de physique pure et appliquées de l'École polytechnique de Bucharest.

Bull. Ac. Romaine: Bulletin mathématique de la Société Romaine des Sciences.

Bull. Intern.: Bulletin international de l'Académie de Bôhème.

Bull. Sci. M.: Bulletin de sciences mathématiques. (Darboux).

Cal. Univ. Publ.: Publications of the University of California.

Camb. Ph. Soc. Proc.: Cambridge Philosophical Society. Proceedings.

Časopis: Časopis pro Pěstováni Mathematiky a Fisiky.

Cat. Acc. Gio. Atti: Atti della Accademia Gioenia di Scienze Naturali in Catania.

Comm. Helv. Math.: Commentarii Mathematici Helvetici.

C. R.: Comptes Rendus. Académie des sciences de l'Institut de France. Paris.

Deutsche Math.-Ver.: Deutsche Mathematiker-Vereinigung. Jahresbericht.

Edinb. M. S. Proc.: Edinburgh Mathematical Society. Proceedings.

Enseign. Math.: Enseignment mathématique.

Gaz. Mat. Buk.: Gazeta matematica. Bukarest.

Giorn. di Mat.: Giornale di Matematiche.

Hamb. M. G. Mitt.: Hamburg. Mathematische Gesellschaft. Mitteilungen.

Heidelb. Sitzber: Sitzungsberichte der Heidelberger Akademie der Wissenschaften. Mathematisch-naturwissenschaftliche Klasse.

Ind. M. S.: Indian Mathematical Society. Journal.

Isis: Isis

J. de Math.: Journal de mathématiques pures et appliquées. Liouville.

J. Éc. Polyt.: Journal de l'École polytechnique. Paris.

J. f. Math.: Journal für die reine und angewandte Mathematik.

Jap. M. J.: Japanese Journal of Mathematics.

Jour. Hirosh Univ.: Hiroshima University Journal of Science. Section A.

Lening Ann. Ist. Polyt.: Izvestija Teknologiceskogo Instituta imeni Leningrad kogo Seveta (Annales de l'Institut polytechnique de Leningrad.)

Liège S. R. Mém.: Liège. Société royale des sciences. Mémoires.

Lomb. Ist. Mem.: Reale Istituto Lombardo di Scienze e Lettere. Milano. Memorie.

Lomb. Ist. Rend.: ————————. Rendiconti.

London M. S. Proc.: London Mathematical Society. Proceedings.

London M. S. J.: ————————. Journal.

London R. S. Trans.: Royal Society of London. Philosophical Transactions.

Math. Ann.: Mathematische Annalen.

Math. Zschr.: Mathematische Zeitschrift. Berlin.

Mathesis: Mathesis. Belgium.

Mem. Acc. It.: Reale Accademia d'Italia. Memorie della Classe di Scienze Fisiche, Matematiche e Naturali.

Mem. Univ. Jap.: Memoirs of the Faculty of Science and Agriculture, Taihoku Imperial University.

Mess. of Math.: Messenger of Mathematics.

Monatsh. f. Math.: Monatshefte für Mathematik und Physik.

Münch. Ber.: München Königliche Bayerische Akademie der Wissenschaften. Mathematisch-physikalische Klasse. Sitzungsberichte.

Nap. R. A. Atti: Napoli. Reale Accademia delle Scienze, Fisiche e Matematiche. Atti.

Nap. R. A. Rend.: ————————. Rendiconti.

Nieuw Arch.: Nieuw Archief voor Wiskunde Uitgegeven door het Wiskundig Genootschap.

Nieuw Tidschr.: Nieuw Tijdschrift voor Wiskunde.

Norske Vid. Ak. Avh.: Avhandlinger utgift av det Norske Videnskaps Akademie.

Note e Eserc. Cat.: Note ed esercitazioni matematiche.

Nouv. Ann.: Nouvelles annales de mathematiques.

Pad. Sem. Rend.: Rendiconti del Seminario matematico della Reale Università di Padova.

Pal. Circ. Mat.: Palermo. Circolo Matematico. Rendiconti.

Per. di Mat.: Periodico di Matematiche.

Phil. Mag.: Philosophical Magazine.

Prac. mat. fiz.: Prace matematyczne fizyczne. Warsaw.

Prag. Ber.: Prager Abhandlungen der königlich Böhmischen Gesellschaft der Wissenschaften. Sitzungsberichte.

Ras. di mat.: Rassegna di matematica e fisica. Rome.

Rec. Math.: Receuil mathématique. Moscow.

Rev. de Math. Spec.: Revue de mathématiques speciales (1890-1914).

Rom. Acc. L. Mem.: Roma. Reale Accademia dei Lincei. Memorie.

Rom. Acc. L. Rend.: ——————. Rendiconti.

Roum. Ac. Bull.: Bulletin de l'Académie Roumaine.

Sächs. Ber.: Berichte über die Verhandlungen der Sächsischen Akademie der Wissenschaften zu Leipzig.

Soc. M. F. Bull.: Société mathématique de France. Bulletin. Paris.

Tôhoku M. J.: Tôhoku Mathematical Journal.

Tôhoku Sc. R.: Tôhoku Imperial University. Science Reports.

Torino Mem.: Torino. Reale Accademia delle Scienze. Memorie.

Toulouse Ann.: Toulouse. Faculté des sciences de l'Université. Annales.

Univ. Masaryk Pub.: Publications de la Faculté des Sciences de l'Université Masaryk.

Wash. Nat. Ac. Sc. Proc.: Washington, National Academy of Sciences. Proceedings.

Wien. Ber: Wien. Kaiserliche Akademie der Wissenschaften. Mathematisch-naturwissenschaftliche Klasse. Sitzungsberichte.

Wis-en Nat. Verh.: Wis-en Natuurkundige Verhandlingen.

Zent. blatt f. Math.: Zentralblatt für Mathematik und ihre Grenzgebeite.

Zurich Congr. Int.: Verhandlungen des internationalen Mathematiker-Congresses, Zurich, 1932.

BOOKS

Appell, P., and Goursat, E. Théorie des fonctions algébriques et de leurs integrales. Vol. I. 2nd ed. Paris, Gauthier-Villars, 1929. Repr. New York, Chelsea Publishing, in prep. xxxvii + 524 p.

Baker, H. F. Principles of geometry. Vol. IV. Higher geometry, 1925. xvi + 250 p. Vol. V. Theory of curves. 1933. x + 247 p. Vol. VI. Algebraic surfaces. 1933. ix + 308 p. Cambridge, Cambridge Univ. Press. Repr. New York, Ungar, 1965.

Beltrami, E. Opere matematiche. 4 vols. Milan, Hoepli, 1902, 1904, 1911, 1920. Repr. New York, Chelsea Publishing, in prep.

Berzolari, L. Algebraische Transformationen und Korrespondenzen. Vol. III_2, part 12 of Enzyklopädie der mathematischen Wissenschaften. 1933. 437 p.

Bouligand, G. Notions sur la géométrie réglée et sur la théorie du complex quadratique. Paris, Vuibert, 1928. 84 p.

Brill, A. Vorlesungen ueber ebene algebraische Kurven und algebraische Funktionen. Braunschweig, Vieweg, 1925. x + 340 p.

Cauchy, A. Oeuvres complètes. 8 vols. (First series.) Paris, Gauthier-Villars, 1884-1911. 15 vols. (Second series.) Paris, Gauthier-Villars, 1897-1958.

Cayley, A. Collected Mathematical Papers. 13 vols. and supplement. 1889-1897. Repr. New York, Johnson Reprint, 1968.

Clifford, W. K. Mathematical papers. Cambridge, 1882. Corr. repr. New York, Chelsea Publishing, 1968. lxx + 658 p.

Coble, A. B. Algebraic geometry and theta functions. Am. Math. Soc. Coll. Publ. 10. 1929. vii + 282 p.

Comessatti, A. Lezioni di geometria analitica e proiettiva. Part I. x + 743 p. Part II. xii + 439 p. Padua, Cedam, 1926, 1931.

Coolidge, J. L. A treatise on algebraic plane curves. Oxford, Clarendon Press, 1931. Repr. New York, Dover Publications, 1959. xxiv + 513 p.

Dresden, A. Solid analytical geometry and determinants. New York, Wiley, 1930. Repr. New York, Dover Publications. x + 310 p.

Edge, W. L. The theory of ruled surfaces. Cambridge, Cambridge Univ. Press, 1931. viii + 324 p.

Enriques, F., and Campedelli, L. Lezioni sulla teoria delle superficie algebriche. Padua, Antonio Milani, 1932. iv + 481 p.

Enriques, F., and Chisini, O. Lezioni sulla teoria geometrica delle equazioni e delle funzioni algebrici. Vol. III. Bologna, Zanichelli, 1924. 593 p.

Fermat, P. Oeuvres de Fermat. 5 vols. Paris, Gauthier-Villars, 1891-1922.

Ganguli, S. The theory of plane curves. Vol. I. 2nd ed. Calcutta, Univ. of Calcutta, 1925. xxi + 387 p.

Godeaux, L. Les transformations birationelles du plan. (Mem. des Sci. Math., fasc. XXII.) Paris, Gauthier-Villars, 1927. 58 p.

Graustein, C. W. Introduction to higher geometry. New York, Macmillan, 1930. xv + 486 p.

Hadamard, J. Oeuvres de Jacques Hadamard. 4 vols. Paris, Editions du Centre National de la Recherche Scientifique, 1968.

Halphen, G. H. Oeuvres d'Halphen. 4 vols. Paris, Gauthier-Villars, 1916, 1918, 1921, 1924.

Hensel, K., and Landsberg, G. Theorie der algebraische Funktionen einer Variabeln. Leipzig, Teubner, 1902. Repr. New York, Chelsea Publishing, 1965.

Hilbert, D., and Cohn-Vossen, S. Geometry and the imagination. [Trans. of Anschauliche Geometrie.] New York, Chelsea Publishing, 1952. 358 p.

Hudson, J. P. Cremona transformations in plane and space. Cambridge, Cambridge Univ. Press, 1927. xx + 454 p.

Hurwitz, A. Mathematische Werke. 2 vols. Basel, Birkäuser, 1963.

Jacobi, C. G. J. Gesammelte Werke. 8 vols. Berlin, Georg Reimer, 1881-1891. Corr. repr. New York, Chelsea Publishing, 1969.

Jessop, C. M. A treatise on the line complex. Cambridge, Cambridge Univ. Press, 1903. Corr. repr. New York, Chelsea Publishing, 1968. xv + 364 p.

Kempe, A. B. How to draw a straight line; a lecture on linkages. London, Macmillan, 1877. Reprinted in: Squaring the circle, and other monographs. New York, Chelsea Publishing, 1953; New York, Chelsea Publishing, 1969.

Klein, F., and Fricke, R. Vorlesungen ueber die Theorie der elliptische Modulfunktionen. 2 vols. Leipzig, Teubner, 1890, 1892. Repr. New York, Johnson Reprint, 1966.

Kowalewski, G. Determinantentheorie, einschliesslich der Fredholmschen Determinanten. 3rd ed. Berlin, Walter de Gruyter, 1942. Repr. New York, Chelsea Publishing, 1948. vii + 320 p.

Krazer, A. Lehrbuch der Thetafunktionen. Leipzig, Teubner, 1903. Repr., Chelsea Publishing, in prep. xxiv + 509 p.

Kronecker, L. Leopold Kronecker's Werke. 5 vols. Leipzig, Teubner, 1895-1930. Corr. repr. New York, Chelsea Publishing, 1968.

Lefschetz, S. Selected papers. Vol. I [includes: L'Analysis Situs. . . .] New York, Chelsea Publishing, in prep.

Lie, S. Gesammelte Abhandlungen. 6 vols. in 11. Oslo, Aschehoug, 1922-1939.

Lie, S. Theorie der Transformationsgruppen. 3 vols. Leipzig, Teubner, 1888, 1890, 1893. Corr. repr. New York, Chelsea Publishing, 1970.

Liè, S. and Scheffers, G. Geometrie der Berührungstransformationen. Leipzig, Teubner, 1896. Corr. repr. New York, Chelsea Publishing, in prep. xi + 693 p.

Loria, G. Curve piani speciali algebriche e trascendenti, Teoria e storia. Vol. I. Curve algebriche. xvi + 574 p. Vol. II. Curve trascendente. Curve dedotte da altre. xi + 439 p. Milan, Hoepli, 1930.

Loria, G. Curve sghembe speciali algebriche e trascendenti. Vol. I. Curve algebriche. xii + 376 p. Vol. II. Curve spheriche. Curve definite da una relazione fra flessione e torsione. Curve particolari sopra superficie assegnate. ii + 256 p. Bologna, Zanichelli, 1925.

Minkowski, H. Geometrie der Zahlen. Leipzig, Teubner, 1896. Repr. New York, Chelsea Publishing, 1953. viii + 256 p.

Möbius, A. F. Gesammelte Werke. 4 vols. 1885-1887. Repr. 1969.

Morley, F., and Morley, F. V. Inversive geometry. London, G. Bell and Sons; New York, Ginn, 1933. Repr. New York, Chelsea Publishing, 1954. xi + 273 p.

Picard, E., and Simart, G. Théorie des fonctions algébriques de deux variables indépendantes. 2 vols. Paris, Gauthier-Villars, 1897, 1906. Repr. 2 vols. in one. New York, Chelsea Publishing, in prep. 796 p.

Pluecker, J. Gesammelte wissenschaftliche Abhandlungen. 2 vols. Leipzig, Teubner, 1895-1896. Repr. New York, Chelsea Publishing, in prep.

Rademacher, H., and Toeplitz, O. Enjoyment of Mathematics. [Trans.] Princeton, Princeton Univ. Pres, 1957. 204 p.

Ruffini, P. Opere matematiche. 3 vols. Palermo, 1915, 1953, 1954.

Salmon, G. Analytic geometry of three dimensions. Vol. I. 7th ed. xxiv + 470 p. Vol. II. 5th ed. xvi + 334 p. New York, Chelsea Publishing, 1958, 1965.

Steiner, J. Gesammelte Werke. 2 vols. 1st ed. Berlin, Georg Reimer, 1881, 1882. 2nd ed. New York, Chelsea Publishing, 1970.

Weyl, H. Concept of a Riemann surface. [Trans.] New York, Addison-Wesley, 1964.

White, H. S. Plane curves of the third order. Cambridge, Harvard Univ. Press, 1925. xii + 168 p.

Wieleitner, H. Algebraische Kurven. [Sammlung Goeschen, Bd. 435.] Berlin, W. de Gruyter, 1930. 146 p.

CHAPTER I

QUADRATIC CREMONA TRANSFORMATIONS

A. Early Examples of Quadratic Transformations

The purpose of this introductory note is to point out the most important phases and their originators in the early development of quadratic transformations, and to conclude with Cremona's famous memoir on the transformations which now bear his name. A report on the subsequent developments of Cremona transformations and their applications will be given in Chapter IV.

The first conspicuous instance of a quadratic transformation process appears in the discovery of the stereographic projection. According to bishop Synesius of Cyrene([221]) it was Hipparchus([98]) who established the law of this projection, although the first writer about the subject is Ptolemy. Only an Arabic translation of Ptolemy's work is in existence. A poor Latin translation by Jordanus was reedited and retranslated by Commandinus([37]), ([178]).

P. G. Dandelin([43]) establishes the principal properties of stereographic projection and their application to the solution of certain problems, as to Apollonius' problem of the tangent-circles of three given fixed circles. He also discovers the relation of the foci of a plane section of a right circular cone with the tangent spheres to the cone and its plane section, and comes across the relation $\rho\rho' = R^2$ between the vectors of corresponding points of lemniscates and conics.

A. Quetelet([179]) refers to Dandelin's investigations which he supplements in some important particulars. Quetelet defines polar (inverse) of a curve, and establishes the well known analytic form of inversion.

In 1836 G. Bellavitis wrote a paper on inverse figures([11]) in which he refers to the papers by Quetelet, Dandelin, Sturm on stereographic projection, and establishes the theory of "equipolenze." Two vectors are said to be equivalent when their projections a, b and a', b' on the $x-$ and $y-$ axes are such that $a+ib=a'+ib'$. Bellavitis gives clear-cut definitions of inversion, introduces *inverse figures, center of inversion, radius of inversion,* and establishes all principal properties of inversion, like conformity, and invariance of circularity. He also extends inversion to space. Bellavitis([9]) establishes certain equalities between segments cut out on lines and curves by lines through a fixed point O (center of gravity of point-system on secant lines) and calls these "equipolenzes."

The transformation by reciprocal radii was established in special form and for special purposes by F. Vièta([236]) for the plane and by P. Fermat([76]) for space.

The discovery of inversion in its full and general bearing, according to information which has recently come to light, should be credited to J. Steiner, who established the laws of inversion in 1824([20]), which, in his picturesque language, he called "Wiedergeburt und Auferstehung." There can exist no doubt that Steiner([214]) used this principle in his solution of Malfatti's problem and a great number of other problems of inversion geometry. Pluecker was familiar with the principle of inversion and its analytic form([173]).

L. J. Magnus'([144]) mere inversions of given figures into new figures, according to Steiner's viewpoint, do not lead to essentially new properties.

In the course of time inversion was rediscovered by a number of investigators; for example, by W. Thomson([227]), ([228]). See also F. Buetzberger([20]).

A. F. Moebius([156]), ([157]) made an important generalization of inversion, or transformation by reciprocal radii, in his *Theorie der Kreisverwandtschaft* by establishing in a purely geometrical manner the most general transformation in a plane, which transforms circles into circles.

The rudiments of analytic formulation([83]), ([84]) of this transformation in the complex plane are contained in C. F. Gauss' study of the stereographic projection by means of the representation of points by complex numbers.

The first systematic treatment of the Kreisverwandtschaft, circle-transformation, in the complex plane may be found in F. H. Siebeck([208]). The transformation appears in the form $Y=\dfrac{A+BX}{C+DX}$. Siebeck also studies a more general transformation (involutorial) $F(X, Y)=0$, where F is symmetric in X and Y.

A more detailed study of inversion and the circle transformation will be made under Section E.

The rudiments of quadratic transformations are noticeable in I. Newton's "de curvarum descriptione organica"([167]). Let A and B be the vertices of two angles of given magnitudes, P the intersection of one side of the first angle with one of the second and D the intersection of the remaining two sides. Then when P describes a conic, D describes a quartic, with double points at A and B and the point E as the limiting position of D when AP and BP coincide with AB. The quartic degenerates into a cubic when the conic passes through A and B.

C. McLaurin([142]) actually introduced the idea of a quadratic transformation by stating that in Newton's relations between D and P, to the lines described by D correspond conics through three fixed points. To a curve of order n corresponds a curve of order $2n$.

W. Braikenridge([18]) describes a geometric mechanism which produces a quadratic transformation. If three lines A_1A_2, A_2A_3, A_3A_1 turn about three fixed points B_3, B_1, B_2, while A_3 always moves on a fixed line and A_1 describes a curve of the nth order, then A_2 describes a curve of order $2n$. He, however, did not perceive the importance of the relation between A_1 and A_2 as a quadratic transformation.

V. Poncelet([177]) gave examples of involutorial quadratic transformations and studied their properties. The first instance is the transformation defined by the relation between a point P and the point of concurrence P' of its polars with respect to two circles. It is immediately recognized that this becomes identical with McLaurin's correspondence between P and D, in the case of right angles PAD and PBD, by assuming A and B as zero circles.

Poncelet([176]) generalized this transformation by replacing the circles by any two conics. The polars of a point P with respect to a pencil of conics are concurrent at a point P'. Again P and P' define the involutorial transformation.

The same transformation was also defined by Steiner([215]) as the correspondence between points which are simultaneously harmonic with respect to the three pairs of lines of a quadrangle. This is, of course, identical with Poncelet's definition, since the three pairs of lines belong to the pencil of conics through the quadrangle. This transformation is frequently referred to as the Steinerian tiansformation (Steinersche Verwandtschaft).

Another method of establishing a quadratic transformation, as pointed out by Steiner, consists in assuming a fixed triangle ABC, and a fixed triangle $A_1B_1C_1$ inscribed in and perspective to ABC. Now let $A_2B_2C_2$ be inscribed in and perspective to $A_1B_1C_1$ with center of perspective P. Then $A_2B_2C_2$ is also perspective to ABC with center of perspective P'. P and P' are thus corresponding points in a quadratic transformation. Choosing $A_1B_1C_1$ as the coordinate triangle, this latter transformation in modern form is identical with

$$x_1'=x_2x_3, \quad x_2'=x_3x_1, \quad x_3'=x_1x_2.$$

The first statement of a real general quadratic transformation for real distinct fundamental points between two planes is due to Steiner, who called it "schiefe Projektion," skew projection([212]). He assumes two fixed planes e and e' and two fixed lines l and g in a general position. The common secants of l and g cut e and e' in couples of points P and P' which establish a general quadratic correspondence between e and e', with real and distinct fundamental points. Steiner discovered the fundamental points and their bearing on the transformation. The usefulness and importance of the "projection" is shown by a systematic development of a great number of new and organically connected propositions. The special transformation mentioned above as well as the general quadratic transformation were known to Steiner as early as 1828.

In the footnote on page 295 of the *Systematische Entwickelung,* immediately preceding the appendix, Steiner makes a significant statement which makes it obvious that Steiner anticipated the possibility of more general transformations than quadratic. This case was also considered by G. J. Verdam (**235.2**).

Magnus (**143**) must be credited with the first analytic treatment of the general quadratic Cremona transformation, based upon two bilinear relations, and the establishment of its essential properties by Cartesian coordinates.

J. Pluecker (**174**) established a quadratic transformation in a form which he considers, mistakenly, as the general expression of such a transformation.

Much later G. V. Schiaparelli (**193**) conceived the problem of transformations under a more general aspect. He establishes a transformation between the two planes (x, y) and (ξ, η) defined by two equations

$$f'(x, y; \xi, \eta) = 0, \qquad f''(x, y; \xi, \eta) = 0$$

so that to a "primitive point" (x, y) correspond, one, two, or even an infinite number of "transformed points" (ξ, η) according to the nature of the functions f' and f''. He discusses only the cases in which to a point (x, y) corresponds only one point (ξ, η), and conversely, and calls such transformations of the first order. By two postulates Schiaparelli further restricts the $(1, 1)$ character of this transformation to the case where f' and f'' are bilinear relations.

Discussing the "general" transformation of the "first order," Schiaparelli shows that the three quadratic expressions in ξ and η must represent three conics which all pass through three fixed points. This of course is the quadratic Cremona transformation.

Schiaparelli extends the investigation of transformations of the "first order" to space of three dimensions by defining them by a manner entirely analogous to that for the plane.

Even earlier, in 1859, a decided step in advance was made by E. De Jonquières * (**44**), (**45**) by the establishment of transformations of a cer-

* This memoir was submitted to the Académie des Sciences in Paris, Oct. 10, 1859, and in definite form Jan. 23, 1860; but no report of it *was ever* made to the Academy.

tain type of higher order than the second now known as De Jonquières transformations.

L. Cremona([38]), ([39]), ([40]) was the first (1862) who solved the problem of algebraic one-to-one correspondences between two planes in its general form: *To every point of a first figure corresponds a single point in the second, and reciprocally to every point in the latter a single point in the former.* Cremona refers to the articles by Magnus and Schiaparelli, which are particular cases of a much more general transformation which, following Schiaparelli, he calls of the first order. For this and subsequent developments see Chapter IV.

B. General Quadratic Transformation

1. *Definition and General Properties of Quadratic Transformations.*— The discovery that there exist transformations between two planes $\Sigma(x)$ and $\Sigma(x')$, or simply (x) and (x') besides the earlier known collineation and inversion (transformation by reciprocal radii), such that, in general, to a point in (x) corresponds only one point in (x'), and conversely to a point in (x') only one point in (x), is of fundamental importance. The first step in this discovery was the establishment of quadratic transformations with this property as described in the preceding section, in which to a line in (x) must correspond a conic in (x'); to a line in (x') a conic in (x).

This is accomplished by a system of conics through three points, the simplest case of what Cremona, in the study of the general (1, 1) transformation, calls a homaloidal net of rational algebraic curves, that is, of rational curves through the same system of base-points consisting of singular (including ordinary) points of these curves.

Thus, a homaloidal system of conics defines a quadratic transformation between two planes.

The exceptional points of the transformation, the base points of the net, are commonly called the fundamental points of the transformation, and the lines which correspond pointwise to these points the *fundamental*, or *principal lines* of the transformation. Thus *a pencil of lines through a fundamental point A_i is transformed into a projective pencil of lines through the fundamental point A'_i.* This proposition is the basis for Seydewitz's construction of the quadratic transformation, which will be discussed later on.

The quadratic transformation, and the behavior of various singularities of algebraic curves on the fundamental system are fully discussed by K. Doehlemann([55]) by a graphic method.

2. *Geometric Description of General Quadratic Transformations.*—Magnus([143]), as noted before in the historical introduction, gives a discussion of the general quadratic transformation in Cartesian coordinates but did not notice the important fact that to a pencil of lines with A_i as a vertex corresponds properly a projective pencil of lines with A'_i as a vertex.

F. Seydewitz([207]) was the first to establish a quadratic transformation by means of projective pencils and ranges. In a plane Σ assume two points A_1 and A_2 and the two pencils $A_1(s_1)$ and $A_2(s_2)$ with these points as vertices. Likewise in a plane Σ' establish two pencils $A'_1(s'_1)$ and $A'_2(s'_2)$. Then establish the projectivities

$$A_1(s_1) \overline{\wedge} A'_1(s'_1),$$
$$A_2(s_2) \overline{\wedge} A'_2(s'_2).$$

Any point P in Σ joined to A_1 and A_2 determines two rays s_1 and s_2 in the two pencils in Σ. To these correspond in Σ' two rays s'_1 and s'_2, by the given projectivities, which intersect in a point P'. Conversely to a point Q' in Σ' corresponds, by these projectivities, a point Q in Σ. To a line l in Σ, or a line g' in Σ', correspond evidently a conic L' in Σ', and a conic G in Σ, respectively. Thus, Seydewitz establishes a quadratic transformation between the two planes.

After Seydewitz the next important step was taken by Th. Reye([185]), ([187]). After a short critical review of what has been done before, Reye enumerates the possible kinds of quadratic transformations between two planes Σ and Σ'.

As a first example of a quadratic transformation Reye mentions the case of a quadric F with two points S and S' on F, and two planes Σ and Σ'. Every point P of F is joined to S and S'. SP cuts Σ in P; $S'P$ cuts Σ' in P' so that P and P' form corresponding points of a quadratic transformation between Σ and Σ'.

According to Reye the simplest method to establish a general quadratic transformation between two planes is to set up two correlations between Σ and Σ', so that to every point in $\Sigma(\Sigma')$ correspond two lines (called polars by Reye) in $\Sigma'(\Sigma)$. Thus there exists a correspondence between the points of one plane and the intersections of the lines of pairs of corresponding correlative lines in the other plane, which is precisely a general quadratic correspondence.

T. A. Hirst([104]), also in 1866, derives the quadratic transformation with real fundamental triangles by two couples of projective pencils without being aware of the fact that the same method had been established before by Seydewitz. He discusses fundamental points and lines, and merely indicates special cases. He finds that, in case of two superposed planes, there are, in general, four invariant points.

In his paper Hirst refers to A. Transon, who in 1865 ([231]) establishes a quadratic transformation which is identical with Steiner's skew projection. Transon, apparently, did not know that Steiner had done this more than thirty years before. A year later ([232]) Transon discusses the skew projection in connection with Hirst's quadric inversion which will be explained in the next section.

E. Weyr ([242]) in 1869, covers analytically in the main the same ground as did Reye three years before by synthetic methods.

Weyr defines as quadratically or birationally related two plane systems when the coordinates of corresponding points are connected by bilinear relations so that to a point of one system corresponds, in general, only one point in the other.

It is a curious fact that this mistaken belief, that the quadratic transformation was the only birational transformation between two planes outside of collineation and correlation, was held by the majority of geometers up to the time of Cremona's classic memoir.

Weyr proves, what is, of course, only true for a quadratic transformation, that a (1, 1)-correspondence between two planes is determined by seven pairs of corresponding points.

H. Schubert ([201]) discusses the degeneration of quadratic transformations and finds that altogether, there are 18 cases of quadratic transformations, of which 17 are special.

G. Lazzeri ([131]) discusses birational correlation of order n in a plane. In particular, Lazzeri proves the theorem that a correlation in which every line passes through its pole (corresponding point) can only be a quadratic correlation:

$$\rho u'_1 = a_1 x_2 x_3, \ \rho u'_2 = a_2 x_3 x_1, \ \rho u'_3 = a_3 x_1 x_2; \ a_1 + a_2 + a_3 = 0,$$

so that evidently $\Sigma u'_i x_i \equiv 0$.

E. Czuber ([41]) considers the (1, 1)-point-transformations of the plane, defined by two bilinear equations

$$f(x, y; \ x', y') = 0,$$
$$g(x, y; \ x', y') = 0,$$

and the special cases.

He makes the old mistake by stating that this quadratic transformation is the most general (1, 1) transformation between two planes. He also discusses cases of involutorial transformations, but misses Hirst's quadric inversion.

E. Duporcq([58]) studies quadratic and rational correspondences between two plane figures.

He finds among other results that if five couples of a quadratic transformation are contained in a collineation, then they determine an infinite number of couples of conjugate points in the quadratic transformation, which are couples of corresponding points in the collineation of two conics. Duporcq extends the investigation to space. Among the results may be mentioned the notable theorem: If five points of a plane P move on five fixed spheres whose centers lie on a fixed plane P', then there exists on P a sixth point which also describes such a sphere.

M. Fréchet([81]) establishes certain normal forms for birational quadratic transformations. He shows that the three cases of the fundamental points A, B, C, namely A, B, C distinct, A and B consecutive, A, B, C consecutive, may all be represented under the same form

$$\frac{X}{x\cdot F'_z}=\frac{Y}{y\cdot F'_z}=-\frac{Z}{y\cdot F'_x+y\cdot F'_y}$$

where

$$F=xy-z^2,\quad F=x^2-z^2,\quad F=x^2-2yz,$$

according to the three cases. By the same method he also proves the well known fact, that every quadratic Cremona transformation may be obtained as the product of an involutorial quadratic transformation of the plane into itself, followed by a collineation.

O. Terquem([223]), in a paper *Sur les méthodes métamorphiques* (de transformation) is led to the study of quadratic transformations by first attacking the problem of birational transformations between two surfaces.

Suppressing one dimension, Terquem obtains as special cases the quadratic transformation and the homographic transformation (collineation).

In a similar manner, J. G. H. Swellengrebel([222]) discusses the transformations defined by two algebraic functions $f(u, v, t', w')=0$, $F(u, v, t', w')=0$; gives various geometric interpretations, including quadratic and collinear transformations.

C. Wiener([244]) gives a short historic account of Cremona transformations, in particular quadratic transformations, and then makes a study of some multiply valued algebraic correspondences between two planes. He gives a number of examples in which is explained the derivation of properties of curves from those of lower order by the transformation.

C. Servais([205]) writes on a geometric interpretation of birational quadratic transformations and has a note on the theory of transformations in ([206]).

P. Muth([162]) in a paper on ternary bilinear forms, establishes invariant conditions on two bilinear forms $f_i(x, y)=0$, $i=1, 2$, that the quadratic reciprocity determined by them degenerates into a linear reciprocity.

V. Weiss([240]) describes a construction of a quadratic transformation between two planes when seven pairs of corresponding points are given, by means of the pencil of correlations.

In a note on a new method of transformation([135]) by a quadratic reciprocity, G. Leinekugel studies properties of rational quartics.

P. Del Pezzo([46]) considers quadratic transformations with either four invariant points, or a conic of invariant points, or intermediate cases.

P. Cassani([24]) constructs the quadratic transformation in a plane α by means of two skew lines r_1, r_2 and a quadric Q. Given x in α draw the transversal xr_1r_2. The polar of this transversal with respect to Q cuts α in x'. An elaborate study of the quadratic correspondence was made by E. Catalan([24.1]). Accounts of the general quadratic transformation may be found in a number of text-books and treatises on algebraic geometry. As such may be mentioned: F. Aschieri, *Geometria projettiva,* 2d ed., 1888, p. 376-383. Th. Reye, *Geometrie der Lage,* vol. 2, 4th ed., 1907, p. 209-214. R. Sturm, *Die Lehre von den geometrischen Verwandtschaften,* vol. 4, 1909. K. Doehlemann, *Geometrische Transformationen,* vol. 2. Salmon-Chemin-Halphen, *Traité de Géométrie Analytic,* Paris, 1903, p. 430-463. Salmon-Fiedler, *Analytische Geometrie der Kegelschnitte,* vol. 2 (1903), 789-796. H. J. Verhagen, *Verwantschappen, etc., Quaddratische Transformatie,* Utrecht, 1910. German translation, *Verwandtschaften, vermittelt Jurch die Fundamentalpunkte und die Koinzidenzpunkte der quadratischen Transformation.* Diss. Groningen (1910)—Utrecht. H. Laurent, *Traité d'Analyse,* vol. 4, Paris, 1889. Clebsch-Lindemann, *Vorlesungen über Geometrie,* vol. 1 (1876), p. 474-496. H. Hilton, *Plane Algebraic Curves* (1920), p. 120-136. H. Wieleitner, *Theorie der ebenen algebraischen Kurven* (1905), p. 137-172.

H. Durège([60]): *Die ebenen Curven dritter Ordnung* (1871), p. 121-127.

H. Schroeter (Steiner), *Vorlesungen über synthetische Geometrie,* vol. 2 (1876), p. 301-304.

Charlotte A. Scott, *Introduction to Modern Analytic Geometry* (1892), 218-225.

F. Klein, *Vorlesungen über höhere Geometrie,* 3d edition (1926).

G. Loria, *Il passato ed il presente delle ...,* 2d ed. (1896).

J. L. Coolidge, *Geometry of the Circle and the Sphere,* (1916).

F. Kötter, in the *Jahresberichte d. D. Math. Ver.,* 5_2: (1901).

C. Special Quadratic Transformations

1. *Involutorial Quadratic Transformation.*—Two conics $a_x^2=0$ and $b_x^2=0$ determine a pencil of conics $a_x^2+\lambda b_x^2=0$, with the four points of intersection $B_1B_2B_3B_4$ of the two conics as base-points. Let $A_1A_2A_3$ be the diagonal triangle of the quadrangle (B). The polars of a generic point (x') with respect to the pencil of conics form a pencil $a_{x'}a_x+\lambda b_{x'}b_x=0$, whose vertex (x) is uniquely determined by (x'). Conversely, the polars of (x) all pass through (x'). In this manner is established an involutorial quadratic transformation, which has been established by Poncelet and Steiner, as stated in section A.

The diagonal triangle $A_1A_2A_3$ is the fundamental triangle of the transformation, and the vertices of the quadrangle (B) are the invariant points. It is evident that various special cases of this involutorial transformation may be obtained by choosing the two conics, or the quadrangle (B), in some special intersecting relations, which are abstractly identical with the various cases of the solutions of a biquadratic equation.

This involution also arises from Weyr's two bilinear equations between two sets of variables $x, y; x', y'$, when the equations are symmetrical with respect to the two sets; and in case of a *real quadrangle* (B) of distinct points, determined by two pairs of real lines, becomes identical with Steiner's involutorial quadratic transformation described in Section B. As this mode of establishing the transformation lends itself very easily to geometric and particularly to constructive applications, and has been used extensively for this purpose, it does not seem to be improper to use for it the name of *Steinerian transformation,* as proposed by H. Durège([60]). The construction of the transformation itself is as follows: Let (a_1, b_1); (a_2, b_2); (a_3, b_3) represent the three line pairs of the pencil of conics with the vertices $A_1A_2A_3$, respectively, and P a generic point, connect P to A_1 by a line c_1, and construct the fourth harmonic line d_1 conjugate to c_1 with respect to (a_1, b_1) as the other pair of conjugate lines. Repeat the construction for A_2, (a_2, b_2); then d_2 will cut c_2 in a point P' which corresponds to P in the Steinerian transformation. As a check, d_3 must also pass through P'; i. e., $d_1d_2d_3$ are, of course, concurrent at the point P' corresponding to P in the transformation.

When the three line-pairs of the quadruple (B) are rectangular, i. e., when one of the points of (B) is the orthocenter of the remaining three, the construction becomes particularly simple. In this case c_i and d_i are symmetric with respect to both a_i and b_i. This special assumption of orthogonality does, of course, not invalidate projectively the results obtained.

The involutorial quadratic transformation may be defined somewhat differently as follows: Assume again a pencil of conics which cut every generic line of the plane in an involution, whose double points correspond to each other in the involutorial quadratic transformation determined by the pencil of conics, or by the quadrangle of base-points. M. Disteli(50) uses this transformation in a memoir on the metrics of circular cubics in connection with geometric theorems of Jacob Steiner.

Again, the quadruple (B) and the triple (A) form the base of a net of cubics, such that any two cubics of the net intersect in a pair P, P' of corresponding points of the transformation. A. Emch, in a paper on the invariant net of cubics in the Steinerian transformation(67), makes use of this property. See also Jacob Steiner's Vorlesungen ueber Synthetische Geometrie (by H. Schroeter)(213), where the quadratic transformation based on a quadrangle is also called Steinerian transformation.

C. Servais(205) gives a "geometric interpretation of the birational quadratic transformation," which practically amounts to the correspondence existing between a pole and the point of concurrence of its polars with respect to a pencil of conics.

J. Neuberg(165) shows that the quadratic transformations determined by involutions in the pencils on B, C determine a third in the pencil on A. There exist three couples of lines, one in each pencil, which touch a conic. The involution of lines on A mark an involution on BC; Neuberg shows that three couples of points one on BC, one on CA, one on AB lie on a conic.

W. Massny(149), in a paper of a more or less expository character, investigates a particular case of quadratic transformations.

L. Rippert(188) treats of the dual of Neuberg's method of obtaining involutorial quadratic transformations, and extends investigation to space.

S. Loria(141), studies pedal and antipedal transformations in the plane and in ordinary space. If (x, y) is the foot of the perpendicular from the origin upon a line whose Pluecker coordinates are u and v, Loria considers the quadratic correlations between the point (x, y) and the line (u, v) and its inverse and then extends the discussion to space.

D. N. Lehmer(133) writes on the theory of inversion and the quadratic reciprocal transformation, and shows by a purely geometric method, how the latter gives rise to inversion.

The same author studies(134) combinations of involutorial quadratic transformations.

Ch. Tweedie(233) considers the general form of the involutorial (1, 1)-quadratic transformation in a plane. Tweedie points out Czuber's omission of Hirst's inversion and mistake of considering the quadratic transformation as the most general Cremona transformations. E. Beltrami([12]) treats of the Steiner transformation in connection with the "conics of nine points."

A. Emch([62]) in his Introduction to Projective Geometry, establishes the Cartesian form of the Steiner transformation, based upon an orthocentric quadrangle. In this case the circular points at infinity are corresponding points of the transformation, and to the line at infinity corresponds the circle through the fundamental points. On this is based an effective theory of circular cubics.

2. *Quadric Inversion.*—Although this is also an involutorial quadratic transformation, its historic and theoretical interest and usefulness for the applications are of sufficient importance to deserve a special treatment.

This transformation was discovered simultaneously by T. A. Hirst([107]) and F. Geiser([85]) in 1865. G. Bellavitis([10]) is credited with the distinct suggestion of the quadric inversion. Two years before, in 1836, Bellavitis([11]), developed very fully the principles of ordinary cyclic inversion. Hirst also refers to J. W. Stubbs' paper on the application of a new method (inversion) to the geometry of curves and curved surfaces([216]).

Hirst defines as inverse points, p and p' conjugate with respect to a fixed "fundamental conic" F and collinear with a fixed point A. The several pairs of inverse points p, p' on any line R through the origin A form an involution, the double points of which are the real or imaginary intersections of R with F. For convenience Hirst assumes A in such a position that two real tangents may be drawn to F, whose points of tangency, B and C, together with A, form the fundamental points, and the polars BC, CA, AB of A, B, C the principal (fundamental) lines.

From a constructive stand-point, Hirst's transformation is extremely simple. To construct the corresponding point p' of a generic point p, connect p with A by a line R; construct the polar of p with respect to F, which will cut R in the required point p'.

Geiser([85]) extends the transformation to space, by assuming a fixed point P and a fixed quadric F. The join of a generic point p to P cuts F in two points k_1 and k_2. The fourth harmonic p_1 of p with respect to k_1 and k_2 corresponds to p in an involutorial quadratic transformation in space.

Accounts of the quadric inversion by T. A. Hirst may also be found in ([99]), ([100]), and ([101]).

3. *Special Transformations Obtained by Various Geometric Devices.*—One finds in the literature a large number of special quadratic transformations based on special geometric constructions, of which some typical cases deserve to be mentioned in detail.

In ([52]), M. d'Ocagne gives an account of the following extremely simple construction: To each point P of the plane one lets correspond a point P' so that the segment PP' appears under right angles from two fixed points A and B. Thus, to obtain P', connect P with A and B, and at A and B erect perpendiculars to AP and BP, which will intersect in the required point P'. G. Battaglini([5.1]) constructs T_2 by two correlations. A geometric treatment of the method of Magnus was given by P. Bretschneider([18.1]). A metrical T_2 is given by G. Cordoso-Laynes([21.2]). See also F. Hofmann([107.1]), B. Igel([109.1]), A. Jacobi([110.1]), G. Kilbringer([120.1]), L. Lipkin([137.1]) and P. M. D'Ocagne([53.1]).

D'Ocagne generalizes this transformation by collineation.

D'Ocagne does not state the fact that this involutorial quadratic transformation is a special case of that determined by a pencil of conics defined by two zero-circles, say $A[(x-1)^2+y^2=0]$ and $B[(x+1)^2+y^2=0]$. Under the latter assumption the transformation assumes the simple form,

$$x'=-x, \qquad y'=\frac{x^2-1}{y}.$$

The couple of corresponding points on a generic line l is obtained by joining the point of intersection M of l with the y-axis with A, and describing a circle K with M as center and MA as radius. K cuts l in the required couple P, P'.

D'Ocagne extends the transformation to space.

M. J. Neuberg([164]) discusses the transformation established by d'Ocagne and determines the fundamental triangle of the transformation, and then makes a further generalization.

Neuberg also shows how a certain case may be reduced to the quadric inversion of Hirst and Geiser. The paper contains a number of geometrical applications of these transformations.

P. H. Schoute([195]) establishes two involutorial Cremona transformations of order 2 and 5.

The first is based on a triangle ABC. Let P be a generic point in the plane of the triangle, and join P to A, B, C, so that $\sphericalangle PAB=\alpha$, $\sphericalangle PBC=\beta$, $\sphericalangle PCA=\gamma$. Construct the lines $P'A, P'B, P'C$, symmetric to PA, PB, PC with respect to the bisectors of the interior angles of the triangle. The three symmetric lines intersect in a point P' which corresponds to P in an involutorial quadratic transformation with ABC as the fundamental triangle. There is $\sphericalangle P'AB=A-\alpha$, $\sphericalangle P'BC=B-\beta$, $\sphericalangle P'CA=C-\gamma$; if A, B, C also denote the angles of the triangle; and

$$\frac{\sin \alpha}{\sin (A-\alpha)}\cdot\frac{\sin \beta}{\sin (B-\beta)}\cdot\frac{\sin \gamma}{\sin (C-\gamma)}=1.$$

In ([196]), Schoute applies his special quadratic transformation (by symmetric lines) to the proof of a theorem of Steiner: The series of similar conics circumscribed to a triangle ABC envelopes a quartic with double points, at $A, B, C,$ and which has, therefore, only 4 double tangents, of which one is the line at infinity. Schoute finds several other properties of this curve and its special cases.

The same involutorial quadratic transformation is described by J. J. A. Mathieu([150]), who calls it trilinear inversion. He also studies a case of quadratic correlation and applies the theory to the proof of several theorems on systems of conics (equilateral hyperbolas).

E. Vigarié([237]) gives an analytic treatment of Schoute's transformation with applications to the geometry of the triangle.

A. Cayley([27]) studies the transformations,

$$(1) \qquad \xi=\sqrt{x}, \qquad \eta=\sqrt{y}, \qquad \zeta=\sqrt{z},$$

$$(2) \qquad \xi=x^2, \qquad \eta=y^2, \qquad \zeta=z^2,$$

$$(3) \qquad \xi=\frac{1}{x}, \qquad \eta=\frac{1}{y}, \qquad \zeta=\frac{1}{z}.$$

The latter is an analytic interpretation of the Steiner transformation.

In an elaborate memoir([191]) L. Saltel studies the following quadratic transformation: Join a generic point μ to $A, B, C,$ and these to a fixed point E, cutting the opposite sides at A', B', C'. To $A\mu$ determine the conjugate in the involution having AB, AC as a pair and AA' as a double line. Repeat the same construction on the other two vertices; then the three conjugates meet in a point M which corresponds to μ in an involutorial quadratic transformation.

P. Mansion([148]) discusses Saltel's second arguesian transformation, and shows that it is identical with a general quadratic transformation with real fundamental points.

If ABC is the triangle of reference, AA', BB', CC' are three lines concurrent at $O(1, 1, 1)$, moreover AM, BM, CM three other lines movable with $M(x, y, z)$, one may determine the lines corresponding to AM, BM, CM in the three involutions determined by (AB, AC, AA'); (BC, BA, BB'); (CA, CB, CC'), with AA', BB', CC' as double lines; they intersect in a point $M'(x', y', z')$, related to $M(x, y, z)$ by a quadratic transformation.

Finally, it is proved that Saltel's triangular arguesian transformation, is reducible by a linear transformation to the one defined above.

D. Amanzio([2]) discusses some special Cremona transformations and their applications to cubics and quartics, and derives properties of such curves corresponding to the theorems of Pascal and Brianchon by the process of these transformations.

A quadratic transformation based upon a cuspidal cubic is established by K. Zahradnik(245). Given a cuspidal cubic C_3. From a generic point P draw three tangents PT_1, PT_2, PT_3 to C_3. The centroid S of $T_1T_2T_3$ corresponds to P in a quadratic transformation. If C_3 is chosen as a cissoid, the transformation becomes quadratic cyclic.

G. De Longchamps(139) considers a "reciprocal transformation," defined by joining a generic point M to a fixed point P and erecting a perpendicular QI, at another fixed point Q to QM. The correspondence between M and I is an involutorial quadratic transformation.

In a "study of comparative geometry"(140), he also establishes a quadratic transformation with application to a class of sextics, and quartics of class four.

F. Ferrari notes(77) that if of two triply perspective triangles five vertices are fixed, the sixth is constructible linearly, and uses this property to construct a quadratic transformation.

E. Cavalli(25) discusses reciprocal figures and the quadratic transformation in cinematics.

R. Vercellin(235.1) writes on trilinear birational transformations, quadratic transformations with coincident fundamental triangles.

N. G. W. H. Beeger(8) studies the quadratic reciprocity between pole and polar with respect to a triangle.

J. Klima(125) considers a special quadratic reciprocal correspondence in a plane.

On this, and similar transformations also see M. D'Ocagne(51), on "polar transformations" of curves and surfaces by M. D'Ocagne(53) and H. Le Pont(136).

E. Dewulf(47) writes on a general geometric transformation of which a particular case is applicable to cinematics, by attaching movements in a plane to projective geometry.

E. Berné(13) gives an elaborate exposition of what he calls "transformation par inversion symétrique," partly connected with inverse properties.

L. E. Dickson(48) studies a quadratic Cremona transformation defined in connection with a fixed conic. On a conic K choose four points $ABCD$. To a point P we make correspond P' as follows: Join AP and DP cutting K in F and G. Join FB and GC; they intersect in the point P' which corresponds to P in an involutorial quadratic transformation. The join PP' passes through O.

A projective treatment of this transformation is given by the same author in (48.1).

H. Wiener(243) considers the transformations which may be compounded by two reflexions: (1) projectivity on a line, (2) circular transformations of the plane, or on sphere, (3) plane (space) collineations which leave a curve (surface) of the 2. order invariant, (4) movements of rigid space-systems, and transformations of equisymmetrical space-systems.

G. Darboux([43.1]) considers space constructions for T_2. M. W. Haskell([94.1]) reduces the equations of T_2 to standard form.

A. J. Third([226]) investigates the quadratic transformation,

$$x' : y' : z' = x(by+cz) : y(cz+ax) : z(ax+by),$$

and makes applications to the geometry of the triangle and certain conics connected with it.

In ([54]) D'Ocagne has a note on two transformations of plane curves. From a generic point M drop perpendiculars MP and MQ to OX and OY, then the join PQ corresponds to $M(x, y)$ in the involutorial quadratic correlation $u = -\frac{1}{x}$, $v = -\frac{1}{y}$. D'Ocagne studies the envelope-curve, when M describes a given curve.

K. Ono([169]) studies some properties of the quadratic correlation which exists between a point P and its so-called triangular polar with respect to a fixed triangle. This polar is obtained by constructing the fourth harmonic points α', β', γ' of the intersections α, β, γ of PA, PB, PC with the opposite sides, and the corresponding pairs of vertices of the triangle. Then $\alpha'\beta'\gamma'$ are collinear on the "triangular polar."

D. Inversion. Circle Transformation

1. Circle transformations (Kreisverwandtschaften) are of such importance in many fields of mathematical research, that, although belonging to the class of quadratic transformations, a particular section must be reserved for them.

Inversion, a special case of circle transformation, was discovered and applied to the solution of certain geometric problems long before the establishment of the general circle transformation by Moebius. As its historic aspects have been considered before in section A of this report, it is not necessary to repeat an account of them in this place. It is, however, pertinent to point out the fact that the introduction of complex quantities and their geometric representation by Wessel-Argand-Gauss gave the theory of circle transformations in its further development a tremendous impetus. Their application in the theory of functions of a complex variable as developed by Cauchy, Riemann, Weierstrass, and expounded by a large number of mathematicians, is evident to every beginner. The crowning results of the brilliant efforts in connecting the theory of circle transformations with the theory of substitution groups, or the corresponding division of the complex plain into cells bounded by circular arcs; then with the theory of automorphic functions, with non-Euclidean geometry, and with the uniformization of analytic functions are a sufficient attestation for the importance of circle transformations, especially when considered from the standpoint of function theory. An historical account of this development is outside the limits set for this report, and we must content ourselves with mentioning a few landmarks in this connection: F. Klein(122); Klein-Fricke(123), (124); Poincaré(175).

In what follows, an account will be given of the purely geometric aspects of circle transformation and its applications. Much of the work that has been done in this direction is mathematically of little consequence, while in a number of cases the results obtained deserve detailed accounting.

2. *Inversion.*—J. Liouville(137) in commenting on W. Thomson's method introduces the term "transformation par rayons vecteurs réciproques."

H. S. Hart(94) writes on transformations of curves. He refers to Salmon's mention of inversion introduced (not for the first time) by Dr. Ingram(110) and Mr. Stubbs(216). Hart finds a number of properties of inversions of conics, as the limaçon, cartesian ovals, etc., by choosing the center of inversion at a focus.

N. M. Ferrers(78) refers to Hart's method of transforming curves, consisting in writing $\frac{k}{r}$ for r in the polar equation of the curve, and discusses some metrical properties of inverse figures.

G. H. O. Boeklen(15) considers three geometric transformations: (1) Transformation of reciprocal radii-vectors. (2) Transformations by proportionally divided coordinates. (3) Transformation by bending of surfaces; triangulation of surfaces; infinitely small triangles must remain congruent by deformation.

B. Tortolini(230) makes a study of space-inversions, and some metrical relations on inverse figures.

A. Mannheim(146) shows how properties of cyclides may be obtained very simply by inversion. Mannheim solves by inversion a certain problem on circle-envelopes.

A little later Mannheim(147) investigates self-inverse curves and studies "anticaustiques."

M. Moutard(160) studies the inverse of surfaces which contain the imaginary sphero-circle as a multiple line. If m is the degree of the surface, p the order of multiplicity of the center of inversion, q the multiplicity of the sphero-circle, m', p', q' are the corresponding numbers of the transformed surface, then $m'=2m-p-2q$, $p'=m-2q$,

$q' = m - p - q$, and conversely, $m = 2m' - p' - 2q'$, etc. When $m' = m$, then $p' = p$, $q' = q$. Some of these surfaces are transformed into themselves. Moutard proposes to call such self-inverse figures *analagmatic* (α privatif, $\alpha\lambda\lambda\alpha \cdot \tau\sigma\omega$, je change), a rather bombastic word now in general use. In ([161]) Moutard derives certain metric properties of bicircular quartic surfaces.

H. E. Habich([92]) writes on the transformations of plane curves by the method of reciprocal radii-vectors and derives relations between the curvature and centers of curvature of a curve and its inverse.

A. Mannheim([145]) gives a geometric demonstration of a property of the transformation by reciprocal radii. He considers two inversions in space, the first transforming F into F_1, the second F_1 into F_2, and shows geometrically that it is possible to obtain a figure F_3 symmetric to F_2 with respect to a plane, by transforming F directly.

J. C. Maxwell([151]) applies inversion to the study of cyclides.

A. Cayley([29]) obtains equations of cyclides by inversion of quadric surfaces.

J. Casey([23]) applies inversion on circle and sphere, to establish certain properties of cyclides, for example, the fact that the bicircular quartic surface can be generated in five different ways as the envelope of a variable sphere, which cuts a given sphere orthogonally, and whose center moves on a given quadric.

In ([22]) Casey gives an elaborate exposition of the generation and classification of cyclides.

In a short note([171]) A. Peaucellier describes the linkage which now goes by his name, a mechanism without sliding movements realizing inversion of a circle, and consequently the mechanical description of a straight line. Peaucellier had announced his discovery before([170]). He calls a linkwork a "compas composé," and shows how by such mechanisms, lines, circles, conics, chonchoids, and cissoids may be described.

Lipkin, a student of Tschebichefs, rediscovered Peaucellier's inversor in 1871.

A. B. Kempe([119]), ([120]) describes "general" methods of producing exact rectilinear motions by linkworks.

H. S. Hart([94]) constructs a number of linkages and shows by their means the mechanical description of various algebraic curves.

A. Emch([65]), ([70]) illustrates the elliptic integral of the first kind by a certain linkage consisting of a combination of Peaucellier's inversors, and is thereby able to prove a number of problems of closure on Poncelet polygons and circle-series.

H. Schroeter([200]) follows, presumably, the procedure of Steiner, i. e., the method of inversion, to solve Malfatti's problem of three circles touching each other, and each two sides of a triangle, as described by Steiner([214]) without proof. The solution, obtained without analysis, disproves Pluecker's insinuation, that Steiner very likely had no proof for his solution.

H. E. Grassmann([90]) gives a geometric proof of a metric theorem concerning the inversion of a sphere.

G. Hessenberg([97]) describes linkages for the circle transformation.

P. H. Schoute([197]) shows that the six tangents from a point P to the lemniscate $(x^2+y^2)^2=2a^2(x^2+y^2)$ have their contacts on a conic. This conic cuts the lemniscate in two residual points, P_1, P_2. The line P_1P_2 cuts the lemniscate in two further points, P_3, P_4 which are residual points of a point P', a quadratic transform of P. This transformation is the inversion with respect to the circle with center at origin and radius a.

The same author([198]) applies inversion to the figure of one triangle symmetric with respect to another, and to the figure of three circles symmetric with respect to a triangle, and makes extension to space.

F. Meyer([153]) uses inversion and the quadratic transformation as aids in the study of nodes of algebraic curves.

M. Cochez([34]) gives a general account of the geometric theory of inversion, or transformation by reciprocal radi; likewise, A. Daguillon([42]).

C. Grolleau([91]) solves a number of problems, as, for example, the inversion of two arbitrary circles into two equal circles.

R. Bouvaist([17]) applies inversion for the derivation of certain properties of the strophoide.

F. Fouret([80]) shows that there is no curve which admits ∞^1 inversions whose centers form a curve locus (though curves with ∞^1 axes of symmetry and their transforms admit (discontinuous) ∞^1 inversions).

J. Hadamard([93]) finds the surfaces which admit ∞^1 inversions whose centers form a curve locus. See also F. Fouret([80.1]) on the principal poles of inversion of Dupin's cyclide; and G. Humbert([109]), who makes a comprehensive study of cyclides, following Moutard, Laguerre and Darboux.

R. Lachlan([127]) abstract in ([128]), uses inversion as a working method in an extensive study of systems of circles and spheres.

S. L. Ravier([182]) writes on the transformation by reciprocal radii and the mechanical description of quadrics, and J. Reveille([134]) uses inversion methods to generate anallagmatic curves.

J. Kleiber([121]) contributes to the cinematic theory of linkages, in which, in part, inversions play a rôle, and refers to the extensive bibliography contained in L. Burmester's Lehrbuch([21]) der Kinematik, Leipzig, 1888. See also, F. J. van den Berg([235]); P. H. Schoute([198]). M. J. Neuberg([164]), v. (1886), V. Fuhrmann([82]).

A. Gob ([86]) establishes some properties of Neuberg circles.

In an interpretation of transformations by reciprocal radii, J. Amaldi ([1]) uses a mapping process of circles in a plane upon points of an S_3.

G. Lazzeri ([131]) gives a new geometric definition of inversion on a circle (sphere) which leads to a chain of theorems, and eventually to the usual definition.

R. Bricard ([19]) gives a proof that under inversion lines of curvature of a surface pass into lines of curvature of the transform.

L. Sire ([209]) studies some relations existing between inverse and polar reciprocal curves.

G. Valiron ([234]) shows how the theorem that lines of curvature are preserved by inversion, as a consequence of Dupin's theorem, can also be proved by a theorem of Joachimsthal.

Th. Leconte ([132]) gives an elementary proof of the theorem that inversion preserves the character of lines of curvature.

K. Ogura ([168]) introduces differential forms of a surface which are useful in the investigation of invariantive properties of surfaces by inversion.

R. A. Johnson ([111]) studies inversion with reference to directed angles and gives a proof of Schoute's theorem, and proves a number of propositions in connection with the geometry of the triangle.

D. Coelingh ([35]) gives an expository account of inversion and its applications. He considers first the point transformation obtained by inversion and second a line transformation which is the dual of the first, and combines these two operations to derive properties of systems of circles.

J. J. Quinn ([180]), ([181]) describes new mechanisms for inversion.

A. Larmor ([130]) extends the contact problem of eight circles touching three given circles to circles on a sphere, and shows how the principle of polarity is made complete and the method of inversion on the sphere is then seen in its true light. This generalization also enables Larmor to deduce the descriptive geometry of a quadric with reference to its plane sections.

J. W. Clawson ([32]) studies an inversion of the complete quadrilateral, and derives some of its properties by inversion.

A. Voss ([238]) in a memoir on the theory of reciprocal radi, investigates the metric properties of this transformation and makes application to differential geometry.

G. Bouligand ([16]) connects the theorem that the only real surface with all its points umbilics, is the sphere, with Liouville's theorem on the compositions of conformal space-transformations.

A few notes on inversions are given by T. Bhattacharyya([7]). E. Cavalli([26]) studies reciprocal figures and the application of quadratic transformations in cinematics.

3. *Circle Transformation* (*Kreisverwandtschaft*).—After the establishment of the general circle transformation by Moebius as reported in section A, A. Cayley([28]) proves a theorem on circles which amounts to a description of the same correspondence. Given $ABCP$ and $A'B'C'$. Through BCP, CAP, ABP, ABC pass circles α, β, γ, ω respectively. Construct three circles α' through $B'C'$ cutting ω' (through $A'B'C'$) at same angle as α cuts ω; β' through $C'A'$, γ' through $A'B'$ with similar cutting of ω'. Then α', β', γ' meet in P' which corresponds to P in a circle transformation. α', β', γ', ω' pass through $B'C'P'$, $C'A'P'$, $A'B'P'$, $A'B'C'$.

The discovery that the circle transformation in a special case of the general quadratic transformation, dates back to Magnus, Reye, Weyr, contained in papers previously mentioned in section A. When two of the fundamental points, say B and C of a quadratic transformation coincide with the circular points I and J of the plane Σ, then the homaloidal net consists of the net of circles through a fixed point A. The lines of the fundamental triangle consist of the isotropic lines through the point A, and the line at infinity. An analogous situation exists in the plane Σ. Lines and circles in Σ are transformed into lines and circles in Σ'. When A and A', and Σ and Σ' coincide, the involutorial circular transformation arises.

In case of an involutorial circle transformation

$$z' = \frac{az+b}{cz-a},$$

and superposed planes Σ and Σ', the pencil of conics, with respect to which pole and point of concurrence of polars are corresponding points, consists of a concentric pencil of equilateral hyperbolas, through two fixed real points R_1 and R_2, whose mid-point is A, and two fixed imaginary points I_1 and I_2, obtained as finite imaginary intersections of zero-circles around R_1 and R_2. AIJ is now the diagonal triangle of the quadrangle $R_1R_2I_1I_2$ of invariant or base-points of the pencil. The involution cut out on the line at infinity by the pencil of equilateral hyperbolas has I and J as double points. In particular when the coordinates of R_1 and R_2 are $(1, 0)$ and $(-1, 0)$, those of I_1 and I_2 $(0, i)$ and $(0, -i)$, respectively, the transformation becomes identical with complex inversion

$$z' = \frac{1}{z}.$$

This theory and its generalization for transformations of the type $z'=f(z)/g(z)$; $z'=z+\phi(z)/\psi(z)$, and in particular,

$$z'=z-\frac{(n+1)f(z)}{f'(z)},$$

where $f(z)$ is a polynomial of degree n, and applications to projectivity between pencils of stelloids and pencils of their first polars, is given by A. Emch([68]). See also C. M. Hebbert([96]).

In the theory of functions of a complex variable, the circle transformations of the second kind,

$$z'=\frac{a\bar{z}+b}{c\bar{z}+d},$$

where $\bar{z}=x-iy$ are also of great importance. They differ from the direct circle transformations (of the first kind) in that they are connected with generalized reflexions, i. e., geometric inversion or transformation by reciprocal radii. Contours of closed figures in Σ' corresponding to contours of closed figures in Σ have the same—or opposite—directions to those in Σ, according as the circle transformation is of the first or second kind.

H. Durège([61]) applies Siebek's representation $y=(p+qx)/(r+sx)$ of the circle transformation to the description of curves by y, and their properties, when x describes a given curve.

M. Laissant([129]) expresses the desirability that the term transformation by reciprocal radii should be replaced by inversion. He studies various transformations of similar character, for example, those that preserve areas.

G. Battaglini([6]) considers a non-Euclidean circle transformation established by two projective pencils of double contact conics, such that to a conic having a double contact with a fixed conic corresponds a conic with double contact with another fixed conic.

A general account of circle transformations in the complex plane is given by A. Emch([64]) with various applications, for example, to the Steiner circle series.

F. N. Cole([36]) discusses the linear functions of a complex variable and their geometric interpretation by points of complex planes, from a geometric point of view. In particular he studies the relation between the circle transformation of the complex plane, the stereographic projection of the sphere and the equivalent collineations of the sphere into itself. The arrangement follows more or less closely Klein's lectures on the theory of functions in 1880-81.

Ed. von Weber([239]) gives a general account of and discusses circle transformations, in the main by use of the complex plane.

E. Kasner([118]) develops the invariant theory of the inversion group with application to the geometry upon a quadric surface. Kasner's paper is concerned with the projective geometry on a quadric, in particular with those properties of algebraic curves on a proper quadric which are invariant under the collineation group which leaves the quadric invariant. When the quadric becomes a sphere the stereographic projection of the sphere, and on it the 6-parameter group of collineations which leaves the sphere invariant, gives in the plane the group of circular transformations; in general, inversion geometry in the plane. Kasner also studies the form problems of these geometries and their bearing upon the theory of curves on quadrics and in the plane of inversion.

In ([117]) E. Kasner considers a relation between the circular and the projective transformations of the plane. Kasner establishes a collineation as follows: Let τ be a circular transformation and O a finite point, not the pole of τ. Let O' correspond to O in τ. Any circle through O is transformed into a circle through O'. The centers of such circles correspond to each other in a collineation H. Conversely, to any collineation H, not an affinity, there correspond two combinations (τ, O) each of which gives rise to H.

A. Emch([63]) gives simple demonstrations of some geometric properties of circular transformations which are of value in the study of ring-shaped domains as used in the definitions of certain automorphic functions of the Schottky type.

M. Pieri([172]) points out the fact that up to now there are two main lines of approach to the study of inversion geometry. The first is by means of the circular transformations (6-parameter group) in the plane; the second by tetracyclic and pentaspheric coordinates.

To avoid the intermediary of a projective geometry in an S_4, in his memoir, Pieri proposes to establish inversion geometry and its principal propositions by its natural elements, the point and the circle, and a set of twenty postulates between them, without making particular use of previous modes of treatment.

A. Emch([66]) studies the rectilinear congruence (2, 2) obtained by joining corresponding points of a circular transformation between two fixed planes.

W. Scheibner([192]) writes on Moebius' Kreisverwandtschaft and inversion; likewise C. Klobasa([126]) on the principles of inverse transformations in the plane.

A. Simionov([208.1]) uses inversion and reciprocal polars for the transformation of geometric propositions.

The dissertation of A. Neumann([166]) deals with quadratic transformations in the plane and in space, in particular circle- and sphere-transformation.

E. Meyer([152]) investigates involutorial conformal point-transformations of space.

In the last paragraph of a paper on space-sextics of genus 2, M. Stuyvaert([220]) shows that such sextics can be transformed by inversion into a quintic of genus 2, a fact already noted by D. Montesano([158]). For further references on circle transformations, etc., see Holzmueller([108]).

4. *Geometry of the Circle and the Sphere.*—The natural continuation of elementary inversion geometry of the circle and the sphere leads to an extensive doctrine embodied in the geometry of the 10-parameter group of conformal transformations of S_3, and the higher sphere geometry of the 15-parameter group of linear transformations of six homogeneous variables, and their extension to higher spaces. In this is included the geometry of tetracyclic and pentaspherical coordinates, and the connection with line-geometry. For all this the reader is referred to the Geometry of the Circle and the Sphere by J. L. Coolidge, and F. Klein's Vorlesungen ueber Hoehere Geometrie, 3. edition 1926. A detailed account and bibliography of this geometry should form the object of a separate report.

E. Various Applications. Groups of Quadratic Transformations

1. *Geometric Applications. Invariant Algebraic Curves.*—G. Jung and A. Armanente([112]) give a general exposition of Cremona transformations and their application to birational reduction of curves of given genus p. For example, it is shown how the general quintic may be obtained from a certain septimic with 9 double points by a quadratic transformation. See also Clebsch-Lindemann([33]).

2. In some applications of involutorial quadratic transformations, E. Duporcq([59]) proves that the four tangents from a point on a lemniscate have contacts on a line which envelopes the conic touching the six flex-tangents.

V. Retali([183]) shows how to get cusps of the second kind by quadratic transformations.

E. Beltrami in ([12]) makes use of quadratic transformations to study the properties of such curves as Steiner's tricuspidal hypo-cyclide.

F. P. Ruffini([189]) studies properties of conics which are polar reciprocals of each other.

K. Doehlemann([56]) uses quadratic transformations to study metric properties of the net of conics on three lines.

K. Doehlemann([55]), ([57]) investigates involutorial forms which can be contained in a plane Cremona, particularly in a quadratic, transformation. He defines first what he calls "abgeleitete," derived transformation. To (X, Y') correspond X' in $(\Sigma \to \Sigma')$, and Y in $(\Sigma' \to \Sigma)$. There exists a Cremona transformation between X' and Y, which is of degree n^2. When X' describes a line in Σ', to it corresponds in $\Sigma(\Sigma' \to \Sigma)$ a curve ϕ described by X. But, considering X in place of Y' in Σ, then Y in $(\Sigma' \to \Sigma)$ describes a curve of order n^2. This transformation between X' and Y is called the *derived* transformation, and is represented by,

$$x'_i = \phi_i[\phi_1(x), \phi_2(x), \phi_3(x)], \qquad i=1, 2, 3.$$

A quadratic transformation may contain a general, or a rational cubic, with central or non-central involution, as an involutorial curve. If the involution on the C_3 is not central, the quadratic transformation thus defined, has no double points. If the involution is central and M its central point, then A_1B_1, $A'_1B'_1$, $A'_2B'_2$ meet in M_1, the tangential of M. Moreover the points of contact of the tangents from M are the invariant points $D_1D_2D_3D_4$ of the transformation. C_3 is the isologue of M. In this case the derived transformation is of order 4 and has C_3 as a pointwise invariant curve.

Doehlemann then considers quadratic transformations which have as an involutorial C_3 a line and a conic, and finally three lines; and some other special quadratic transformations.

He finds that a rational C_3 with non-central involution can be contained only in an involutorial quadratic transformation of the plane.

M. Disteli (49) gives a simple plane representation of the forms of plane cubics, their construction and problems of closure; utilizing the Steiner transformation. The method leads to a constructive classification of cubics.

In (50) M. Disteli applies the Steiner transformation for the proof of a large number of theorems which Steiner published mostly without proof.

Ch. Michel (154) studies transformations of cubics into themselves by a Steiner transformation. He finds two classes: first, those whose joins of corresponding points meet in a point of the cubic; second, those whose joins envelope a class cubic. See also B. Bydzovsky (21.1).

S. C. Baghi (5) proves certain properties of four pairs of inverse points on a self inverse cubic.

P. H. Schoute (194) uses quadratic transformations as an auxiliary in the investigation of quartics with three inflexional nodes.

A. Emch (70), pp. 275-284, by means of the Steiner transformation gives simple constructive methods for the Newtonian five types of cubics.

W. Binder (14) maps plane rational cubics and quartics on a conic by means of a Steiner transformation, and studies some of their properties in their relation to the conic and the point correspondences on it.

A. Emch([62]) by means of the Steinerian transformation attached to an orthocentric quadrangle, develops a simple method for the study of circular cubics and its special cases.

A. Emch([67]) studies the properties of the invariant net of cubics in the Steinerian transformation.

The same author([69]) discusses possible classes and types of algebraic curves which are invariant under an involutorial quadratic Cremona transformation. In particular, hyperelliptic curves are considered, and a classification of hyperelliptic sextics and special quadratic transformations are studied. In the second paper the couples of corresponding points (P_1P') of the transformation, $\rho x'_1 = x_2x_3$, $\rho x'_2 = x_3x_1$, $\rho x'_3 = x_1x_2$ are mapped on a Cayley cubic surface by means of the invariant functions, $\rho y_1 = x_1(x_2^2+x_3^2)$, $\rho y_2 = x_2(x_3^2+x_1^2)$, $\rho y_3 = x_3(x_1^2+x_2^2)$, $\rho y_4 = 2x_1x_2x_3$. The Cayley cubic thus obtained is $2y_1y_2y_3-(y_1^2+y_2^2+y_3^2)y_4+y_4^3=0$, and the mapping process between its points and the couples $(x)(x')$ or lines (n) in the plane of transformation is studied in detail.

The characteristic difference between two classes of invariant curves is exemplified by the two classes of invariant cubics,

$$a_1x_1(x_2^2-x_3^2)+a_2x_2(x_3^2-x_1^2)+a_3x_3(x_1^2-x_2^2)=0,$$

and

$$a_1x_1(x_2^2+x_3^2)+a_2x_2(x_3^2+y_1^2)+a_3x_3(x_1^2+y_2^2)+2a_4x_1x_2x_3=0.$$

Only the square of the first may be rationally represented on the Cayley cubic.

The classification of invariant curves by means of this mapping process, which is, in general, possible for any involution in a plane is extended to quartics, quintics, and sextics.

The investigation of invariant algebraic curves and surfaces is extended by A. Emch to general involutorial Cremona transformation in the plane and higher spaces([69, 71]). It is found that in any involutorial Cremona transformation of any dimension, there are, in general, two classes of invariant forms (A) and (B), so that the square of (A) belongs to class (B).

The problem of invariant curves in a Cremona transformation is also considered by A. Myller([163]). See also F. G. Taylor([222.1]) on birationally related cubics.

E. Ciani([30]) studies some constructions of the plane quartic with a double point, which he generates by the use of quadratic transformations.

The problem of mapping couples of corresponding points of an involutorial quadratic transformation upon certain cubics has also been considered by L. Godeaux([88]). See also ([87]).

O. M. Thalberg (224) investigates the geometric relation between the tangents at multiple points of certain algebraic curves by means of involutorial quadratic transformations.

In a previous paper (225) the same author considers problems of this sort of a more general type by means of plane involutorial Cremona transformations, in particular properties of curves which are invariant in such transformations.

Concerning birational transformations of cubics into themselves see also E. Weyr (241), and C. Segre (204).

K. Ogura (168) applies involutorial quadratic transformation to cubics which are invariant, and studies their relations with the geometry of the triangle.

E. Ouivet (169.1) applies Cremona transformations to certain types of differential equations.

H. E. Timerding (229) investigates quadratic Cremona transformations also in connection with certain important classes of curves. Reference is made to G. B. Guccia's investigations (91.1) and first an account is given of curves and their properties connected with the general Cremona transformation. Timerding applies the results to the quadratic transformation, thus leading to a treatment which already W. Godt explained in (89). Timerding's object is to return once more to this analytic treatment and to add some new results.

3. *Repeated Quadratic Transformations.*—S. Kantor (113) determines the number of cyclic groups in a plane quadratic Cremona transformation. He first establishes the fact that in every quadratic transformation there is, in general, one pair of points which are in involutorial correspondence. The four invariant points of the transformation form the base points of a pencil of conics in the net of conics determined by three pairs of corresponding sides of the fundamental triangles. There are three cyclic groups of three points each.

Kantor finds the following formula for the number of cyclic groups of N points. If we put $N = a_1^{m_1} a_2^{m_2} \ldots a_\nu^{m_\nu}$, where $a_1 a_2 \ldots a_\nu$ are prime factors, then the number of groups, in any quadratic Cremona transformation, for which every point of a group returns to its original position after applying the transformation N times is,

$$Z_n = \frac{2 - \Sigma 2^{\frac{N}{a_r}} + \Sigma 2^{\frac{N}{a_r a_s}} + \ldots + (-1)^\nu 2^{\frac{N}{a_1 a_2 \ldots a_\nu}}}{N}$$

$$= \frac{1}{N} \cdot \sum_{\mu=0}^{\mu=\nu} \sum_{r=1}^{r=\nu} (-1)^\mu 2^{\frac{N}{a_{r_1} a_{r_2} \ldots a_{r_\mu}}},$$

in which for $r_1 r_2 \ldots r_\mu$ any permutation of the numbers $1, 2, \ldots, \nu$ must be set.

Thus there is one involutorial couple, there are two periodic triples, three periodic quadruples, 6, 9, 18 periodic quintuples, sextuples, septuples, etc.

For $N=a$, a part of Fermat's theorem results. Kantor's paper is dated Rome, February 13, 1880, and its results are generalized for a Cremona transformation of any order, in the same volume(114).

The same problem and some additional questions are considered by S. Kantor(115).

T. A. Hirst(105) studied the same problem of periodic quadratic transformations in 1881. The formula for the number of periodic transformations of cycles of m points, which is identical with that given by Kantor, was derived for Hirst by Professor H. J. Smith.

V. Snyder(211) refers to Kantor's and Wiman's results on periodic transformations and points out the omission by Kantor of an important category of such transformations. Snyder's purpose is to obtain the equation, define the system of fundamental elements, and discuss some of the properties of the transformation, and to show the existence of new ones not heretofore considered.

4. *Groups of Quadratic Transformations.*—L. Autonne(4) was one of the first to investigate quadratic Cremona substitution groups of three homogeneous variables.

H. E. Slaught(210) investigates the "cross-ratio group" of 120 quadratic Cremona transformations of the plane. This is a particular case of E. H. Moore's cross-ratio group(159). In the second part of the paper Slaught studies the complete form system of invariants of the cross-ratio group of order 120.

P. Field(79), in a note on certain groups of transformations of the plane into itself, considers the case of Slaught's diagram of five points and the 120 regions of the cross-ration group of the 120 quadratic Cremona transforms, when 4 of the 5 points are imaginary.

E. Ciani(31) investigates some notable groups of quadratic transformations in a plane. They have the property that they transform into themselves two groups of four lines each, which correspond to each other in two non-concentric projective pencils in the same plane. Ciani studies and classifies the groups formed by such transformations. The group is of order 8 when the two quadruples of lines are generic; of order 48 when they are harmonic; and of order 96 when they are equianharmonic. The most notable of these transformations are the quadric inversions, others are involutorial with four invariant points, and the remaining ones are of period 3 or 4.

Ciani applies the results to the elliptic quartic, which has the property that the quadruples of tangents drawn from the two double points are projective.

Earlier, G. Fano([74]) investigates continuous groups of Cremona transformations in a plane, and certain groups of projective transformations.

The determination of Cremona transformations which form continuous groups has been carried out by F. Enriques([72]).

The same problem for space was solved by Enriques-Fano([73]).

A good account of continuous Cremona transformations may be found in G. Fano([75]).

F. Systems of Correlations and Collineations in Connection with Quadratic Cremona Transformations

1. *Hirst's Investigations on the Correlations of Two Planes.*—T. A. Hirst([106]) takes as a starting-point the correlations determined by four points, $A_1B_1C_1D_1$ in Σ and four corresponding lines $a_2b_2c_2d_2$ in Σ'. If a line corresponding to A_1 passes through A_2, then the line corresponding to A_2 passes through A_1. Two points like A_1 and A_2 are called conjugate; likewise, a_1 and a_2 are conjugate lines if the point corresponding to a_1 lies on a_2, and conversely. There is *one* condition that in a correlation two points shall be conjugate, or two lines shall be conjugate. The problem to determine a correlation which shall satisfy *any* eight given conditions admits, in general, a limited number of solutions. Hirst finds six types.

The systems determined by α points in Σ with α given polars in Σ; β lines in Σ with β polar in Σ'; γ points and δ lines in each plane with given conjugates in the other, are as follows: if $(\alpha\beta\gamma\delta)$ denotes signature of system of correlations, and noting that $(\alpha\beta\gamma\delta)\equiv(\beta\alpha\gamma\delta)$; $2\alpha+2\beta+\gamma+\delta=7$,

Type			Type		
I	(3010)	(0301)	V	(1050)	(0105)
II	(2110)	(1201)		(1041)	(0114)
III	(2030)	(0203)		(1032)	(0123)
	(2021)	(0212)	VI	(0070)	(0007)
IV	(1130)	(1103)		(0061)	(0016)
	(1121)	(1112)		(0052)	(0025)
				(0043)	(0034)

altogether 26 distinct fundamental systems.

Of particular importance are the cases (3010), (2030), (1130), (1050), (0070), each of which leads to a pair of conjugate connexes of the first class and second order. In each of these five systems, a point in either plane has a conjugate in the other plane, while every line is represented by a conic passing through three fixed singular points of the three exceptional correlations which each system includes.

We have here the quadratic transformation in its most general form; the three pairs of singular points being identical with the three pairs of fundamental points.

Hirst in ([103]) derives new numerical relations connecting exceptional correlations and shows how by means of these the above classification may be obtained in a much simpler manner.

Before this simplification Hirst extended the investigation, contained in the first paper, to correlations in space([102]), which led to much more complicated questions.

2. *Further Investigations on Correlations and Collineations.*—Th. Reye([186]) studies reciprocal transformations (correlations), polar, and null-systems; F^3— and ϕ^2— systems; and apolar relations.

R. Sturm([218]) investigates correlative pencils, and in ([217]) correlative bundles, where he extends the idea of Hirst's signature to space-relations.

R. Sturm([219]) considers correlations and correspondences connected with them, and first remarks that Hirst showed that in every ∞^1 system of correlations there are two kinds of singularities according as there is a singular point in each plane or a singular line in each plane.

In the system (0070) of which the remaining four mentioned by Reye and Hirst are special cases, the three pairs of singular points are those from which the seven points in one field and the seven corresponding points in the other field are projected by projective pencils.

Sturm also discusses the quadric inversion by Hirst-Geiser, as generated by ∞^1 correlations (1130). W and w as pole and polar in both senses (Σ and Σ'), and 3 points of K^2 as self conjugate. K^2 is common to all correlations of the system.

Sturm also points out an error made by A. Millinowski([247]) in a paper on correlations of the second order, obtained by two planes collinear in two ways. The latter thought, mistakenly, that the correspondence established by lines between Σ and Σ' was (1, 1).

After extending the investigation to correlative bundles, Sturm studies infinite numbers of quadratic transformations between Σ and Σ' with the same seven pairs of corresponding points. In order that this may be possible, the fundamental points in $\Sigma(\Sigma')$ must lie on a cubic $K^3(K'^3)$, as pointed out in connection with Millinowski's paper. In this case there are ∞^1 singular correlations among the ∞^2.

Also, the null-system of the second order, formed by A. Ameseder (3), is discussed. This may be put in simple form,

$$\rho u_1 = a_1 x_2 x_3,$$
$$\rho u_2 = a_2 x_3 x_1,$$
$$\rho u_3 = -(a_1 + a_2) x_1 x_2,$$

from which there is obviously $u_1 x_1 + u_2 x_2 + u_3 x_3 = 0$.

R. Sturm (219.1) proves a theorem due to Clebsch, published without proof, which, in terms of correlation, may be stated as follows: All ∞^3 correlations of two fields A, B which have five pairs of conjugate pairs of conjugate points $A_1, B_1; \ldots; A_5, B_5$ in common, have a sixth pair A_0, B_0 in common. Two points in two correlated fields (planes) are said to be conjugate, when each lies in the corresponding line of the other. Sturm discusses investigations by Rosanes and Voss on collinear and correlated systems and their connections with quadratic transformations.

S. Kantor (116) discusses the relation between nets of collineation and quadratic transformations.

F. London (138) discusses important constructive problems connected with the theory of correlation and the theory of quadrics and quadric systems. He succeeds in finding a simple construction of a quadratic transformation when seven corresponding pairs are given. Then the problem is solved to construct the common null-pair of three correlations when six of them are given. This is important because the construction of space quartics of the first kind, when eight of its points are given, may be reduced to a special case of that problem. Similarly, the construction of a quadric, when nine points are given, becomes identical with the construction of a quadratic transformation when seven pairs of corresponding points are given.

Other solutions of the latter problem are given by H. Schroeter (199), and by Th. Reye (187).

3. *Trilinear Correspondence.*—H. Schubert (202), (203) establishes trilinear relations between three one-dimensional forms g, g', g'', whose parameters are connected by an equation,

$$xx'x'' + ax'x'' + a'x''x + a''xx' + bx + b'x' + b''x'' + c = 0.$$

It has six singular pairs. Such trilinear relations lead to forms of order 3; for example, 3 trilinear pencils of planes generate a cubic surface. By the properties of this relation Schubert is able to construct the cubic surface linearly.

G. Hauck (95) develops a theory of trilinear correspondence between plane systems $\Sigma_1, \Sigma_2, \Sigma_3$, and threefold (dreibuendig-eindeutig) quadratic Cremona transformations with three pairs of fundamental conics.

BIBLIOGRAPHY

1 Amaldi, J., Nap. R. A. Rend., 5: 238-240 (1891).

2 Amanzio, D., Math. Gaz., 14: 48-53 (1876).

3 Ameseder, A., Wien. Ber., 83: 385-401 (1881).

4 Autonne, L., see reference in Chap. IV.

5 Baghi, S. C., Math. Qu., 8: 54-55 (1905).

6 Battaglini, G., Giorn. di Mat., 16: 256-263 (1878).

7 Bhattacharyya, T., Calc. Bull., 10: 29-34 (1919).

8 Beeger, N. G. W. H., Wisk. Tijds., 7: 42-47 (1910).

9 Bellavitis, G., Nap. Soc. It. Sc. Mem., (3), 3: no. 4, 43 p. (1879).

10 ———, Padova N. Saggi, 4: 243-288 (1838).

11 ———, R. L. Ven. Ann., 6: 126-141 (1836).

12 Beltrami, E., Bologna Mem., (2), 2: 361-395 (1862); (3), 5: 543-566 (1874). °

13 Bernès, E., J. de M. E., 15: 121, 145, 175, 197, 217, 244, 265 (1891); 16: 3, 25, 49, 73, 97, 126, 145, 169, 192, 217, 241 (1892); 17: 3, 25, 49, 76 (1893).

14 Binder, W., Z. Math. Phys., 36: 78-98 (1891).

15 Boeklen, G. H. O., Arch. Math. Phys., 32: 83-94 (1859).

16 Bouligand, G., Nouv. Ann., (5), 1: 266-267 (1922-23).

17 Bouvaist, R., Rev. de Math. Spec., 8: 162 (1904-06).

18 Braikenridge, W., Exercitatio geometrica de descriptione linearum curvarum, London, (1733).

18.1 Bretschneider, P., Prog. Plauen in V., 24 (1876).

19 Bricard, R., Nouv. Ann., (4), 3: 16-17 (1903).

20 Buetzberger, F., Ueber bicentrische Polygone, Steinersche Kreisreihen und die Erfindung der Inversion, Leipzig, (1913).

21 Burmester, L., Lehrbuch der Kinematik, Leipzig, (1888).

21.1 Bydzovsky, B., Prag. Cesk. Ac. Bull., (1899).

21.2 Cardoso-Laynes, G., Il Pit., (1), 5_2: 34-38 (1898).

22 Casey, J., Lond. R. S. Trans., 161: 585-721 (1872).

23 ———, Lond. R. S. Proc., 19: 495-497 (1870-71).

24 Cassani, P., R. I. Ven. Atti., (7), 9: 30-34 (1897-98).

24.1 Catalan, E., Brux. Mém. Sc., (2), 2: 88 p. (1910); Brux. Bull., (2), 36: 605-606 (1873).

25 Cavalli, E., Nap. R. A. Atti., (2), 9: no. 12, 28 p. (1899).

26 ———, Nap. R. A. Rend., (3), 4: 411-412 (1898).

27 Cayley, A., J. de Math., (1), 14: 40-46 (1849). °

28 ———, Quart. J., 2: 162 (1858). °

29 ———, Quart. J., 11: 283-288 (1871).

30 Ciani, E., Giorn. di Mat., (3), 53: 192-197 (1915).

31 ———, Pal. Circ. Mat., 42: 323-338 (1917).

32 Clawson, J. W., Am. Math. Month., 24: 71-73 (1917).

33 Clebsch-Lindemann, Vorlesungen ueber Geometrie, 1: 474-496, 690-693 (1876).

34 Cochez, M., J. de M. E., 1: 225, 257, 321, 353 (1877); 2: 1, 33 (1878).

35 Coelingh, D., Nieuw. Arch., 16: 116-159 (1889).

36 Cole, F. N., Ann. of Math., (1), 5: 121-176 (4 plates) (1889-90).

37 Commandinus, New Translation of Ptolemy's Work.

38 Cremona, L., Bologna Mem., (2), 2: 621-630 (1863).

39 ———, Bologna Mem., (2), 5: 3-35 (1865).

40 ———, Giorn. di Mat., 1: 305-311 (1863); 3: 269-280 (1865).

41 Czuber, E., Monatsh. f. Math., 5: 267-286 (1894).
42 Daguillon, A., J. de M. E., 6: 129-133 (1882).
43 Dandelin, P. G., Brux. Mém., 4: 1-10, 11-27 (1827).
43.1 Darboux, G., Soc. Phil., 5: 72-76; 77-80 (1868).
44 De Jonquières, E., Giorn. di Mat., 23: 48-75 (1885).
45 ———, Nouv. Ann., (2), 3: 97-111 (1864).
46 Del Pezzo, P., Nap. R. A. Rend., (3), 10: 366-372 (1904).
47 Dewulf, E., Ann. Éc. Norm., 3: 405-431 (1886).
48 Dickson, L. E., Am. Math. Month., 2: 218-221 (1895).
48.1 ———, Pal. Circ. Mat., 9: 256-259 (1895).
49 Disteli, M., Z. Math. Phys., 36: 138-157 (1891).
50 ———, Z. Vierteljschr., 37: 255-305 (1891).
51 D'Ocagne, M., J. de M. E., (2), 4: 15-16, 33-36 (1885).
52 ———, Liège S. R. S. Mém., (2), 16: no. 7, 10 p. (1890).
53 ———, Mathesis, (1), 4: 73-79 (1884).
53.1 ———, Nouv. Ann., (3), 12: 337-352 (1893).
54 ———, Ann. Polyt. Porto Ac., 9: 5-9 (1912).
55 Doehlemann, K., Geometrische Transformationen, vols, 1, 2.
56 ———, Z. Math. Phys., 32: 120-127 (1887).
57 ———, Z. Math. Phys., 36: 356-378 (1891).
58 Duporcq, E., C. R., 126: 1405-1406 (1898).
59 ———, Rev. de Math. Spec., 6: 177-178 (1900-02).
60 Durège, H., Die ebenen Curven dritter Ordnung, 121-129 (1871).
61 ———, Arch. Math. Phys., 42: 1-41 (1864).
62 Emch, A., Introduction to Projective Geometry, New York, 185-216 (1905).
63 ———, Am. Math. Month., 21: 139-146 (1914).
64 ———, Ann. of Math., (1), 12: 141-160 (1898-99).
65 ———, Ann. of Math., (2), 1: 81-92 (1900); 2: 60-63 (1901).
66 ———, Ann. of Math., (2), 13: 155-160 (1912).
67 ———, Bull. Am. M. S., 24: 327-330 (1918).
68 ———, Pal. Circ. Mat., 34: 333-344 (1912).
69 ———, Tohoku M. J., 21: 310-326 (1922); 24: 68-87 (1925); 25: 63-76 (1925).
70 ———, Colo. Stud., 1: 111-127, 275-284 (1904).
71 ———, Am. J. M., 48: 21-44 (1926).
72 Enriques, F., Rom. Acc. L. Rend., (5), 2: 468-473; 532-538 (1893).
73 Enriques-Fano, Ann. di Mat., (2), 26: 59-98 (1897).
74 Fano, G., Pal. Circ. Mat., 10: 16-29 (1896).
75 ———, Monatsh. f. Math., 9: 17-29 (1898).
76 Fermat, P., Varia Opera Matematica, Toulouse, (De contradibus sphaericis, p. 75-88) (1679).°
77 Ferrari, F., Soc. Phil., (9), 1: 93-103 (1899).
78 Ferrers, N. M., Quart. J., 1: 32-35 (1857).
79 Field, P., Bull. Am. M. S., 12: 234-236 (1905-06).
80 Fouret, F., Nouv. Ann., (3), 2: 259-262 (1883).
80.1 ———, Nouv. Ann., (3), 7: 113-116 (1888).
81 Frechet, M., Nouv. Ann., (4), 3: 503-507 (1903).
82 Fuhrmann, V., Mathesis, (1), 10: 105-111 (1890).
83 Gauss, C. F., Werke, 8: 354-355.
84 ———, Werke, 8: 102-105.
85 Geiser, F., Bern. Mitt., 97-107 (1865).

86 Gob, A., Liège S. R. S. Mém., (2), 16: no. 5, 12 p. (1890).
87 Godeaux, L., Brux. Bull., no. 7, 527-534 (1926).
88 ———, Mathesis, 26: 19-23 (1922); 30: 353-360 (1926).
89 Godt, W., Dissertation, Goettingen, (1873).
90 Grassmann, H. E., Leipz. Ber., 29: 133-134 (1877).
91 Grolleau, C., Rev. de Math. Spec., 5: 545-548 (1898-1900).
91.1 Guccia, G. B., Pal. Circ. Mat., 1: 119-132 (1887).
92 Habich, H. E., Nouv. Ann., (2), 5: 399-407 (1866).
93 Hadamard, J., Bull. Sci. M., (2), 12: 118-121 (1888).
94 Hart, H. S., Camb. Dub. Math. J., 8: 47-50 (1853); Mess. of Math., 4: 82-88 (1875).
94.1 Haskell, M. W., Cal. Ac. Proc., (3), 1: 1-12 (1898).
95 Hauck, G., J. f. Math., 98: 304-332 (1885).
96 Hebbert, C. M., Tohoku M. J., 13: 1-14 (1918).
97 Hessenberg, G., Wuerttenberg Ges., (1924).
98 Hipparchus, Account in Almagest of Ptolemy.
99 Hirst, T. A., Ann. di Mat., 7: 49-65 (1865).
100 ———, Brit. Ass. Rep., 34: 1st part, 3-4 (1864); 35: 2d part, 6-7 (1865).
101 ———, Giorn. di Mat., 4: 278-293 (1866).
102 ———, London M. S. Proc., 6: 7-9 (1874).
103 ———, London M. S. Proc., 8: 262-273 (1877).
104 ———, Nouv. Ann., (2), 5: 213-218 (1866).
105 ———, Quart. J., 17: 301-311 (1881).
106 ———, London M. S. Proc., 5: 40-70 (1873-74).
107 ———, London R. S. Proc., 14: 91-106 (1865).
107.1 Hofmann, F., Z. Math. Phys., 31: 283-295 (1886).
108 Holzmueller, G., Einfuehrung in die Theorie der isogonalen Verwandtschaften, Leipzig, 21-23 (1882).
109 Humbert, G., J. Éc. Polyt., 55: 127-252 (1885).
109.1 Igel, B., Z. Math. Phys., 17: 516-518 (1872).
110 Ingram, J. R., Dublin R. S. Trans., 1: 57-63 (1842-43).
110.1 Jacobi, A., J. f. Math., 23: 243-254 (1842); 31: 76-82 (1846).°
111 Johnson, R. A., Am. Math. Month., 24: 313-317 (1917).
112 Jung, G., and Armamante, A., Giorn. di Mat., 7: 235-253 (1869).
113 Kantor, S., Ann. di Mat., (4), 10: 64-70 (1880).
114 ———, Ann. di Mat., (4), 10: 71-73 (1880).
115 ———, Wien. Ber., 82: 237-259 (1880).
116 ———, Math. Ann., 20: 297-298 (1882).
117 Kasner, E., Ann. of Math., 5: 99-104 (1903-04).
118 ———, Am. M. S. Trans., 1: 430-498 (1900).
119 Kempe, A. B., How to draw a straight line, London, (1877).°
120 ———, London R. S. Proc., 23: 565-577 (1874-75).
120.1 Kilbringer, G., Z. Math. Phys., 33: 14-21 (1888).
121 Kleiber, J., Z. Math. Phys., 36: 296-301 (1891).
122 Klein, F., Vorlesungen ueber das Ikosaeder, (1884).
123 Klein-Fricke, Vorlesungen ueber die Theorie der elliptischen Modulfunktionen, 1: (1890); 2: (1892). °
124 ———, Vorlesungen ueber die Theorie der automorphen Funktionen, 1: (1897); 2: (1912). °
125 Klima, J., Casopis, 52: 64-72 (1922-23).

126 Klobasa, C., Grundsaetze der inversen Transformationen in der Ebene, Troppau, Prog., 12 p. (1907).
127 Lachlan, R., London R. S. Trans., 177: 481-625 (1886); Pure Geometry, (1893).
128 ———, London R. S. Proc., 40: 242-245 (1886).
129 Laissant, M., Mathesis, 10: 224-230 (1900).
130 Larmour, A., Brit. Ass. Rep., 57: 607 (1887).
131 Lazzeri, G., Per. di Mat., (2), 2: 137-143 (1900).
132 Leconte, Th., Nouv. Ann., (4), 12: 542-544 (1912).
133 Lehmer, D. N., Am. Math. Month., 17: 135-136 (1910).
134 ———, Am. Math. Month., 18: 52-57 (1911).
135 Leinekugel, G., Nouv. Ann., (3), 14: 391-406 (1895).
136 Le Pont, H., J. de M. E., (2), 4: 207-210, 224-225 (1885).
137 Liouville, J., J. de Math., (1), 12: 265-290 (1847).
137.1 Lipkin, L., Pet. Ac. Sci. Bull., 16: 57-60 (1871).
138 London, F., Math. Ann., 38: 334-368 (1891).
139 Longchamps, De. G., J. de M. E., (2), 6: 49-53, 77-82, 97-103, 121-126, 145-149, 193-195 (1882); J. de M. E., (2), 10: 109-127, 154-177, 198-229, 243-270 (1886).
140 ———, Nouv. Ann., (2), 5: 118-128 (1866).
141 Loria, G., Per. di Mat., (3), 4: 214-224 (1907).
142 McLaurin, C., Geometria organica sive descriptio linearum curvarum universalis, London, (1720).
143 Magnus, L. J., J. f. Math., 8: 51-63 (1832).
144 ———, Sammlung von Aufgaben ... (Part 3), 229-240 (1833).
145 Mannheim, A., J. de Math., (2), 16: 317-320 (1871).
146 ———, Nouv. Ann., (1), 19: 67-79, 115-117 (1860).
147 ———, Nouv. Ann., (1), 20: 218-222 (1861).
148 Mansion, P., Brux. Bull., (2), 36: 625-633 (1873); Nouv. Corr., 1: 54-61 (1874); 2: 129-135 (1876).
149 Massny, W., Programm. Gym. Cross-Strehlitz, no. 4, 20 p. (1887).
150 Mathieu, J. J. A., Nouv. Ann., (2), 4: 393-407, 481-493, 529-537 (1865).
151 Maxwell, J. C., Nouv. Ann., (2), 10: 162-182 (1871).
152 Meyer, E., Arch. Math. Phys., (3), 13: 149-152 (1908).
153 Meyer, F., Edinb. Proc., 13: 931-946 (1884-86).
154 Michel, Ch., J. de M. E., (5), 21: 111-114 (1897).
155 Millinowski, A., J. f. Math., 79: 140-158 (1875).
156 Moebius, A. F., J. f. Math., 52: 218-228, 229-242 (1856).
157 ———, Leipz. Ab., 2: 529-595, (1855); Werke, 2: 243-314.
158 Montesano, D., Torino Atti, 27: 660-690, 1053-1069 (1892).
159 Moore, E. H., Am. J., 22: 279-291, 336-342 (1900).
160 Moutard, M., Nouv. Ann., (2), 3: 306-309 (1864).
161 ———, Nouv. Ann., (2), 3: 536-539 (1864).
162 Muth, P., Math. Ann., 42: 257-272 (1893).
163 Myller, A., Nouv. Ann., (4), 14: 126-130 (1914).
164 Neuberg, M. J., Liège S. R. S. Mém., 16: no. 8, 12 p. (1890).
165 ———, Mathesis, (1), 6: 5-7 (1886); (1), 8: 177-183 (1888).
166 Neumann, A., Ueber quadratische Verwandteschaften in Ebene und Raum, insbesondere Kreis und Kugelverwandtschaft. Dissert. Koenigsberg, 82 p. (1908).

167 Newton, I., Enumeratio linearum tertii ordinis; Opera, Sam. Horsley's edition, 1: 156-157, 527-560 (1779).
168 Ogura, K., Tohoku M. J., 4: 132-137 (1913-14).
169 Ono, K., Tohoku M. J., 10: 31-45, 144-154, 198-206 (1916).
169.1 Ouivet, E., C. R., 150: 1036-1038 (1910).
170 Peaucellier, A., Nouv. Ann., (2), 3: 414-415 (1864).
171 ———, Nouv. Ann., (2), 12: 71-78 (1873).
172 Pieri, M., Giorn. di Mat., 49: 49-96 (1911); 50: 106-140 (1912).
173 Pluecker, J., Analytisch geometrische Entwickelungen, Essen, 1: (1828); *Also* J. f. Math., 11: 117-129, 219-225, 356-360 (1834).
174 ———, System der analytischen Geometrie, Berlin, (1835).
175 Poincaré, H., Acta Math., 1: 1-62, 193-249 (1882); 3: 49-92 (1883).
176 Poncelet, V., Traité des propriétés projectives des figures, 1: 185-248 (1865); First Edition, (1822).
177 ———, Traité, 1: (1865); sect. 1, ch. 2, 43-44.
178 Ptolemy, Ptolomaei Planisphaerium, Venice, (1558).
179 Quetelet, A., Brux. Mém., 4: 79-113 (1827).
180 Quinn, J. J., Am. Math. Month., 12: 105-106 (1905); 13: 143-144 (1906).
181 ———, Am. Math. Month., 16: 1 (1909).
182 Ravier, S. L., Nouv. Ann., (3), 10: 371-373 (1891).
183 Retali, V., Interméd., 16: 274-280 (1909).
184 Reveille, J., Nouv. Ann., (3), 12: 180-182 (1893).
185 Reye, Th., Geometrie der Lage, 4th ed., 2: 209-214 (1907).
186 ———, J. f. Math., 82: 173-206 (1877).
187 ———, Z. Math. Phys., 11: 280-310 (1866).
188 Rippert, L., Mathesis, (2), 9: 185-190, 217-221 (1899).
189 Ruffini, F. P., Bologna Mem., 6: 383-394 (1875).
190 Saltel, L., Brux. Bull., (2), 34: 51-52 (1872).
191 ———, Brux. Mém., 27: no. 5, 66 p. (1877).
192 Scheibner, W., Leipz. Ber., 58: 217-227 (1906).
193 Schiaparelli, G. V., Torino Mem., (2), 21: 227-319 (1864).
194 Schoute, P. H., Arch. Math. Phys., (2), 2: 113-128 (1885); 3: 113-137 (1886); 4: 308-322 (1886); 6: 113-156 (1888).
195 ———, Bull. Sci. M., (2), 6: 152-168, 174-188 (1882).
196 ———, Bull. Sci. M., (2), 7: 314-324 (1883).
197 ———, Wien. Ber., 89: 1252-1267 (1884).
198 ———, Nieuw Arch., 9: 117-140 (1882).
199 Schroeter, H., J. f. Math., 62: 215-231 (1863).
200 ———, J. f. Math., 77: 105-142, 230-244 (1874).
201 Schubert, H., Hamb. M. G. Mitt., 1: 31-36 (1882).
202 ———, Hamb. M. G. Mitt., 1: 1-3 (1882).
203 ———, Math. Ann., 17: 457-472 (1880).
204 Segre, C., Math. Ann., 27: 296-314 (1886).
205 Servais, Ch., Mathesis, (1), 7: 110-114, 129-134, 187-190 (1887).
206 ———, Mathesis, (1), 8: 105-109 (1888).
207 Seydewitz, F., Arch. Math. Phys., 7: 113-148 (1846).
208 Siebeck, F. H., J. f. Math., 55: 221-253 (1858).
208.1 Simionov, A., Buk. Gaz. Mat., 13: 71-74 (1907).
209 Sire, L., Rev. de Math. Spec., 9: 153-154 (1907-08).
210 Slaught, H. E., Am. J., 22: 343-380 (1900); 23: 99-138 (1901).

211 Snyder, V., Ann. of Math., 13: 140-148 (1911-12).

212 Steiner, J., Systematische Entwickelung der Abhaengigkeit geometrischer Gestalten voneinander, 251-295 (1832).

213 Steiner-Schroeter, Vorlesungen ueber synthetische Geometrie, 2: 2d ed., 301-304 (1876); 3d ed. by R. Sturm, (1898).

214 Steiner, J., J. f. Math., 1: 161-184, 252-288 (1826).

215 ———, J. f. Math., 3: 207-212 (1828); Werke, 1: 173-180.

216 Stubbs, J. W., Phil. Mag., (3), 23: 338-347 (1843).

217 Sturm, R., London M. S. Proc., 7: 175-194 (1875-76).

218 ———, Math. Ann., 12: 254-368 (1877).

219 ———, Math. Ann., 19: 461-486 (1882).

219.1 ———, Math. Ann., 22: 569-588 (1883).

220 Stuyvaert, M., Giorn di Mat., (3), 50: 67-99 (1912).

221 Synesius of Cyrene, Synesi Cyrnea ad paconium, etc., Paris, (1864).

222 Swellengrebel, J. G. H., J. f. Math., 43: 245-282 (1852).

222.1 Taylor, F. G., Edinb. M. S. Proc., 34: 163-175 (1915-16).

223 Terquem, O., Nouv. Ann., (1), 1: 403-404 (1842); (1), 5: 497-508 (1846).

224 Thalberg, O. M., Den Sjette Skandinaviske Matematiker Kongress, 473-477 (1925).

225 ———, Oslo Selsk. Skr., no. 3, 51 p. (1924).

226 Third, J. A., Edinb. M. S. Proc., 18: 11-24 (1899-1900).

227 Thomson, W., Brit. Ass. Rep., 17: 2d part, 6-7 (1847).

228 ———, J. de Math., (1), 10: 364-367 (1845).

229 Timerding, H. E., Math. Ann., 53: 191-219 (1900).

230 Tortolini, B., Ann. di Math., (1), 2: 189-195, 316-329 (1859).

231 Transon, A., Nouv. Ann., (2), 4: 385-393 (1865).

232 ———. Nouv. Ann., (2), 5: 63-70 (1866).

233 Tweedie, Ch., Edinb. R. S. Trans., 40: 253-262 (1901-02).

234 Valiron, G., Rev. de Math. Spec., 11: 238-239 (1911-12).

235 Van den Berg, F. J., Nieuw Arch., 7: 78-90 (1881).

235.1 Vercellin, R., Per. di Mat., (3), 7: 24-35 (1910).

235.2 Verdam, G. J., Amst. K. Ak. Wet. Verh., 12: 67-93 (1846).

236 Vièta, F., Francisis Vietae Apollonius Gallus sen excuscitata Apolloni Pergae, Paris, (1600).

237 Vigarié, E., J. de M. E., (3), 5: 101-105, 129-131, 153-157, 181-183, 207-211 (1891).

238 Voss, A., Muench. Ber., 49: 229-259 (1920).

239 Weber, E., Muench. Ber., 31: 367-408 (1901).

240 Weiss, V., Wien. Ber., 111_{2a}: 1489-1495 (1902).

241 Weyr, E., Wien. Ber., 87: 837-872 (1883).

242 ———, Z. Math. Phys., 14: 445-477 (1869).

243 Wiener, H., Leipz. Ber., 43: 644-673 (1891).

244 Wiener, C., Math. Ann., 3: 11-33 (1871).

245 Zahradnik, K., Wien. Ber., 75: 437-440 (1877).

CHAPTER II

ANALYSIS OF SINGULARITIES OF PLANE ALGEBRAIC CURVES

1. *Account of the Development of the Theory of Singularities of Plane Algebraic Curves.*—In the Lezioni Sulla Teoria Geometrica delle Equazioni e delle Funzioni Algebriche, by F. Enriques, edited by O. Chisini, Vol. 2, the whole theory of singularities is divided into three chapters: (1) Singularities and expansions in Puisseux series, (2) singularities with respect to quadratic transformations, (3) singularities with respect to differential calculus. The three aspects and their connections are critically illuminated. The treatment concludes with a historic survey of the whole problem.

As there are important connecting links between the various points of view, the historic part of the report will not be limited to the treatment of singularities by quadratic transformations, and will follow in the beginning mainly Enriques' Lezioni([50]).

The theory of algebraic curves or functions has its origin in the study of successive approximations, and as such is connected with the work of Newton. In reality Newton did not occupy himself particularly with singular points, but in([99]) giving examples, studied the problem of successive approximations of the implicit function defined by the equation of $f(x, y)=0$, from which results the expansion of y in powers of x. Newton considered this problem as an extension of the determination of the successive decimal places of a root of an equation $f(x)=0$. Considering $f(x, y)$ in the neighborhood of a point (o, a), he puts $y=a+z$, and obtains the resulting equation in z, which is supposed to be satisfied by putting $z=bx+z_1$, thus giving rise to a new equation in z_1, which is then solved by putting $z_1=cx^2+z_2$, and so forth.

To write effectively the equation which determines z, and analogously the succeeding equations, Newton arranges the terms of $f(x, a+z)$ in rows and columns, such that the terms of each row or column have the same degree with respect to one of the variables. Thus *Newton's parallelogram* arises([100]).

De-Gua([43]) applied this diagram to the problem of the separation of the branches of the algebraic function $y(x)$ at a singular point. For this purpose he assumes the singular point at the origin, and thus determines the groups of characteristic terms to which the equation reduces, by means of infinitesimals of higher order. In De-Gua's treatment, Newton's parallelogram assumes the form of "algebraic triangle," which Cramer later called "analytic triangle." De-Gua thought that singularities at the origin could be approximated by $y^\nu = ax^\mu$. This error was corrected by Euler, 1749, who observed that a branch may be imaginary, when x is negative, but still admits of a real approximation, and illustrated this by the cusp of the second kind $y = ax^2 + bx^{\frac{5}{2}}$, already discovered by De L'Hopitale.

The complete separation of the branches at a singular point by means of the calculus of successive approximations is found in Cramer([38]). The same method is used in Chapter 7 for the asymptotic approximation of infinite branches. According to Enriques, Cramer's method, for practical purposes, has not been surpassed even today, although other methods have been added since that time.

For Cramer the purpose of the investigation of a singularity is attained, when it is possible to separate the branches forming the singularity, and the approximation may be carried as far as desired. But Cramer's method does not examine the infinite power series to which the branches give rise and the possibility of an analytic representation of the branches of an implicit function $y(x)$. This problem was considered by Newton's school, and Newton himself stated how one may obtain an "exact" value of a root of an equation by an infinite expansion.

B. Taylor([145]) followed the path of Newton and found as his main contribution to the theory of implicit functions, the series bearing his name. C. MacLaurin([94]), based on the method of undetermined coefficients, found the particular case of Taylor's series, known as MacLaurin's series.

The question concerning the validity of Taylor's series goes back to Lagrange([85]), who gives as a definition of an arbitrary function $f(x)$ of one variable an expression in which the variable enters in any manner. According to this a function $f(x+i)$ may be expanded into a power series of integral powers of i, with the exception of particular values of x. This expansion becomes illusory when the consecutive derivatives of f, for the point x, up to a certain order become infinite. Then $f(x+i)$ may be expanded in power series with fractional and also negative exponents when $f(x) = \infty$. These conclusions suppose a particular form of the function, constructed with a small number of symbols of calculation. Lagrange was the first to give an expression for the remainder of Taylor's series.

A complete justification of this expansion was accomplished only by the extension to functions of a complex variable by the efforts of Cauchy, Laurent, and Puisseux. Cauchy([28]) as early as 1826, gave a precise answer to the problem of the convergence of Taylor's series and the possibility of expanding a function by such series. In particular he determined rigorously the expansion of functions into entire power series, under the conditions of regularity which exclude singular points.

Laurent showed, in 1843, how in the algebraic domain, the case of poles gives rise to negative exponents. Puisseux([123]) admitted the case of branch points, and shows how the branches of an algebraic function in the neighborhood of a singular point may be represented by power series with fractional powers of the independent variable, and by using Newton-Cramer's method also gives effective calculations. Puisseux' method has been treated and in some points perfected by Fischer([55]). See also the treatises by Clebsch-Lindemann([33]) ; Baltzer([5]) ; Netto([98]) ; Hensel-Landsberg([70]) ; E. Picard([113]) ; Jordan([73]) ; and other references in the Encyklopädie.

With this ends the first period of the theory of singularities of which the " Théorie des fonctiones elliptiques " by Briot et Bouquet([26]) may be mentioned as a representative treatise.

The second, or modern period of the theory of singularities of algebraic curves, may be characterized by the problem of intersections, to which one is led also by those researches whose object is the absolutely general formulation of Pluecker's relations. Cayley([31]) in 1865 found the multiplicity of the intersections of two curves by taking the definition of the resultant as the product of the differences of the roots as a starting point. This multiplicity is the sum of those of the couples of branches, and, when considering the singularity at the origin and the expansions of x and y by entire powers of a parameter t, its determination is reduced to the evaluation of the minimal degree in which t figures in the resultant of Weierstrass([151]).

The problem of evaluating the intersections of two tangent branches when one knows the respective Puisseux expansions, is solved by H. J. Smith, in ([139]) and by G. Halphen([65]).

These investigations put in evidence the purpose which certain " characteristic terms " and the exponents of the variable have in the Puisseux series.

A more original and very fruitful point of view is taken by Noether([101]), by introducing quadratic transformations. This method is able to serve equally well for the analytic calculation of the Puisseux expansions and for the geometric study of the nature of the singularity.

As for the first task, the idea of using quadratic transformations was conceived by Hamburger([67]) and afterwards by Weierstrass([151]). See also Koenigsberger([81]).

An extensive development with respect to the geometric aspect is given by Noether in a fundamental memoir([103]).

The possibility of transforming a curve with any singularities into a curve with multiple points with distinct tangents goes back to Kronecker, who communicated the theorem verbally to Riemann and Weierstrass in 1858. Kronecker([84]) based his argument on Puisseux's expansions and imposed a rational substitution on a single variable x, which is rationally invertible for the points of the curve $f(x, y)=0$, as may also be seen in his researches on discriminants of algebraic functions of one variable, in which he distinguishes between an essential and non-essential factor. The first factor which corresponds to the branch points of the function, turns out to be invariant in rational transformations of the curve.

It is in the introduction of this paper in which Kronecker writes of his contact with Riemann and Weierstrass on this matter.

Kronecker's method is based upon the fact that the discriminant of y, i. e., the resultant $R(x)$ of the elimination of y between $F(x, y)=0$ and $\frac{\partial F}{\partial y}=0$ is composed of two factors D and D_1, of which one is an essential divisor, the other an unessential divisor which enters as a square, so that $R=D\cdot D_1^2$. The first factor depends on the branch points x of y and their multiplicities only. The second factor depends on the multiple points of the curve, and the transformation permits the separation of its quadratic factors from each other and from the factors of D. Riemann([125]) accomplished the same result by transcendental means. In this connection see further G. Landsberg([86]); E. Lasker([87]); L. W. Thomé([146]); S. Kantor([76]). G. A. Bliss([18.1]) gives a valuable critical review of what has been done concerning the reduction of singularities of plane curves by birational transformations, and makes a study of birational transformations simplifying singularities of algebraic curves. The method followed consists in the construction of a pair of rational functions $\xi(x, y)$, $\eta(x, y)$ satisfying an irreducible equation $\phi(x, y)=0$ birationally related to $f(x, y)=0$, having certain properties described subsequently. Also, the distinction between the theorems for the projective and the function theoretic planes is explained.

V. Snyder([139.1]) shows that curves of given order and genus with ordinary double points exist for every value of the genus p not exceeding $\frac{1}{2}(n-1)(n-2)$, n being the order of the curve, and determines the equation in each case.

S. Lefschetz([88.1]) considers the loci with given singularities and shows the existence of a curve having assigned Plueckerian characters. The question is equivalent to the determination of the maximum number of cusps that a curve of order n and genus p may have. He also points out the extension of the theory to surfaces and higher varieties.

J. L. Coolidge([36.1]) discusses the characteristic numbers of real algebraic plane curves and proves the theorem that the only algebraic identities involving any combination of real or total singularities, which are valid for all real algebraic plane curves, are those which are deducible from the known equations of Pluecker and Klein([79]).

T. R. Hollcroft([70.1]) considers problems such as to find the greatest number of cusps that may be added to the singularities of a plane curve of given order and number of cusps and nodes.

In ([70.2]) Hollcroft studies maximal cuspidal curves, i. e., curves which for a given order and genus have the maximum number of cusps, and as such are unique.

Operating with birational transformations on a curve one arrives at a much more expressive result, namely, the possibility of reducing an arbitrary curve to one possessing only double points. See Hensel-Landsberg([70]).

Klein seems to have been the first to point out this consequence of Kronecker's theorem, by recalling an observation made by Clebsch that the curves with multiple points with distinct tangents can be reduced to curves with double points only, by the medium of space curves and the plane mapping of the cubic surface. More precisely, Halphen, in Étude referred to above([58]), points out, that any algebraic singular curve can be transformed into a space curve without singular points, from which by projections one obtains a plane curve with nodes only. A simple geometric proof of the theorem is given by Bertini([13]). Other proofs are given by Picard-Simart([116]); Halphen([58]); Appell-Goursat([2]); Simart([138]); A. R. Forsyth([56]); E. Vessiot([150]); A. Del Re([47]).

Noether([101]) used Cremona transformations, in particular quadratic Cremona transformations. By passing from a given curve to a series of transforms by quadratic transformations, one obtains a concept of the composition of the singularity by means of consecutive multiple points, which may be considered as the most important acquisition in the theory of singularities. Noether is thus led to conceive any extraordinary singularity as a limit-singularity due to the reunion of several multiple points which approach each other indefinitely, and of simple branch points.

Noether's theory has been given a simpler geometric treatment by Bertini in several papers([12]), and in the systematic expositions in the appendix of his Introduzione alla geometria projettiva degli iperspazi([9]); German translation by A. Duschek([48]).

The reduction of singularities by transformations also permits one to extend algebraically to algebraic curves with any sort of singularities the concept of that invariant characteristic which constitutes the genus, which on the other hand may be defined with equal generality by the order of connectivity of the corresponding Riemann surface.

From the relation of invariance of the genus applied to the correspondence between point curves and envelope curves, Noether([102]) deduces the extension in case of any singularity of one of Pluecker's formulas; on the other hand, by calculating the intersections of the curve with a polar, he determines the reduction of the class due to any singularity, which by another method was also evaluated by Halphen.

The complete extension of Pluecker's formulas requires us, moreover, to pass from the characteristics of singularities of a point-curve to those of the dual curve, which is accomplished by making use of the observation of Cayley([31]) and Halphen([66]) that in a dual transformation order and class of a branch are permuted.

Other relations between the characteristics of a curve capable of valuable applications were established by H. J. Smith([139]) and Zeuthen([159]), ([161]).

With the extension of Pluecker's formulas is also associated the question of Plueckerian equivalents. Cayley([31]) in 1866, is led inductively to the concept that every singularity in its effects on Pluecker's formulas may be replaced by a certain number of double points and cusps, double tangents and flexes.

A purely arithmetical justification for this was given by H. J. Smith([139]). There remains the essential problem to recognize in the Plueckerian equivalents the trace of a passage to the limit, by which every singularity results from the superposition of elementary singularities. This question may be illuminated as well by the expansions into series, as by the method of quadratic transformations.

The first method is adopted by A. Brill([23]) who substitutes for a branch of the curve an "osculating hyperparabola" which he lets vary conserving rationality. The second method is implicitly contained in Noether's definition of a singularity as a limit of ordinary multiple points which approach each other. This was done by Miss Scott([129]).

The relation between the theory of singularities by Noether and that of Smith-Halphen, based on Puisseux expansions comes to light when one proposes the effective calculation of those expansions by quadratic transformations with Hamburger's method mentioned above. The explicit determination of the successive multiple points of a branch in terms of the exponents of the development into series, following from the inversion of the preceding procedure, is contained in a memoir by

Noether([105]), and is obtained by the use of quadratic transformations, substantially as applied by Noether in his preceding investigations.

The various methods are coordinated and supplemented by F. Enriques([52]), and again systematically presented in his Lezioni([51]). The importance of the new concept, in which the "satellites" and "free" points connected with a singularity are arithmetically illuminated, consists in the completion of Noether's analysis. For Noether the singularity is simply defined as reunion of consecutive multiple points; for Enriques the idea of the position of these points must be added, which is expressed by the concept of "satellitisme," on which depends the existence of superlinear branches.

The notion of "satellite points" is based upon the following considerations on the intersection of a branch of the second order

$$(1) \qquad y=a_1x+a_2x^2+\ldots+a_sx^s+a_{s+1}x^{s+\frac{1}{2}}+a_{s+2}x^{s+1}+\ldots,$$

$$s\geqq 1, \qquad a_1\neq 0, \qquad a_{s+1}\neq 0,$$

and the linear branch

$$(2) \qquad y=a_1x+a_2x^2+\ldots+a_ix^i+b_{i+1}x^{i+1}+\ldots.$$

When $i=s-1$, the branch (1) has $s-1$ double points $O_1\ldots O_{s-1}$ consecutive to O, and succeeding each other on the linear branch

$$y=a_1x+a_2x^2+\ldots+a_{s-1}x^{s-1}.$$

When $i\geqq s$, to the $(s-1)^{\text{st}}$ double point O_{s-1} consecutive to O on the branch (1) follows an sth point O_s which is simple on (1). In this case (1) is defined as a branch of *species* $s-1$.

Now on a branch of the second order and of species $s-1$, there exist $s-1$ double points consecutive to O, and then the simple points O_s, O_{s+1}, O_{s+2}, ..., then the second simple point O_{s+1} is defined as the *satellite* of O_s, so that every branch of the second order passing properly through O_s (i. e., through $O\ O_1\ldots O_s$) passes in consequence also through O_{s+1}.

Another new feature of Enriques' treatment is the study of the differential conditions which characterize the passage of a curve through consecutive points and the relative multiplicities. In this manner is obtained a more precise determination of the concept of extraordinary singularities as limit-singularities, from which follow interesting applications, in fact a complete theory of algebraic singularities from the point of view of infinitesimal analysis. Enriques gives some historic data leading to this development.

Hamburger([67]), in 1871, calculated the Puisseux expansions for linear branches, by determining the coefficients as successive derivatives of the algebraic function $y(x)$, and Stolz([142]) in 1874 reconsidered and continued Pluecker's analysis in the form later generalized by Enriques, and, like Hamburger, reduced by proper transformations the case of superlinear branches to that of linear branches. See also Koenigsberger(81); De La Gournerie(45); Darboux(40).

The Enriques investigation begins with the demonstration and extension of Stolz's formulas for the successive derivatives of implicit functions and of composite functions. The symbolism introduced permits of a treatment in terms of the theory of operations together with the expansion of the power of a polynomial. The application of the formulas of the calculus thus obtained leads in a simple manner to the required result to characterize with differential conditions the composition of the points which make up the elements of a singularity. The result seems particularly important when taking into account the case of satellite points and hence of the superlinear branches, whose origin appears thus in its true light; likewise the classical method of Newton-Cramer, which thus attains a more profound significance. Also, the relation with the arithmetical point of view of the theory becomes apparent in the diagram by which Enriques has illustrated his differential conditions, thus moving in an order of ideas which have contact with some recent arithmetical theories, as, for example, contained in Minkowsky's "Geometrie der Zahlen"(97).

On the theory of singularities of an algebraic curve and the neighboring points by means of differentials of various orders of the coordinates, or the "elements of first and higher order," which is of importance in the treatment of differential equations of first and higher order, one may find comment by F. Engel(49); E. Study(144); M. Noether(106), (107); Zeuthen(161).

The double points with a single tangent may be put in two classes, into proper nodes (tacnodes, oscnodes) and cusps (of the first, second, ... kind), according as they may be resolved into two simple points or a single point after a finite number of quadratic transformations. The characteristic form of the equation of a curve with such a singularity at the origin is given by B. Levi(90). Let Δ be the order of the branch at A, i. e., the number of intersections at A of a generic line r, sufficiently close to A, so that the sum of the orders of all branches through A is equal to the order s of the multiplicity. There exists a line a, however, one of the tangents to f at A, such that when r forms a sufficiently small angle with a, r will cut the branch in more than Δ points. a is then called the singular tangent of said branch through A. The class Δ' of the branch is the number of tangents to the branch from a point different from A whose distance from a remains below a certain limit. Thus for an ordinary point, a cusp of the first, and a cusp of the second kind, Δ and Δ' have the values 1, 1; 2, 1; 2, 2. The idea of order and class of a branch (in case of reality) goes back to Pluecker([119]). Further references on branches and corresponding expansions may be found in E. Picard([114]); Appell-Goursat([2]); C. Jordan([73]); Halphen([58]), ([63]) with abstract of this in ([64]). For real branches see E. Cosserat([37]); E. Vessiot([149]). An analytic treatment of a curve in the neighborhood of a finite or infinite point is given by Ch. Bieler([17]). M. de Franchis gives a synthetic treatment of higher singularities in ([42]). A. Cayley([31]), and in particular G. Halphen([63]), ([64]) have investigated the mapping of singularities between correlated curves. To a branch (Δ, Δ') of one curve corresponds a branch (Δ', Δ) of the other. The number of intersections of the branch with a is equal to the number of tangents from A coincident with a, i. e., equal to $\alpha=\Delta+\Delta'$.

When x_1y_1 are the coordinates of A, and assuming the singular tangent different from $x-x_1=0$, which is always possible, from the series of quadratic transformations which resolves the singularity, and the possibility of expansion of the coordinates x, y of $f(x, y)=0$ in series of integral exponents of a parameter t in one-to-one correspondence with the points of the branch, i. e., t a rational function of x, y, follows the representation of the branch

$$y-y_1=(x-x_1)^{\frac{a}{\Delta}}\cdot\phi[(x-x_1)^{\frac{1}{\Delta}}],$$

in which, according to Halphen ϕ is a series of ascending positive powers of $(x-x_1)^{\frac{1}{\Delta}}$. In this expansion $\alpha\geqq\Delta$, for $y=y_1$ is a singular tangent $\alpha>\Delta$, in fact according to Cayley-Halphen $\alpha=\Delta+\Delta'$. Compare O. Stolz([142]); Weierstrass([153]); L. Koenigsberger([81]); O. Biermann([18]); Pluecker([119]).

A. Brill in ([25]) and ([19]) gives a rigorous function-theoretic treatment of algebraic functions and its geometric interpretation for algebraic curves. Much use is made of the Weierstrassian "Vorbereitungssatz" which in the hands of Brill becomes a powerful instrument of analysis and by which he proves Noether's fundamental $A\phi+B\psi$-theorem. By these methods is also shown how a reduced polynomial may be resolved explicitly into linear factors concerning y; thus,

$$f(x,y)=c(x-a)\prod_{i=1}^{n}\left\{y-\frac{1}{(x-a)^{pi}}-Q^{(i)}[(x-a)^{\frac{1}{\Delta i}}]\right\}$$

in which the Q's are power series. Making use of these developments, the discussion of the branches of the curve offers no particular difficulty.

Weierstrass' investigations concerning the aforesaid theorem are contained in ([152]). See also W. F. Osgood([108]).

2. *Consecutive Points of a Singularity; Characteristics. Birational Transformation of Plane Algebraic Curves into Non-Singular Space Curves.*—On account of the importance of certain terms used in the theory of singularities, a more detailed presentation of their definitions will perhaps be helpful in reading the report.

In the first place Noether's definition of consecutive singular points must be mentioned. The behavior of curves and its ordinary singularities in a quadratic transformation has been discussed previously in Chapter 1. Now consider an r-fold point O of a curve f. Assuming O as a fundamental point of a quadratic transformation in Σ, the r-fold point is transformed into the points O'_1, O'_2, ... of multiplicities r_1, r_2, ..., $(r_k \geqq 1)$ in Σ'. Every one of these points whose order is >1, may be treated similarly by new quadratic transformations, and so forth; so that finally after a finite number of quadratic transformations, a transformed curve with only ordinary singularities (distinct tangents) is obtained. This constitutes Noether's famous theorem. That by birational transformation the reduction may be made to a curve with ordinary double points only, was first proved implicitly by Kronecker([84]) as pointed out before; then in a purely geometric manner by Bertini([13]) and reproduced by Picard in his Traité([115]). For other proofs see Halphen([60]), Appell-Goursat([2]), Simart([138]). See also Hensel-Landsberg([69]).

That the series of quadratic transformations is limited is proved by Hamburger([67]). See also Bertini([12]); Severi([137]); Picard([115]). Reversing the order of transformation by their inverses, the original curve f is obtained with an r-fold point at O, which now appears as surrounded by consecutive points $O_1O_2\dots O_i$ in the first neighborhood, O_{11}, O_{21}, ... in the second neighborhood, O_{111}, O_{211}, ... in the third neighborhood, etc. whose multiplicities are r_1, r_2, ..., r_i; r_{11}, r_{21} ...; r_{111}, r_{211}, ... respectively, and which satisfy the fundamental neighborhood relation

$$r=r_1+r_{11}+r_{111}+r_2+r_{21}+r_{211}+\dots.$$

This is the fundamental idea used by Noether in his classic memoir of 1871.

The multiplicities of the points of a branch whose Puisseux expansion is $y=ax+a_1x^{\frac{\nu+\nu'}{\nu}}+\dots+a_{m+1}x^{\frac{\nu+\nu'+\dots+\nu^{(m+1)}}{\nu}}+\dots$, depend essentially on those terms in which $\nu^{(i)}$ is not divisible by the highest common divisor of the preceding ν's. The positions of the successive multiple points of the branch depend on the coefficients a, a_1, ..., a_m, and on the positive integers $\nu(>1)$, ν', ν'', ..., ν^{m+1} in which m denotes the smallest index for which $\nu=\rho\rho_1\ldots\rho_m(\rho_m>1)$. It is assumed that $a_1\neq 0, \ldots a_{m+1}\neq 0$, and $\frac{\nu'}{\nu}=\frac{\rho'}{\rho}, \ldots, \frac{\nu^{(m+1)}}{\nu}=\frac{\rho^{(m+1)}}{\rho\rho_1\ldots\rho_m}$; $\rho\geqq 1, \rho_1\geqq 1$ $\ldots$, and ρ^{i+1}/ρ_i are irreducible fractions. The orders of multiplicity of said points are the remainders of the successive divisions: $\nu':\nu; \nu:\nu_1; \ldots, \nu_{\sigma-2}:\nu_{\sigma-1}; \nu'':\nu_\sigma; \ldots; \nu''':\nu_{\sigma+\sigma'}, \ldots$, which lead to the determination of the highest common divisor of $\nu\nu', \nu\nu'\nu'', \ldots, \nu\nu'\nu''\ldots\nu^{(m+1)}$. Under the hypotheses made

$$\text{h.c.d. }(\nu\nu')=\nu_\sigma=\frac{\nu}{\rho}=\rho_1\rho_2\ldots\rho_m,$$

$$\text{h.c.d. }(\nu\nu'\nu'')=\nu_{\sigma+\sigma'}=\frac{\nu}{\rho\rho_1}=\rho_2\ldots\rho_m,$$

$$\cdots\cdots\cdots\cdots\cdots\cdots\cdots\cdots\cdots\cdots$$

$$\text{h.c.d. }(\nu\nu'\nu''\ldots\nu^{(m)})=\rho_m,$$

$$\text{h.c.d. }(\nu\nu'\nu''\ldots\nu^{m+1})=1.$$

The characteristic terms introduced by Smith and Halphen mentioned in Section 1 are substantially $a_i x^{\frac{\nu+\nu'+\ldots+\nu^{(i)}}{\nu}}$ and the characteristic couples of numbers

$$\text{h.c.d. }(\nu\nu'\ldots\nu^{(i-1)})=\rho_{i-1}\ldots\rho_m,$$

$$\nu^{(i)}-\rho_{i-1}\ldots\rho_m\left|\frac{\nu_i}{\rho_i\ldots\rho_m}\right|.$$

By another notation, to the branch, as represented above, correspond the characteristic couples $\nu, \nu_1; \nu_\sigma, \nu_{\sigma+1}; \ldots$ where

$$\nu_1=\nu'-\nu\left|\frac{\nu'}{\nu}\right|, \quad \nu_\sigma=\text{h.c.d. }(\nu\nu')=\rho_1\ldots\rho_m; \quad \nu_{\sigma+1}=\nu''-\nu_\sigma\left|\frac{\nu''}{\nu_\sigma}\right|, \text{ etc.}$$

The number of characteristic couples is called genus of the branch by Enriques.

When a singular point is at the origin O, then a generic line a through O cuts f in an intersection of multiplicity Δ. The branch is linear or superlinear according as $\Delta=1$, or >1. When the multiplicity of the intersection of a is $\Delta+\Delta'$, $\Delta'>0$, then a is a singular tangent to the branch at O, and f may be represented parametrically as $x=ct^\Delta+\ldots$, $y=c't^{\Delta+\Delta'}+\ldots$, where $c, c'\neq 0$ are constants. Δ' is called the class of the branch. The tangents at the points of a branch are dual to the branch (described by points) and Δ and Δ' are dually corresponding. A. Cayley([30]) conjectures the theorem that Δ' is the number of tangents coinciding with a which pass from a point of a (except O) to the branch and $\Delta+\Delta'$ the number of tangents from O coinciding with a. This was proved by Halphen([63]), ([64]). See also Zeuthen([160]); Halphen([59]); Bertini([8]); Segre([133]).

As already indicated in Section 1, Noether([105]) reconsiders the question of singularities in order to establish the connection between his former results obtained by quadratic transformations and the characteristics of Halphen and Smith. It is known that analytically these developments must be derived from a succession of quadratic transformations of the curve which resolves the singularity step by step. But this analytic process leads to a series of continued fractions by which numbers defining the singularity result directly. Noether asks and answers the question, which of the combinations of these numbers of transformations are the characteristic numbers of Smith and Halphen.

A very important aspect of the problem of singularities is gained by the method of projection between higher spaces. Veronese([147]) proved the theorem that any algebraic curve whatsoever may be obtained as a projection of a hyperspace curve without singularities. This is implicitly contained in Brill-Noether's theory of linear point-series on an algebraic curve. The theorem concerning the birational transformation of an ordinary space curve into a non-singular curve of the same space was proved by Poincaré([122]).

See in this connection also Del Pezzo([46]); Vessiot([148]); Segre([134]); Levi([89]); Hensel-Landsberg([70]).

As a matter of fact, by the theory of linear point series it is easily proved, (Severi-Loeffler([137])), that every plane algebraic curve may be obtained by a birational transformation from a space curve without singularities in a space of $r \geqq 3$ dimensions. Moreover, every plane algebraic curve, with any singularities whatsoever, may be obtained as projection of a space curve without singularities in a space of four or more dimensions. Again, as stated before, an algebraic curve with any singularities, may always be transformed birationally into a plane curve with only ordinary double points.

M. De Franchis([42]) develops a geometric theory of the singularities of a plane algebraic curve by taking the definition of a cycle as a starting point. If C is a curve and P a point on it, which by a certain Cremona transformation is transformed into a curve C_1, so that to a point P correspond simple points $P'_1 P''_1 \ldots P_1^{(s)}$ of C_1, then de Franchis calls cycle of C the totality of points of C consecutive to the same point P_1.

The graphical features of higher singularities are investigated by Ch. A. Scott([129]) who gives a graphical analysis of the singularities of algebraic curves, by means of the quadric inversion, and is thus enabled to ascribe a definite geometric meaning to the numbers δ and κ in Pluecker's formulas, also in cases of higher singularities.

A. B. Basset([6]) by means of birational transformations discusses compound singularities of quintic curves, and in the second paper uses special quadratic transformations to resolve certain types of singularities of plane algebraic curves.

Ch. A. Scott([130]) gives an account of quadric inversion in an introductory study of certain modern ideas and methods in plane analytic geometry.

H. W. Jung([74]), ([75]) treats singular points of plane algebraic curves by the arithmetical theory of algebraic functions, and establishes and clarifies the connection of his theory with that of the geometric-algebraic method.

3. *Multiplicities of Intersection; Additional References on Singularities.*—The question of the multiplicities of intersections of two curves with given singularities at a given point is closely connected with that of singularities in general, and deserves particular discussion. The importance of this will not be diminished by some repetitions of references given in the preceding sections.

Weierstrass([154]) investigates the multiplicities I of the intersection of two curves f, f' at a common point, which may be determined by adding the multiplicities of the intersections of one of the curves with the branches of the other. I may also be determined by the formation of the resultant of f, f' as a product of differences, as was done by Cayley([31]). Compare also Smith([139]); Halphen([61]), ([66]); De La Gournerie([44]), ([45]); Painvain([109]).

Segre([136]) (see also Bertini([10])) points out that it is not sufficient for the characterization of a singular point merely to establish the orders of the branches and the multiplicities of the consecutive infinitely close branches. The nature of the branches passing through the multiple point must be ascertained also.

Another method for the determination of the multiplicities of intersections was established by Brill([24]). If f, and f_1 are curves of order n and n_1 with arbitrary singularities, but no common parts, the common intersections are obtained by the linear factors of the resultant R. Cayley([31]) and H. J. Smith([139]) give as definitions of the resultant the product of the differences of the values of y which from $f=0$, $f_1=0$ correspond to one and the same value of x. For an intersection $A(x_1, y_1)$ it is sufficient to use the expressions for y which assume the value y_0 for $x=x_1$. Substituting in R for the y's their expansions in terms of $x-x_1$, then the least exponent of the result will give the required multiplicity. Concerning this see also O. Stolz([143]), Noether([104]), Segre([135]). Halphen([66]) gives a geometric form to the definition of multiplicities of intersections, as the sum of the orders of the infinitely small segments which are cut out between the two curves by a transversal whose distance from A is infinitely small of the first order. For this see also Zeuthen([157]), ([158]).

Noether studies multiplicities of intersections by means of quadratic transformations which resolve the singularities of f and f_1 at A. If $r_i, r_{ik}, \ldots$ and $s_i, s_{ik}, \ldots$ denote the multiplicities appearing in the succession of quadratic transformations of f and f_1 with multiple intersections at A, then the multiplicity of intersection of f and f_1 at A is expressed by

$$rs+\sum_i r_i s_i+\sum_{i,k} r_{ik}s_{ik}+\ldots.$$

Macaulay([92]) gives a method for determining the multiplicities of intersection in connection with Noether's fundamental theorem. See also Ch. A. Scott([131]), and Macaulay([93]).

Bacharach([4]) gives a geometric form to the fundamental theorem of intersections in the form: If mp of the intersections of a C_m with a C_n lie on a C_p, then the remaining $m(n-p)$ lie on a C_{n-p}. Zeuthen([162]) proves this theorem directly and derives from it the analytic representation of such related curves. Bacharach also shows that when a C_p is made to pass through the intersections of C_m and a $C_n(p\geqq m, p\geqq n)$, then for $p>m+n-3$ the mn conditions to be imposed on C_p are independent; however, for $p\leqq m+n-3$, of these conditions exactly

$$d=\tfrac{1}{2}(m+n-p-1)(m+n-p-2)$$

are a consequence of the remaining ones. In other words, a C_p through $mn-d$ of those intersections also passes through the remaining d, except in the case that these lie on a $C_{m+n-p-3}$.

See also Jacobi([71]); Pluecker([121]); Cayley([30]), ([32]); Clebsch([35]); Meray([96]).

B. M. Walker([150.1]) gives a detailed account of the proof suggested by Bertini by mapping the curve by cubics on six points upon a space-curve and projecting the space-curve back into the plane.

In ([156.1]), O. Zariski extends Puisseux's theorem on the expansion in series of analytic functions in the neighborhood of a critical algebraic point to the case of algebraic functions $y(x)$ in a field which contains several critical points. In particular he asks the question whether the relation between x and y may be uniformized by the introduction of a parameter z which is an algebraic function of x, and succeeds in establishing the condition for such a possibility, i. e., the substitutions of $y(x)$ in the critical points of C must belong to an algebraic function of genus $p=0$.

Additional references concerning algebraic curves, their singularities, correspondences, and their properties in general will be found in the following alphabetical list.

BIBLIOGRAPHY

1 Appell, P., et Goursat, E., Théorie des fonctions algébriques, (1895).
2 Appell-Goursat, Théorie des fonctions algébriques, Ch. 6 (1895).
3 ———, Théorie des fonctions algébriques, 276-284 (1895).
3.1 Autonne, L., C. R., 121: 673-676 (1895).
4 Bacharach, J., Erl. Ber., (1879); Diss. Erl., (1881); Math. Ann., 26: 275-299 (1885).
5 Baltzer, H. R., Analytische Geometrie, 295-324 (1882).
6 Basset, A. B., Quart. J., 36: 359-372 (1905); 37: 313-328 (1905-1906).
7 ———, Quart. J., 43: 151-160 (1912).
8 Bertini, E., Introduzione alla geometria projettiva degli iperspazi, 355-369 (1907).
9 ———, Introd., 370-390 (1907).
10 ———, Introd., 385 (1907).
11 ———, Giorn. di Mat., 7: 105-106 (1869).
12 ———, Lomb. Ist. Rend., (2), 21: 326-333, 413-423 (1888).
13 ———, Riv. di Mat., 1: 22-24 (1891).
14 ———, Ann. di Mat., (2), 22: 1-40 (1894).
15 ———, Math. Ann., 44: 158-160 (1894).
16 Berzolari, L., Allgemeine Theorie der hoeheren algebraischen Kurven; Encyk. 111 C, 4: 313-455 (1903).
17 Bieler, Ch., Nouv. Ann., (2), 19: 492-507 (1880); 20: 97, 489, 537, (1881); (3), 2: 354, 397 (1883); 3: 367 (1884); 4: 153, 223, 249 (1885).
18 Biermann, O., Theorie der analytischen Funktionen, 193 ff. (1887).
18.1 Bliss, G. A., Am. M. S. Bull., 29: 161-183 (1923); Am. M. S. Trans., 24: 274-285 (1922).
19 Brill, A., Vorlesungen ueber algebraische Kurven und algebraische Funktionen, (1925).
20 Brill, A., and Noether, M., Die Entwickelung der Theorie der algebraischen Funktionen in aelterer und neuerer Zeit; Jahresbericht, 3: 107-566 (1894).
21 Brill, A., Vorlesungen ..., 39-68, 238-284 (1895).
22 Brill, A., and Noether, M., Math. Ann., 7: 269-310 (1874).
23 Brill, A., Math. Ann., 16: 348-408 (1880); *also* Vorlesungen ueber algebraische Funktionen und Kurven, (1925).
24 ———, Münch. Ber., 18: 81-94 (1888).
25 ———, Münch. Ber., 21: 207-220 (1891).
26 Briot et Bouquet, Théorie des fonctions elliptiques, 2 ed. (1875).
27 Brocard, H., Interméd., 18: 18-19 (1911).
28 Cauchy, A., Exercises de Mathématique, 1: 25; sur le calcul des résidus et le calcule des limites, Turin, (1831-1832); exercises d'analyse et de physique mathématiques, 2: 41, 109 (1841).
29 Cayley, A., Article on "curve," Encycl. Brit., 9. ed., 6: 716-728 (1877); Papers, 11: 460-489.
30 ———, Camb. Math. J., 3: 211-213 (1843); Papers, 1: 25-27.
31 ———, Quart. J., 7: 212-223 (1866); J. f. Math., 64: 369-371 (1865); Papers, 5: 424-426, 520-528.
32 ———, Math. Ann., 30: 85-90 (1887); Papers, 12: 500-504.
33 Clebsch-Lindemann, Vorlesungen ueber Geometrie, 319-359 (1875-1876).

34 ———, Vorlesungen ueber Geometrie, 1: (1875-1876); French translation by A. Benoist, 1: (1879); 2: (1880); 3: (1883).

35 Clebsch, A., J. f. Math., 63: 189-243 (1864).

36 Clebsch-Lindemann, Vorl., 1: 474-496 (1876).

36.1 Coolidge, J. L., Pal. Circ. Mat., 42: 260-266 (1917).

37 Cosserat, E., Toulouse Ann., 4: 1-16 (1890).

38 Cramer, G., Introduction à l'Analyse des lignes courbes algébriques, ch. 10, 148-351 (1750).

39 Cromona, L., Introduzione ad una teoria geometrica delle curve plane, Bologna Mem., (1), 12: 305-436 (1861). German ed. by M. Curtze, Greifswald (1865).

40 Darboux, G., C. R., 36: 204-208 (1868).

41 Dedekind, R., and Weber, H., J. f. Math., 92: 181-290 (1882).

42 De Franchis, M., Pal. Circ. Mat., 11: 104-153 (1897).

43 De Gua, J. P., Usage de l'Analyse de Descartes, etc., 457 p. (1740).

44 De La Gournerie, J., J. de Math., (2), 14: 425-435 (1869).

45 ———, C. R., 77: 573-577 (1873).

46 Del Pezzo, P., Nap. R. A. Rend., (2), 7: 15-21, 45-50 (1893).

47 Del Re A., Modena R. Acc. Mem., (2), 10: 415-448 (1895).

48 Duschek, A., Einfuehrung in die projektive Geometrie mehrdimensianaler Raeume: 429-551 (1924). German translation of E. Bertini's "Introd."

49 Engel, F., Leipz. Ber., 45: 468-476 (1893); 54: 17-51 (1902); Jahresbericht, 11: 187-188 (1902).

50 Enriques-Chisini, Lezioni sulla teoria geometrica delle equazioni e delle funzioni algebriche, 2: 535-544 (1918).

51 Enriques, F., Lezioni, 2: 327-534.

52 ———, Rom. Acc. L. Rend., (5), 25: 607-613 (1916).

53 Euler, L., Introductio in analysin infinitorum, Lausannae, 2: (1748).

54 Fiedler, W., Geometrie der Lage, 2: (1885).

55 Fischer, H. V., Puisseux's Untersuchungen ueber algebraische Functionen, Halle (1861).

56 Forsyth, A. R., Theory of functions of a complex variable, 2 ed.: 544-583 (1900); Mess. of Math., 30: 1-7 (1900).

57 Guccia, G. B., Lezioni di geometria superiore, lit., Palermo, (1890).

58 Halphen, G., Appendix to the French translation of Salmon's Higher Plane Curves, Paris, (Étude...) 619-636 (1884).

59 ———, Étude, p. 544.

60 ———, Étude, p. 630.

61 ———, Par. Ac. Mem. Sav., (2), 26: (1878); Bull. Soc. M., 1: 130-142 (1873); 2: 34-52 (1873); 3: 76-92 (1875).

62 ———, C. R., 78: 1105-1108 (1874).

63 ———, C. R., 80: 97 (1875).

64 ———, Oeuvres, 1: 212-215 (1916).

65 ———, Soc. M. F. Bull., 3: 76-92 (1874-1875); 4: 29-41 (1875-1876); J. de Math., (3), 2: 257-290, 371-408 (1876); Étude, p. 575-648.

66 ———, Oeuvres, 1: 216-311 (1916).

67 Hamburger, M., Zschr. Math. Phys., 16: 461-491 (1871).

68 Hensel, K., und Landsberg, G., Theorie der algebraischen Funktionen einer Variabeln..., ° (1902).

69 Hensel, K.-Landsberg, G., Theorie der algebraischen Funktionen einer Variabeln, 418-419 (1902).°

70 Hensel-Landsberg, Theorie der algebraischen Funktionen einer Variabeln, 39-116, 365-419 (1902). °

70.1 Hollcroft, T. R., Am. M. S. Bull., (2), 29: 407-414 (1923).

70.2 ———, Ann. of Math., (2), 26: 37-46 (1924-25).

71 Jacobi, C. G., J. f. Math., 15: 285-312 (1836); Werke, 3: 331-354.°

72 Jordan, C., Cours d'Analyse, 2 ed., 1: (1893).

73 ———, Cours d'Analyse, 2 ed., 1: 397-411, 561-612 (1893).

74 Jung, H. W., Einfuehrung in die Theorie der algebraischen Funktionen einer Veraenderlichen, (1923).

75 ———, Math. Ann., 84: 161-201 (1921).

76 Kantor, S., Monatsh. f. Math., 11: 193-286 (1900).

77 Klein, F., Einleitung in die geometrische Funktionentheorie, lit., (1880-81); (1892). °

78 ———, Riemann'sche Flaechen, 2 vols., lit., (1891-92).

79 ———, Math. Ann., 10: 199-219 (1876); Erl. Ber., (1875).

80 Koenigsberger, L., Vorlesungen ueber die Theorie der elliptischen Funktionen, part 1, lecture 9 (1874).

81 ———, Vorlesungen ueber die Theorie der elliptischen Functionen, 167-198 (1874).

82 Koetter, E., Grundzuege einer rein geometrischen Theorie der algebraischen ebenen Kurven. Preisschrift Berlin Ac. Abh., (1887).

83 ———, Die Entwickelung der synthetischen Geometrie.... Jahresbericht, 5: 1-486 (1898-1901).

84 Kronecker, L., J. f. Math., 91: 301-334 (1881).

85 Lagrange, J., Théorie des fonctions analytiques (1797); 3 ed., sec. 25 (1847); Oeuvres, 9: 83-85 (1881).

86 Landsberg, G., J. f. Math., 131: 152-164 (1906).

87 Lasker, E., Math. Ann., 60: 20-116; 607-608 (1905).

88 Laurent, H., Traité d'Analyse, 2: (1887); 4: (1889).

88.1 Lefschetz, S., Am. M. S. Trans., 14: 23-41 (1913).

89 Levi, B., Rom. Acc. L. Rend., (5), 1: 111-113 (1898).

90 ———, Torino Atti., 40: 139-167 (1905).

91 Lucas, F., Études analytiques sur la théorie générale des courbes planes, Paris, (1864).

92 Macaulay, F. S., London M. S. Proc., 26: 495-544 (1895); 29: 673-695 (1898); 31: 15-30, 381-423 (1899); 32: 418-430 (1900).

93 ———, Am. M. S. Bull., (2), 4: 540-544 (1898).

94 MacLaurin, C., Treatise of Fluxions (1742), book 2, ch. 2, art. 751.

95 ———, De linearum geometricarum proprietatibus generalibus tractatus, which is an appendix to A treatise of algebra (1748), (5 ed. 1788). French edition by E. De Jonquières: Melanges de geometrie pure, 197-261 (1856).

96 Meray, Ch., Ann. Éc. Norm., (1), 4: 159-193 (1867).

97 Minkowsky, H., Geometrie der Zahlen, (1896).°

98 Netto, E., Vorlesungen ueber Algebra, 2: 50-62 (1900).

99 Newton, I., Methodus funtionum et serierum infinitarum, Opuscula, edit. Castillion, Losannae Ginevra, 1: 37-41 (1774).

100 ———, Analysis per aequationes numero terminorum infinitas (1771); Methodus fluxionum (1736), Opera edit. Horsley, 1: 257-282; 391-518.

101 Noether, M., Math. Ann., 5: 635-639 (1872); Goett. Nachr., 267-278 (1871).

102 ———, Math. Ann., 9: 166-182 (1876).

103 ———, Math. Ann., 9: 166-182 (1876); 23: 311-358 (1884).

104 ———, Math. Ann., 23: 311-358 (1884).

105 ———, Pal. Circ. Mat., 4: 89-108 (1890).

106 ———, Chicago Congr. M. Papers, 253-257 (1896).

107 ———, Math. Ann., 56: 677-684 (1902).

108 Osgood, W. F., Encykl., II B 1, no. 45.

109 Painvain, L., Bull. Sci. M., 4: 131-142 (1873); 5: 138-144 (1873).

110 Pascal's Repertorium, 2 ed., 291-322 (1910).

111 Peschka, von, G. A., Darstellende und projektive Geometrie, 2: (1884).

112 Picard, É., Traité d'Analyse, 2 ed., 2: (1905).

113 ———, Traité, 2: 348-367 (1893).

114 ———, Traité, 2 ed., 2: ch. 13 (1905).

115 ———, Traité, ..., 2 ed., 2: 404-411 (1905).

116 Picard-Simart, Théorie des fonctions algébriques de deux variables indépendantes, 2: 1-49 (1906).°

117 Pieri, M., Riv. di Mat., 4: 40-42 (1894).

118 Pluecker, J., Theorie der algebraischen Kurven, Bonn, (1839).

119 ———, Theorie der algebraischen Kurven, sec. 2, p. 205 (1839).

120 ———, Theorie der algebraischen Kurven, 155 ff. (1839).

121 ———, J. f. Math., 16: 47-57 (1837); Abh., 1: 323-333; J. de Math., 2: 11-15 (1837); Abh., 1: 334-338 (1895).

122 Poincaré, H., C. R., 117: 18-23 (1893).

123 Puisseux, B., J. de Math., (2), 15: 365-480 (1850); 16: 228-240 (1851).

124 Reuschle, G., Praxis der Kurvendiskussion, part 1 (1884).

125 Riemann, B., J. f. Math., 54: 115-155 (1857); Werke, 2 ed., 88-144 (1892).

126 Salmon, G., A treatise on the higher plane curves, Dublin, (1852); (3 ed., 1879); German edit. by W. Fiedler (2 ed., 1882); French ed. by O. Chemin, and followed by Étude sur les points singulièrs by G. Halphen, Paris (1884).

127 Schmeidler, W., Math. Ann., 81: 223-234 (1920).

128 Schubert, H., Kalkuel der abzaehlenden Geometrie, (1879).

129 Scott, C. A., Am. J., 14: 301-325 (1892).

130 ———, Am. J., 15: 221-243 (1893).

131 ———, Am. M. S. Bull., (2), 4: 260-273 (1898).

132 ———, Am. M. S., Trans., 43: 216-263 (1902).

133 Segre, C., Ann. di Mat., (2), 22: 41-142 (1893).

134 ———, Ann. di Mat., (2), 25: 2-54 (1896).

135 ———, Giorn. di Mat., 36: 1-50 (1898).

136 ———, Torino Atti., 36: 645-651 (1901).

137 Severi, F., Vorlesungen ueber algebraische Geometrie, 46-52, 134-139 (1921); German translation by E. Loeffler.

138 Simart, G., C. R., 116: 1047-1050 (1893).

139 Smith, H. J., London M. S. Proc., 6: 153-182 (1876); partly presented to Soc. in (1873-1874); Papers, 2: 101-131.°

139.1 Snyder, V., Am. M. S. Bull., (2), 15: 1-4 (1908-09).

140 Stahl, H., Theorie der Abel'schen Functionen, (1896).

141 Stolz, O., Grundzuege der Diff. u. Integr. Rechnung, part 1, 177-198 (1893).

142 ———, Math. Ann., 8: 415-443 (1875).

143 ———, Math. Ann., 15: 122-160 (1879).

[144] Study, E., Leipz. Ber., 53: 338-403 (1901).
[145] Taylor, B., Methodus incrementorum, London, (1717).
[146] Thomé, L. W., J. f. Math., 126: 52-70 (1903).
[147] Veronese, G., Math. Ann., 19: 161-234 (1882).
[148] Vessiot, E., Soc. M. F. Bull., 22: 208-216 (1894).
[149] ———, Bull. Sci. Math., 20: 29-31 (1896).
[150] ———, Toulouse Ann., 10: 1-14 (1896).
[150.1] Walker, B. M., Diss. Chicago, (1906) 54 pages.
[151] Weierstrass, K., Vorlesungen ueber die Theorie der Abel'schen Transcendenten (1869); Werke, 4: 1-46.
[152] ———, Abhandlungen aus der Functionenlehre, 107 (1886); Werke, 2: 135.
[153] ———, Werke, 1: ch. 1.
[154] ———, Werke, 4: ch. 2, 6.
[155] Wieleitner, H., Bibliographie der hoeherenn algebraischen Kurven (1890-1904); Jahresbericht Gym. Speyer (1904-05).
[156] ———, Theorie der ebenen algebraischen Kurven, (1905).
[156.1] Zariski, O., Rom. Acc. L. Mem., (6), 1: 482-495 (1926).
[157] Zeuthen, H. G., Lehrbuch der abzaehlenden Geometrie, 17-28 (1914).
[158] ———, Encykl., 111 C 3, sect. 5 and 6.
[159] ———, Math. Ann., 10: 210-220 (1876).
[160] ———, Math. Ann., 10: 210-220 (1876); Acta Math., 1: 171-188 (1882).
[161] ———, Acta Math., 1: 171-188 (1882).
[162] ———, Tidskr f. Math., (5), 5: 65-79 (1887); Math. Ann., 31: 235-251 (1888).

CHAPTER III

LINEAR SYSTEMS OF PLANE CURVES

1. *Definitions.*—The system of curves represented by an equation of the form

$$\sum_{i=1}^{k+1} \lambda_i f_i(x_1, x_2, x_3) = 0 \tag{1}$$

where the f_i are of the same order n and linearly independent, is a linear system of curves of dimension k. A point which is at least r fold for each curve of the system is called an r fold basis point. It was shown by E. Bertini([1]) that if a linear system of curves has a variable r fold point, the system consists of a fixed curve counted $r-1$ times and a variable curve. He also showed that if every curve of a linear system is composite, either there is a fixed component or else each curve consists of the same number of curves belonging to the same pencil. For another proof see J. Lüroth([14]) and ([15]). In studying linear systems of curves it is convenient to consider systems of maximum dimension with a given set of basis points. An r fold point imposes $\frac{r(r+1)}{2}$ conditions on a curve so that counting the conditions separately we have the virtual dimension

$$K = \frac{n(n+3)}{2} - \Sigma \frac{r(r+1)}{2}. \tag{2}$$

In certain cases the conditions are not independent so that the effective dimension is

$$k = K + s, \tag{3}$$

where s is the superabundance.

A system for which $s=0$ is said to be *regular.*

A curve of the system is determined by k arbitrary points. The virtual genus of a general curve of the system as calculated from the basis points is

$$P = \frac{(n-1)(n-2)}{2} - \Sigma \frac{r(r-1)}{2}. \tag{4}$$

The effective genus p is the number of linearly independent adjoints of order $n-3$ having an $r-1$ fold point at each r fold point of the system. For an irreducible system $p=P$; G. Castelnuovo([5]). For reducible systems $p=P+c-1$ where c is the number of component curves which make up a curve of the system, G. Biermann (3). The number of variable intersections of 2 curves of the system is the grade

$$D=n^2-\Sigma r^2. \tag{5}$$

The numbers K, D, and P satisfy the relation

$$K=D-P+1 \tag{6}$$

so that

$$k=D-p+s+1 \tag{7}$$

for an irreducible system, where s is $\leqq p$ because $D \geqq k-1$, V. Martinetti (20).

Since $D \geqq 2(k-1)$, s is $\leqq p-k+1$; G. Castelnuovo (5).

The numbers D, p, and from (1) k are invariant under birational transformation, see G. Jung (11). See also G. B. Guccia (9).

2. *Superabundance.*—C. Küpper (13) determined the maximum number of simple basis points such that through them pass a system of curves of dimension K and order N.

F. S. Macaulay (16) and (17) showed that the superabundance of the complete intersection of C_l, C_m for C_n is $(l+m-n-1)(l+m-n-2)$ for $n \geqq l \geqq m < l+m$ and is zero for $n \geqq l+m-2$. The special case $n=l+m-2$ is given by C. Küpper (13). Macaulay also found an expression for the superabundance of the residual intersection of C_l and C_m through $N(<lm)$ points for $C_{l+m-n-3}$, and applied the theorem to construct point groups with required superabundance for curves of given orders. V. Martinetti (18) determined the superabundant systems for $p=2$, $k>1$, and M. de Franchis (7) and (8) found those with $p=2$, $k=1$ and also for $p=3$. H. Krey (12) showed that through $\frac{n(n+3)}{2}-4$ arbitrary points and 2 points properly chosen pass ∞^3 curves of order n for $n>4$.

3. *Characteristic Series.*—C. Segre (21), by considering the G_D^{k-1} cut on a curve of the system by the other curves of the system, showed that if $D>2p-2$ or $k>p$, $k=D-p+1$ so that $s=0$ and if D is $>2p$, or $k>p+1$ the system is *simple*, that is, if a curve of the system passes through a point it does not therefore pass through one or more associated points.

4. *Maximum Dimension.*—G. B. Guccia (10) and V. Martinetti (19) had determined the linear systems of least order for $p=1$, namely cubics, and C_4 with $2P_2$, and C. Segre (21) had found for $p=2$, the C_4 with P_2, the C_5 with P_3 and P_2, and the C_6 with $8P_2$. Hence if $p=1$, the maximum $K=D=9$, and for $p=2$, the maximum $K=11$, and the maximum $D=12$.

G. Castelnuovo[4], by adding additional double points, each of which reduces p by 1, k by 3, and D by 4, found the maximum K to be $3p+5$ and the maximum D to be $4p+4$, limits actually attained by hyperelliptic curves.

5. A *fundamental curve* is one which does not meet a general curve of the system except at basis points. If we impose the condition that a point on a fundamental curve F lies on a curve of the system we have a system of dimensions $k-1$ consisting of F and a residual curve. E. Bertini[2] proved that if $s=0$, then the genus of F is 0, and G. Castelnuovo[5] showed that the genus of F is $\leq s$. He also proved that a superabundant system has at least 9 basis points. E. Ciani[6], continuing Castelnuovo's work, distinguished between a proper fundamental curve and an improper one which can be transformed into a point which is not a basis point of the system. G. Jung[11] showed that the difference between the number of fundamental points and curves is invariant under a birational transformation. C. A. Scott[20] considered cases in which a fundamental curve is a multiple factor of partial systems of a linear system.

6. *Adjoints.*—G. Castelnuovo[5] studied linear irreducible systems $|C|$ by the aid of the adjoints of order $n-3$ which contain as a fixed component any fundamental curve for which $s=0$. The residual system $|C'|$ is termed the pure system of adjoints and cuts on a curve of $|C|$ the series G^{p-1}_{2p-2}.

7. *Virtual Characteristics of a Composite System.*—G. Castelnuovo[5]. If the system $|C|$ has a residual $|C''|$ with respect to $|C'|$, with $|C'|$ and $|C''|$ meeting $|C|$ in Δ' and Δ'' points and $|C'|$ meeting $|C''|$ in I points then

$$\begin{aligned} K &= K'+K''+I \\ p &= p'+p''+I-1 \\ D &= D'+D''+2I \\ K'' &= K-\Delta'+p'-1 \\ \Delta'-D' &= \Delta''-D''=I \\ \Delta'+\Delta'' &= D \\ p-p' &= \Delta''-K''. \end{aligned}$$

8. *Adjoints and Dimension.*—G. Castelnuovo[5]. If $|C'|$ is the pure adjoint system of $|C|$ then $\Delta'=2p-2$ and $K''=K-2p+p'+1$. If now k is $>p+1$, $K=k$ and $D>2p$.

Hence Δ'' is >2 and $\Delta''-K''$ is ≤ 2.

Hence $p'<p-1$ so that s' which is $\leq p'-k'+1=0$ and therefore if k is $>p+1$, $|C'|$ is regular and p' is $<p-1$.

Since $p-p''=\Delta'-K'$ therefore $p''=1$ so that K'' is ≤ 9. Hence the maximum value of k is $2p-p'+8$ when $|C|$ can be transformed into all the curves of a certain order. With this exception we have $k\leq 2p-p'+7$. The minimum value of p' is $2-p$ so that the maximum value of K is $3p+5$.

BIBLIOGRAPHY

1 Bertini, E., Lomb. Ist. Rend., (2), 15: 24 (1882). [1.

2 ———, Pal. Circ. Mat., 3: 5-21 (1884). [5.

3 Biermann, G., Monatsh. f. Math., 10: 373-375 (1899). [1.

4 Castelnuovo, G., Ann. di Mat., (2), 18: 117-128 (1890). [4.

5 ———, Torino Mem., (2), 42: 3-43 (1891). [1.

6 Ciani, E., Giorn. di Mat., (2), 33: 57-79 (1895). [5.

7 de Franchis, M., Pal. Circ. Mat., 13: 1-27 (1899). [2.

8 ———, Pal. Circ. Mat., 13: 130-160 (1899). [2.

9 Guccia, G. B., Pal. Circ. Mat., 1: 139-156 (1884-1887). [1.

10 ———, Pal. Circ. Mat., 1: 169-187 (1884-1887). [4.

11 Jung, G., Ann. di Mat., (2), 15: 277-312 (1887). [5.

12 Krey, H., Math. Ann., 19: 82-91 (1881). [2.

13 Küpper, C., Monatsh. f. Math., 6: 5-11, 127-156 (1895). [2.

14 Lüroth, J., Math. Ann., 42: 457-470 (1893). [1.

15 ———, Math. Ann., 44: 539-552 (1894). [1.

16 Macaulay, F. S., London M. S. Proc., 26: 495-544 (1895). [2.

17 ———, London M. S. Proc., 28: 673-695 (1895). [2.

18 Martinetti, V., Pal. Circ. Mat., 1: 202-216 (1884-1887). [2.

19 ———, Lomb. Ist. Rend., (2), 20: 264-269 (1887). [4.

20 Scott, C. A., Quart. J., 32: 209-238 (1901). [5.

21 Segre, C., Pal. Circ. Mat., 1: 217-221 (1884-1887). [3, 4.

CHAPTER IV

PLANAR CREMONA TRANSFORMATIONS

The more important general properties of these transformations have been developed for the plane by Luigi Cremona in the two memoirs (86) (1863); (88) (1865). He remarks(86) that earlier writers had regarded the quadratic transformation as the most general birational transformation from plane to plane and had himself apparently accepted this view two years earlier (85) (1861-62). He notes, however, that the sequence or product of two or more quadratic transformations is usually of higher order. In the present account sections 1, 2, 3, 4 are devoted to those properties of the transformations which appear when the two planes in correspondence are not necessarily superposed; sections 5, 6, 7, 8, 9, 10 to properties for superposed planes; and sections 11, 12 to applications of the transformations in geometry and algebra. General accounts of this subject have been given by Clebsch-Lindemann(67), Doehlemann(97), Sturm(272), Malet (192), Coble(77), and Mlodziejowski(201). Historical résumés are given by Pascal[(242.1), pp. 356-365] and by Loria[(188.1), pp. 253-259]. A comprehensive exposition of the theory is given by Hudson[(144.1), Part I].

1. *Definitions and General Properties.*—We use the terms *pencil, net,* and *web* for linear systems of dimension one, two, and three respectively. Sylvester(276) calls a flat space *homaloidal* or a *homaloid* (Gr. ομαλοσ=level); Cremona(89) extends the term to cover a rational surface, and calls a web of surfaces, such that the variable net in the web has a single base point variable throughout the space, a *homaloidal* web. In (90) Cremona defines a *homaloidal net* of plane curves to be a net whose variable pencil has a single base point variable throughout the plane.

Cremona(88) establishes the general transformation of *order n* by setting up a projective correspondence between the net of lines on a plane E_x and a homaloidal net H_y of order n on a plane E_y by making four lines (no three on a point) of E_x correspond respectively to four curves (no three in a pencil) of H_y. If the pencil of lines on x in E_x corresponds to a pencil of H_y with variable base point at y then y is the image of x in the Cremona transformation T from E_x to E_y. The net is determined by fixed points of given multiplicity called *fundamental points* or *F-points* (also base-points, principal points). If these consist of β_k k-fold points ($k=1, \ldots, n-1$) then the linear conditions they impose on the net and the intersections they absorb lead to([86]) the relations

$$(1) \qquad \sum_{k=1}^{n-1} \frac{k(k+1)}{2} \beta_k = \frac{(n-1)(n+4)}{2}, \quad \sum_{k=1}^{n-1} k^2 \beta_k = n^2 - 1.$$

Since the genus of a curve of H_y must be zero the independence of the linear conditions is confirmed. Lines on E_y correspond under the inverse transformation T^{-1} to curves also of order n of the *inverse* (or conjugate) homaloidal net H_x on E_x which is determined by α_k k-fold ($k=1, \ldots, n-1$) F-points on E_x where the numbers α_k satisfy the relations (1) ([88]). We assume unless expressly stated otherwise that the F-points in E_x, E_y are all distinct, i. e., the homaloidal nets have ordinary multiple points all of whose tangents are variable in the net.

Cremona([88]) shows that the number of F-points in E_x is the same as the number ρ of F-points in E_y. Clebsch([68]) arranges the F-points in E_x, F_i, in order of decreasing multiplicity r_i, i. e., $r_i \geqq r_{i+1}$; and the F-points in E_y, F_j of multiplicity s_j in similar order, i. e., $s_j \geqq s_{j+1}$ ($i, j=1, \ldots, \rho$). Cremona([88]) shows that the pencil of curves in E_y which is the transform of a pencil of lines on F_i in E_x has for common factor a curve P_i of order r_i—a curve which we call a *principal* (or fundamental) *curve* or *P-curve*. He proves that the various points of P_i on E_y correspond to the points on E_x consecutive to F_i in various directions or more briefly to directions at F_i. An algebraic proof by r-th polars of this theorem is given by Cayley([58]); see also Rosanes([251]). If the singularities of the net at the F-points are compound, or if the F-points are in special position, the P-curves may break up; cf. Autonne([23.1]), Biasi([38.1]) and Hudson[([144.1]), Chap. II].

With respect to the relative behavior of the P-curves and F-points, Cremona([88]) proves the following: Every curve which is a fixed part of a composite pencil of the homaloidal net is a P-curve and conversely; every composite curve of the homaloidal net has at least one P-curve as a part; a curve which cuts the curves of the net only at the F-points is a P-curve and conversely; a P-curve is of genus zero, has no multiple points except at the F-points, and is completely determined by its multiplicities at the F-points; the number, α_{ij}, of times a P-curve P_i passes through the F-point, F_j, on E_y is the same as the number of times the P-curve P_j passes through the F-point, F_i, on E_x; the aggregate of the P-curves of a homaloidal net is the Jacobian of the net [proved earlier by T. A. Hirst in "The Reader," Oct. 1 (1864)]. The class, the number of flexes, and the number of double tangents of P_i is transferred to the number of curves of the net with a given behavior at F_i. Cremona([87])

also notes that the sum of the orders of the P-curves of a homaloidal net is equal to the sum of the orders of the F-points. For the particular case of a Jonquières transformation for which $a_{n-1}=\beta_{n-1}=1$ and $a_1=\beta_1=2(n-1)$ de Jonquières [(145) (1859); (146) (1860); (147) (1864); in full (148) (1885)] had noted the F-points and P-curves. The first algebraic treatment is due to Rosanes(251) (1871). For such treatment the following theorems of Bertini(32) concerning linear systems in any space are important. If every member of a linear system has an r-fold point outside the base of the system, the locus of these r-fold points is an $(r-1)$-fold factor of the system; if every member of the system is composite, then either the members have a common factor or the members are composed of j members of a fixed pencil [cf. Lüroth(190)]. Pasch(243) proves that between three ternary forms of order n and no common factor with vanishing jacobian there exists an irreducible equation of genus zero and degree $\nu(n=\nu\mu)$; and the three given forms can be expressed as binary forms of order ν whose arguments are ternary forms of order μ. Darboux(91) (1872) gives a review of the above and Dewulf(94) an excellent résumé of progress up to 1873.

The theorems stated above lead to the following relations on the positive integers n, r_i, S_j, a_{ij}:

$$
(2)\qquad
\begin{array}{ll}
\Sigma_i\ r_i^2=n^2-1 & \Sigma_j\ s_j^2=n^2-1\\
\Sigma_i\ r_i=3(n-1) & \Sigma_j\ s_j=3(n-1)\\
\Sigma_i\ a_{ij}^2=s_j^2+1 & \Sigma_j\ a_{ij}^2=r_i^2+1\\
\Sigma_i\ a_{ij}=3s_j-1 & \Sigma_j\ a_{ij}=3r_i-1\\
\Sigma_i\ a_{ij}r_i=ns_j & \Sigma_j\ a_{ij}s_j=nr_i\\
\Sigma_i\ a_{ij}a_{ik}=s_js_k & \Sigma_j\ a_{ij}a_{kj}=r_ir_k
\end{array}
\qquad (i,j=1,\ldots,\rho).
$$

Clebsch(68) (1871) notes that the absolute constants of T are the $2\rho-8$ absolute projective constants of the ρ F-points in either plane. He finds that

$$
\Delta=|a_{ij}|=\pm n,\qquad \pm r_i=\Sigma_j s_j\frac{\partial\Delta}{\partial a_{ij}},\qquad \pm s_j=\Sigma_i r_i\frac{\partial\Delta}{\partial a_{ij}},
$$
$$
r_is_j=na_{ij}\mp\frac{\partial\Delta}{\partial a_{ij}}.
$$

He proves that the matrix $||a_{ij}||$ for a group of a_l equal r_i's and a group of β_m equal s_j's has elements a_{ij} which are all equal except that for the group of $a_l>1$ points there is one and only one square matrix $(\beta_m=a_l)$ in which the elements a_{ij} are all equal to a except for a_l elements which by proper arrangement of the β_m points can be brought into the principal diagonal and these a_l elements are all equal to $b\neq a$. Thus *the numbers* a_l, β_m *coincide except as to order.* By subtracting one column of Δ from the others he finds that $(a-b)^{a_{l-1}}$ is a factor of $\Delta=\pm n$. Bertini([30]) shows, however, that $a-b=\pm 1$; and that $a_{ij}>a_{ij'}$ if $s_j>s_{j'}$.

The P-curves of a Cremona transformation are rational, and are completely determined by their multiplicities at the F-points which themselves may be in general position. Thus not every rational curve can be a P-curve. Bertini([27]) calls a rational curve of order t for which the positions of all its multiple points of multiplicities

$$t_1, t_2, \ldots (t_1 \geqq t_2 \geqq \ldots)$$

may be arbitrarily assigned an L-curve. He proves that $t_1+t_2+t_3>t$ and that an L-curve can be transformed by quadratic transformation into an L-curve of lower order and eventually into the directions about a point. Since any Cremona transformation is a product of quadratic transformations (cf. §7) Bertini's L-curves are the P-curves of Cremona transformations. The L-curves for which $t_1+t_2+t_3=t+1$ belong to seven types: $t=5$, $a_2=6$; $t=6$, $a_3=1$, $a_2=7$; $t=8$, $a_3=7$; $t=11$, $a_4=7$, $a_3=1$; $t=17$, $a_6=8$; $t=20$, $a_7=8$, $a_3=1$; $t=t$, $a_{t-1}=1$. Brahana and Coble([42]) note that all the P-curves, including the directions at a point, completely defined by a set of ρ F-points (which are infinite in number for $\rho \geqq 9$) give rise to a discontinuous division of the plane and study the simple finite map for $\rho=4$.

H. W. E. Jung([159]) treats the general Cremona transformation with singularities of any kind whatever at the F-points. The theorems obtained differ in some respects from those which are valid for the transformations with ordinary discrete F-points. Both the exposition and the results are given in the notation of the Hensel-Landsberg theory of algebraic functions.

2. *Determination of the Types.*—The type of a Cremona transformation is determined when one of the two homaloidal nets is given. The net is determined when the position of the F-points is given as well as the order n and multiplicity r_i of the curves of the net. We consider here the determination of the positive integers n, r_i which satisfy the relations

$$(3) \qquad r_1+r_2+\ldots+r_\rho=3(n-1), \qquad r_1^2+r_2^2+\ldots+r_\rho^2=n^2-1.$$

This is not merely a diophantine problem. The solutions are subject to the inequalities which express the fact that a curve of order m cannot cut all the curves of the net in more than mn points. Thus de Jonquières([150]) distinguishes between the algebraic solutions for which some of the integers are negative; the arithmetic solutions for which the integers are positive but the net does not exist; and the geometric solutions for which the integers are positive and the net exists [cf. also Palatini([237])].

The types may be listed either according to the order n or according to the number ρ of F-points, the latter classification being perhaps more fundamental. Cremona([86]) gives the first list: $n=2$, $\alpha_1=3$; $n=3$, $\alpha_2=1$, $\alpha_1=4$; $n=4$, $\alpha_2=3$, $\alpha_1=3$; $n=4$, $\alpha_3=1$, $\alpha_1=6$; and the de Jonquières type, $n=n$, $\alpha_{n-1}=1$, $\alpha_1=2(n-1)$. He adds([88]) the inverse nets and P-curves for all solutions from $n=2$ to $n=10$ inclusive; for $n\equiv 0$, or 1 (mod. 2) one pair of inverse nets; and for $n\equiv 0$, 1, 2, or 3 (mod. 4) four pairs of inverse nets. Roberts([250]) determines the numbers α_{ij} for the types given by Cremona as reproduced by Cayley([58]). Guccia([128]) gives algebraic expressions for the homaloidal nets and their inverses of four of the general types given by Cremona. Bianchi([38]) first derives new types of nets by transforming others. By transforming a Jonquières type of order m by one of order n he finds new self-inverse types for any order. Ruffini([252]) made little effective progress by arithmetic methods. Guccia([135]) derives the type of a product from that of the factors. Thus if T_k (order k) has inverse F-points of orders $s_1, \ldots, s_\rho$ and T_l of orders $s'_1, \ldots, s'_{\rho'}$ then $T_kT_l=T_{kl}$ has F-points of orders

$$s_1, \ldots, s_\rho, ks'_1, \ldots, ks'_{\rho}.$$

He notes that a type given by de Jonquières([149]) is the product of two Jonquières types. G. Jung([158]) determines the jacobian of the product of two T's and the relations on the integers r_i, α_{ij} of the product. Palatini([238]) finds a variety of new types by combining in various ways the known types with some general values in their integers α_i, β_k. De Jonquières([150]), ([149.2]) devises a rather complicated transition from solutions for one order n to another $n+k$, gives the geometric solutions for $n=12$, and for $k=1$ deduces those for $n=13$. It is a method of trial with high orders used as illustrations. Earlier([149.1]) he discussed composition as a method for finding types. Guccia([129.1]) gave the type which arises from a product of quadratic transformations which are not particularly related. Ruffini([252.1]) gives types with simple and double F-points.

Clebsch([67.1]) and Diekmann([96]) recognize that two different mappings of a rational surface on a plane lead to a Cremona transformation of the plane—the extension of the construction of a quadratic transformation by two projections of a quadric surface. The latter applies this to two mappings of a cubic surface by cubic curves on six points. Sturm([275]) maps the cubic surface on a plane first by transversals of two skew lines u, v and second by bisecants of a twisted cubic K on the surface. According as (σ, τ), the respective numbers of points at which u, v meet K, are (0, 0), (1, 0), (1, 1), (2, 0), (2, 1) and (2, 2), T's of orders 4, 6, 8, 8, 10, 12 are obtained which are included in the known tables. Ines Larice([184]) gives an arithmetic method for deducing types of order n

when those of previous orders are known. The method is suggested by projection of the Veronese F_2^4 in S_5. As examples the types for $n=14$, 15 are determined. She also gives([185]) two types already listed by Ruffini([252]).

Montesano([209]) considers the relative values of the a_i and makes a decided advance, e. g., the theorem

$$(n-1)\{\rho(n+1)-9(n-1)\}=\Sigma(r_i-r_k)^2$$

when $n>2$. He proves* that there are no symmetric types other than $n=2$, $a_1=3$; $n=5$, $a_2=6$; $n=8$, $a_3=7$; $n=17$, $a_6=8$. In ([211]) he constructs new types by taking the product of a given type by a quadratic transformation. He finds that the four symmetric types and the Jonquières type are the only exceptions to $r_1+r_2+r_3=n+k(k>1)$. He also finds *asymmetric* types for which $r_i\neq r_k$ contrary to the opinion expressed by Clebsch([68]), that of least order being $n=23$, $r_i=12$, 10, 9, 8, 7, 6, 5, 4, 3, 2. Many particular properties of the integers r_i are given, e. g., $r_1+r_2+r_4>n$ (except for the case $n=2$); $n-1\geqq r_1\geqq\frac{n+1}{3}$, the limits being attained only in the Jonquières case and the symmetric cases, respectively. He shows that if k of the r_i's are equal, then $k<9$ and gives for $\rho=9$ an infinite sequence of types. He also adds new types for $n=14$, 15 to the lists of Larice([184]). Montesano([213]) determines also the *semi-symmetric* types, i. e., those in which the F-points fall into two groups of like order [cf. also ([214])]. Miss Hudson([143]) considers solutions for n a number of the form $n=x\mu+\nu(\nu=0, 1, \ldots, x-1)$. For $x=2$, 3 the general types which can be stated in terms of such a division of n have been given in part by Cremona and Ruffini and in larger measure by Palatini. For $x=4$ Miss Hudson adds new types to those already known and finally determines types for general x. Mlodziejowski([202]) derives the types for larger n from previous types by quadratic transformation and compiles tables for the first 21 orders. Later ([202.1]) he discusses the arithmetic solutions of the equations of condition and allied linear systems. For a very complete discussion consult Hudson[([144.1]), Chap. IV].

S. Kantor([168]) first recognizes the importance of ρ, the number of F-points, in the classification of types. He states (p. 273) that the number of types for $\rho=9$ is infinite and gives (p. 280) a table of the finite number of types for $\rho<9$ which includes the non-existent type $n=11$, $a_3=5$, $a_5=3$. His theorem (p. 297) that the numbers n, r can be expressed as rational functions of $\rho(\rho+1)/2$ parameters is not practically useful since the formulae "n'ont bésoin que d'une restriction

* See p. 121.

relativement au caractere entier des coefficients." Nevertheless the only effective method thus far found for obtaining solutions n, r_i with actually existent nets is that of combining known transformations. For this method the group of characteristics introduced by Kantor[(168) Part IV] and treated in the next section is indispensable. In this way Coble(74), (p. 375) has proved that the infinite number of types for $\rho=9$ divide into $960.3^7.2$ classes such that the types in each class depend upon the unrestricted variation of 8 integers.

3. *Linear Transformations with Integer or Rational Coefficients Associated with Cremona Transformations. The Geometry of the Cremona Group.*—If the transformation T of order n has F-points in E_x of multiplicities $r_i(i=1\ldots\rho)$ and if T^{-1} has F-points in E_y of multiplicities $s_j(j=1,\ \ldots,\ \rho)$; also the P-curve of the point F_i in E_x passes a_{ij} times through the point F_j on E_y, then a curve C_x in E_x of order x_0 with multiplicity x_i at F_i is transformed by T into a curve C_y in E_y of order y_0 with multiplicity y_j at F_j where

$$(1)\qquad G:\quad \begin{array}{l} y_0=nx_0-r_1x_1-r_2x_2\ldots-r_\rho x_\rho \\ y_1=s_1x_0-a_{11}x_1-a_{21}x_2\ldots-a_{\rho 1}x_\rho \\ y_2=s_2x_0-a_{12}x_1-a_{22}x_2\ldots-a_{\rho 2}x_\rho \\ \cdots\cdots\cdots\cdots\cdots\cdots\cdots\cdots \\ y_\rho=s_\rho x_0-a_{1\rho}x_1-a_{2\rho}x_2\ldots-a_{\rho\rho}x_\rho \end{array}$$

If the behavior of the transforms C_x, C_y at pairs of ordinary corresponding points should also be indicated, the equations (1) may be extended by adding

$$\begin{array}{l} y_{\rho+1}=-(-1)x_{\rho+1} \\ y_{\rho+2}=-(-1)x_{\rho+2}. \end{array}$$

Since C_x is the transform of C_y by T^{-1} the transformation G^{-1} associated with T^{-1} has a form similar to (1) and arises from (1) by interchanging the coefficients r, s and the a_{ij} with the a_{ji}. Hence the determinant of the coefficients, D, has the value

$$(2)\qquad D=\begin{vmatrix} nr_i \\ s_ja_{ij} \end{vmatrix}=\pm 1.$$

If C_y is transformed by a new Cremona transformation T_1 into C_z and if G_1 is the linear transformation (1) associated with T_1 then C_x is transformed into C_z by the product TT_1 whose associated transformation is the product GG_1. Naturally the products TT_1 and GG_1 are not completely defined unless the distribution of the F-points of T_1 among the F-points (or ordinary points) of T^{-1} is given. Since T itself is a product of quadratic transformations, the linear transformation G may be generated by the element

$$(3)\qquad \begin{aligned} y_0 &= 2x_0 - x_1 - x_2 - x_3 \\ y_1 &= \ \ x_0 \qquad\ \ - x_2 - x_3 \\ y_2 &= \ \ x_0 - x_1 \qquad\ \ - x_3 \\ y_3 &= \ \ x_0 - x_1 - x_2 \end{aligned}$$

and permutations of $x_1, \ldots, x_\rho$.

The linear transformation G appears first in G. Jung[(156)1887-8] but the first comprehensive investigation of it is due to S. Kantor(168) (pp. 293-316). He finds that G has the invariant linear and quadratic forms,

$$(4)\qquad \begin{aligned} &3x_0 - x_1 - x_2 \ldots - x_\rho \\ &x_0^2 - x_1^2 - x_2^2 \ldots - x_\rho^2 \end{aligned}$$

and observes that all the numerical relations, §1 (2), are a consequence of the invariance of these forms [cf. also Montesano,(210); Coble(74).] He calls G (or its system of coefficients) the "characteristic" of T and $x_0, x_1, \ldots, x_\rho$ the "singularity-complex" of C_x. Two curves C_x, C'_x with complexes x, x' meet in

$$x_0x'_0 - x_1x'_1 - \ldots x_\rho x'_\rho$$

ordinary points so that the invariance of the polarized form of the quadratic (4) is evident. A cubic curve on the F-points of a quadratic T is transformed into a cubic curve on the F-points of the quadratic T^{-1} whence the particular complex

$$(5)\qquad 3, 1, 1, \ldots, 1$$

is invariant under T itself. Hence its linear polar form (4) as to the quadratic form (4) is invariant. Hence if the complex x is transformed by G into the complex y then the complex

$$x_0 - 3,\ x_1 - 1,\ x_2 - 1,\ \ldots,\ x_\rho - 1$$

is transformed into the complex

$$y_0 - 3,\ y_1 - 1,\ y_2 - 1,\ \ldots,\ y_\rho - 1,$$

i. e., if C_x is transformed into C_y the canonical adjoints of C_x are transformed into the canonical adjoints of C_y. Kantor notes the existence of the generator (3) and gives an extended account of the characteristic equation of the linear transformation G with particular reference to the periodicity of G in connection with the periodicity of T. He erroneously ascribes to D the value $(-1)^\rho$. Since the complex (5) is invariant the characteristic equation of G always has one root $+1$. If it has ν roots $+1$ and ν linearly independent complexes Kantor says that G is of "rank ν."

Thus (3) is of rank 3 with invariant complexes 1, 1, 0, 0; 1, 0, 1, 0; 1, 0, 0, 1 upon which (5) is now dependent.

Montesano([210]) shows that if a group of equal r_i's is coordinated to a like group of equal s_j's and if F_{i_1}, F_{i_2} are two F-points of the first group coordinated to the points F_{j_1}, F_{j_2} respectively of the second group by the scheme of Clebsch (§1) then $x_{i_1}-x_{i_2}=\pm(y_{i_1}-y_{i_2})$. Montesano([212]) also discusses the behavior of the matrices n, r_i, s_j, a_{ij} in the operation of forming various products of Cremona transformations. The inequality

$$r_i+s_j\leqq m+a_{ij}$$

is proved by a hyperdimensional argument by Coble([74]).

The linear transformation, G, used by Kantor almost exclusively in connection with periodic transformations and finite groups for which the F-points must satisfy restrictive conditions has been applied by Coble([74]) to the study of general sets of discrete points of the plane. He defines two ordered sets of m points, $P_m{}^2$ in E_x with points $p_1, \ldots, p_m$ and $Q_m{}^2$ in E_y with points $q_1, \ldots, q_m$ to be *congruent* under the Cremona transformation T if ρ of the pairs p_i, q_i are the F-points of T, T^{-1} and the remaining $m-\rho$ of the pairs are corresponding pairs of ordinary points. Necessarily then $m\geqq\rho$. If $m=\rho$ the conditions that $P_m{}^2$ and $Q_m{}^2$ be congruent under T are the $2\rho-8$ projective conditions on the F-points of T; if $m=\rho+1$ these conditions are the $2\rho-6$ projective conditions which amount to a projective construction for T. In order to put into evidence the projective invariants of an ordered set $P_m{}^2$ the first four points are placed at the reference and unit points and the factors of proportionality in the remaining $m-4$ points are so adjusted that the last coordinate of each is the same. Then the set has $2(m-4)+1$ variable homogeneous coordinates (whose ratios are absolute projective constants of $P_m{}^2$) which are taken as the coordinates of a point P in a space $\Sigma_{2(m-4)}$. To the $m!$ ways of reordering the set $P_m{}^2$ there correspond $m!$ points P in $\Sigma_{2(m-4)}$ which are conjugate under a Cremona group $G_{m!}$ in $\Sigma_{2(m-4)}$ which is discussed in ([73]). To the passage from a set $P_m{}^2$ to a congruent set $Q_m{}^2$ there corresponds in $\Sigma_{2(m-4)}$ a Cremona transformation from P to Q. Thus all sets $Q_m{}^2$ congruent in any order to a given set $P_m{}^2$ are represented by points Q in Σ which form a conjugate set under a Cremona group $G_{m,\,2}$ in Σ for which the identity corresponds in the plane to the passage from $P_m{}^2$ to any projectively equivalent set. This discontinuous group $G_{m,\,2}$ (infinite for $m\geqq 9$) is isomorphic with the group of transformations G of type (1) on variables $x_0, x_1, \ldots, x_m$; a group $g_{m,\,2}$, generated by (3) and the permutations of $x_1, \ldots, x_m$. The isomorphism is simple unless congruence implies projectivity which is true for only three cases (for general position of P_m^2): $m=5$, under quadratic transformation; $m=7$, under the Geiser transformation, $n=8$, $\alpha_3=7$; $m=8$, under the Bertini transformation, $n=17$, $\alpha_6=8$. The structure of the arithmetic group $g_{m,2}$ has been used for the study of $G_{m,2}$, and the properties of the related ternary Cremona transformations. The group $g_{7,2}$ isomorphic with the group of the double tangents of a plane quartic curve is discussed by Bramble(43).

There are 72 sets of points P_6^2 congruent in some order to one of the 72 sets; these are the sets such that cubic curves on P_6^2 map the plane on the *same* cubic surface; there are 288 sets P_7^2 congruent to a given set; these are the Aronhold sets of seven double points of a quartic envelope of class four. Kohn(182) had defined a "Wurf" of m points to be a set P_m^2 or any projective transform. He had considered the projective relations among the 72 Würfe P_6^2 attached to a cubic surface and the 288 Aronhold Würfe P_7^2 attached to a quartic envelope but did not recognize the existence of the Cremona transformations under which these Würfe are congruent. Conner(82) derives some of these transformations by projection from the space figure determined by the eight base points of a net of quadrics. Kasner(179) obtains the $G_{6!}$ in Σ_4 determined by the permutations of P_6^2 from the irrational invariants of a single-six of lines on a cubic surface. A group preliminary to $G_{m,2}$ is the cross-ratio group, due to E. H. Moore(216), a Cremona group which arises from the permutations of a set P_m^1 of m points on a line. Tinto(282.1) finds the transformations which lead to congruent sets P_6^2 as transformations of chords of two twisted cubic curves on a cubic surface. Further applications of the notion of congruence are found in §§ 9, 12.

Coble(74) (p. 370) has obtained a group $e_{m,2}$ isomorphic with $g_{m,2}$ and $G_{m,2}$ as follows. If P_m^2 is on an elliptic cubic C^3 with elliptic parameter u (such that $u_1+u_2+u_3\equiv 0$ is the collinear condition) a set Q_m^2 congruent to it under the quadratic transformation indicated by (3) is projective to a set Q_m^2 on C^3 itself. If $u_1, \ldots, u_m$ are the parameters of P_m^2 on C^3 the parameters $u'_1, \ldots, u'_m$ of Q_m^2 on C^3 are

$$(6)\qquad \begin{aligned} u'_i &\equiv u_i - \tfrac{2}{3}(u_1+u_2+u_3) \quad (i=1, 2, 3), \\ u_j &\equiv u_j + \tfrac{1}{3}(u_1+u_2+u_3) \quad (j=4, 5, \ldots, m). \end{aligned}$$

The third-periods implied in $\frac{1}{3}(u_1+u_2+u_3)$ may be removed by a collineation on C^3 or they may be avoided altogether by taking P_m^2 on a cuspidal cubic with collinear condition $t_1+t_2+t_3=0$ [cf. Bramble(43)]. The transformation (6) together with permutations of $u_1, \ldots, u_m$ generate $l_{m,2}$. The elements have rational coefficients with at worst the denominator 3. It is in effect the form which $g_{m,2}$ takes when x_0 is eliminated by the use of the linear invariant (4). This group is especially useful in discussing the first infinite case P_9^2.

Klein in his Erlangen Program[([181]) §8, 1 (1893)] remarked that a geometry, in the sense in which he used the term, of the Cremona group had not been developed. Fano[([111]) § 21] calls attention to the discontinuous character of the Cremona group and questions whether there are properties of curves invariant under the Cremona group which are not also invariant under the birational group. Marletta([193]) finds conditions under which two curves in birational correspondence are also in Cremona correspondence. The conditions are expressed in terms of the existence of the canonical adjoints of successive indices. He also finds conditions for the reduction by Cremona transformation of the order of a given curve and gives types of curves of least order. Coble([77]) compares the meagre development of Cremonian geometry with that of the related birational and projective geometry. He suggests as a geometric element the set of points P_m^2 rather than the point. As dual element there may be used either the linear systems of curves defined at P_m^2 or the aggregate of P-curves defined by P_m^2([42]). As the basis for an invariant theory he uses the conditions on the set P_m^2 which express that two of the P-curves coincide. These are called the *discriminant conditions* attached to P_m^2. If P_m^2 and Q_m^2 are congruent under T then T transforms the discriminant conditions of P_m^2 into those of Q_m^2. Ota([231]) considers the projective geometry of a homaloidal system (net, web, etc.) as a more general type than the geometry of the original space. As a rule, however, the theorems are merely translations from one linear system to another and the difficulties with respect to F-points are but lightly touched.

4. *Construction of the F-Points; Construction and Description of Particular Transformations.*—Clebsch([68]) had noted that if the ρ F-points of T in E_x are given and four of the ρ F-points of T^{-1} in E_y are given the remaining $\rho-4$ F-points in E_y are determined. Thus for $n>2$ and therefore $\rho\gtreqless 5$ the 2ρ F-points of T must satisfy $2\rho-8$ projective conditions and T is then determined. For $n=2$, T is determined by the two sets of F-points p_1, p_2, p_3; q_1, q_2, q_3 and a given corresponding pair, p_4, q_4. The construction for T then arises from the fact that if p_5, q_5 is any other corresponding pair the five points p_1, p_2, p_3, p_4, p_5 in E_x are projective in order to the five points q_1, q_2, q_3, q_5, q_4 in E_y. In general then two problems arise: first, to obtain some statement for the conditions on the F-points; and second, when these are satisfied some statement of the conditions on a corresponding pair. All of these conditions may of course be implied in a definite geometric construction.

Many special types are considered under involutorial and periodic transformations, and under geometric applications.

De Jonquières([147]) describes the transformation ($n=n$, $a_{n-1}=1$, $a_1=(2n-1)$) named after him and finds the inverse F-points and the P-curves [cf. also ([145])]. Cremona([86]) gives the following construction for this type. In space a line is drawn from a point x on a plane E_x across two fixed curves to meet a plane E_y in y. If one of the two curves is of order $n-1$ and the other a line $(n-2)$-secant to the curve, e. g., a plane curve of order $n-1$ with an $(n-2)$-fold point and a line on the $(n-2)$-fold point, or a $(1, n-2)$ curve on a quadric and a generator of the first kind, then the x, y are images in a special Jonquières transformation. Doehlemann([99]) gives the latter construction for $n=3$. Coble([76]) connects the two sets of $2(n-1)$ simple F-points with a set of $2(n-1)$ points in higher space. Schuttenhelm([256.1]) discusses the Jonquières transformation particularly for $n=2$, 3.

The cubic transformation has a double F-point p_1 and four simple F-points $p_2, \ldots, p_5$ on E_x and similar F-points $q_1, q_2, \ldots, q_4$ on E_y such that the simple F-point $p_i(q_i)(i=2, \ldots, 5)$ has for P-curve the line $\overline{q_1q_i}(\overline{p_1p_i})$ and the double F-point $p_1(q_1)$ has for P-curve the conic on all the points $q(p)$. The five points p are projective in order to the five points q. Bertini([31]) shows that any cubic transformation between two planes E_x, E_y can be generated by transversals of a line and conic in space with one common point; or by a variable conic required to pass through a fixed point in each of the given planes and to meet a fixed conic in space in two points. Saltel([253]) defines the "Arguesienne" (after Desargues) transformation as the pairs x, x' on a line through a fixed point P and on a conic of a fixed pencil. This cubic involution has a double F-point at P and simple F-points at the base points of the pencil. Various particular cases are discussed. This involution followed by a collineation is the general cubic transformation. Cardoso-Laynes([50]) and Lacour([183]) discuss metrically defined cubic transformations. Retali([248]) considers the cubic transformation $z=re^{i\phi}$, $z'=r\cos^2\phi e^{i\phi}$ in the complex plane of z; and Monaco-Aprile([206]) gives an algebraic account of it. Maschke([198.1]) studies the determination of the transformation by assigned F-points.

The quartic transformation ($a_2=3$, $a_1=3$) is the product of two quadratic transformations in general position. Cotterill([83]) gives the following construction. Conics on three points F_i and a point x cut a line l in pairs of an involution; conversely, conics on F_i which cut l in pairs of an involution meet in a point x. Let then F_i be a triangle in a plane π, F'_i a triangle in a plane π' which meets π in l; if conics on

F_i and x and conics on F'_i and x' meet l in pairs of the same involution, then x, x' are images in the quartic transformation. The F-points and P-curves are found and connected with Cremona's theory. Veronese([284]) constructs the transformation by his V_2^4 in S_5 (the map of the plane by the linear system of conics) from two planes each of which cuts V_2^4 in three points. Doehlemann([98]) gives the following construction: from x on a fixed plane α there is a unique chord across a fixed space cubic curve which meets a fixed plane β in y. If the curve breaks up into a conic and a secant line the transformation is cubic. Schmidt([253.2]) discusses the determination of the transformation when fixed points are assigned.

The quintic transformation ($\alpha_2=6$) has F-points $p_1, \ldots, p_6$ in E_x and $q_1, \ldots, q_6$ in E_y such that the P-curve of $p_i(q_i)$ $(i=1, \ldots, 6)$ is the conic on all the points $q(p)$ except $q_i(p_i)$. Müller([219]) gives the following construction. If S is the collineation which sends $p_1, \ldots, p_4$ into $q_1, \ldots, q_4$; and σ the collineation which sends $p_2, \ldots, p_5$ into $q_2, \ldots, q_5$, the pencils of conics on these respective sets are projectively related by S and σ. If x is on a conic of each pencil in E_x, y is on a corresponding conic of each pencil in E_y. Kohn([182]) obtains the points p, q as sections of halves of a double six of lines on a cubic surface; and also as the projections of six of the eight base point of a net of quadrics from each of the other two base points. Coble([72]) gives other properties and ([73]) considers the ways in which the "associated" P_6^2, Q_6^2 may coincide. He notes also([76]) that if p_7, p_8, p_9 form with P_6^2 the base points of a pencil of cubics a quadratic transformation with F-points at p_7, p_8, p_9 will send P_6^2 into Q_6^2. Schoute([255]) defines lines AO and AO' as symmetric with respect to a triangle ABC if $\angle BAO=\angle O'AC$; given O the three lines on it from A, B, C have symmetric lines which meet at O where O, O' are pairs of a quadratic involution. If the lines on O are replaced by circles on O and a fixed point then O, O' are pairs of an involution of the fifth order. He determines the F-points and P-curves and notes that it is the transform of the quadratic involution by an inversion.

Le Paige([234]) gives constructions for the F-points of the de Jonquières involutions $n=5, 6$; and later([235]) for the transformations. These constructions are extended by Schoute([256]) to transformations of higher order.

Veronese([284]) states the theorem that every Cremona transformation of order n can be determined by two projections of the surface $F_2^{n^2}$ obtained by mapping the plane by all curves of order n. This follows immediately from the fact that if to the net of lines on E_x we add a fixed curve of order $n-1$, the two homaloidal nets on E_x, E_y each of order n, are in projective correspondence. Coble([74]) gives constructions for all congruent sets P_m^2, Q_m^2 ($m \leqq 8$). This implies the construction of the F-points for $\rho=8$ and both the F-points and the transformation for $\rho<8$. In the two further symmetric cases ($n=8$, $a_3=7$; $n=17$, $a_6=8$), the two sets of F-points are projective.

Godeaux([122]) discusses the transformation from x to x' where x, x' are the further intersections of a cubic of the pencil on $p_1, \ldots, p_8$ with a conic of the pencil on $p_1, \ldots, p_4$. He also([124]) defines a transformation in terms of a net of order n with an $(n-1)$-fold point at O which contains a pencil with an $(n-3)$-fold point at O. If the line $\overline{xx'}$ is on O and the curve of the net on xx' touches $\overline{xx'}$ at x' then x, x' are images under a Cremona transformation of order $n-\nu-2$ where ν is the number of inflexional tangents at O of a general curve of the original net. Again Godeaux([125]) takes a pencil of cuspidal cubics with necessarily fixed cusp and cusp tangent and discusses the transformation of order 7 between a point x and its tangential point x' on the same cubic.

5. *Isologous Curves; Fixed Points and Curves; Cyclic Sets.* When the planes E_x and E_y coincide and the point y becomes a point x' on E_x the Cremona transformation T from x to x', which we write in direct and inverse form as

$$(1) \qquad T: x'_i = f_i(x); \quad T^{-1}: x_i = \phi_i(x') \qquad (i=1, 2, 3),$$

acquires new properties. The F-points of T^{-1} may coincide in part or altogether with the F-points of T. For certain points x, x' will coincide with x. These are the fixed points of T and of T^{-1}. Two given transformations T_1, T_2 have now a definite product T_1T_2. We may also speak of the powers T^m of a given transformation T. A fixed point of T^m is either a fixed point of T, or a member of a set of m_1 (m_1 a factor of m) points permuted cyclically into each other by T and called a *cyclic set* of T of order m_1. If the general point of E_x is in such a cycle of precisely m points, then T^m is the identical transformation and T itself is periodic with period m. Finite groups of Cremona transformations now occur as well as infinite groups both discontinuous and continuous. These matters are considered in §§ 5, ..., 9.

De Jonquières([147]) defines the *isologous* curve of a point $P(p_1, p_2, p_3)$ to be the locus of points x which are on a line with P and the image x' of x and shows that the Jonquières transformation of order n has $n+2$ fixed points. Cremona([88]) sets up the isologous net (for variable P) of the transformation T, a net of curves of order $n+1$ with an r-fold point at $F_r (r=1, \ldots, \rho)$ and on the $n+2$ fixed points of T, whose equation is $|pxf|=0$. Guccia([129]) considers certain points and lines whose existence puts two conditions on the isologues and therefore are finite in number, e. g., a point R at which the isologue of P has a cusp. Guccia([130]) develops a variety of points and lines projectively related to the isologues and states a number of theorems concerning them into which the fixed points and cyclic sets enter. He also([136]) investigates the two isologous nets, $|pxf|=0$, $|qx'\phi'|=0$ which occur for T and T^{-1}. Doehlemann([100]) finds that a curve of fixed points is a factor of each net. He distinguishes between fixed points at which every direction is fixed and fixed points at which but two directions are fixed.

A fixed curve of T is one which T transforms into itself. Thus Rosanes([251]) finds that the quadratic involution $x'_i=1/x_i$ has a net and a web of invariant cubics. T may also have a curve of fixed points. Doehlemann([100]) shows that, if T of order n has a curve of fixed points of order μ, then $n \gtreqless (3\mu-1)/2$. He also finds the class of the envelope of lines joining x, x' as x' approaches x as a limit. Castelnuovo([53]) proves that if T has an irreducible curve of fixed points of genus $p>1$ then either T is periodic of period 2, 3, or 4, or T is the transform of a Jonquières type. Conner([82]) discusses the general quintic transformation $(a_2=6)$ which has a cubic curve of fixed points. Most of the involutions discussed in the next section have curves of fixed points of considerable interest.

Rosanes([251]) notes that a quadratic transformation has two cyclic triads and remarks that the number of cyclic sets might be stated in general but that this number would have many exceptions for particular transformations. S. Kantor([163]) was the first to give the formula

$$\pi_m=\frac{1}{m}\,[n^m-\Sigma n^{\overline{\frac{m}{p_1}}}+\Sigma n^{\overline{\frac{m}{p_ip_j}}}\ldots(-1)^\nu n^{\overline{\frac{m}{p_1\cdots p_\nu}}}]$$

for the number π_m of cyclic sets of order $m=p_1^{m_1}p_2^{m_2}\ldots p_\nu^{m_\nu}$ (p_i a prime) of a transformation T of order n. For $n=2$ he had given this formula earlier([162]). This result for $n=2$ had been read by Hirst at the Birmingham meeting of the British Association in 1865 but it remained unpublished until([140]) 1881, a year after the appearance of Kantor's results. Doehlemann([101]) considers the case for which T has ∞^1 cyclic pairs which lie on a curve of fixed points of T^2. This curve is of order at most $2n-1$ and in general this limit it attained only for $n=2$, a case treated in more detail. Miss Hudson([144]) following out a particular case noted by Doehlemann([100]) proves that if under non-involutorial transformation two curves of fixed points coincide their general points are of the second kind (i. e., every direction at a point is a fixed direction); and that if all the general points of a curve are of the second kind the curve can be regarded as a pair of adjacent curves of fixed points. Such a transformation may also have an isolated double invariant point.

The content of this section is given in detail by Hudson([144.1], Chap. V).

6. *Involutorial Cremona Transformations.*—If the transformation T sends a general point x into x' and at the same time sends x' into x, i. e., if $T^2=1$, then T is *involutorial,* or of *period two,* or an *involution.* Then $T=T^{-1}$ and the two inverse nets coincide. Thus only those transformations T whose inverse nets are of the same type can, under proper conditions, become involutorial. One of these conditions is that the two sets of F-points coincide but this is not usually sufficient.

The most important contribution to our knowledge of involutions T is due to Bertini who, recognizing that the number of projective types T is unlimited, sought to reduce them to a finite number of types by *transformation,* i. e., by the formation of products $T_1^{-1}TT_1$ where T_1 is any Cremona transformation. He shows that by proper choice of T_1 this transform reduces to one of the four types:

(a) the harmonic homology;
(b) the Jonquières involution;
(c) the Geiser involution;
(d) the Bertini involution.

The Jonquières involution is composed of pairs x, x' on a line with O and harmonic to the two further intersections of this line with an irreducible curve of order $p+2$ and genus p with a p-fold point at O. The Geiser involution is composed of pairs x, x' which form with seven given points $p_1, \ldots, p_7$ the base points of a pencil of cubics. The Bertini involution is composed of pairs x, x' which form with eight given double points the base points of a net of sextics.

Bertini([28]) first proved that every involution of the Jonquières type $(n=n, \alpha_{n-1}=1, \alpha_1=2(n-1)$ could be transformed into either (a) or (b) and gave the general properties of such involutions. Later([29]) he proved by the use of an invariant linear system (L) of curves derived from a P-curve which passes α_{ii} (an $\alpha_{ii}>0$ always exists) times through its own F-point that an involution can be transformed into one of the four types (a), ..., (d). Here he first recognized the property of a web of sextics with eight given double points that the net of the web which passes through a point x also passes through a point x'. Two cases were not covered: first the case where the P-curves break up; and second the case $\alpha_{ii}>3$ and every curve of (L) not invariant. For a review of these two papers see Darboux([92]). Finally Bertini([30]) shows that an involution with isolated F-points either has an $\alpha_{ii}\leqq 3$ or is transformable into one of lower order and this can be transformed into one of his four types. He shows also that if $\alpha_1=0$, $\alpha_2=0$ the P-curves *of greatest order* r_1 go through all the F-points; that if $\alpha_1\neq 0$ then $\alpha_1=3$ and the 3 P-curves of order r_1 go through all the F-points except one simple F-point; and that if $\alpha_1=0$, $\alpha_2\neq 0$, then $\alpha_2=6$ and the 6 P-curves of order r_1 go through all the F-points except one double F-point. Bertini's study of the birationally distinct types of involutions forms the model for the study by S. Kantor of the birationally distinct types of periodic transformations and of finite groups (cf. § 8). They recur there and are discussed in the general accounts of these topics.

Geiser(120) first noted the involution (c) composed of the two further intersections x, x' of pencils of the net of cubics on seven points of the plane, P_7^2, with points $p_1, \ldots, p_7$. He observed that if x is at p_1, x' is on the cubic of the net with node at p_1; if x is on the line $\overline{p_1p_2}$, x' is on the conic through $p_3, \ldots, p_7$; if x is on a line, x' is on an octavic with triple points at P_7^2; and further that x, x' coincide on a sextic with nodes at P_7^2. Clebsch(69) showed that this sextic is the Jacobian of the net of cubics and mapped it by means of the net of cubics upon a general plane quartic curve; cf. also Milinowski(199) and Williams(287). Goring(127) finds that the involution is of class one; that its isologous curves are the cubics of the net; and that as x describes a line a, the line $\overline{xx'}$ envelopes a rational cubic envelope whose double tangent a has contacts at the pair x, x' on a. Emch(103) discusses problems of closure connected with the involution. Wallstaff(286.1) considers a degenerate case of order six. Bennett(25) discusses invariant linear systems under the four types of involutions.

The Bertini involution determined by 8 points P_8^2 is of order 17. The F-point p_1 has for P-curve the sextic with triple point at p_1 and nodes at the other points. Wiman(289) maps it into an involution on a cubic surface. It has a fixed point at p_9 the ninth base point of the pencil of cubics on P_8^2, each of which is an invariant cubic. It has a curve of fixed points, the locus of the ninth node of sextics with nodes at given P_8^2 [cf. Halphen(139)]. For general accounts see Doehlemann(97) (165-74), Sturm(272) (105-7), Coble(75) (251-4). Snyder(267) gives an algebraic exposition. Conner(82) shows that the Geiser and Bertini involutions in the plane are the projections of pairs of points of a space involution, which lie on an invariant quadric. The space involutions are the Geiser involution determined by six points and the Kantor involution determined by seven points.

The birational classification of involutions due to Bertini is supplemented by a projective classification due to Caporali. For an involution of order n Caporali(49) introduces the idea of the *class* ν, the number of pairs x, x' on a line. The locus of fixed points has the order $n-2\nu$ and the genus p. An isologous curve has order $2\nu+1$, genus π and triple points at the isolated fixed points. If K is the class of the envelope of lines on which a pair x, x' coincide, then $K+p+\pi=n$. Caporali notes that at a point on the locus of fixed points there are two fixed directions, one along the curve and one on a line of the envelope of class K whereas at isolated fixed points every direction is fixed. Bertini([33]) proves the following theorems: Involutions of class zero are of Jonquières type. Involutions of Jonquières type and class not zero have two, one, or no lines of fixed points and are of classes $(n-2)/2$, $(n-1)/2$, or $n/2$ respectively. The construction of these types is given. Involutions of class one are the Geiser involution and its particular cases of lower order (9 types in all). Involutions of class two are either of Jonquières type; or of type $n=11$, $\alpha_5=4$, $\alpha_2=5$ with a pencil of fixed cubics on these 9 F-points or a specialization of this type (11 types in all). All involutions of class one except the general Geiser involution are transforms of involutions of class zero by one (or in a single case two) quadratic transformation. Involutions of class two are transforms of involutions of class zero by one quadratic transformation. Martinetti([194]) lists the types of involutions of classes 3 and 4; and further([195]) gives the construction for and thereby verifies the existence of the types listed. Berzolari([35]) gives numerous theorems relating to the F-points and isologous curves for an involution of class ν and tabulates the involutions of class 5. Okken([229]) adds one type to and eliminates two types from the list of Berzolari. Okken([228]) uses the net of isologous curves to discuss the involutions of class 6. Berzolari([36]) proves that if an involution has an r-fold F-point whose P-curve passes through it $r-3$ times and if $r\geqq 8$ then the involution belongs to one of 15 types whose class $\nu\leqq 7$. Martinetti([198]) obtains lower limits for the genus of the general curve of the isologous net for involutions of various types. He uses a linear system with the same behavior at the F-points as the isologous net but otherwise unrestricted. Morgan([217]) sets up equations of the four types of involutions given by Bertini and shows how the involutions of class $\nu\leqq 5$ can be transformed into these types. Conversely, he shows how involutions of any class can be obtained by transformation. Aprile([4]) discusses some involutions of class 3.

If the points of a plane E_x are in (2, 1) rational correspondence with the points of a plane E_y so that to a point x of E_x there corresponds one point y of E_y, whereas to one point y of E_y there corresponds two points x, x' of E_x, then x, x' is a pair of a Cremona involution. Following Clebsch([69]), de Paolis([240]), ([241]), ([242]) studied these correspondences (for a résumé see ([93])) which are reported in Chap. V § 1. The relation of the (2, 1) correspondence to involutions is recognized by Noether([226]) and again by Lüroth([189]) who maps the pairs of the various types of involutions on the points of a rational surface and thereby on the points of another plane. The latter object is accomplished by Bertini([34]) by simpler geometric considerations. Castelnuovo([54]) proves more generally that the sets of k points in any plane involution ($k=2$, the pairs of a Cremona involution) can be put into (1, 1) rational correspondence with the points of a plane. Boyd([41]) discusses the (2, 1) correspondence associated with the Jonquières involution.

Palatini([236]) uses the mapping of point pairs of a plane upon the points of an S_4 to derive a method for setting up involutorial Cremona transformations in the plane. Steinmetz([269]) considers the properties and simple cases of the involutions determined by a net of curves with two variable intersections. Bobek([39]) defines an aggregate of involutions by means of a pencil of cubic curves on the base points $P_9{}^2$. A rational curve C^ν of order ν with y_i-fold points at p_i is completely determined if

$$\sum_{i=1}^{9} y_i = 3\nu - 1, \qquad \sum_{i=1}^{9} y_i(y_i - 1) = (\nu-1)(\nu-2).$$

The C^ν meets a cubic curve in one point q_ν outside $P_9{}^2$. If then the cubic curves are projected each into itself from the point q_ν on it, an involution is determined of order $3\nu+5$, with an F-point of order $(\nu-2y_i+1)$ at p_i, and with a curve of fixed points of order $\nu+5$. These involutions are studied in detail. Stiner([270]) defines an involution by means of a pencil of line conics and a point p. Pairs x, x' are such that the three lines $\overline{xx'}$, $\overline{xp}$, $\overline{x'p}$ touch the same conic of the pencil. This involution is used to construct a rational quintic curve with six given double points and two given simple points; and is applied also to other problems. Godeaux([123]) considers an involution as cut out on the plane by a system L of lines and a system C of conics, an L and a C determining a pair x, x' of the involution. He divides involutions into four types according as the dimension of the systems L, C are 1, 1; 1, 2; 2, 1; 2, 2. F-points and P-curves for the first three types are given but the fourth is not completely discussed. Van den Berg([26.1]) constructs a quintic involution with six double F-points by symmetric circles; cf. also Servais([260.1]) and Vigarie([284.1]). Crocchi([90.1]), Schoute([254.1]), and Chizzoni([63.1]) study involutions constructed by means of pencils and nets. Emch([103.1]) considers curves invariant under involutions.

In addition to the types of involutions described in § 4 others appear in the finite groups reported in § 8. A general account of involutions is given by Sturm([272]) (pp. 95-130); and Hudson([144.1]) (Chap. V, II).

For the determination of the birationally reduced types of periodic transformations and finite groups (§ 8) and of infinite continuous groups (§ 10) certain results in connection with linear systems are necessary and such systems are considered in the next section.

7. *The Reduction of Linear Systems of Curves.*—In the reduction of linear systems of curves to simpler typical forms by Cremona transformations an important theorem is that which states that a homaloidal net H_x of curves of order n, can be transformed by a sequence of quadratic transformations, $Q_1, Q_2, \ldots, Q_j$ into the net of straight lines. For if T is the Cremona transformation which converts lines into H_x, then the product $TQ_1, Q_2, \ldots, Q_j$ is a collineation C, and T itself is a product $(CQ_j^{-1}) \cdot Q_{j-1}^{-1}, \ldots, Q_1^{-1}$ of j quadratic transformations. If then any linear system can be transformed into a simpler form by T, it can in general be transformed into simpler form by quadratic transformations alone.

Cayley(58) states that Clifford surmised that every T can be expressed as a product of quadratic transformations and he verified this for $n=2, \ldots, 8$. Clifford's notes(71) indicate an examination of these cases. After an announcement(223.1) of the theorem, Noether(224) shows that the relations $\sum_{i=1}^{\rho} r_i^2 = n^2-1$, $\sum_{i=1}^{\rho} r_i = 3(n-1)$ attached to a homaloidal net with F-points of orders $r_1, \ldots, r_\rho$ lead to $r_1+r_2+r_3>n$ and the reducibility of the net by quadratic transformation follows. Rosanes(251) gives a proof by induction from n to $n+1$ that $r_1+r_2+r_3>n$ and draws the same conclusion; cf. also Bertini(27) and Mlodziejowski(200). Noether(225) extended the proof of the reducibility of the homaloidal net by a product of quadratic transformations to the case of coincident F-points. Nearly thirty years later Segre(260), after an analysis of Noether's work, pointed out that a quadratic transformation would not necessarily reduce the order of the net if, of the three F-points of highest order, the last two approach the first in different directions; nor if the last two approach the first in the same direction but not on the same linear branch; nor in certain cases on the same superlinear branch. Segre also made the same objection to the existing work on the reduction of linear systems. Castelnuovo(57) then proved that every Cremona transformation can be expressed as a product of Jonquières transformations, each itself a product of quadratic transformations. Alexander(1) raises objection to Castelnuovo's proof and gives a new proof of reducibility by means of quadratic transformations alone; cf. also Chisini(63). At certain stages of Alexander's process the order of the net may be raised so that Jonquières transformations may be necessary for repeated lowering of the order. Veronese(284) tried to prove the reducibility by using the projections of his V_2^4.

In the following account of the reducibility of linear systems to systems of least order by Cremona transformation the notation of Chapter III is used.

For rational curves ($p=0$), Noether([224]) proves that a pencil is reducible to a pencil of lines. Guccia([131]) shows that any system ($p=0$) of grade D is reducible to one of four types: (a) $n=(D+2)/2$, with an ordinary base-point of order $D/2$ and an isolated simple base-point; (b) $n=(D+\sigma+1)/2$ ($0\leqq\sigma\leqq D-1$) with a base-point of order $(D+\sigma-1)/2$ with σ (distinct or coincident) fixed tangents; (c) a linear system of conics without base-points; (d) the net of lines. He also proves that a system ($p=0$) with ordinary base-points is reducible if $n>(D+2)/2$ (D even) or $n>(D+1)/2$ (D odd). Guccia([132]) observes that Picard's proof of the theorem that the rational ruled surfaces and the Steiner quartic are the only surfaces with rational plane sections does not consider possible relations among the base-points; and he applies the results of the preceding paper to demonstrate the theorem.

For elliptic curves ($p=1$) Bertini([29]) proves that a pencil ($k=1$) is reducible to a pencil of order $3r$ with 9 r-fold points; and a net ($k=2$, $D=2$) to a net of cubics on 7 points. For $k>1$ Guccia([133]) shows that a system ($p=1$) can be reduced to one of the two types: (a) a system of cubics with 0, 1, . . ., 7 simple base-points distinct or coincident; (b) a system of quartics with two double base-points distinct or coincident; cf. also Segre([259]) and Martinetti([197]). Guccia([137]) defends the validity of his methods against a criticism of G. Jung. Nalli([220]) determines the types of least order for a pencil of elliptic curves which is transformed into itself by a Cremona involution.

For systems of genus 2 Bertini([29]) proves that a web ($k=3$, $D=4$) is reducible to a web of sextics with 8 nodes, or a web of quartics with one node and 8 simple points. Segre([259]) shows that every system ($p=2$, $k>2$) is reducible to a system of quartics with one node, or of quintics with one triple point and one node, or of sextics with 8 nodes. To this Guccia([134]) notes the exception of a system of sextics with a common four-fold point with two pairs of coincident branches. This exception can be removed by Jonquières transformation. Martinetti([196]) considers superabundant systems ($p=2$, $s>0$, $k>1$) and finds three irreducible types for all of which $k=2$, $s=1$ as follows: (a) quartics with a node O and 10 simple points, two on a line with O, the other 8 with O the base of a pencil of cubics; (b) quartics with a node and 8 simple points all on one cubic; (c) sextics with 8 nodes and 2 simple points all on a cubic. To this list de Franchis([117]) adds the type: quintics with triple point p_1, double point p_2, and 11 simple points $p_3, \ldots, p_{13}$ such that p_2 is consecutive to p_1, and that $p_1, p_3, \ldots, p_{13}$ are consecutive; and that p_1, p_3, p_4 are on a line. De Franchis([115]) shows that pencils of genus 2 are reducible to one of six types: (a) quartics with a node and 12 simple points; (b) the type described just above with an additional simple point p_{14} consecutive to p_{13}; (c) sextics with 8 nodes and 4 simple points; (d) septimics with one triple point and 10 nodes; (e) nonics with 8 triple, 2 double, and 1 simple point; (f) 13-ics with 1 five-fold and 9 four-fold points. The existence of these types is proved.

For systems of genus 3, both regular and superabundant, of any dimension de Franchis([116]) lists the irreducible types.

Bertini([29]) proves a general theorem on reducibility: a system $[k=a, p=(a-1), D=2(a-1)]$ is reducible if $a-1$ is less than the greatest order r_1 of a base-point; if $a-1=r_1$ the system is reducible except in the cases of (a) the linear system of curves of order n, genus $n-2$, with one common $(n-2)$-fold point and $2n$ simple points, and (b) the web of sextics with 8 nodes. The most comprehensive general contribution to reducibility is given by G. Jung([156]). Complete and general systems for which $s=0$ are divided by Jung into *normal* and *derived* types. A derived type arises from a normal type by adding a simple base-point—thus normal types have no simple base-points. Normal types are divided into *monomial* types, (A) with no multiple base-point, (B) with one multiple base-point; *binomial* types (C) with just two multiple base-points; and *trinomial* types (D) with at least three multiple base-points for which $r_1+r_2+r_3 \leq n$. For every value of p of the form $(n-1)(n-2)/2$, there are systems of type (A); for given p and every solution in positive integers $r_1>1$ and i of $p=(i-1)r_1+\frac{1}{2}(i-1)(i-2)$ there are systems of type (B); and for each type (A), (B) there are derived types which in some cases are enumerated. The binomial types (C) are divided into two kinds: (1st) $r_1+r_2<n$; (2d) $r_1+r_2=n$; the trinomial types (D) into three kinds: (1st) $r_1=r_2=r_3$; (2d) $r_1>r_2=r_3$; (3d) $r_1>r_2>r_3$. For these similar diophantine equations are given which in the case (D) are subject to the condition of *geometric* compatability. A number of theorems resulting from this classification are given. Incomplete and special systems are then taken up and divided into: type 1° for which some arbitrary base-points determine the others *uniquely* (as 8 base-points of a pencil of cubics determine the ninth); and type 2° for which this is not so (as in the pencil $n=3r$ with 9 r-fold points). The memoir concludes with tables of general systems for $p=0, 1, 2, 3, 4$. In ([157]) Jung extends the considerations above to the case where the base-points are higher singularities and gives tables of types for $p=0, \ldots, 4$. He finds as upper limits for the degree and dimension for given p, $D \lesseqgtr 6p+2$, $k \lesseqgtr 5p+3$. Castelnuovo([51]) reduces these limits to $D \lesseqgtr 4p+4$, $k \lesseqgtr 3p+5$. If these maxima are attained the system is reducible to either a system of quartics $(p=3)$ or to a hyperelliptic system of type: (a) curves of order $p+3$ with an ordinary $(p+1)$-fold point and a double point; or (b) curves of order $p+\mu+2$ $(\mu=0, 1, \ldots, p)$ with a $(p+\mu)$-fold point and μ double points consecutive to it. He modifies results of Segre([259]) to read as follows: If a linear system determined by its base points has irreducible curves and if $k>p$ the system is complete; if $k>2$ and the plane involution of order D is composite (i. e., if curves on one point must pass through other points), then $k \overline{<} p+1$. In ([52]) Castelnuovo introduces a variety of new and effective ideas into the theory of linear systems such as the "virtual" or "effective" order of a base-point, and "virtual" and "effective dimension." Important theorems are obtained by the use of adjoint systems and the *characteristic* linear series cut out on one curve of the system by the others [cf. Chap. III].

Armenante([5]) had early used a Cremona transformation to reduce to a minimum order the mapping curves of a rational surface. A résumé of the results on linear systems which are immediately applicable to the mapping of the plane upon surfaces in S_3 or S_n and their consequences in the theory of surfaces is given by Castelnuovo and Enriques([55]) with a rather complete list of references to that time [cf. also Chap. XII].

The use of "successive canonical adjoints" in the study of linear systems is based on the theorem that if a system Σ is transformed by Cremona transformation into a system Σ', then the adjoints of order $n-3$ of a general curve f of Σ are transformed into those of the general curve f' of Σ'. Such adjoints were used by Brill-Noether([44]), ([227]) but only with respect to the unique linear series g_{2p-2}^{p-1} which they cut out on f and f'. The extension of the covariance of the system of canonical adjoints to the entire plane under Cremona transformation is proved by Castelnuovo([52]) though the fact had been stated and effectively used by S. Kantor([167]), ([168]). Other instances of the value of this method are found in Castelnuovo([54]), ([53]); in the determination of the types of finite and of continuous groups, cf. § 8 and § 10; and in the study of the rational double plane by Castelnuovo and Enriques([56]); cf. also G. Jung([153]).

After the revision by Segre and Castelnuovo of the proof of the reducibility of the homaloidal net to the net of straight lines by quadratic transformation, Ferretti([113]) revised the discussion of linear systems and confirmed the results of Bertini, Guccia, Martinetti and de Franchis which were given above. He derived theorems which connect the order n of a reduced system, the dimension x_j of the j-th successive canonical adjoint, and the order of the highest base-point. Accounts of the work of Noether, Segre, Castelnuovo, and Ferretti are given along somewhat different lines by Franciosi([118]) and Nencini([221]).

A general account of the reduction of linear systems has been given by Fano([111]).

S. Kantor remarks that the base-points of a homaloidal net or more general linear systems may not be given. The groups of points of like order may then be separated by rational processes but the determination of the points within such a group would involve algebraic irrationalities. He obtains([174]) the reduced types of rational, elliptic, and hyperelliptic linear systems for which the reduction does not involve such irrationalities; and ([175]) the similar problem involved in the expression of a Cremona transformation as a product of simpler types. Kantor also([173]) considers the expression of T as a product of simpler types when only integer coefficients occur in the homogeneous equations of T, T^{-1}. He finds that the factors may be of sixteen different types and draws some conclusions as to the irrationalities involved in the expression of T as a product of quadratic factors.

8. *Periodic Transformations and Finite Groups.*—A transformation T is periodic and of period r if T^r is the lowest power of T for which $T^r=1$, the identical collineation. A finite group is composed of periodic elements and the powers of a periodic element form a cyclic finite group so that the methods applicable to periodic transformations and to finite groups must be in the main the same. Groups may be classified according to Autonne with respect to the maximum order n of the transformations T which compose them. This would lead to an infinite number of types since this maximum order would rise indefinitely under continued random Cremona transformation of the group. Or they may be classified according to S. Kantor as equivalent under Cremona transformation to one of a finite number of types. They may be regarded as known if such an aggregate of types has been exhibited. A particular type may involve one or more arbitrary integers as in the case of the Jonquières type among the four types of involutions derived by Bertini.

S. Kantor([164]) gives an account of periodic quadratic transformations. Let A, B, C be F-points of Q; A', B', C' be F-points of Q^{-1}; A'_i, B'_i, C'_i $(i=1, 2, \ldots)$ be the transforms of A', B', C' by $Q, Q^2, \ldots$; and A_i, B_i, C_i the transforms of A, B, C by $(Q^{-1})^i$. If Q^r is the identical collineation, then A'_i, B'_j, C'_k coincide eventually with A or B or C. For points on the line BC pass successively into directions at A', at A'_1, at A'_2, etc. If eventually A'_i is not at A, B, or C, this set of directions cannot become again the points of a line. This necessary condition Kantor later calls the "principle of *enchainement* (*Verkettung*) of the F-points." Obviously, all the points $A'_i, \ldots, A_i, \ldots$ are F-points of Q or of one of its powers. Five particular cases are considered: (1) $A=A'$, $B=B'$, $C=C'$, the usual involution with four fixed points; (2) $A=A'$,

$B=C'$, $C=B'$, an inversion with pole A and conic of fixed points, or such an inversion combined with a rotation (center A and circular points B, C) of period n; (3) $A'=C$, $B'=A$, $C'=B$, of period 6 with one fixed point f and a cycle of three points which with f is a 4-point with diagonal triangle ABC; (4) $A'=A$, $B'=B$, of period r, equal to the least multiple of r_1, r_2 if the pencils of lines on A, B are cyclically transformed into themselves with periods r_1, r_2 respectively; (5) $A'=B$, $B'=A$, $C'_r=C$, then $Q^{r+2}=1$ (r even) or $Q^{(2r+4)}=1$ (r odd). The case $A'_2=A$, $B'_2=B$, $C'_2=C$ is not periodic since no reduction in the order of successive transforms of a line occurs. The case $A'=A$, $B'_k=B$, $C'_r=C$ is pronounced periodic but the intermediate cases are not examined so that a given necessary and sufficient condition for periodicity is evidently inadequate. Snyder([266]) discusses algebraically the case (4) and gives canonical forms more specific than those of Autonne([6]) both for distinct and for coincident F-points.

Autonne[([9]), abstracts ([6]), ([8])] defines a quadratic group to be one all of whose elements are quadratic. He divides these into three classes: (1) a G_{120} with F-points at four points; (2) groups with two points a, a', F-points both of Q and of Q^{-1}, which may or may not be interchanged, which arise from finite groups of projectivities in the pencils on a, a'; (3) a class with coincident a, a'. The orders of the possible groups are not indicated. G. Jung([158]) contributes an addition to this paper. The G_{120} is included in the series of cross-ratio groups of E. H. Moore([216]) and is exhaustively treated by Slaught([262]) who gives fundamental regions for it and discusses its invariants. P. Field([114]) gives the modifications of Slaught's work which occur when two or four of the points are conjugate imaginaries. Autonne([7]) defines the order of a Cremona group to be the maximum order of a transformation. A semi-cubic group is a cubic group for which the product of any two cubic elements is quadratic or linear. He gives canonical forms of semi-cubic groups in two categories of which the first contains groups of orders n, 10, 8, 24, 24; and the second groups of orders 16, 64, 96, 160. Autonne[([11]), abstract ([10])] considers cubic groups with distinct F-points and fixed double F-point O. The group G contains an invariant subgroup Γ which effects the identical permutation on the lines through O. Seven types of groups Γ are listed which are enlarged to 15 additional groups G. In ([14]) the discussion is extended to the case of coincident simple F-points. Autonne([12]) in a preliminary article considers substitutions on the homogeneous contragredient variables x_1, x_2, x_3; u_1, u_2, u_3 and requires that they be *contact* transformations, i. e., that $(xu)=0$, $(dxu)=0$, $(xdu)=0$ shall entail $(x'u')=0$, $(dx'u')=0$, $(x'du')=0$. If linear they are collineations or correlations. In an extended series of memoirs [([17]), ([20]), ([21]); abstracts ([13]), ([15]), ([16]), ([22])] he considers rational and reversible "cremonienne" transformations of the form

$$x'_i = \phi_i(x^a, u^b), \qquad u'_i = \psi(x^c, u^d)$$

of the orders a, b, c, d indicated in x, u which are subject to the conditions of contact above. The ∞^4 elements x, u of the plane are transformed among themselves in such a way that the ∞^3 *principal elements* for which $(xu)=0$ are invariant. Thus the transformations indicated are not Cremona transformations in general but rather *birational* transformations of the quadric variety $(xu)=0$. If this variety be mapped upon the points of an S_3 [cf. ([21]) pp. 451-2 or 409-11], the cremonienne gives rise to a particular type of Cremona transformation which satisfies a differential condition whose geometric equivalent is that the aggregate of curves whose tangents belong to a fixed linear complex be invariant. In [([23]), ([24]); abstracts ([18]), ([19])] this theory is applied to differential equations of the first order with rational coefficients in connection with the view of Clebsch of such an equation as the principal coincidence of a connex $(ax)^m(u\alpha)^n=0$.

S. Kantor([165]), developing methods of his prize memoir (not then published), observes that if a rational surface, mapped on a plane, admits a (1, 1) periodic correspondence (usually a periodic collineation) then this correspondence is the map of a periodic Cremona transformation in the plane and that the various mappings of the surface on the plane lead to various transforms of this periodic plane transformation. In ([166]) he notes types of periodic quadratic transformations. In ([167]) Kantor proves that every curve which admits a (1, 1) birational correspondence B can be birationally transformed into a curve on which B is effected by a Cremona transformation which is either (a) a homography; or (b) a Jonquières transformation with $O'=O$ or (c) a transformation with 6, 7 or 8 points in its characteristic. He further([171]) considers the possible number of correspondences B on a hyperelliptic curve of order n with an $(n-2)$-fold point and shows that if B is periodic it is effected by a periodic Jonquières transformation. Montesano([207]) defines a finite group by means of four points $p_1, \ldots, p_4$ and a line α. The four inversions I_k with center p_k and fixed conic with self polar triangle p_l, p_m, p_n and p_k, α as pole and polar satisfy the relations $I_iI_j=I_jI_i=I_kI_l$ and generate a G_8.

S. Kantor in the memoir([168]) crowned by the Naples Academy in 1883 seeks to determine the types of periodic transformations which are not reducible by Cremona transformation to simpler types. Caporali([49.1]) gives a résumé of this memoir. A preliminary discussion of periodic collineations and their invariant curves and of the various correspondences existent on particular cubics and pencils of cubic curves is applied later to the construction of transformations. The various "enchainments" of the F-points are then discussed as in ([164]) and those whose characteristics (cf. § 3) are not periodic are discarded. The periodic characteristics are then tested for geometric compatibility, and of the resulting transformations those only are retained which are not reducible by transformation of the type $Q_1^{-1}QQ_1$ to periodic collineations or to quadratic types already listed. There result 10 periodic types with periods 6, 9, 12, 12, 14, 15, 18, 20, 24, 30. A similar process for $n=3$ and $n=4$ yields 5 cubic types of periods 6, 6, 6, 10, 12; and 3 biquadratic types of periods 3, 8, 6. An arithmetic study of the general characteristic then leads to the conclusion that all periodic ones are birationally equivalent to a Jonquières type or to one with a characteristic that contains at most 8 points. The final conclusion is that all periodic Cremona transformations can be transformed into either: (1) a periodic collineation; or (2) a Jonquières transformation of one of three kinds, each kind having two integer parameters; or (3) one of 28 particular types of order $n \leqq 17$ of which 18 are mentioned above. A lengthy excerpt of this memoir has been given by Kantor himself([170]). The arithmetic discussion of the characteristics is replaced later([172]) by a consideration of the invariant systems of curves, a method which also ensures the geometric existence of the characteristic. However, Castelnuovo([54]) had proved earlier that every cyclic Cremona transformation can be reduced by transformation to one that transforms into itself either (a) the net of lines (a collineation); or (b) a pencil of lines (a Jonquières transformation); or (c) a system of $\infty^r (r>1)$ of cubics with $(9-r)$ base-points; or (d) a web of sextics with eight double base-points. This theorem applied to finite groups rather than to cyclic elements is fundamental in the determination by Kantor([161]) of the types of finite groups. Kantor lists 53 types of which, apart from the collineation groups, six are ascribed to Autonne and 40 are new. A more elegant exposition along precisely the same lines is given by Wiman([289]) who makes a number of additions and corrections to Kantor's list. It appears by the use of successive canonical adjoints that every finite group is equivalent to: either (1) a collineation group; or (2) a Jonquières group with fixed line pencil; or (3) a group of quadratic transformations with an invariant pair of line pencils; or (4), (5), (6), (7), (8) a group with an invariant linear system of cubic curves on 3, 4, 5, 6, 7 points respectively; or (9) a group with an invariant web of sextics with 8 nodes. Of the later cases the type (7) is handled by mapping the plane on a cubic surface by means of the invariant web of cubics. The possible collineation groups of the surface furnish the required groups in the plane. In type (8) the invariant net of cubics maps the plane on a double plane with a quartic curve of branch points, C^4. The collineation groups which C^4 may admit furnish the Cremona groups desired. Cox([84]) shows how the transition from collineation group to Cremona group may be carried out practically. In type (9) the invariant web of sextics maps the plane upon a quadric cone in S_3 on which there is a space sextic curve, $C^6(p=4)$, of branch points. The collineation groups which C^6 may admit furnishes the required Cremona groups. Sharpe([261]) discusses the character of these Cremona groups.

Wiman([288]) discusses the groups of birational transformations, some of which are Cremona transformations, of curves of genera $p=4$, 5, 6; cf. for $p=4$ Miss Van Benschoten([26]) and for $p=5$, McKelvey([191]). Snyder([264]) obtains the groups of collineations and quadratic transformations which leave a curve of genus 6 unaltered. Miss Tappan([277]) determines the finite groups, at most quadratic, which leave sextic curves not transformable to lower order unaltered. Miss Allen([2]) in listing involutions of degree 3 determined by nets of genus 0, 1, 2 finds certain cyclic sets of quadratic transformations of period 3. Coble([72]) considers a group G_{120} of quintic transformations determined by five points of the plane whose invariant theory (in dual form) had been given by Clebsch([70]). The group is a transform of Autonne's G_{120}. Ciani([65]) discusses ternary quartic curves which admit finite groups of quadratic transformations. Ciani([64]) studies the quadratic groups of orders, 8, 48, 96 which arise when two projective pencils of four lines each are unaltered according as the four lines are general, harmonic, or equianharmonic.

Cherubino([59]) considers finite groups of the form

$$x'=\frac{a_i x+b_i}{c_i x+d_i},\qquad y'=\frac{y}{(c_i x+d_i)^{p+1}}\qquad (i=0,\ 1,\ \ldots,\ k-1)$$

and their hyperelliptic invariant curves largely from the view-point of function-theory and group-theory. These transformations are of Jonquières type with an invariant line pencil. In ([60]) these groups are extended to include elements of the form $y'=y\alpha_i(x)/\delta_i(x)$ which are called *reduced* types with respect to the more general *semi-projective* group, $y'=\{\alpha_i(x)y+\beta_i(x)\}/\{\gamma_i(x)y+\delta_i(x)\}$, i. e., one in which the transformation on one variable y is projective. An additional transformation on z', z similar to that on y', y extends the transformation to S_3, etc. In S_3 the transformations are of the generalized Jonquières type studied by Enriques and Fano([106]) and by Fano([110]) (cf. § 10). Cherubino studies the curves and surfaces which admit the reduced and semi-projective group. He further([61]), ([62]) extends these ideas to include the anti-birational elements of Comessatti[(79); cf. § 11] obtained by adjoining $x'=\bar{x}$, $y'=\bar{y}$. The extended groups are said to be of the second kind.

9. *Infinite Discontinuous Groups.*—S. Kantor(171) considers the infinite group of Jonquières transformations which have a hyperelliptic curve of order n with an $(n-2)$-fold point as a fixed curve with particular reference, however, to the construction of such transformations and with no mention of group properties. He also(176) calls attention to the existence of sub-groups of the Cremona group which are of the same general character as the group itself, e. g., the group which has an invariant elliptic curve and the Jonquières group which has a given line pencil for invariant. Bydzovsky(45) discusses the transformations which have a given cubic curve as an invariant and finds that for any system of F-points of T on C^3 there are 18 transformations which carry C^3 into itself. Indeed, if T' has F-points on C^3 it transforms C^3 into C'^3 which can be transformed back into C^3 by 18 collineations K so that $T'K=T$. This idea is used by Coble(74) to obtain the group isomorphic with the group of characteristics (cf. § 3). Coble[(74), p. 178] finds that a pencil of plane cubics can be transformed into only 960 projectively distinct types; a pencil of plane sextics with 9 nodes into only $2^8.960$ projectively distinct types; in general a pencil of order $3r$ with 9 r-fold points into only a finite number of types. Hence in all of these cases there is an infinite discontinuous group of Cremona transformations under which the pencil is invariant. Invariants of such groups are also discussed (pp. 384-5). Coble(75) finds that a general rational sextic can be transformed into only $2^{13}.31.51$ projectively distinct rational sextics. Thus there is an infinite discontinuous group which transforms the sextic into itself. The group is generated by the Bertini involutions determined by any 8 of the 10 nodes. The associated group of characteristics is congruent to the identity modulo 2. The group determines a group of linear transformations on the parameter of the sextic which is generated by the involutions whose double pair is the Jacobian of two pairs of nodal parameters. A method for exhibiting infinite series formally invariant under the group is set forth by Coble[(77), pp. 359-61]. Coble[(76), p. 16] remarks that a Veronese surface V_2^4 in S_5 is transformed into a Veronese surface by a regular transformation (cf. Chap. 7, § 3) which in the plane corresponds to the quintic transformation $(n=5,\ a_2=6)$. Thus there arises in the plane an infinite discontinuous group generated by one such quintic transformation and collineations.

10. *Continuous Groups.*—Though it was long known that the Cremona group contained continuous subgroups such as the group of inversions, Enriques(104) was the first to study the continuous c-parameter groups

G of Cremona transformations. He showed that G must have an invariant linear system of curves of order m and dimension $k \gtrless 1$. By taking adjoints of order $m-3$ a new invariant linear system is obtained, etc.; until finally it appears that G has an invariant linear system of dimension $k \gtrless 1$ of rational or elliptic curves which may be transformed into one of the normal types of Guccia (cf. § 7). Of these types only three admit continuous groups as follows: (a) the group (∞^8) of collineations or a subgroup of it; (b) the group (∞^6) of quadratic transformations which transforms into itself the system of conics on two points (or the two pencils of lines on the two points—the inversive group) or a subgroup of it; (c) the group (∞^{n+5}) of Jonquières transformations of order n which transform into itself the linear system ($k=n+1$) of curves of order n with a fixed $(n-1)$-fold point and $(n-1)$ fixed tangents thereat or a subgroup of it. By using these invariant linear systems for each type as a mapping system Enriques proves that the geometry in the plane which has as a principal group [cf. Klein([181])] a continuous group of Cremona transformations depending upon a finite number of parameters coincides with the projective geometry on (a) the plane; (b) a ruled quadric; or (c) a rational normal cone of order n in S_{n+1}; or with a particular case of one of these geometries. Enriques then([105]) studies the invariant subgroups of his type (c) since Lie had already proved that (a) is simple and (b) is known to contain only two invariant subgroups, each (∞^3) of which leaves the lines on one of the two points unaltered. From the normal cone in S_{n+1} he finds three invariant collineation groups in G_{n+5} all of whose elements have the vertex O of the cone as a fixed point. The first group G_{n+4} consists of the elements of G_{n+5} which have a second fixed point at O. The second group G_{n+2} consists of the elements of G_{n+5} perspective at O (homologies) each of which has a plane of fixed points. The third group G_{n+1} is the subgroup of permutable elations whose plane of fixed points is on O. These groups translate to the plane as follows. If μ is the invariant linear system of G_{n+5} with $(n-1)$-fold point at O the general element of G_{n+5} has a fixed curve in μ which does not contain fixed lines on O. The elements of G_{n+4} have fixed curves in μ which have at least two fixed lines on O as a part. The elements of G_{n+2} leave every line on O unaltered. The elements of G_{n+1} effect on each line through O a parabolic collineation. By taking the dual of the hypercone in S_{n+1} as a curve N of order n at infinity, Enriques finds that the geometry of the plane under G_{n+5} can be interpreted as a new extension to S_{n+1} of ordinary metric euclidean geometry in S_3, for which the curve N is the absolute.

Fano([108]) notes that all the continuous Cremona groups of S_k will be determined if all the projective groups in $S_i (i \gtrless k)$ which have a rational invariant V_k have been determined. Having obtained ([107]) these projective groups he finds in this way for $k=2$ the types given by Enriques. Mohrmann in his dissertation ([203]), and more fully in ([204]), studies by algebraic methods the collineation group of the normal cone of order n in S_{n+1} attached to Enriques group (c). He finds an essential difference in the cases n odd, n even; and gives finite equations both for the collineation group and for the Jonquières group. Later ([205]) he studies the geometry upon the normal surfaces of the groups of Enriques by suitable algebraic methods. Bohlmann ([40]) verifies that every continuous group of quadratic transformations is Enriques' group (b) or group (c) for $n=2$. Kasner ([177]) develops a comprehensive theory of algebraic invariants under the inversive group as a medium for the study of projective geometry on a quadric surface. This theory of double binary forms in digredient variables is pursued in ([178]). Newson ([222]), ([223]) gives an account of the continuous inversive group using group-theoretic and geometric methods. Fano [([112]), pp. 339-43] has given a general report on continuous Cremona groups.

11. *Geometric Applications.*—We consider first certain applications in which the Cremona correspondence is not specifically from plane to plane. Cremona ([88]) establishes a correspondence between two bundles of lines by setting up a Cremona transformation from plane E_x to plane E_y in space. The lines ax, $a'y$ on two fixed centers a, a' correspond. He shows that the locus of the points of intersection of corresponding lines is a curve of order $n+2$, genus $n-1$, with certain Cayley numbers. Reye ([249]) discusses the quadratic transformation between bundles of lines, and planar fields of either points or lines, and develops related theorems concerning quadrics. He uses these in the determination of the transformation itself by pairs of F-points or of corresponding points. Darboux ([90.2]) uses a quadratic transformation for the construction of a quadric on nine points. Hirst ([141]) discusses the congruence of order $n+2$ and class n obtained by drawing lines through corresponding points of a Cremona transformation T of order n between two planes α, β in space. The two planes are singular planes of the congruence and the F-points of T, T^{-1} are singular points of the congruence. The focal surface is of order $4n$ and class $4(n-1)$. G. Jung ([151]) shows that if a bundle of lines on O and a bundle of planes on O' are put in Cremona correspondence of order n they generate a monoid of order $n+1$; or if $O'=O$ the lines in their corresponding planes form a cone of order $n+1$, the planes envelop a curve of class $n+1$. This idea is developed further by Jung ([152]) and by Visalli ([285]). Lo Piano ([245]) sets two bundles of lines in Cremona correspondence and through a fixed point P passes a line across two corresponding lines. The locus of the fourth harmonic

of P as to the two crossings is a rational surface of order $n+2$ which is studied in general and for $n=2, \ldots, 5$. Tinto(282) remarks that the chords of a space cubic curve form a ternary domain and considers the effect in space of a Cremona transformation between two such domains. Visalli(286) discusses the complex of lines which pass through a point x of a plane α and meet a line ξ of a plane β when α, β are connected by a Cremona correlation. Zahradnik(293) uses a quadratic correlation to construct the rational cubic and quartic. Lazzeri(186) discusses the Cremona correlation and the two curves of respectively order $n+1$ and class $n+1$ for which image point and line are incident. He proves that only for $n=2$ can incidence always occur; cf. Ameseder(3) and Sturm(274). The only correlation, other than the polarity in a conic ($n=1$), which is a polarity occurs for $n=2$, point and polar line as to a triangle; cf. Hain(138). Oppenheimer(230) uses Ameseder's null-system to construct triangles inscribed in a given cubic curve C^3 whose sides pass through a point of C^3. Beyel(37) uses two quadratic null-systems to generate the plane cubic of order 3 and class 3 and brings out certain properties of it. Tuck(283) uses the quadratic reciprocity to construct triangles inscribed and circumscribed to a cubic curve. Ota(232) uses the quadratic correlation for geometric purposes.

Certain geometric studies relate to the transformations themselves. Thus G. Jung(154) takes a Cremona transformation T of order n which sends x into x', x' into x'' where $\overline{xx'}$ is ξ, $\overline{x'x''}$ is η, and $\overline{xx''}$ is ζ, and studies the two $(1, n)$ correspondences between x', η and between x', ξ. In (155) he discusses the $(1, 2n)$ correspondence between x', ζ and makes applications to the transformations of orders up to six. This article connects closely with that of Guccia[(130), cf. § 5]. Pannelli(239) extends the idea of Jung to the l-th power of T and discusses the n^l+2 fixed points which break up into the cycles of Kantor. Kanda(160) makes some general remarks on Cremona transformations and their product in a Grassmann nomenclature. He defines two products T_1T_2 and $T_1 \times T_2$, of which the first is the usual product, and the second sends x into the line $\overline{x_1x_2}$ when T_i sends x into $x_i (i=1, 2)$. Kantor(169) recapitulates some of the geometric results of his prize memoir relating to birational transformations of a cubic into itself [cf. Geuss(121); and Martinetti(196.1)]. Each such transformation is in general contained in ∞^2 quadratic transformations. Montesano(208) considers the possibility of putting two planes in Cremona correspondence in ∞^1 ways so that a point of the one plane corresponds to the ∞^1 points of a rational curve in the other. He obtains a space construction for the required figure.

Cremona transformations are frequently used in geometric constructions both of particular sets of points and of loci. Müller(219) applies the transformation of the fifth order ($a_2=6$) to the problem of projectivity formulated by Sturm ([273]). Müller also gives a construction for the 9-th base-point of a pencil of cubics [cf. Oppenheimer ([230])]. The construction of a quadric on 9 given points is given by Reye ([249]), Sturm ([273]), and London ([188]). London also gives a construction for the space quartic curve on 8 points and for the 8 base-points of a net of quadrics. Saltel ([253]) uses the Arguesienne transformation to obtain constructions for various curves of higher order which are transforms of constructible curves. Pecl ([244]) uses the cubic transformation,

$$x'=ax^3/(bx^2+cy^2), \qquad y'/x'=y/x$$

to construct rational curves. Zahradnik ([291]), ([292]), ([290]) uses the cubic transformation $x'=xy^2/(x^2+y^2)$, $y'/x'=x/y$ or $(x'+iy')(x+iy)=ixy$ to construct and discuss rational curves and certain special curves such as the lemniscate and cissoid [cf. Morgan ([218]), and Saltel ([253.1])]. Thalberg ([279]) applies involutorial transformations to the study of plane curves, particularly the hyperelliptic curves of Jonquières type and special types of quartics.

Cremona transformations sometimes occur incidentally in connection with birational transformations. Miss Scott ([257]) shows that the maximum reduction in the order of the transform of a curve by birational transformation is secured by the use of adjoints of order ν where $n-3 \leqq \nu < n$ and that the transforming net becomes a homaloidal net if these adjoints are rational. Snyder ([263]) shows that curves of sufficiently high genus for given order can be transformed into curves of the same order by collineations only. In ([265]) he considers the lowest order of possible transforms of a given curve and finds that this transform of lowest order can usually be obtained by Cremona transformation.

Many papers deal with the plane quartic curve of general or of special type. Müller ([219]) obtains a quartic as the locus of points from which two given sets of four points appear to be projective. Godt ([126]) indicated the connection of the quadratic transformation in a single plane with the theory of the plane quartic; cf. Clebsch-Lindemann [([67]), vol. 3, pp. 442-52]. Timerding ([281]) elaborated this theme by pointing out that with a general Cremona transformation T in a single plane, there is determined a form $\theta(u, v)$ of degree $2(n-1)$ in each of u, v which for given u is the line equation of the transform of u by T and for given v the transform of v by T^{-1}. For $u=v$ there is a locus of lines of class $4(n-1)$, which touch their corresponding curves. For $n=2$ this is a quartic of class 4 with double-points of the three F-points and the four fixed points of T. From this Timerding derives the usual Aronhold theory of the double tangents. Pyrkosch ([247]) discusses a pencil of quartics with three fixed nodes as the transform of a pencil of conics. Ciani(65) discusses quartics which admit quadratic transformations. Bydzovsky(47) discusses the quadratic transformations which leave an elliptic plane quartic unaltered. There are nine, seven of which form with the identity a group. He inquires(46) as to the conditions under which this number may reduce because some coincide or reduce to collineations; and later(48) considers special cases with more quadratic transformations. Thalberg(278) derives a variety of geometric properties of quartics with at least one node by the use of Cremona transformations.

Of isolated geometric applications we note the following: W. Stahl(268) observes that, if p_7, p_8 are two of the base-points of a net of quadrics, a line λ_x on p_7 is on a unique quadric of the net and meets a cross-generator λ_y which is on p_8. The correspondence between λ_x, λ_y is of type $(n=4, a_3=1, a_1=6)$. E. H. Moore(215) discusses in a general way the manner of introducing or removing either the parameters of, or the conditions inherent in, a given geometric figure by the use of Cremona transformations and illustrates his remarks by considering various transforms of six points in multiple perspective relation. Dewulf(95) employs a special Jonquières transformation of order n to obtain results of metric interest. Miss Scott(258) uses Cremona transformations to prove certain theorems concerning the number of circuits (complete branches) of a plane curve of genus 0 or 1. Poincaré(246) uses Cremona transformations in the study of rational points on given curves—for the most part the elliptic cubic. Leib(187) uses a quadratic involution to determine and interpret more complicated invariants of two triangles from simpler ones. Conner(81) makes free use of various Cremona transformations in the study of multiple correspondences determined by the rational plane quintic curve; cf. also (80). Miss Hudson(142) considers the possible forms into which a rational plane sextic curve can be transformed when one of the discriminant conditions on its nodes vanishes; Coble(75) considers the general case. Emch(102) discusses the curves of order $3n$ invariant under a quadratic involution and gives a particular account of the sextics. Thalberg(280) uses the same involution to prove that the $4r$ corner tangents of a curve of order $3r$ with r-fold points at the four corners and three diagonal points of a quadrangle touch a curve of class $r=2, 3, 4$. Sturm[(272), p. 20-31; (271), p. 508] gives applications of ternary transformations in space geometry.

Comessatti(79) makes an important departure. He defines an *antibirational* transformation in a plane π to be the transformation obtained by combining the conjugate transformation $z'=\bar{z}$, $y'=\bar{y}$ with a Cremona transformation. In the mapping of a real rational surface F upon π, imaginary elements will in general occur so that two conjugate points of the real surface do not in general map upon conjugate points of π but rather upon pairs of an involutorial anti-birational T. If two real surfaces F, F' thus gives rise to involutions T, T', the surfaces F, F' are *similar* (the one a transform of the other under a *real* transformation) if, and only if, the T, T' are *similar* (the one a transform of the other by a plane Cremona transformation). The involutions T are reduced to five types: (a) the conjugate transformation; (b) the anti-quadratic transformation without fixed points; and (c), (d), (e) products of (a) and real Jonquières, Geiser, and Bertini involutions. Therefrom real rational surfaces divide into three types: (a) those with a pencil of real curves; (b) those not under (a) but with a real net of elliptic curves, $D=2$; (c) those not under (a), (b) but with a real web of curves of genus 2, $D=4$. These surfaces have (a) any number of real folds; (b) four real folds; (c) five real folds. The groups of real transformations of the surface into itself are given as groups of antibirational transformations in π.

12. *Algebraic and Other Applications.*—One would suppose that Cremona transformations would frequently occur in the great variety of algebraic applications in which rational representations are desired since by proper procedure the denominator of such expressions may be removed. Nevertheless such applications are rare. Coble(72) is the first to use finite Cremona groups in the solution of equations. After a preliminary study of a G_{120} defined by five points in the plane, the group is transformed into Autonne's quadratic group g_{120}. In this form, and as a particular case of E. H. Moore's(216) series of cross-ratio groups, Slaught(262) had determined a set of fundamental regions and had given the invariantive forms. Coble obtains these invariantive forms by using Clebsch's[(66), p. 353] typical representation of the binary quintic and states the form-problem of the Cremona group. The solution of the quintic in terms of the solution of this form-problem is given. To solve the form-problem it is noted that, under the even subgroup g_{60} of g_{120}, the simplest invariant linear system of curves (the system of cubics on four points) divides into two invariant nets which experience under g_{60} the linear transformations of Klein's[(180) Chap. 4, Part 2] group of the A's and group of the B's respectively. The solution of the form-problem of the Cremona group is given in terms of the solution of the form-problem of the linear group of the A's which Klein had indicated in terms of the ikosahedral irrationality. The Cremona group appears as the natural bond between the collineation group of the A's and the original group of the permutation of the roots of the quintic [cf. (73), p. 198]. Further applications of this kind are given in Chap. 8, § 12.

Schottky(254) shows that, when one of the even abelian theta functions ($p=4$) vanishes for the zero values of the arguments (which happens when the normal curve is on a quadric cone rather than a general quadric), there will exist certain three (rather than four) term relations among the products of odd theta's which serve to define a set of 8 points in the plane. The properties of the functions are then used to set up an especially simple form of the curve of fixed points of order nine and genus four of the Bertini involution defined by the eight points and to obtain the equations which map this curve upon the normal space sextic as Bertini and Noether had done. The coordinates of the eight points thus appear as modular functions of genus four. Frobenius(119) carries through a similar procedure with functions of genus three for the double tangents of the planar quartic determined by seven points of the plane. Neither recognizes that under integer linear transformation of the periods these planar sets of points pass into other sets which are *congruent* to the original set of points under proper Cremona transformation as defined in § 3. The discriminant conditions of the ternary sets of points are given by the vanishing of the even functions for the zero arguments.

Ouivet(233) applies Cremona and in particular quadratic transformations to the determination of the general integral of a differential equation of degree and order one.

BIBLIOGRAPHY

1 Alexander, J. W., Am. M. S. Trans., 17: 295-300 (1916). [7.
2 Allen, Florence, Quart. J., 45: 258-88 (1913-1914). [8.
3 Ameseder, A., Wien, Ber., 83: 385-402 (1881). [11.
4 Aprile, G., Cat. Note e Mem., 1: 102-110 (1921). [6.
5 Armenante, A., Ann. di Mat., (2), 4: 50-72 (1870-71). [7.
6 Autonne, L., C. R., 97: 567-570 (1883). [8.
7 ———, C. R., 99: 646-649 (1884). [8.
8 ———, C. R., 98: 565-567 (1884). [8.
9 ———, J. de Math., (4), 1: 431-454 (1885). [8.
10 ———, C. R., 101: 53-55 (1885). [8.
11 ———, J. de Math., (4), 2: 49-103 (1886). [8.
12 ———, C. R., 102: 313-315 (1886). [8.
13 ———, C. R., 103: 1176-1178 (1886). [8.
14 ———, C. R., 105: 267-270 (1887). [8.
15 ———, C. R., 104: 767-770 (1887). [8.
16 ———, C. R., 104: 1422-1425 (1887). [8.
17 ———, J. de Math., (4), 3: 63-85 (1887). [8.
18 ———, C. R., 105: 850-854 (1887). [8.
19 ———, C. R., 105: 929-932 (1887). [8.
20 ———, J. de Math., (4), 4: 177-247 (188). [8.
21 ———, J. de Math., (4), 4: 407-464 (1888). [8.

22 ———, C. R., 106: 262-265 (1888). [8.
23 ———, J. Éc. Polyt., 61: 35-122 (1891); 62: 47-180 (1892). [8.
23.1 ———, C. R., 121: 673-676 (1895). [1.
24 ———, C. R., 122: 1043-1045 (1896). [8.
25 Bennett, T. L., Am. J., 48: 257-276 (1926). [6.
26 Van Benschoten, Anna L., Am. J., 31: 213-252 (1909). [8.
26.1 Van den Berg, F. J., Nieuw. Arch., 7: 78-90 (1881). [6.
27 Bertini, E., Giorn. di Mat., (1), 15: 329-335 (1877). [1, 7.
28 ———, Ann. di Mat., (2), 8: 11-23, Corr. p. 40 (1877). [6.
29 ———, Ann. di Mat., (2), 8: 244-286 (1877). [6, 7.
30 ———, Lomb. Ist. Rend., (2), 13: 443-451 (1880). [1, 6.
31 ———, Lomb. Ist. Rend., (2), 15: 154-157 (1882). [4.
32 ———, Lomb. Ist. Rend., (2), 15: 24-28 (1882). [1.
33 ———, Lomb. Ist. Rend., (2), 16: 89-101, 190-208 (1883). [6.
34 ———, Lomb. Ist. Rend., (2), 22: 771-778 (1889). [6.
35 Berzolari, L., Ann. di Mat., (2), 16: 191-275 (1888-89). [6.
36 ———, Pal. Circ. Mat., 3: 145-159 (1889). [6.
37 Beyel, C., Zschr. Math. Phys., 38: 65-83 (1893). [11.
38 Bianchi, L., Giorn. di Mat., 16: 263-266 (1878). [2.
38.1 Biasi, G., Gallizzi (Sassari), 27 p. (1905). [1.
39 Bobek, K., Wien. Ber., 91_2: 476-518 (1885). [6.
40 Bohlmann, G., Goett. Nachr., 44-54 (1896). [10.
41 Boyd, P. P., Am. J., 34: 291-324 (1912). [6.
42 Brahana, H. R., and Coble, A. B., Am. J. M., 48: 1-20 (1926). [1, 3.
43 Bramble, C. C., Am. J. M., 40: 351-365 (1918). [3.
44 Brill, A., and Noether, M., Goett. Nach., 116-132 (1873); Math. Ann., 7: 269-310 (1874). [7.
45 Bydzovsky, B., Rozpravy, 18: no. 34, 8 p. (1909); Bulletin Internationale, 15: 1-5 (1910). [9.
46 ———, Rozpravy, 29: no. 23, 10 p. (1920); Bulletin Internationale, 23: 68-71 (1920). [11.
47 ———, Rozpravy, 29: no. 17, 10 p. (1920); Bulletin Internationale, 23: 72-75 (1920). [11.
48 ———, C. R. Strasbourg Congrès, 383-387 (1920). [11.
49 Caporali, E., Nap. R. A. Rend., 18: 212-219 (1879). [6.
49.1 ———, Nap. R. A. Rend., 22: 314-320 (1883). [8.
50 Cardoso-Laynes, Per. di Mat., (2), 4: 33-41 (1902). [4.
51 Castelnuovo, G., Ann. di Mat., (2), 18: 119-128 (1890). [7.
52 ———, Torino Mem., (2), 42: 43 p. (1891); Torino Atti., 26: 595-602 (1890-91). [7.
53 ———, Rom. Acc. L. Rend., (5), 1_1: 47-50 (1892). [5, 7.
54 ———, Math. Ann., 44: 125-155 (1894); Rom. Acc. L. Rend., (5), 2_2: 205-209 (1893). [6, 7, 8.
55 Castelnuovo, G., and Enriques, F., Math. Ann., 48: 241-316 (1896-7). [7.
56 ———, Pal. Circ. Mat., 14: 290-302 (1900). [7.
57 Castelnuovo, G., Torino Atti, 36: 861-874 (1900-01). [7.
58 Cayley, A., London M. S. Proc., 3: 127-180 (1870). [1, 2, 7.
59 Cherubino, S., Torino Atti, 49: 705-726 (1913-14). [8.
60 ———, Lomb. Ist. Rend., (2), 47: 959-982 (1914). [8.
61 ———, Torino Atti, 50: 39-55 (1915). [8.

62 Cherubino, S., Lomb. Ist. Rend., (2), 48: 347-361 (1915). [8.

63 Chisini, O., Modena R. Acc. Mem., (5), 6: 7-13 (1921). [7.

63.1 Chizzoni, F., Rom. Acc. L. Rend., (3), 19: 301-343 (1883). [6.

64 Ciani, E., Pal. Circ. Mat., 42: 323-338 (1917). [8.

65 ———, Giorn. di Mat., (3), 10: 31-75 (1919). [8, 11.

66 Clebsch, A., Binäre Formen; Leipzig, (1872). [12.

67 ———, (Lindemann, Benoist) Leçons sur la Géométrie, Paris, 2: (1880); 3: (1883). [1, 12.

67.1 ———, Math. Ann., 1: 253-316 (1869). [2.

68 ———, Math. Ann., 4: 490-496 (1871). [1, 2, 4.

69 ———, Math. Ann., 3: 45-75 (1871). [6.

70 ———, Math. Ann., 4: 284-345 (1871). [8.

71 Clifford, W. K., Works, London, 538-542 (1882).[c] [7.

72 Coble, A. B., Am. M. S. Trans., 9: 396-424 (1908). [4, 8, 12.

73 ———, Am. M. S. Trans., 16: 155-198 (1915). [3, 4, 12.

74 ———, Am. M. S. Trans., 17: 345-385 (1916). [2, 3, 4, 9.

75 ———, Am. J., 41: 243-265 (1919). [6, 9, 11.

76 ———, Am. M. S. Trans., 24: 1-20 (1922). [4, 9.

77 ———, Am. M. S. Bull., 28: 329-364 (1922). [1, 3, 9.

78 Coble, A. B., with Brahana, H. R., (see Brahana).

79 Comessatti, A., Math. Ann., 73: 1-72 (1912-3). [8, 11.

80 Conner, J. R., Am. J., 32: 115-176 (1910). [11.

81 ———, Am. M. S. Trans., 13: 265-278 (1912). [11.

82 ———, Am. J., 38: 155-176 (1916). [3, 5, 6.

83 Cotterill, T., London M. S. Proc., 2: 119-125 (1866-9). [4.

84 Cox, L. C., Am. J., 39: 59-74 (1917). [8.

85 Cremona, L., Bologna Rend., 88-91 (1861-2). [1.

86 ———, Bologna Mem., (2), 2: 621-630 (1863); Giorn. di Mat., 1: 305-311 (1863). [1, 2, 4.

87 ———, British Assn. Report, 34: 3 (Bath, 1864). [1.

88 ———, Bologna Mem., (2), 5: 3-35 (1865); Giorn. di Mat., 3: 269-280, 363-376 (1865). [1, 2, 5, 11.

89 ———, Lomb. Ist. Rend., (2), 4: Nota 1^a: 269-279, Nota 2^a: 315-324 (1871). [1.

90 ———, Math. Ann., 4: 213-230 (1871). [1.

90.1 Crocchi, L., Giorn. di Mat., 12: 378-380 (1874). [6.

90.2 Darboux, G., Ann. Éc. Norm., 6: 61-68 (1869). [11.

91 ———, Bull. Sci. M., 3: 251-256 (1872). [1.

92 ———, Bull. Sci. M., (2), 2: 212-213 (1878). [6.

93 ———, Bull. Sci. M., (2), 2: 209-211 (1878). [6.

94 Dewulf, E., Bull. Sci. M., 5: 206-240 (1873). [1.

95 ———, Ann. Éc. Norm., (3), 3: 405-431 (1886). [11.

96 Diekmann, J., Math. Ann., 4: 442-475 (1871). [2.

97 Doehlemann, K., Geom. Trans., Leipzig, II: (1908). [1, 6.

98 ———, Zschr. Math. Phys., 32: 315-320 (1887). [4.

99 ———, Zschr. Math. Phys., 33: 243-245 (1888). [4.

100 ———, Math. Ann., 39: 567-597 (1891). [5.

101 ———, Zschr. Math. Phys., 36: 356-378 (1891). [5.

102 Emch, A., Tôhoku, M. J., 21: 310-326 (1922). [11.

103 ———, Am. M. S. Bull., 30: 527-535 (1924). [6.

103.1 ———, Am. J., 48: 21-44 (1926). [6.
104 Enriques, F., Rom. Acc. L. Rend., (5), 2_1: 468-473 (1893). [10.
105 ———, Rom. Acc. L. Rend., (5), 2_1: 532-538 (1893). [10.
106 Enriques, F., and Fano, G., Ann. di Mat., (2), 26: 59-98 (1897). [8.
107 Fano, G., Pal. Circ. Mat., 10: 1-15 (1896). [10.
108 ———, Pal. Circ. Mat., 10: 16-29 (1896). [10.
109 Fano, G., with Enriques, F., (see Enriques). [8.
110 Fano, G., Torino Mem., (2), 48: 221-278 (1898). [8.
111 ———, Encykl. III AB4b, 21, 22. [3, 7.
112 ———, Encykl. III AB4b, 339-343. [10.
113 Ferretti, G., Pal. Circ. Mat., 16: 236-279 (1902). [7.
114 Field, P., Am. M. S. Bull., (2), 12: 234-236 (1905-06). [8.
115 de Franchis, M., Pal. Circ. Mat., 13: 1-27 (1899). [7.
116 ———, Pal. Circ. Mat., 13: 130-160 (1899). [7.
117 ———, Pal. Circ. Mat., 13: 200-201 (1899). [7.
118 Franciosi, V., Giorn. di Mat., (3), 9: 141-190 (1918). [7.
119 Frobenius, G., J. f. Math., 99: 275-314 (1886). [12.
120 Geiser, C. F., J. f. Math., 67: 78-89 (1867). [6.
121 Geuss, A., Diss. Erlangen, 43 p. 8vo (1909). [11.
122 Godeaux, L., Brux. Bull., 898-900 (1907). [4.
123 ———, Brux. Bull., 5: 217-225 (1911). [6.
124 ———, Brux. Bull., (5), 9: 521-529 (1923). [4.
125 ———, Nouv. Ann., (5), 3: 12-24 (1924). [4.
126 Godt, W., Diss., (1873). [11.
127 Goring, L., Diss. Strassburg, (1878). [6.
128 Guccia, G. B., Pal. Circ. Mat., 1: 17-19, 20-23, 24-25, 50-53 (1884-7). [2.
129 ———, Pal. Circ. Mat., 1: 56-57 (1884-87). [5.
129.1 ———, Pal. Circ. Mat., 1: 68 (1884-87). [2.
130 ———, Pal. Circ. Mat., 1: 119-132 (1884-87). [5, 11.
131 ———, Pal. Circ. Mat., 1: 139-156 (1884-87). [7.
132 ———, Pal. Circ. Mat., 1: 165-168 (1884-87). [7.
133 ———, Pal. Circ. Mat., 1: 169-189 (1884-87). [7.
134 ———, Pal. Circ. Mat., 1: 388-389 (1884-87). [7.
135 ———, C. R., 101: 808-809 (1885). [2.
136 ———, C. R., 101: 866-869 (1885). [5.
137 ———, Pal. Circ. Mat., 3: 233-235 (1889). [7.
138 Hain, E., Arch. Math. Phys., 63: 407-412 (1879). [11.
139 Halphen, G. H., Soc. M. F. Bull., 10: 162-172 (1882). ° [6.
140 Hirst, T. A., Quart. J., 17: 301-311 (1881). [5.
141 ———, London M. S. Proc., 14: 259-301 (1882-83). [11.
142 Hudson, Hilda P., London M. S. Proc., (2), 15: 385-400 (1916-17). [11.
143 ———, London M. S. Proc., (2), 22: 223-247 (1923-24). [2.
144 ———, Pal. Circ. Mat., 50: 219-228 (1926). [5.
144.1 ———, Cremona Trans. in Plane and Space, Camb. Univ. Press, (1927), [1, 2, 5, 6.
145 de Jonquières, E., C. R., 49: 542 (1859). [1, 4.
146 ———, C. R., 50: 187 (1860). [1.
147 ———, Nouv. Ann., (2), 3: 97-111 (1864). [1, 4, 5.
148 ———, Giorn. di Mat., 23: 48-75 (1885). [1.
149 ———, C. R., 101: 720-724 (1885). [2.

149.1 de Jonquières, E., C. R., 101: 921-922 (1885). [2.
149.2 ———, C. R., 101: 857-861 (1885). [2.
150 ———, Giorn. di Mat., 24: 1-11 (1886). [2.
151 Jung, G., Rom. Acc. L. Rend., (4), 1: 762-767, 773-774, 810-812 (1885). [11.
152 ———, Rom. Acc. L. Rend., (4), 2_1: 85-89 (1886). [11.
153 ———, Rom. Acc. L. Rend., (4), 2_2: 302-306 (1886). [7.
154 ———, Rom. Acc. L. Rend., (4), 2_2: 339-344 (1886). [11.
155 ———, Lomb. Ist. Rend., (2), 19: 812-818 (1886). [11.
156 ———, Ann. di Mat., (2), 15: 277-312 (1887-88). [3, 7.
157 ———, Ann. di Mat., (2), 16: 291-326 (1888-89). [7.
158 ———, Lomb. Ist. Rend., (2), 19: 161-166 (1886). [2, 8.
159 Jung, H. W. E., J. F. Math., 138: 255-318 (1910). [1.
160 Kanda, A., Monatsh. f. Math., 22: 279-302 (1911). [11.
161 Kantor, S., Theorie der endlichen Gruppen, Berlin, (1895). [8.
162 ———, Ann. di Mat., (2), 10: 64-70 (1880). [5.
163 ———, Ann. di Mat. (2), 10: 71-73 (1880). [5.
164 ———, Wien. Ber., 82_2: 237-259 (1880). [8.
165 ———, C. R., 100: 42-44 (1885). [8.
166 ———, C. R., 100: 95-97 (1885). [8.
167 ———, C. R., 100: 343-345 (1885). [7, 8.
168 ———, Premiers fondements, Naples, (1891). [2, 3, 7, 8.
169 ———, Torino Atti, 29: 9-16 (1893). [11.
170 ———, J. f. Math., 114: 50-108 (1895). [8.
171 ———, Pal. Circ. Mat., 9: 65-78 (1895). [8, 9.
172 ———, Acta Math., 19: 115-193 (1895). [8.
173 ———, C. R., 126: 946-949 (1898). [7.
174 ———, Monatsh. f. Math., 10: 18-53 (1899). [7.
175 ———, Monatsh f. Math., 10: 54-74 (1899). [7.
176 ———, Wien. Ber., 112_{2a}: 667-754 (1903). [9.
177 Kasner, E., Am. M. S. Trans., 1: 430-498 (1900). [10.
178 ———, Am. M. S. Trans., 4: 86-102 (1903). [10.
179 ———, Am. J., 25: 107-122 (1903). [3.
180 Klein, F., Ikosaeder, Leipzig, (1884). [12.
181 ———, Math. Ann., 43: 63-100 (1893). [3, 10.
182 Kohn, G., Wien. Ber., 114_{2a}: 1431-1459 (1905). [3, 4.
183 Lacour, E., Nouv. Ann., (4), 2: 169-177 (1902). [4.
184 Larice, Ines, Ven. Ist. Atti, (8), 11: 731-756 (1908). [2.
185 ———, Per. di Mat., (3), 6: 234-236 (1909). [2.
186 Lazzeri, G., Rom. Acc. L. Rend., (4), 2_2: 61-67 (1886). [11.
187 Leib, D. D., Am. M. S. Trans., 10: 361-390 (1909). [11.
188 London, F., Math. Ann., 38: 334-368 (1891). [11.
188.1 Loria, G., Il passato ed il presente . . ., Torino, (1896). [1.
189 Lüroth, J., Diss. Freiburg i. B., 25 p. (1889). [6.
190 ———, Math. Ann., 42: 457-470 (1893); 44: 539-552 (1894). [1.
191 McKelvey, J. V., Am. J., 34: 115-146 (1912). [8.
192 Malet, H., Transformations birationelles, Paris, (1921). [1.
193 Marletta, G., Pal. Circ. Mat., 24: 229-242 (1907). [3.
194 Martinetti, V., Ann. di Mat., (2), 12: 73-106 (1883-84). [6.
195 ———, Ann. di Mat., (2), 13: 53-80 (1884-85). [6.
196 ———, Pal. Circ. Mat., 1: 205-216 (1884-87). [7.

196.1 ———, Giorn. di Mat., 23: 37-47 (1885). [11.
197 ———, Lomb. Ist. Rend., (2), 20: 264-269 (1887). [7.
198 ———, Pal. Circ. Mat., 4: 30-42, 126-142 (1890). [6.
198.1 Maschke, H., Diss. Breslau, (1879). [4.
199 Milinowsky, H., J. f. Math., 77: 263-268 (1874). [6.
200 Mlodziejowski, B., Mosk. Math. Samml., 29: 269-275 (1913-15). [7.
201 ———, Mosk. Math. Samml., 31: 7-34 (1922). [1.
202 ———, Mosk. Math. Samml., 31: 34-57, 58-77 (1922). [2.
202.1 ———, Mosk. Math. Samml., 31: 341-344 (1924). [2.
203 Mohrmann, B., Habil.-Schrift Karlsruhe, 30 p. (1911). [10.
204 ———, Pal. Circ. Mat., 31: 170-200 (1911). [10.
205 ———, Pal. Circ. Mat., 32: 158-187 (1911). [10.
206 Lo Monaco-Aprile, L., Mat. P. A., 2: 188-192 (1902). [4.
207 Montesano, D., Ven. Ist. Atti, (6), 6: 1425-1444 (1887-88). [8.
208 ———, Giorn. di Mat., (2), 10: 181-189 (1903). [11.
209 ———, Nap. R. A. Rend., (3), 11: 259-303 (1905). [2.
210 ———, Pal. Circ. Mat., 31: 363-368 (1911). [3.
211 ———, Nap. R. A. Atti., (2), 15: no. 7, 34 p. (1914). [2.
212 ———, Nap. R. A. Rend., (3), 21: 30-38, 69-79, 113-119 (1915). [3.
213 ———, Nap. R. A. Rend., (3), 21: 248-251 (1915). [2.
214 ———, Nap. R. A. Rend., (3), 24: 31-78 (1918). [2.
215 Moore, E. H., Am. J., 10: 243-257 (1888). [11.
216 ———, Am. J., 22: 279-291 (1900). [3, 8, 12.
217 Morgan, F. M., Am. J., 35: 79-104 (1913). [6.
218 ———, Ann. of Math., (2), 16: 134-137 (1914-5). [11.
219 Müller, H., Math. Ann., 2: 281-292 (1870). [4, 11.
220 Nalli, P., Pal. Circ. Mat., 31: 92-108 (1911). [7.
221 Nencini, D., Ann. di Mat., (3), 27: 259-292 (1918). [7.
222 Newson, H. B., Am. M. S. Bull., (2), 4: 107-121 (1897-98). [10.
223 ———, Am. M. S. Bull., (2), 7: 259-266 (1900-01). [10.
223.1 Noether, M., Goett. Nach., 1-6 (1870). [7.
224 ———, Math. Ann., 3: 161-227 (1871). [7.
225 ———, Math. Ann., 5: 635-639 (1872). [7.
226 ———, Erl. Ber., 10: 81-86 (1878). [6.
227 ———, Math. Ann., 23: 311-358 (1884). [7.
228 Okken, P. A., Diss. Groningen, 92 p. (1913). [6.
229 ———, Lomb. Ist. Rend., (2), 47: 226-230 (1914). [6.
230 Oppenheimer, H., Diss. Jena, 30 p. (1892). [11.
231 Ota, T., Tôhoku Sc. R., 10: 155-232 (1921). [3.
232 ———, Tôhoku Sc. R., 10: 233-272 (1921). [11.
233 Ouivet, E., C. R., 150: 1036-38 (1910). [12.
234 Le Paige, C., Brux. Bull., (3), 3: 760-762 (1882). [4.
235 ———, Brux. Bull., (3), 4: 415-431 (1882). [4.
236 Palatini, F., Saggio di un metodo, Palermo, 12 p. (1892). [6.
237 ———, Ven. Ist. Atti, (7), 8: 1555-1568 (1896-97). [2.
238 ———, Per. di Mat., (3), 9: 129-143 (1912). [2.
239 Pannelli, M., Giorn. di Mat., 29: 154-162 (1891). [11.
240 de Paolis, R., Rom. Acc. L. Mem., (3), 1: 511-544 (1876-77). [6.
241 ———, Rom. Acc. L. Mem., (3), 2: 31-50 (1877-78). [6.
242 ———, Rom. Acc. L. Mem., (3), 2: 851-878 (1877-78). [6.

242.1 Pascal, E., Repertorium, Leipzig, Teubner, (1910). [1.
243 Pasch, M., Math. Ann., 18: 93-94 (1881). [1.
244 Pecl, P., Pr. Gymn. in Raudnic, 14 p. (1905). [11.
245 Lo Piano, D., Nap. R. A. Rend., (3), 6: 130-135 (1900). [11.
246 Poincaré, H., J. de Math., (5), 7: 161-233 (1901). [11.
247 Pyrkosch, Progr. (Nr. 226) Gymn, Breslau, 16 p. (1907). [11.
248 Retali, V., Mat. p. a., 2: 192-197, 222-227 (1902-03). [4.
249 Reye, T., Zschr. Math. Phys., 11: 280-310 (1866). [11.
250 Roberts, S., London M. S. Proc., 4: 121-139 (1871-73). [2.
251 Rosanes, J. f. Math., 73: 97-110 (1871). [1, 5, 7.
252 Ruffini, F. P., Bologna Mem., (3), 8: 457-525 (1877). [2.
252.1 ———, Bologna Mem., (3), 9: 199-215 (1878). [2.
253 Saltel, L., Brux. Mém. Cour., 22: Mem. 5, 53 p. (1872). [4, 11.
253.1 ———, Brux. Bull., (2), 34: 51-52 (1872). [11.
253.2 Schmidt, Diss. Breslau, (1882). [4.
254 Schottky, F., J. f. Math., 103: 185-203 (1888). [12.
254.1 Schoute, P. H., Assoc. Fr. C. R., 8: 194-205 (1879). [6.
255 ———, Bull. Sci. M., (2), 6: 152-168, 174-188 (1882). [4.
256 ———, Assoc. Francaise p. l'avanc. d. S., Rouen, 131-152 (1883). [4.
256.1 Schuttenhelm, Diss. Strassburg, 47 p. (1901). [4.
257 Scott, Charlotte A., Quart. J., 28: 377-381 (1896). [11.
258 ———, Am. M. S. Trans., 3: 388-398 (1902). [11.
259 Segre, C., Pal. Circ. Mat., 1: 217-221 (1884-7). [7.
260 ———, Torino Atti, 36: 645-651 (1900-1). [7.
260.1 Servais, C., Mathesis, 8: 105-109 (1888). [6.
261 Sharpe, F. R., Am. J., 37: 55-64 (1915). [8.
262 Slaught, H. E., Am. J., 22: 343-80 (1900); 23: 99-138 (1901). [8, 12.
263 Snyder, V., Am. J., 30: 10-18 (1908). [11.
264 ———, Am. J., 30: 325-336 (1908). [8.
265 ———, Am. J., 30: 337-346 (1908). [11.
266 ———, Ann. of Math., (2), 13: 140-148 (1911-2). [8.
267 ———, Am. J., 33: 327-336 (1911). [6.
268 Stahl, W., J. f. Math., 95: 297-316 (1883). [11.
269 Steinmetz, C. P., Am. J., 14: 39-66 (1892). [6.
270 Stiner, G., Zürich Naturf. Gess., 40: 217-229 (1895). [6.
271 Sturm, R., Liniengeometrie, Leipzig, 3: (1896). [11.
272 ———, Geometrische Verwandtschaften, Leipzig, 4: (1909). [1, 6, 11.
273 ———, Math. Ann., 1: 533-574 (1869). [11.
274 ———, Math. Ann., 19: 461-486 (1882). [11.
275 ———, Math. Ann., 26: 304-308 (1886). [2.
276 Sylvester, J. J., Camb. Dub. Math. J., 6: 1-17 (1851). [1.
277 Tappan, Anna H., Am. J., 37: 309-336 (1915). [8.
278 Thalberg, O. M., Quart. J., 50: 33-54 (1924). [11.
279 ———, K. Vid. Sels. Skr., no. 3, 51 p. (1924). [11.
280 ———, Proc. Congress, Copenhagen, 473-477 (1925). [11.
281 Timerding, H. E., Math. Ann., 53: 193-219 (1900). [11.
282 Tinto, J. F., Edinb. M. S. Proc., 34: 133-145 (1915-16). [11.
282.1 ———, Edinb. M. S. Proc., 36: 17-21 (1917-18). [3.
283 Tuck, T., Diss. Jena, 51 p. (1890). [11.
284 Veronese, G., Rom. Acc. L. Mem., (3), 19: 344-371 (1883-84). [4, 7.

284.1 Vigarie, E., J. d. M. Elem., (3), 5: 101-105, 129-131, 153-157, 181-183, 207-211, 256 (1891). [6.

285 Visalli, P., Rom. Acc. L. Rend., (4), 2_2: 80-87 (1886). [11.

286 ———, Rom. Acc. L. Rend., (5), 4_1: 480-487 (1895). [11.

286.1 Wallstaff, Diss. Breslau, 73 p. (1902). [6.

287 Williams, A. R., Univ. Cal. Pub., 1: 211-222 (1920). [6.

288 Wiman, A., Sven. Vet. Ak., Bih. 21_1: 1-23, 24-41 (1895-6). [8.

289 ———, Math. Ann., 48: 195-240 (1896-97). [6, 8.

290 Zahradnik, K., Wien. Ber., 133_{2a}: 973-986 (1904). [11.

291 ———, Casopis, 34: 105-123, 329-341 (1905). [11.

292 ———, Wien. Ber., 114_{2a}: 669-691 (1905). [11.

293 ———, Wien. Ber., 117_{2a}: 1167-1190 (1908). [11.

It is stated on page 84 that Montesano proved (1905) that there exist only four symmetric types of plane Cremona transformations, namely those of order 2, 5, 8, 17. This was previously established by Bertini (Richerche sulle transformazioni nel piano, Annali di Matematica (2), Vol. 8 (1877), p. 271, in 1877 in a footnote of the page cited. The first line of the table given in the footnote is incorrect. The equations (8), (9), page 252 from which the table is derived are, however, correct. Moreover, the correction in the solution was evidently made by Bertini that same year, inasmuch as another table giving the characteristics of these four symmetric types correctly occurs as an erratum at the bottom of the second index page of the same volume [T. R. HOLLCROFT, Bulletin of A.M.S.]

CHAPTER V

MULTIPLE CORRESPONDENCES BETWEEN TWO PLANES

1. *The* (1, 2) *Correspondences between* S'_2 *and* S_2.—The first generalization of the (1, 1) correspondence as established by the Cremona transformations between two planes is the (1, 2) case, which will now be considered. A general synthetic discussion is given by R. Sturm(**99**).

Let the simple plane be denoted by (x), the double plane by (x'). To a point P_1 in (x) corresponds P' in (x'), but to the latter correspond P_1 and P_2 in (x), either of which uniquely determines the other. Thus, associated with the (1, 2) correspondence is an involutorial birational transformation I_2 in the plane (x). The earlier history of plane involutions will be found in Chapter IV, but the memoirs most closely related to the double plane will be mentioned here: C. F. Geiser(**41**) considered the involution defined by a net of cubic curves with seven basis points. A pencil formed by an additional point P also has another basis point Q. This involution is of order 8. It has the seven basis points for fundamental points of order 3, the fundamental cubic belonging to P_i has a node at P_i and belongs to the net. The invariant points define a curve of order 6, genus 3, having double points at the basis points. This is known as the Geiser involution; it will be designated by (G). A. Cayley(**31**) outlined a (1, 2) transformation between two planes.

E. de Jonquières(**49**) constructed another involution by means of a fixed curve $C_n: O^{n-2}$ and a pencil of lines through O. A point P is transformed into P', the harmonic conjugate of P on OP as to the two residual points in which OP meets C_n. This involution is of order n, has O as a basis point of order $n-1$, and the points of tangency of the tangents to C_n from O as simple fundamental points. This is known as the perspective Jonquières type; it will be designated by (J).

The product of (J) by a certain harmonic homology (H) may be an involution if proper restrictions are imposed on (J). The product is called the non-perspective Jonquières type. It may also be transformed into a harmonic homology, as was shown by E. Bertini(**6**). This led Bertini to inquire whether all plane involutions could be reduced to a finite number of types when transformed through Cremona transformations(**7**). In addition to the three types (H), (G), (J), already known, one more was found. It is of order 17 and has eight fundamental points of order 6. The locus of invariant points is a curve of order 9, having the fundamental points for triple points. Through the eight points as double points a ∞^3 system of sextic curves can be passed, but a net within the web determined by one additional basis point has a second, which is the conjugate of P in the involution. This is called the Bertini involution; it will be denoted by (B). The equations of the involution of type (B) were obtained by V. Snyder([93]) who also gives various particular cases. E. Caporali([23]) proposed to classify involutions by their *class,* that is, the number of pairs of distinct conjugate points lying on an arbitrary straight line. The class is not invariant under birational transformation. Constructions of involutions of any given class were given by K. Bobek([12.1]).

For the early literature on attempts to classify involutions according to class see Doehlemann([38]). An extensive more recent study of those of class six was made by P. A. Okken([79]). H. Liebmann([67]) showed that the four types can be obtained by projecting a quadric upon a plane stereographically, and then from points not on the quadric by means of a linear congruence of lines.

The first systematic treatment of (1, 2) correspondences between the points of two planes was given by A. Clebsch([33]). The locus of points $L'=0$ in (x'), the two images of which coincide, is called the curve of *branch-points* (or branch curve) and the image curve in (x), $K=0$ is called the curve of *coincidences* (also called double curve).

The equations of the transformation may be written in the form

$$x'_i=R_i(x)$$
$$x_i=M_i(x')+N_i(x')\sqrt{L'},$$

wherein R_iM_i, N_i are rational polynomials in (x'). A non-self-conjugate locus in (x) is transformed into a curve touching $L'=0$ at every common point except fundamental points. The case in which $L'=0$ is a conic is treated in detail.

When P' describes a locus in (x'), its images P_1, P_2 in (x) may describe the same locus, or different loci, conjugate to each other in the associated involution defined by P_1, P_2. The necessary and sufficient condition that P_1, P_2 describe separate loci is that the locus described by P' shall touch $L'=0$ at every common point except the fundamental points. The problem arose incidentally in connection with that of the bisection of the periods of an abelian function. If $L'=0$ is a conic, I_2 is H. If $L'=0$ is the general quartic, I_2 is G. See also L. Cremona and G. Battaglini([36]).

R. de Paolis([82]) studied the (1, 2) correspondences directly, using only algebraic methods. When P_1, P_2 describe the same curve image of a rational curve of (x'), the curve is hyperelliptic, and is invariant (not point by point) under I. The points P_1, P_2 describe the characteristic $g_2{}^1$ of the curve.

The limiting position of the line P_1P_2, as $P_1 \to P_2$ at a point on K is called the *principal direction* at P_1. It is not, in general, the tangent to K.

If a curve C in (x) has $P_1{}^r$ and its conjugate P_2 to multiplicity s, $s \neq r$, $P_2{}^s$, it is not self-conjugate, and its image C' in (x) has P'^{r+s}, and the conjugate of C has $P_1{}^s P_2{}^r$. If a non-conjugate C meets K in r points, not tangent to the principal directions at those points, C' touches L' in the corresponding points. If a non-self-conjugate C has r-point contact with K, C' has $2r$-point contact with L' at the corresponding point. If a non-self-conjugate C intersects K in the principal direction, C' has a cusp at the corresponding point on L'. If a non-self-conjugate C has a multiple point of order r on K, C' has r branches all touching L' at the corresponding point.

To a net of rational curves in (x') corresponds a net of self-conjugate hyperelliptic curves in (x).

The basis points of the net $|\phi|$, images of the ∞^2 lines of (x'), are the fundamental points of (x). If P_i is of multiplicity i, then

$$\Sigma i^2 = n^2 - 2,$$

in which n is the order of a curve ϕ. The number of conditions imposed by the basis points is

$$h = \frac{n(n+3)}{2} + p - 3.$$

For $p > 1$, the basis points are not independent. Here p, the genus of a general curve ϕ, is called the genus of the transformation.

The lines of (x) are transformed into a non-linear system of rational curves of order n in (x').

The order of the associated involution of conjugate points P_1, P_2 in (x) is $N = n^2 - 1 - \Sigma i'^2$ in which i' is the order of the basis point P'_i, of the system $[\phi']$, images of the lines of (x).

The order of K is $\nu = 3(n-1) - \Sigma i'$.

Each curve of the system $[\phi']$ has $\dfrac{N-v}{2}$ double points.

2. *Fundamental Elements.*—A basis point P_i of multiplicity i of the system $|\phi|$ has for image in (x') a rational curve of order i.

This same point may also be fundamental in I, i. e., conjugate to a rational curve, [First kind]; or it may be conjugate to another fundamental point of the same multiplicity, having the same image curve in (x'), [Second kind]; or it may be self-conjugate [Third kind]. If a fundamental curve meets each ϕ' in i points, the image point in (x) is i-fold in I.

Fundamental curves of the second kind in (x') do not meet curves ϕ' except at fundamental points. In (x), fundamental curves of the first and second kinds are rational, those of the third kind are hyperelliptic.

The fundamental curves of the simple plane together with K constitute the jacobian of the net $|\phi|$. The g_2^1 on each ϕ has $2(p+1)$ coincidences, hence L' is of order $2(p+1)$.

If any C' touches L', its image C has a node on K at the image point.

To tangent lines of L' correspond nodal curves of the net $|\phi|$. The inflexional tangents to L' correspond to curves ϕ having cusps on K. A basis point P_i of order i on $|\phi|$ is of multiplicity $3i$ on the jacobian of $|\phi|$. R. de Paolis([83]) considers the following particular case. Given a conic $A(x)=0$, and a point (y) in its plane. The polar of (y) as to $A(x)=0$ is $A(x, y)=0$. Put $x'_i=y_iA(x)-2x_iA(x, y)$. Between (x'), (x) this defines a $(1, 2)$ correspondence. The jacobian of the net $|\phi|$ consists of K, of $A(x)=0$ and of the line $A(x, y)=0$. The involution I is the general involutorial quadratic inversion.

This transformation is then interpreted in non-euclidean geometry.

In a third paper R. de Paolis([84]) studies the general cubic $(1, 2)$ transformation of genus 1. The net $|\phi|$ now consists of the cubics through 7 arbitrary basis points. L' is the general quartic, and the seven lines, images of the cubics of the net having nodes at the basis points are bitangents of L', forming an Aronhold septuple. The twenty-eight bitangents and the sixty-three systems of contact conics and the system of contact cubics of L' are all obtained. The involution I is the Geiser involution of order 8; all the basis points are fundamental of the first kind. The curve of coincidences, invariant points of I, is a sextic of genus 3, having the basis points for double points.

3. *Derivation of the Types.*—E. Bertini([10]) obtained all the irreducible forms by means of the types of plane involutions, and by projecting from space.

H is obtained by projecting from a cone stereographically; J by projecting ∞^3 systems of curves from a quadric; G by projecting F_3 on a double plane, and by mapping it on another plane birationally; B by $\infty^3 C_6$ on a quadric cone mapped stereographically. See also E. Bertini([7]).

M. Noether([76]) showed that the $(1, 2)$ transformations not associated with harmonic homology in the simple plane are of three types, according as the curve of branch points is $C_{2n}: P^{2n-2}$ or a general C_4 or $C_6: 2P^3$. An outline of the same procedure was given by the same author([77]). The method is intimately connected with that of mapping rational surfaces on a simple and on a double plane.

The (1, 2) point correspondence between the planes (x') and (x) associated with (H) may be expressed in the form $x'_2x_3 - x'_1x_2 = 0$, $x'_1x_2x_3 - x'_2x_1^2 = 0$. The invariant or coincident points constitute the line $x_1 = 0$. That associated with (G) has the form

$$x'_1C_1 + x'_2C_2 + x'_3C_3 = 0,$$
$$x'_1x_1 + x'_2x_2 + x'_3x_3 = 0,$$

in which each $C_i = 0$ is a general conic. Each cubic of the net formed by linear combinations of $x_iC_k - x_kC_i = 0$ is transformed into itself. The curve of invariant points is the jacobian of the net $K_6(r_1^2 \ldots r_7^2)$, in which r_i is a simple basis point. The curve of branch points in (x') in (1, 1) correspondence with K_6, is the general quartic L'_4, which expresses the condition that the line in (x) touches the associated conic. Every line of (x) contains one pair of conjugate points. From this standpoint the system of the bitangents, the contact conics, cubics, etc., of L'_4 can all be determined directly by elementary methods. Their two images in (x) are conjugate as to (G). That associated with (B) may be expressed by the equations

$$x'_1\phi_1 + x'_2\phi_2 = 0,$$
$$x'_1\mathrm{f} + x'_3\phi_1(k_1\phi_1 + k_2\phi_2) = 0,$$

in which $\phi_i = 0$ is a general cubic curve, and $f = 0$ is a proper sextic having eight of the points h_i common to $\phi_1 = 0$, $\phi_2 = 0$ for double points. The curve of invariant points is the jacobian of the system $\phi_1 = 0$, $\phi_2 = 0$, $f = 0$. It is of order 9, and has eight h_i as triple points, $K_9(h_1^3 \ldots h_8^3)$. The curve of branch points is also a sextic L'_6 having three branches touching each other at a common point. An arbitrary line in (x) contains four pairs of conjugate points; i. e., the simplest form of (B) is of class 4. Finally, the (1, 2) correspondence of form (J) can be expressed by $x'_1x_2 - x'_2x_1 = 0$, $x'_3M_1 - u'M_2 = 0$, in which $M_i = 0$ is a curve of order m with an $(m-2)$-fold point at $(0, 0, 1)$; u' is linear in (x'). The jacobian is the curve of coincidences, a hyperelliptic curve having an $(n-2)$-fold point at $(0, 0, 1)$. This involution is of class zero. This is the easiest way to find and to classify the plane involutions of order 2; to find all possible ways in which the curves of a net can have two variable points of intersection. The question of reducibility or of the equivalence of two given cases is not more difficult by this method than by that of Bertini. The seven basis points r_i have for images in (x') seven bitangents of L'_4; a point of a bitangent has two images, one of which is the basis point r_i, and the other describes the cubic of the net having a node at r_i. The point and the nodal cubic are conjugate in the involution (G).

An interesting case of (J) is that in which all the curves of the net have contact of the second order on each branch at the multiple point. Its equation is

$$\lambda_1(-u_{n-1}x_3+u_n)+\lambda_2 u_{n-1}x_1+\lambda_3 u_{n-1}x_2=0$$

in which u_i is binary in x_1, x_2 of order i. All the basis points are now coincident at $(0, 0, 1)$.

In (41.1) L. Godeaux gives an incomplete construction of the types of plane involutions by means of systems of conics and of lines. When both are pencils, type 1, a particular case of the perspective (J) results; the general (J) is called type 2. The author's type 3 is also reducible to (J). The fourth type, that generated by the field of lines and a net of conics, is not completed. As shown above, this is (G). An outline of the algebraic procedure and the derivation of the equations is given by J. Neuberg (75.1).

Another particular I_2 is interpreted dually by Stiner (93.1). Given four fixed lines, no three concurrent, and a point Q. A point P determines a conic touching the four given lines and PQ. It has a second tangent from P and from Q, and these intersect in P'. The I_2 is P, P'. It is employed to construct $C_5: 6P^2 2P^1$.

A (1, 2) transformation between two planes is outlined in a study of the correspondence between the points of S_4 and the pairs of points of S_2 by F. Palatini (79.1).

The curves of coincident points characterize the type of the involution. Hence it follows that any surface mappable on a double plane is, or is not, rational according as its curve of branch points can be reduced to one of the above curves.

Thus, the mapping of a quartic surface with a double line upon the plane, and the birational equivalence of various maps were developed in detail by J. Cardinaal (24) and (25).

A. Schwarz (87) obtained a (1, 2) correspondence between (x), (x') as follows: let V_1, V'_1 be vertices of two projective pencils of lines in (x), (x') respectively, and V_2, V'_2 be vertices of two pencils in (2, 1) correspondence. A point P_1 in (x) determines a line of (V_1) and of (V_2), hence a point P' in (x'). But P' has two images of (x), collinear with V_1. I is the perspective quadratic inversion. The genus of the (1, 2) transformation is 0. The net $|\phi|$ consists of cubics having a common node at V_2 and three simple collinear basis points. It is a particularization of the Geiser net.

Another construction of a (1, 2) correspondence of genus 0 was given by V. Retali (85) in which the simple and the double plane are superposed. Given a conic C and a fixed point R. With any point P' is associated the double points P_1P_2 of the quadratic involution on the line PR, having given the two pairs of conjugates P_iR_i, PR, C. The net $|\phi|$ consists of conics.

M. Noether([78]) gave the configurations in (x') associated with the (B) and (J) involutions, analogous to what had already been done for the types (G) and (H). The curve $L'=0$ is a sextic having three branches tangent at the same point in B. The lines of (x) are transformed into sextics having this same three branch contact and having four variable double points. In (J), $L'=0$ is a sextic with a 4-fold point, for $p=4$.

P. P. Boyd([18]) did the same thing for type (J), of the general case. F. M. Morgan([74]) gave the necessary and sufficient criteria for reducing a (1, 2) transformation of type (G) to the canonical form. The reductions due to particular positions of the basis elements were given by V. Snyder([96]).

All the involutions found by E. Bertini([7]) are rational; that is, the pairs of points P_1, P_2 can be mapped birationally on the points of a double plane, but certain restrictions were imposed on the basis points of the nets $|\phi|$. These restrictions were removed by G. Castelnuovo and F. Enriques([27]). They showed that whenever a surface can be mapped on a double plane in such a way that the curve of branch points is of the form of $L'=0$ as already found, in the (1, 2) transformations associated with the involutions (G), (B) and (J), the surface was necessarily rational, and conversely. The necessary and sufficient condition that a double plane is rational had been formulated in the same way by M. Noether([76]).

V. Retali([86]) considers a (1, 2) plane transformation of order 3 and genus 0. Compare V. Snyder([95]).

4. (1, k) *Correspondences between Two Planes* $k>2$.—When the points of a plane can be arranged in sets of k, such that any point will determine uniquely the entire set to which it belongs, the sets are said to form an involution of order k. If the sets can be mapped birationally upon the points of a second plane (x'), the involution I_k is said to belong to the plane (x').

Numerous constructions for such involutions were given by J. S. Vanecek and M. N. Vanecek([102]) for surfaces and for planes. Plane involutions may be generated by two projective pencils of curves. F. Chizzoni([32]) considered a pencil $|\phi|$ of curves of order n, and a pencil $|\psi|$ of curves of order n'. A curve ϕ and its associate curve ψ have N variable intersections. Any point P determines a curve ϕ and a curve ψ, hence all N points of intersection. Let R_h be a basis point of order r_h of $|\phi|$, and of order r'_h of $|\psi|$. Then

$$N=nn'-\Sigma r_h r'_h, \quad n^2=\Sigma r_h^2, \quad n'^2=\Sigma r'^2_h.$$

A point R_h basis for both $|\phi|$ and $|\psi|$ is a fundamental point of the first kind; if for either one only, of the second kind.

Suppose ζ_i is a curve having no variable point on any curve of $|\phi|$ or $|\psi|$, unless the latter becomes composite. Let ζ_i have R_h to multiplicity η_{ih}, and be of order z_i. Then

$$nz_i = \Sigma r_h \eta_{ih}, \quad n'z_i = \Sigma r'_h \eta_{ih}.$$

Any point of ζ_i has the whole curve ζ_i for its image. Two curves ζ_i, ζ_k intersect in fundamental points of the third kind, apart from those of the first kind and of the second kind.

A large number of relations is derived, but no new results in reducibility. $N=1$, $N=2$ were known before, and the forms $N=2$ confirmed. Particular cases of a I_5 were given by C. le Paige (66).

Constructions by means of a net of curves of order n through $\frac{n(n+3)}{2}-2$ fixed points were given by J. de Vries (115). The involution is of order $\frac{n(n-3)}{2}+2$. The locus of intersection of corresponding curves of two pencils of the net in (p, q) correspondence is of order $n(p+q)$. A somewhat similar treatment is given by C. P. Steinmetz (97). The paper contains several illustrations, but no new results.

G. Castelnuovo (30) extended his studies of involutions on a given curve (29) to apply to a plane field. Given a plane pencil of lines with vertex O. Any line meets a fixed C_n in n points. Consider the polar group of each set by the procedure of the earlier memoir. The new groups all lie on the first polar of C_n as to O, and they depend on $\infty^1 C_n$, giving rise to I_n in the plane. All the points of each group are collinear.

Particular examples of plane involutions were given by J. de Vries (108), (109), and (112). In (110) an I_4 and a related I_3 are discussed.

A cubic involution I_3 is derived by W. van der Woude (101) and (100). The triads of I_3 are vertices of self-polar triangles as to a fixed conic γ. If P describes a line a, its associates describe a quartic curve having a node at the pole of a as to γ. I_3 has ten singular points; their polars are 10 singular lines. Together they form a Desargues $(10_3, 10_3)$ configuration.

J. de Vries (111) obtains a Cremona transformation between the points and lines of a plane by means of an involution I_3 of class 1. Each line is associated with the opposite vertex of the triad to which it belongs. Associate each point P of a line r with the point P^* in which p meets r. When P^* is given, there are n points P, n being the class of the rational envelope of p, when P describes a line. Further properties were found in (113.1) by the same author. If $n=1$, the Desargues $(10_3, 10_3)$ configuration results. Already discussed by W. van der Woude (101).

An I_3 is determined by the intersections with a fixed plane of the congruence of space cubic curves through 5 fixed points. This is called Reye's cubic involution(86.1). Further properties are given by J. de Vries(114.1). J. de Vries(113) constructs I_4 by means of two projective nets of conics. If they each have a basis point, and these coincide, I_3 results. When P describes a line, its associates describe a curve of order 15, with 8-fold points at the singular points. The curve of coincidence is of order 9. The image of the common basis point is a C_8 with a 5-fold point. The I_3 is of class 3. See also P. Visalli(103) and (104).

G. Jung(52) studied two planes (x), (x') in $(k, 1)$ correspondence, using notation and method similar to that of R. de Paolis(83) and (82).

If the image in (x) of a line of (x') is of genus p, then $p=k-1-c$, where c may be positive, zero or negative, according as the system $|\phi|$ is partial, complete, or the basis points of the net are dependent. If $c_1 \sim \phi'$, ϕ' touches L' in v points, and has $\frac{N-v}{2}$ variable double points. The associated involution I_N of (x) has the order $N=m^2-\Sigma r'-1$, and the locus of coincidences K is of order $\nu=3(m-1)-\Sigma r'$.

In (53) the same author considers involutions associated with a given Cremona transformation of order n of two superposed planes. Let $|\phi|$, $|\phi'|$ be the two Cremona nets, having V_i, V'_i for fundamental points of orders r_i, r'_i. Designate the $n+2$ invariant points by u.

Join a point $(x)=P$ to its image P' by a straight line and regard it as a line of the plane (x'). Its image is a curve of ϕ_n and the line PP' now regarded as in (x). To the points of the multiple plane correspond the net of isologues P_{n+1} of the simple plane. The correspondence is of order $n+1$ and of genus $n-1$. The simple plane contains no fundamental points. To an arbitrary line of (x) corresponds a curve of order $n(n+2)$. A similar transformation exists by interchanging (x) and (x'). The Geiser involution is included in this category when $n=2$.

The author points out that the table given by R. de Paolis is incomplete. Three more forms should be added.

Further details are given by G. Jung(50). The same method is followed further by P. Visalli(107).

C. F. E. Björling(12) raises the question of reducibility, and obtains numerous necessary inequalities, but no final results.

M. Pannelli(80) generalizes the method of G. Jung(53) by replacing the given Cremona transformation k times. The greatest difference between this and the simpler scheme lies in the number and complexity of the fundamental elements in the simple plane.

F. Borghese(13) extends to superposed planes in $(1, \nu)$ correspondence the question answered for Cremona correspondences by S. Kantor(57), namely, to determine the number of invariant points, and how many points return to their initial positions when the correspondence is repeated k times. If the curve of coincidences is of order k and has fundamental points to multiplicity λ_i, while the other plane has these same points to multiplicity θ_i, the number of isolated invariant points is

$$n+\nu+1+k(k-3)-\Sigma\lambda_i(\lambda_i-1)=\Sigma\theta_i$$

in which n is the order of ϕ, the image of a line of the multiple plane.

The number of cyclic groups of N elements is equal to the number of cyclic groups of N points in the two uniform correspondences, one of order n and the other of order ν.

That every involution on a straight line is rational was first proved by J. Lüroth(69) and later(68) the same author proved independently that every plane involution of order 2 and those of order 3 which arise from Cremona transformations of period 3 are rational. The general theorem that every plane involution of any order is rational was proved by G. Castelnuovo(26). That the corresponding theorem in S_3 is not true was confirmed by G. Aprile(1), by means of a concrete example. The general theory of plane involutions was developed by F. Deruyts(37).

P. Visalli(105) determined the properties of the congruence of lines joining corresponding points of two planes in $(1, \nu)$ correspondence. If n is order of ϕ, image of a line of the multiple plane, the congruence is of order $n+\nu+1$, class n. The two given planes are singular planes and all fundamental points of each are singular. Other singular planes may exist. The congruence and its focal surface are both rational. This is a direct generalization of T. A. Hirst(42), (43), (44) and of I. Conti(34) and (35).

C. A. Scott(89), (88), (90) studies $(1, k)$ correspondences from the equations $x'_i=\phi_i(x)$, wherein any two curves of the net $|\phi|$ have k variable intersections. Let the images of P' be $P_1, \ldots, P_k$. A curve $f(x)=0$ is transformed into $f(x')=0$, but when P' describes $f'=0$, P_1 describes $f=0$ and $P_2, \ldots, P_k$ describe its conjugate. K, L' are defined as in $(1, 2)$ correspondences but when $P_1=P_2$ describes K, the points $P_3, \ldots, P_k$ describe the co-jacobian. The simple plane is divided into compartments, any one of which is the image of the (x') plane. Fundamental points are of three kinds, as given by R. de Paolis(82). C. A. Scott(91) considers various cases which cannot arise in the $(1, 2)$ correspondence. Systems with dependent elements of determination are considered and the resulting effect on the fundamental elements determined. The theory is intimately connected with that of linear systems of plane curves. See further G. Castelnuovo(28); E. Bertini(8), (9), (11); G. Jung(51); E. Caporali(21); F. S. Macaulay(70), and (71).

G. Ferretti (40) showed how to construct involutions or order n and of class 0 and 1. All the preceding memoirs concerned with the class of the involution take $n=2$ except G. Castelnuovo (26) who considers n general. By means of projections from hyperspace, it is found that if I_n is of class 0, it is an I_3 or a perspective Jonquières involution. If $\nu=1$, $n=8$ and the curves of lowest order which can generate I_n are cubics.

A particular (1, 2) point correspondence was given by J. Neuberg (75) in which lines are transformed into conics.

The types of I_3 as expressed by two equations implicit in (x), (x') were found by A. M. Howe (48) to be five in number.

(1)	(x) lines	$\Sigma x_i x'_i=0$	(x') lines
	lines	$\Sigma x'_i \phi_{3,\,i}=0$	lines
(2)	lines of pencil	$x_1x'_1+x_2x'_2=0$	lines of pencil
	$C_n: P^{n-3}$	$\Sigma x'_i C_{n,\,i}=0$	lines pers. Jon.
(3)	conics 1 basis pt.	$\Sigma x'_i C_{2,\,i}=0$	lines
	conics	$\Sigma x'_i C_{2,\,i}=0$	lines
(4)	cubics: 6 pts.	$\Sigma x'_i C_{3,\,i}=0$	lines
		$\Sigma x'_i C_{3,\,i}=0$	lines
(5)	pencil $C_3: 9P$	$x'_1 C_{3,\,1}+x'_2 C_{3,\,2}=0$	lines
	$C_9: 8P^3$	$\Sigma x'_i \psi_{9,\,i}=0$	lines

followed by an enumeration of the cyclic types.

A similar enumeration of the non-cyclic I_4 was made by T. R. Hollcroft (47). He found ten irreducible types.

Surfaces having equations of the form $z^n=f(x, y)$ have rational (cyclic) involutions belonging to the plane.

A. Bottari (14) determined when the surface is rational. The results are:

(a) For arbitrary n: $f=0$ consists of $n+2$ lines, n through 0.
For $n=3$:

(b) An arbitrary cubic.

(c) C_6 with two tacnodal branches $=8P^2$.

(d) $n=5$.
An arbitrary cubic and an inflexional tangent.

In the second paper (15), with fewer restrictions on n, twenty-five additional types are found. For $n>30$, only types (a) and (b) are possible. When $f=0$ is irreducible, further results were obtained by M. Botasso (17).

5. *Cyclic Involutions I_n in S_2.*—Cyclic involutions I_n are generated by Cremona transformations of period n. When P' describes a straight line, each point P_k in (x) describes a curve which is invariant under the Cremona transformation, all the points P_k are on the same curve. The curve K, if such exists, is the curve of invariant points of the Cremona transformation. All the points $P_1 = P_2 \ldots P_n$ are points of K.

A. Emch([39]) gives a construction of systems of cubic curves by means of involutorial quadratic transformations.

L. Autonne([2]) obtains the necessary and sufficient conditions that a quadratic plane transformation shall be periodic, and that each power shall be quadratic, except the identity. It is expressed in terms of cyclic projectivities of the two pencils, the vertices of which are fundamental for both planes. See S. Kantor([55]), ([56]), ([58]), ([59]), ([60]) and ([61]), A. Wiman([118]), and V. Snyder([94]). For applications of cyclic quadratic transformations to involutions on quartic curves, see Bydzovsky([20]).

6. *Multiple Correspondences (m, n) between Two Planes.*—When both indices (m, n) are greater than one, the correspondence between two planes may not be rational, but a number of cases should be noted.

Given two planes π, π' in (m, n) point correspondence. Consider both immersed in the same S_4, in such a way that the point common to π, π' is not self-corresponding, and connect by straight lines every point of each plane with all of its corresponding points in the other. Then intersect the joining lines with an arbitrary S_3. The locus of the points of intersection is a surface in $(1, n)$ correspondence with π, and in $(1, m)$ correspondence with π'. Conversely, if a surface is in $(1, n')$ point correspondence with one plane and $(1, m')$ with another, it defines a

$$\left(\frac{m'}{\mu}, \frac{n'}{\mu}\right) (\mu \geqq 1)$$

correspondence between the two planes.

C. Stéphanos([98]) considered a point-line transformation defined by the connex $f(x^m, u^n) = 0$. If the point x is fixed, each u touches its envelope in a point y; similarly, if u is fixed, each x of the locus has a tangent v. The totality (y, v) constitute another connex called the conjugate of the given one.

P. Visalli([106]) considered two harmonic homologies in different planes in projection. The pairs of points P'_1, P'_2 have the same conjugates P_1, P_2 and P'_1, P'_2 as their only images. This is a very particular case of a $(2, 2)$ correspondence. An arbitrary line goes into a conic, but a line through the center of homology is transformed into a line through the center of homology in the other plane. Such a correspondence is called a compound involution.

A particular (n, n') compound involution by means of pairs of projective pencils of curves was given by C. Wiener([116]).

G. Burali-Forti([19]) constructed a (2, 2) correspondence as the product of a (2, 1) correspondence between the planes σ, σ'' and a (1, 2) correspondence between σ'' and σ'. Let the former be called (σ), wherein σ is the simple plane, and the latter (σ'), wherein σ' is the simple plane.

Let P''_r be a fundamental point of σ'' of multiplicity i''_r in (σ) and P_r be a fundamental point of multiplicity i_s in (σ). Similarly, let P''_s be of multiplicity $i''_{r'}$, in (σ'), and $P'_{s'}$ be a fundamental point of σ' of multiplicity $i'_{s'}$ in (σ'). Let N be the order of the associated involution (σ) and N' of (σ').

Let K be of order v in (σ), and p be the genus of (σ). Let K' be of order v' in (σ') and p' be the genus of (σ'). From the two sets of formulas for the components, those for the (2, 2) correspondence between σ and σ' are obtained. This is also a compound involution, but more general than that of Visalli. $\Sigma i_r''^2 = n^2-(N+1)$; $\Sigma i_s^2 = n^2-2$; $\Sigma i_r'' = 3(n-1)$; $\Sigma i_s = 3n+2(p-2)$ and similarly for the (σ').

A more general (2, 2) correspondence was given by G. Marletta([72]), namely, the general (non-involutorial) one in which the images of lines in either plane are conics in the other. A point P in π has P'_1, P'_2 for images in π'. The images of P'_1 are P and P_1; those of P'_2 are P and P_2, hence associated with P is a pair of points P_1, P_2 in the same plane. This symmetric (2, 2) is called the conjugate transformation of the given one. A construction by means of projection in S_4 is outlined. The correspondence may be established by two conics with projective tangents, or by projecting a quadric from two arbitrary points O, O', and representing the points stereographically. When O, O' are conjugate, the Visalli case results.

A similar study is made by Marletta([73]) of (2, 2) correspondences in which straight lines in either plane have cubics for images in the other. If $p=p'=0$, the order of the symmetric (2, 2) in each plane is 5 and is also rational. Curves of coincident points are rational cubics. If $p=p'=1$, the correspondence can be realized by mapping a general cubic surface doubly on a plane from any two points on it. It may also be generated by a pencil of lines and a projective pencil of cubic curves, all in the same plane.

A study of general (m_1, m_2) correspondence was made by R. Baldus([3]). Three kinds of fundamental curves exist: (1) A basis point in k_1 leaves of the m_1-fold plane π_1 has for images k_1 rational curves, every point of which has one image at given basis point, while the other m_1-1 images describe curves. (2) A fundamental curve in π_2 may belong to $h_2>1$

superposed planes, having h_2 distinct points in π_1 as its image. (3) The image points may be coincident in π_1, without defining a principal direction. In this case the fundamental curve in π_2 may not be rational. Such singularities may arise by projecting surfaces having a double curve on a double plane, for example quartics from two double points. Every involutorial (m_1, m_2) correspondence between two planes can be expressed as the product of three correspondences $(m_1, 1)$, $(1, 1)$, $(1, m_2)$.

C. Wiesing(117) generated an involutorial (2, 2) correspondence by means of two projective nets of conics in different planes, each with two fixed basis points.

Further constructions for the compound involution were given by V. R. Barraco(4). Given a pencil μ of curves of order m, and a system v of curves of index 2, that is, such that two curves of the system pass through a given point. Suppose any curve of μ meets any curve of v in two points. Given A, the residual intersections of the two v curves with the μ curves fixed by it are A_1, A_2. Now start with A_1, determining A and A'_1.

The pairs (AA_1), (AA_2), (A_1, A'_1) are *conjugate* points, (A_1, A_2), (A, A'_1) are associated points, both forming involutorial (2, 2) correspondences. The formulas obtained can be derived readily by the method of F. Chizzoni(32).

Using a notation similar to his, and denoting the fundamental curves of the third kind by Z_{ij}, it is found that the conjugate of a straight line is of order $4mn-2\Sigma x_i^2-2$, passes with multiplicity $2r'_hm+2r_hn-2\Sigma\rho_{ik}x_i$ through R_h and with multiplicity $2z_i$ through Z_{ij}. The conjugate of the basis point R_h is of order $2mr'_h+2nr_h-2\Sigma\rho_{ik}z_i$, has R_h to multiplicity $4r_hr'_h-2\Sigma\rho_{ik}+2$, Z_{ij} to multiplicity $2\rho_{ij}$, and other basis points R_k to multiplicity $2r_hr'_k+2r_kr'_h-2\Sigma\rho_{ih}\rho_{ik}$. The order of the associated transformation is

$$(4mn-2\Sigma z_i^2-2)^2-(2nr_k+2mr'_h-2\Sigma\rho_{ik}z_i)^2-\Sigma\Sigma(2\rho_{ik})^2-2,$$

a formula previously found by G. Marletta(72).

The particular case $m=1$, then $n=2$, and finally $m=1$ and $n=2$ is considered in detail, confirming results already obtained by Marletta.

A calculus of composition for (m, n) correspondences was developed by A. Kanda(54).

T. Kubota(62), (63), (64), (65) constructs a (2, 2) correspondence by means of a conic γ and two points A_1, A_2 on it. From any point P in the plane draw PA_1, PA_2 meeting γ in C_1, C_2. Let Q be the pole of C_1C_2. This defines a (1, 2) correspondence. In (65.1) the same author adds one more type of involutorial (2, 2) correspondence. It may be defined by the equations

$$x'_1=x_1, \qquad x'_2=x_2, \qquad x'_3=-x_3\pm 2r\sqrt{x_1^2+x_2^2}.$$

Every quadratic (2, 2) correspondence can be constructed by the product of two such (1, 2) correspondence and collineations.

An involutorial (2, 2) correspondence of class 1 was constructed by J. de Vries ([114]) by means of a pencil of conics and a fixed conic γ. A point P is associated with the two points in which its polar as to γ intersects the conic of the pencil through it; γ is the curve of coincidences. Similarly, given a fixed cubic, a point P, its associates are the intersection of its first and second polars as to the cubic. All the inflexional tangents are singular lines. The curve of coincidences consists of the cubic and its Hessian.

The last paper contains a correction of a statement by H. Bateman ([5]) that the Laguerre inversion corresponds to the Lorenz transformations in the theory of relativity.

The classification of (2, 2) correspondences from the forms of the defining equations was made by F. R. Sharpe and V. Snyder ([92]).

There are five types of non-involutorial involutions. Let the equations be of the form

$$\Sigma a_i(x')u_i(x)=0, \qquad \Sigma b_i(x')v_i(x)=0.$$

The five types may be represented by the scheme:

	$a_i(x')$	$u_i(x)$	$b_i(x')$	$v_i(x)$
1.	conic	line	line	conic
2.	conic	conic	line	line
3.	conic	line	conic	conic
4.	conic	conic	conic	conic
5.	C_n	C_n	line pencil	line pencil

There are six types of compound involutions, obtained by the combinations G, G; G, J; G, B; J, J; B, B; J, B.

The necessary and sufficient conditions that a (2, 2) correspondence is a compound involution is that its defining equations have the form

$$\frac{F_1(x)}{F_1(x')}=\frac{F_2(x)}{F_2(x')}=\frac{F_3(x)}{F_3(x')}.$$

A similar classification of the general (2, 3) types was made by T. R. Hollcroft ([45]); twelve types are given. The types of compound (2, 3) involutions were also given by Hollcroft ([46]): he enumerates fifteen types. In ([47.1]) he completes the other (m, n) correspondences, for $m, n\leqq 4$, both the general ones and the compound involutions. No essentially new features appear in general correspondences with one or both indices greater than 4.

BIBLIOGRAPHY

1 Aprile, G., Note e Mem., 1: 102-110 (1921). [4.
2 Autonne, L., C. R., 97: 567-570 (1883). [5.
3 Baldus, R., Math. Ann., 72: 1-36 (1912). [6.
4 Barraco, V. R., Giorn. di Mat., 54: 42-63 (1916). [6.
5 Bateman, H., Am. J., 34: 325-60 (1921). [6.
6 Bertini, E., Ann. di Mat., (2), 8: 11-23 (1877). [1.
7 ———, Ann. di Math., (2), 8: 244-286 (1877). [1, 3.
8 ———, Lomb. Ist. Rend., (2), 14: 24-28 (1882). [4.
9 ———, Lomb. Ist. Rend., (2), 15: 154-157 (1882). [4.
10 ———, Lomb. Ist. Rend., (2), 22: 771-778 (1889). [3, 4.
11 ———, Pal. Circ. Mat., 3: 5-21 (1889). [4.
12 Björling, C. F. E., Sven. Vet. Ak. Bih; Afd. 150_6, 13: 19 p. (1888). [4.
12.1 Bobek, K., Wien. Ber., 91_2: 476-518 (1885). [1.
13 Borghese, F., Giorn. di Mat., 32: 58-68 (1894). [4.
14 Bottari, A., Ann. di Mat., (3), 2: 277-296 (1899). [4.
15 ———, Giorn. di Mat., 41: 285-325 (1903). [4.
16 Bottasso, M., Torino Atti., 44: 12-37 (1909). [4.
17 ———, Torino Atti., 44: 169-195 (1908-09). [4.
18 Boyd, P. P., Am. J., 34: 290-324 (1912). [3.
19 Burali-Forti, C., Pal. Circ. Mat., 5: 91-99 (1891). [6.
20 Bydzovsky, B., Casopis, 47: 247-262 (1918). [5.
21 Caporali, E., Mem. D. Chelini, 424 p., 144-170 (1881). [4.
22 ———, Nap. R. A. Mem., 116-125 (1888). [4.
23 ———, Nap. R. A. Rend., 18: 212-218 (1879). [1.
24 Cardinaal, J., Amst. K. Ak. Wet. Verh., 1: 1-63 (1892-93). [3.
25 ———, Amst. K. Ak. Wet. Versl., (3), 8: 86-146 (1891). [3.
26 Castelnuovo, G., Math. Ann., 44: 125-154 (1894). [4.
27 Castelnuovo, G., and Enriques, F., Pal. Circ. Mat., 14: 290-302 (1900). [3.
28 Castelnuovo, G., Torino Mem., (2), 42: 1-43 (1891). [4.
29 ———, Ven. Ist. Atti., (6), 4: 1167-1200 (1885-86). [4.
30 ———, Ven. Ist. Atti., (6), 4: 1559-1594 (1885-6). [4.
31 Cayley, A., J. f. Math., 67: 95-96 (1867). [1.
32 Chizzoni, F., Rom. Acc. L. Mem., (3), 19: 301-341 (1884). [4, 6.
33 Clebsch, A., Math. Ann., 3: 45-74 (1871). [1.
34 Conti, I., Pal. Circ. Mat., 1: 230-240 (1884-87). [4.
35 ———, Pal. Circ. Mat., 2: 97-106 (1888-89). [4.
36 Cremona, L., and Battaglini, G., Rom. Acc. L. Rend., (3), 2: 152 (1877). [1.
37 Deruyts, F., Liège S. R. S. Mem., (2), 17: 208 p. (1892). [4.
38 Doehlemann, K., Geom. Trans., 2: 350 p. (1908). [1.
39 Emch, A., Am. M. S. Bull., 24: 327-330 (1917). [5.
40 Ferretti, G., Pal. Circ. Mat., 17: 311-326 (1903). [4.
41 Geiser, C. F., J. f. Math., 67: 78-89 (1867). [1.
41.1 Godeaux, L., Brux. Bull., 5: 217-225 (1911). [3.
42 Hirst, T. A., Mem. D. Chelini, p. 51-73 (1881). [4.

43 ———, London M. S. Proc., 14: 259-301 (1882-83). [4.
44 ———, Rom. Acc. L. Mem., 10: 131-150 (1879). [4.
45 Hollcroft, T. R., Am. J., 41: 5-24 (1919). [6.
46 ———, Am. J., 43: 199-212 (1921). [6.
47 ———, Am. J., 44: 163-171 (1922). [4.
47.1 ———, Am. J., 49: 553-564 (1927). [6.
48 Howe, A. M., Am. J., 41: 25-48 (1919). [4.
49 Jonquières, de E., Giorn. di Mat., 23: 48-73 (1885). [1.
50 Jung, G., Lomb. Ist. Rend., (2), 19: 812-818 (1886-87). [4.
51 ———, Lomb. Ist. Rend., (2), 20: 275-290 (1887). [4.
52 ———, Rom. Acc. L. Rend., (4), 2_2: 302-306 (1886). [4.
53 ———, Rom. Acc. L. Rend., (4), 2_2: 339-344 (1886). [4.
54 Kanda, A., Monatsh. f. Math., 22: 279-302 (1911). [6.
55 Kantor, S., Theorie Ebenentransf., Berlin, 120 p. (1895). [5.
56 ———, Acta. Math., 19: 115-193 (1895). [5.
57 ———, Ann. di Mat., (2), 10-11: 64-70 (1880-82). [5.
58 ———, J. f. Math., 114: 50-108 (1895). [5.
59 ———, Nap. R. A. Atti., (2), 4: 1-335 (1891). [5.
60 ———, C. R., 100: 42-44, 95-97 (1885). [5.
61 ———, Wien. Ber., 82: 237-259 (1880). [5.
62 Kubota, T., Tôhoku Sci. R., 6: 359-374 (1918). [6.
63 ———, Tôhoku Sci. R., 7: 36-42, 113-121 (1918). [6.
64 ———, Tôhoku Sci. R., 8: 11-16 (1919). [6.
65 ———, Tôhoku Sci. R., (1), 14: 155-64 (1925). [6.
65.1 ———, Tôhoku Sci. R., (1), 15: 671-674 (1926). [6.
66 Le Paige, C., C. R., 99: 537-8 (1884). [4.
67 Liebmann, H., Diss. Jena, 52 p. (1895). [1.
68 Lüroth, J., Univ. Freiberg, 25 p. (1889). [4.
69 ———, Math. Ann., 9: 163-165 (1876). [4.
70 Macaulay, F. S., London M. S. Proc., (1), 23: 495-544 (1895). [4.
71 ———, London M. S. Proc., (1), 29: 673-695 (1897-98). [4.
72 Marletta, G., Pal. Circ. Mat., 17: 173-184 (1903). [6.
73 ———, Pal. Circ. Mat., 17: 371-385 (1903). [6.
74 Morgan, F M., Am. J., 35: 79-104 (1913). [3.
75 Neuberg, J., Math. Qu., (2), 10: 86-87 (1906). [4.
75.1 ———, Brux. Bull., 5: 194-196 (1911). [3.
76 Noether, M., Erl. Ber., 10: 81-86 (1878). [3.
77 ———, Gött. Nachr., p. 267-278 (1871). [3.
78 ———, Math. Ann., 33: 525-545 (1889). [3.
79 Okken, P. A., Diss. Groningen, 92 p. (1913). [1.
79.1 Palatini, F., Saggio di un metodo, Pal., 12 p. (1892). [3.
80 Pannelli, M., Giorn. di Mat., 29: 154-162 (1891). [4.
81 Paolis de, R., Rom. Acc. L. Rend., (4), 2: 339-344 (1886). [4.
82 ———, Rom. Acc. L. Mem., (3), 1_2: 511-544 (1877). [1, 4.
83 ———, Rom. Acc. L. Mem., (3), 2_1: 31-50 (1878). [2, 4.
84 ———, Rom. Acc. L. Mem., (3), 2_2: 851-878 (1878). [2.
85 Retali, V., Bologna Mem., (4), 10: 653-671 (1889). [3.
86 ———, Ann. Ac. Polyt. Porto., 9: 10-20 (1914). [3.
86.1 Reye, T., Geom. der Lage, 3: 225 (1893). [4.
87 Schwarz, A., Wien. Ber., 94: 310-340 (1886). [3.

88 Scott, C. A., Am. M. S. Bull., 7: 24-38 (1900-01). [4.
89 ———, Quart. J., 29: 329-381 (1897-98). [4.
90 ———, Quart. J., 32: 209-240 (1901). [4.
91 ———, Nieuw. Arch., (2), 3: 243-252 (1898). [4.
92 Sharpe, F. R., and Snyder, V., Am. M. S. Trans., 18: 402-414 (1917). [6.
93 Snyder, V., Am. J., 33: 327-336 (1911). [1.
94 ———, Ann. of Math., (2), 13: 140-148 (1911-12). [5.
95 ———, Am. M. S. Bull., 30: 101-124 (1924). [3.
96 ———, Am. M. S. Trans., 11: 371-387 (1910). [3.
97 Steinmetz, C. P., Am. J., 14: 39-66 (1892). [4.
98 Stéphanos, C., Soc. M. F. Bull., (2), 4: 318-28 (1880). [6.
98.1 Stiner, Z. Vierteljschr., 40: 217-229 (1895). [3.
99 Sturm, R., Geom. Verwandt., 4: 232-256 (1909). [1.
100 van der Woude, W. K., Amst. Ak. Wet. Versl., 18_2: 751-759 (1910). [4.
101 ———, Amst. K. Ak. Wet. Versl., 18_2: 842-851 (1909-10). [4.
102 Vanecek, J. S., and Vanecek, M. R., C. R., 99: 742-744, 856-857, 909-911 (1884). [4.
103 Visalli, P., Sopra le diverse classi...Messina, 35 p. (1884). [4.
104 ———, Memoria sulle trasf....Messina, 55 p. (1884). [4.
105 ———, Lomb. Ist. Rend., (2), 28: 114-126 (1895). [4.
106 ———, Pal. Circ. Mat., 3: 165-170 (1889). [6.
107 ———, Rom. Acc. L. Rend., (4), 3_1: 124-127 (1887). [4.
108 de Vries, J., Monatsh. f. Math., 5: 1-16 (1894). [4.
109 ———, Amst. K. Ak. Wet. Versl., (3), 4: 307-326 (1888). [4.
110 ———, Amst. K. Ak. Wet. Versl., 19: 52-62 (1910-11). [4.
111 ———, Amst. K. Ak. Wet. Versl., 22_2: 872-884 (1913-14). [4.
112 ———, Amst. K. Ak. Wet. Versl., 22_2: 1385-88 (1913-14). [4.
113 ———, Amst. K. Ak. Wet. Versl., 23: 84-89 (1914-15). [4.
113.1 ———, Amst. K. Ak. Wet. Versl., 26_2: 1256-63 (1917-18). [4.
114 ———, Amst. K. Ak. Wet. Versl., 30: 5-9 (1921). [6.
114.1 ———, Amst. K. Ak. Wet. Versl., 34: 391-392 (1924-25). [4.
115 ———, Arch. Néer., 23: 355-366 (1889). [4.
116 Wiener, C., Math. Ann., 3: 11-33 (1871). [6.
117 Wiesing, C., Diss. Breslau, 83 p. (1906). [6.
118 Wiman, A., Math. Ann., 48: 195-240 (1897). [5.

CHAPTER VI

INVOLUTIONS ON RATIONAL CURVES

The geometry on a rational curve C_n of order n in S_r, $r \geqq n$ is intimately associated with that of linear involutions $I_m{}^s$ of order m and of dimension s. It is only from that standpoint that the subject will be considered here. See W. F. Meyer(133), Apolarität und rationale Curven, Tübingen, 1883, and the following articles in the Encyklopädie der mathematischen Wissenschaften:

F. Dingeldey(65) Kegelschnitte und Kegelschnittsysteme.

O. Staude(165) Flächen 2. Ordnung und ihre Systeme und Durchdringskurven, Section X.

G. Kohn and G. Loria(96) Specialle ebene algebraische Kurven, Section IX.

G. Loria(122) Spezielle ebene algebraische Kurven von höherer als der vierten Ordnung, Section BI.

Moreover, valuable summaries are found in E. Pascal's(140) Repertorium der höheren Mathematik, second edition, vol. 2, part I (1910), Chap. XII, F. Dingeldey(66) Kegelschnittsysteme; Chapter XIV, L. Berzolari(16) Die Geometrie auf einer ebenen algebraischen Kurve, Section II; Chapter XIX, E. Ciani and H. Wieleitner(38) Projektive Spezialisierung von Kurven vierter und dritter Ordnung, sections 4 and 5; Chapter XXII, H. Wieleitner(238) Metrisch spezialisierte ebene Kurven, sections 1 and 2. And in vol. 22, part 2, (1922) Chapter XXIX. Also, O. Staude(166) Die Raumkurven dritter und vierter Ordnung; Chapter XXXVII, and L. Berzolari(18) Besondere algebraische Raumkurven.

1. *Involutions on C_1.*—Let x define a point on a straight line.

Given two polynomials $f_n(x)=0$, $g_n(x)=0$ and a parameter λ. The equation $f_n(x)+\lambda g_n(x)=0$ defines an *involution* of order n and of dimension 1. Between the values of λ and of x there exists a $(1, n)$ correspondence. The $2(n-1)$ values of λ for which two of the associated x coincide are called *limiting points* or *branch points;* the associated double points x are called *coincidences.* Some of the properties of (1, 2) correspondences between C_1, C'_1 were given by M. Chasles(34).

I_4 on S_1 cannot be expressed as a square by any rational transformation. P. Th. Pepin(139).

Two involutions of order $n>2$ on the same line do not have a common group unless relations exist among the coefficients in the defining equations.

The curves of a pencil of curves of order n meet an arbitrary line in the groups of an involution.

Two involutions I_n, I'_n are said to be projective when λ, λ' are projective. In two projective involutions there are $n+n'$ groups, each of which has a point in common with its image in the other. E. de Jonquières (**88**) gives numerous applications. Given I_3, I_2 projective, to construct $\phi=0$, the locus of their common elements. L. Cremona (**51**) summarizes results then known. P. Cassani (**29**) generalized the problem of Desargues, by determining the locus of the double points of an I_2 on a line defined by a pencil of conics, as the line describes a pencil. See also E. Waelsch (**183**).

2. *Involutions on a Conic.*—By means of the parametric representation the same properties of I_n can be interpreted on a conic or on any rational curve. Thus, Em. Weyr (**196**) proved that if a complete n-side is circumscribed about a conic and inscribed in a curve of order $n-1$, then there are ∞^1 such n-sides describing a rational I_n. Such an involution is fixed when one group and $n-1$ further pairs of points belonging to the same group g_n are given. The I_3 of osculating circles of an ellipse was found by V. Janni (**83**). The triangle formed by each group has constant area.

Em. Weyr (**225**) proved that given an I_n on a C_2, lines joining pairs of associated points envelop a curve of class $n-1$, called the *involution curve.* For application to (2, 2) correspondences see Em. Weyr (**189**). For particular (m, n) correspondences see M. Chasles (**35**).

Em. Weyr (**192**), (**193**) proved the theorem: Given two C_2 having double contact, an infinite number of triangles can be inscribed in one, circumscribed about the other, forming I_3^1 with points of contact for triple elements. Equilateral triangles and a pair of concentric circles furnish an example.

R. de Paolis (**138**) considered two involutions I_m, I_n on the same C_1 and showed that without knowing the double points of either an I_{m+n}^{m+n-1} can be constructed such that the $m+n$ elements of multiplicity $m+n$ are the double points of the given projective involutions.

Em. Weyr (**229**) studied an I_2 on a parabola by elementary methods. He also (**190**) proved that if I_n^1 has two n-fold points, its elements form projective series with the n-fold points for fixed elements.

Em. Weyr (**226**) gives the two following elementary applications. Given a point P on a conic. Inscribe the right-angled triangles in C_2, right angle at P. The hypotenuse always passes through a fixed point on normal to C_2 at P.

Construct triangles with one vertex at P, such that the normal bisects the angle at P. The base passes through a fixed point on the tangent to C_2 at P.

Em. Weyr(227) contains a summary of known results. Given a pencil of conics. They describe on a fixed C_2 through a basis point an I_3^1. By considering composite conics of the pencil, the I_3^1 is constructed from the two triads.

C. Le Paige(101) defines an *evolution* as follows: Given the pairs aa', bb', cc' of an I_2, the harmonic conjugate of a as to b', c', etc., forms an *evolution*, with the other five points a', b, b', c, c'.

P. Serret(159) gives constructions for involutions of order 2, then generalizes to find the tangents to a rational curve of class n.

C. Le Paige(109) expresses the double points and the branch points of an I_3^1 in terms of the Clebsch-Aronhold notation; the latter being the Hessian of the Jacobian of the two defining cubic forms.

C. Le Paige(108) showed that the branch points of two cubic involutions with common double elements together are three tetrads of an I_4. Had been touched upon by Em. Weyr(205) and (191).

Em. Weyr(219) showed that if two cubically involutorial conics, that is, each of which is the involution conic of an I_3^1 on the other, one involution of points, the other of tangents, touch each other, their radii of curvature are as $1:4$, the conic having the point involution having the larger curvature. If two conics are in double contact and their curvatures are as $1:4$ at one of them, then the same ratio of their curvatures exists at the other.

C. Le Paige(110) gives a construction of the problem: Given three triads of I_3^1 on a C_2, and two points, find the third point of the group determined by them. See Em. Weyr(210) who had already solved the problem for I_n^1. The conics through three fixed points determine an I_3^2 on a conic through one of them. Suggestions for further constructions are given but no complete solution is provided. The method is analytic; the Clebsch-Aronhold notation is used.

C. Le Paige(111) considers *conjugate* involutions. Given an I_3^1, the groups common to two certain I_3^2 derived from it form a conjugate I_3^1. As an application, given three triangles inscribed in a conic; they determine three inscribed conics. These conics have a common tangent.

Em. Weyr(220) considers an I_4^1 of tangents to a fixed C_2. The vertices of the quadrilaterals generate a general cubic curve J_3 as involution curve. The three pairs of opposite vertices are conjugate points in the three non-central involutions on J_3. The six vertices of any tetrad and the vertices of any triangle circumscribed to C_2 are basis points of a pencil of cubics. An I_4^1 is completely and uniquely defined by three triple elements.

Weill([186]) gives further interpretations of I_3^1. Given two conics referred to their common self-conjugate triangle, one C_2 with one variable coefficient. A triangle inscribed in the first, circumscribed to the second defines the I_3. The locus of the Brianchon point is another C_2. Various further properties are mentioned with an outline of possible generalizations.

C. Servais([160]). The branch points of two conjugate I_3, associated with the same double point, are harmonic conjugates of the double elements of the cyclic projectivity defined by the other three double elements.

Em. Weyr([213]) considers an I_n of tangents on C_2. It is determined uniquely by two complete groups, but a finite number of different I_n^1 may have $2(n-1)$ common pairs. A general C_4 is not an involution curve (as a pentagon cannot be inscribed). If one can, ∞^1 can, and the sides of a pentagon are always tangent to a fixed C_2. If C_4 has a node, and the nodal tangents make with the lines joining the node to the points of contact of any bitangents a harmonic set, C_4 is an involution curve. Similarly, C_4 is an involution curve if the tangents at three collinear points form an inscribed triangle.

Em. Weyr([221]) discusses various degenerations of I_4^1 on C_2. If I_4^1 is defined by one tetrad and two double elements, the involution cubic J_3 is nodal. A cuspidal cubic cannot be an involution curve of any I_4^1 of tangents to C_2.

E. Waelsch([184]) discussed the I_4 associated with polar groups on C_2.

Properties of C_2 expressed by means of invariants and covariants of binary forms were given by G. Pittarelli([143]).

If I_4^1 has two pairs of double elements, J_3 is composite. If two simple circumscribing quadrilaterals to C_2 have a common diagonal, their eight vertices lie on a conic C''_2, in I_4 relation to C_2. If I_4^1 contains three groups of only double elements, J_3 consists of three lines, sides of a self-polar triangle as to C_2. It is the diagonal triangle of every tetrad of I_4^1. The seven systems of contact conics with a given rational C_4 are derived.

3. *General Properties of I_n^1, Independent of Carrier.*—de la Gournerie([79]) gives a special I_4^1 defined as follows: Given four points on a line. They determine three quadratic involutions, each of which has a center. These three centers are determined for every group of a quartic involution I_4. These points together with the point at infinity may make a group of an involution, the same for all. One other group determines the involution. Three groups are composed of the two double points of the three quadratic involutions.

Given three collinear points O', O'', O'''. Construct the segment $e'f'$ with its center at O', and harmonic as to O'', O'''. Similarly for $e''f''$,

$e'''f'''$. Take three quadratic involutions with $e^{(i)}f^{(i)}$ as double points. Take any point x, get its conjugate as to each group of the special I_4. Application is made to the spiral lines on the torus and to the solution of the quartic equation.

Em. Weyr(190) proves that if I_n^1 has two n-fold elements, its elements form projective series with the n-fold elements as double elements. Particular applications by Em. Weyr(192), (193) and (195).

It was shown by G. Kohn(91) that the theory of the harmonic center can be founded on that of involutions on S_1. A point m is the harmonic center of degree r as to a point system $a_1 \ldots a_n$ with O as pole, if the a_i all belong to an I_n having O as r-fold point and m is a point of multiplicity $n-r+1$. Then O is harmonic center of the a_i of degree $n-r$ with m as pole. This principle was employed to construct the satellite curve of O as to a plane curve C_n. Draw a C_1 through O meeting C_n in $|a|$. Construct the I_n having $|a|$ for one group, and O^n for a second. Construct the double points $m_1 \ldots m_{n-1}$ and their harmonic centers. As C_1 describes the pencil O, these points describe a curve, one component of which is the first polar of O as to C_n; the other is the $C_{(n-1)(n-2)}$, satellite of O as to C_n. The theory is developed further by G. Kohn(92).

G. Maisano(126) considers a particular I_4^1 with three groups having a triple element. The three simple elements define the cubic covariant of the cubic form given by the three triple elements.

G. Maisano(127). An I_n^1 has $2(n-1)$ double elements arranged in $n-1$ pairs, each of which belongs to a group of I_n. Applications to I_4^1.

C. Le Paige(105) gives various additional theorems on self-conjugate involutions. The determination of the branch points of I_3 when the coincidences are known.

C. Servais(160) provides alternate proofs for the properties of conjugate I_3^1. See C. Le Paige(104) and (112); Em. Weyr(214). The four double elements of a self-conjugate I_3^1 are equianharmonic. Another involution is defined by interchanging branch points and double points.

J. de Vries(174) gives an algebraic proof and a construction for completing I_n defined in various ways.

W. F. Osgood(135) gives an alternate geometric proof of the (1, 1) nature of the correspondence between points on the curve and values of the parameter. An illustrative example is given. See also J. Cardinaal(27).

K. Bartel(7) discusses possible constructions by means of pencils of lines in involution. Two I_n^1 generate C_n with P_n at each vertex. More details for $n=2$. K. Bartel(8) continues the same study for $n=4$. No new results.

Em. Weyr([204]). Four double elements d_i of $I_3{}^1$ form with branch elements v_i pairs d_iv_i. Any d_i is a double element in the projectivity formed by the other three pairs.

4. *Plane Rational Cubic Curves.*—H. Durège([70]) discussed the problem of closure in the following form: Given a point A on a plane rational cubic, find the tangential of A, etc. Parametric form so chosen that the sum of the parameters is zero for collinear points on cuspidal curves, while for nodal cubics, the sum equals the product. Applied to inscribed and circumscribed polygons. Compare A. Clebsch([39]).

In ([198]) the same author (Weyr) constructs a pencil of cubics with the node and five simple basis points given. The nodal tangents form an I_2, hence there are two cuspidal C_3 in the pencil. In the same paper is found an outline of constructing $I_n{}^1$ when two complete groups are given. In ([205]) a pencil of C_3 with fixed nodal tangents is given; the lines from the node to the inflexions describe an $I_3{}^1$ having the fixed nodal tangents for triple elements. A number of known results are proved by a particular parametric representation.

Em. Weyr([191]) studies the nodal cubic parametrically. By choosing 0, ∞ for the values of the parameter at the node three points are collinear when the product of the parameters is constant. Let A, B be two points on C_3. Start with a point P_1 and find P_2 on C_3 such that P_1AP_2 are collinear; P_2BP_3 collinear P_3AP_4, etc. If P_{2n+1} is identical with P_1, the figure $P_1P_2\ldots P_{2n}$ is an inscribed Steiner polygon. The necessary and sufficient condition that a Steiner polygon of $2n$ sides is associated with two points t_1, t_2 is that $t_1{}^n = t_2{}^n$. Elementary properties are given by Em. Weyr([230]). Particular cases were discussed by K. Zahradnik([247]), ([250]), ([248]) and ([245]).

Em. Weyr([210]) maps a nodal C_3 on C_2. Project C_3 from the node upon a conic through the node perspectively. Cotangential points are mapped on an I_2 with images of the node for double points. The C_3 is incidentally mapped on the line joining the images of the node. Parametric representation is given. Another mapping by means of cross-ratio is discussed.

Em. Weyr([212]) studies the systems of contact C_2. A large number of known theorems are proved by other methods. Every contact C_2 of C_3 is an involution C_2 of a definite $I_3{}^1$.

Em. Weyr([215]) considers the system of conics in the triangle of inflexional tangents and touching an arbitrary line. Each has two additional tangents common with C_3, hence there is a (1, 2) correspondence between the conics and the tangents.

Since the fourth basis line is arbitrary, each line of the plane is in (1, 1) correspondence with an involution of tangents on C_3. The discussion is partly analytic.

M. Treibscher ([169]) constructs a curve of class 3 by means of a pencil of C_2 and an I_2 on the line at infinity. The envelopes of the axes and of the asymptotes are determined. Every curve of class 3 having the line at infinity for double tangent can be generated in this way, as the envelope of the asymptotes of a pencil of conics.

H. Drasch ([69]) generates a nodal cubic by a projective pencil of class 2, and one of class 1. Dual in space treated by H. Schroeter ([157]) ; numerous known properties are again proved by this method.

G. Pittarelli ([144]) discusses the invariant form of the parametric representation of the rational cubic curve.

G. Zecca ([253]) replaces each point on a rational plane C_3 by its first tangential and obtains the condition for closure after k applications. The same process is applied to a nodal space C_4, transforming each point into the residual intersection with the osculating plane.

G. Loria ([121]) showed that a nodal cubic has a continuous system of $2n$-gons for n odd, and none for n even. There are none on the cuspidal cubic.

J. de Vries ([171]) and ([172]) gives other methods of generating plane nodal cubic curves.

K. Zahradnik ([252]) showed that various curves can be generated as cissoids, including the right circular cissoid, Sluse's conchoid, the right strophoid, the cissoid of Diocles, the trisectrix, and others.

J. E. Wright ([243]) constructs the nodal cubics through eight given points and discusses their reality. If all basis points are real, not more than two acnodal cubics are in the pencil. In general $R-1=C-A$. R=number real basis points, C=number crunodal cubics, A number acnodal cubics.

K. Zahradnik ([246]) gives a general method of constructing nodal cubics. The method is a combination of inversion and duality.

L. Crelier ([47]) develops a synthetic study of the locus of intersection of corresponding lines of two pencils in (1, 2) correspondence.

E. Cavalli ([33]) gives another method of generating rational plane cubics. W. van de Woude ([242]) discusses quadrilaterals circumscribed about a given conic and inscribed in a cubic. A number of dual interpretations are included in the memoir of C. Servais ([161]) which discusses the plane quartic curve of class 3.

Involutions on C_3, $p=0$ are treated parametrically by R. M. Winger ([240]). The idea of the satellite line of the plane rational cubic is developed by the same author([239]), with an outline of its application to non-singular cubics.

R. M. Winger([241]) obtains further relations among the intersections of a general C_n and a rational C_3.

5. *Space Cubic Curves, and Various Plane Mappings.*—L. Cremona([50]) studies C_3 by means of two lemmas: "if K_2 is the polar conic of a point θ as to a trilateral, vertices *abc*, each tangent to K_2 from θ forms with the lines θa, θb, θc an equianharmonic system" and "two projective pencils, one of lines, one of pairs of conjugate lines of an involution, with common vertex θ, having $\theta\omega_1$, $\theta\omega_2$ for self-conjugate rays and θa, θb, θc common to the two systems, if each of first gives with the second an equianharmonic system, then $\theta\omega_1$, $\theta\omega_2$ in the second correspond to $\theta\omega_2$, $\theta\omega_1$ in the first." They are applied to the section of C_3 by the planes of a pencil, forming an I_3. Through any point of space pass two quadric cones through C_3, and any line touches four such cones.

Let σ be the pole of the section of one of these cones by a given plane, and call the vertex of the cone S. The line $S\sigma$ describes a quadric H. This representation he calls hyperbolic projection of C_3 on the plane.

L. Cremona([53]). Given a space C_3, developable $\Sigma_3{}^4$, any osculating plane of C_3 meets Σ in a C_2 and the generator counted twice. Through any point of the plane pass two other osculating planes, intersecting in A, in general not in the given plane π. Given another osculating plane π_1 and the associated conic $C_2{}^1$, generator P^1, and line A^1. The line A generates a ruled surface as the point (π, A^1) describes a curve. If the given curve is a conic, the ruled surface is of order 4 having C_3 for double curve. Particular cases when C_3 is a cubical parabola are discussed.

Em. Weyr([194]). The system of quadrics through C_2 and three coplanar points of a given space C_3 form a pencil meeting C_3 in $I_3{}^1$. The other basis C_2 of $|H|$ is in the plane of the given triad. The quadrics of the pencil touching C_3 and the 16 having contact of second order are determined. Various particular positions of the basis elements are considered.

R. Sturm([167]), by use of two apolar binary forms, obtains the equation of the linear complex to which C_3 belongs, and derives many of its properties. Continuation of Th. Reye([146]) and ([147]), where the word "apolar" was first used. J. Rosanes([149]) uses "conjugate" instead of "apolar."

L. Cremona([49]) first gave the parametric form

$$x_1 : x_2 : x_3 : x_4 = t^3 : t^2 : t : 1.$$

See also Joachimsthal([84]).

E. d'Ovidio([136]) applies the Clebsch-Aronhold notation and the theory of binary invariants to various involutions on space cubic curves. No new results were obtained nor in ([142]) by G. Pittarelli. Continued by E. d'Ovidio([137]) and by F. Gerbaldi([74]) in which the methods of E. d'Ovidio([137]) are applied to a system of two projectively related space cubics.

C. Le Paige([111]) shows that if C_3 is met by any three planes of an $I_3{}^2$, and the bisecant to C_3 drawn through their common point, its points on C_3 belong to the three simple $I_3{}^1$ determined by the three triads.

Em. Weyr([223]). The I_3 defined by a pencil of planes has its double points at the four points of contact of the tangent planes through the line, axis of the pencil. The lines joining conjugate points meet the axis in points of the I_3. Conversely, every $I_3{}^1$ on C_3 is cut by the planes of a pencil. The dual interpretation allows the null-system to which C_3 belongs to be developed. Weyr shows in a new way that the locus of the vertex of K_2 containing six basis points of C_3 is the Weddle surface, which is invariant (but not point for point) under the cubic involution defined by the bisecants of C_3.

Em. Weyr([224]) considers an $I_n{}^1$ on C_3. The involution (ruled) surface is of order $2(n-1)$, having C_3 to multiplicity $n-1$, and the planes connecting points by threes define the involution curve of order $\frac{(n-1)(n-2)}{2}$. The particular case of $n=4$ is studied, including a number of special cases.

C. Le Paige([113]) showed that if three planes of a tetrahedron turn about a triad of a cyclic I_3 on a space C_3, the fourth vertex describes a quadric surface.

A. Cantone([26]) reproduces by synthetic methods known theorems on harmonic division of bisecants of a space cubic.

E. Caspary([28]) introduces a new symbolic notation to express the condition that points on C_3 are collinear, coplanar, etc. A new proof of the existence of the Weddle surface is given.

A. C. Dixon([67]) gives a synthetic and algebraic proof of known elementary theorems, followed by consideration of special cases, showing the connection with division of elliptic functions. The only involution employed is an I_2. In A. C. Dixon([68]) the only transformation employed is point-plane duality.

E. Waelsch([184]) interprets binary quintic forms on C_3. Given an $I_4{}^1$ of points. The tangents at the points of each tetrad have two transversals, conjugate in the linear complex to which C_3 belongs. As the tetrad describes C_3, the transversals generate a quadric surface. When the transversals coincide, the tetrad is equianharmonic. The polar triads of groups of $I_4{}^1$ form an $I_3{}^1$ which cut C_3 by planes of a pencil.

G. Kohn(95) shows that cubics having contact with the developable of a given cubic in one or more points can be found by means of certain (2, 2) correspondences.

G. Gallucci(73) arranges the points of C_3 on ∞^2 orthocentric tetrahedra.

J. Fairson(71) gives a geometric interpretation of the invariants of binary forms, applied to the space C_3.

S. Jolles(87) presents an alternative synthetic proof of the harmonic pole of a line as to the triangle formed by points of a plane as to C_3; it is the null-point of the plane with respect to the complex determined by C_3.

J. de Vries(180) employs the cubic involutorial transformation to systems of cubic curves, images of lines of a congruence.

E. K. Wakeford(185). Given six mutually skew lines in S_3. There are six C_3 having all as bisecants, and one $C_3{}^1$ having them as axes of two osculating planes. There are ∞^1 tetrahedra inscribed in C_3, circumscribed to $C_3{}^1$. If three C_3 have six common bisecants, they have seven. If seven lines are common bisecants of two cubics, there exists a third having all seven for bisecants.

6. *Plane Rational Quartics.*—From Em. Weyr(197). Given a lemniscate C_4 with P_2 at O. From any point X on C_4 can be drawn four tangents. The points of contact are equianharmonic on $C_1 \perp OX$, and distant $\dfrac{OX}{2}$ from O. The envelope of C_1 is an equilateral hyperbola, similar to that of which C_4 is the pedal. Through each point Y pass three osculating circles, points of osculation on $C_1 \perp OY$. This defines $I_3{}^1$ with two triple elements at O. The osculating circles at four con-cyclic points have con-cyclic residual intersections.

Any triad of parallel tangents to the cardioid meet the double tangent in three points which are projected from M, one-fourth distance from the cusp to the point of intersection of the curve and its cuspidal tangent, by lines making angles of 60° with each other. Pencil in $I_3{}^1$. Em. Weyr(195).

Given a plane C_4 with three nodes, all the nodal tangents inflexional. If a line through a node D meets C_4 in A and B, and the line joining the other nodes in D', these four points $DABD'$ are harmonic. From D' can be drawn six tangents, points of contact of which lie on three lines through D describing an $I_3{}^1$ with tangents at D as triple elements. The theorem was applied to the lemniscate. Em. Weyr(200). Compare Em. Weyr(233) and K. Zahradnik(249) and (251) for applications to the cardioid.

Em. Dewulf and P. H. Schoute in (64) determine the number of rational quartics through two given P_2 and $7P_1$. They consider a pencil of cubics and invert by quadratic inversion into quartics—hence obtain 12 solutions.

C. Bobek(19) gives the generation of plane rational C_4 by means of projective pencils of conics through three common basis points. The treatment is synthetic. He considers the case in which all nodal tangents are inflexional. A table of fifty forms is provided.

Any rational C_4 has $3I_2^1$, $\infty^2 I_4^1$, $\infty^1 I_3^1$. Curve mapped on C_2, equivalent to quadratic inversion through the nodes. Em. Weyr(217).

A. Ameseder(4) shows that the trinodal quartic curve is determined by five conditions besides the positions of the nodes.

C. Le Paige(107) employing a method given by C. Le Paige(106) constructed a C_4 when fourteen points were given.

P. H. Schoute(153). If the nodal tangents to a trinodal C_4 are inflexional, the points of contact of tangents from any point of a curve are collinear. The envelope is a C_4 having nodes at the vertices of a self polar triangle. Same paper in substance in (155). See also Em. Weyr(236), and E. Laguerre(99) and (98).

C. Le Paige(115). Given two perspective triangles. Keep one fixed; as center describes circle, axis envelopes a rational curve, class 4.

See also P. H. Schoute(154) and (151) and (152); G. Kohn(94).

M. Lazarski(100) constructs the C_4 with a triple point by means of a (1, 3) line correspondence.

W. Gross(80) showed that the C_2 through six points of inflexion of a rational C_4, the C_2 through the points of contact of four double tangents, and the C_2 touched by the six inflexional tangents all belong to the same pencil. The method was that of simultaneous invariants of binary forms.

G. Kohn(93) gives another proof of the projective property of the lemniscate, and states that the first proof is due to Em. Weyr(231). Given C_4 with three cusps, cuspidal tangents meeting at P. Any line through P meets C_4 in the four points equianharmonic as to C_1 and as to C_4. Dual theorem. By quadratic inversion, consider a triangle and an inscribed C_2, P as Brianchon point. Any C_2 circumscribing triangle and passing through P meets first conic in points equianharmonic on both.

J. de Vries(173) studies involutions on C_4 with a triple point; also (176).

A. Adler(3) proves that the contour of a cylindroid on a plane normal to the double directrix is the three-cusped hypocydoid.

G. Majcen(125) discusses a particular quartic.

J. N. Rice(148) obtains further properties of inscribed and circumscribed triangles.

7. *Rational C_4 in S_3.*—L. Cremona(48) gives an exhaustive synthetic study of the general case, obtaining many results found by others later. For early history, compare Repertorium, vol. II 2, pp. 946-952(18).

Em. Weyr([199]) employed parametric representation in an invariant form: he gives the condition that four points shall be coplanar, three collinear points forming an I_3^1. Through any point of C_4 passes a plane meeting it in three other points, the osculating planes at which pass through the first point, forming an I_3^1.

Em. Weyr([192]) shows that the curve $x_1=t^4$, $x_2=t^3$, $x_3=t$, $x_4=1$ has two linear inflexions. Four points are coplanar if $\Sigma t_i t_k=0$. The I_3^1 formed by the trisecants has the inflexions for triple elements. E. Bertini([10]) proves that the general C_4 has three bisecants which form the intersection of the osculating planes at their points on C_4, and they meet in a point. A new I_4^1 is determined as follows: Given a point P, pass a plane through P and each of the axes of osculating planes; each determines one point on C_4. These points with P form a group of I_4^1. If C_4 has two linear inflexions, it has ∞^1 axes of osculating planes. If the four stationary points are equianharmonic, so are the residual intersections of the four tangent-secants. In this case C_4 is the intersection of a ruled cubic and a polar quadric. A. Armenante([5]) and ([6]) reproduces the substance of L. Cremona([48]) in algebraic form, by use of the Clebsch-Aronhold method. He also gives an alternate proof that every C_4 has three concurrent axes of osculating planes, E. Bertini([10]). He also proves that the points of a Steiner surface and the bisecants of C_4 are in (1, 1) correspondence, from which many properties of the surface are derived.

Em. Weyr([233]) derives properties of the I_3^1 of trisecants; continuation in ([225]) and ([198]). He also([201]) treats the cuspidal C_4 parametrically. By letting $t=\infty$ represent the cusp and $t=0$ the point of contact of the stationary plane, four points t_i are coplanar when $\Sigma t_i=0$. The points of contact of the double tangent planes are harmonic as to the cusp and stationary point. The lines joining these all pass through the vertex of a second quadric cone containing C_4. The residual points of intersection with C_4 of the osculating planes of four coplanar points are also coplanar.

The general C_4, $p=0$ is mapped on a conic by Em. Weyr([202]) and ([203]). The C_4 lies on a quadric surface F_2, meeting the plane of the given C_2 in a conic $\overline{C}_2$, and C_2 itself in four points, through each of which passes a trisecant t of C_4. From a point on t, project C_4 on the plane of C_2 into $\overline{C}_4$ with P_3 on C_2, hence C_2, $\overline{C}_4$ are perspective. The system (t) gives rise to I_3 on C_2. Many properties of C_4 can be obtained by this process. If the osculating planes of one triad t, C_4 meet on C_4, the same property holds for every triad. The results obtained synthetically are then compared with those obtained analytically by E. Bertini([10]), with frequent reference to Em. Weyr([199]) who gives a new proof of theorem that every C_4, $p=0$ has three concurrent axes of osculating planes. If C_4 is projected from this point upon a plane, the trinodal quartic has inflexional tangents for all its nodal tangents.

Em. Weyr([206]) continues the parametric representation to map a nodal C_4 on C_2. Four points t_i are coplanar if $t_1t_2t_3t_4=K$, the node being 0 and ∞. The osculating planes to C_4 from a point on it form a group of $I_3{}^1$, as the point describes the curve. The node represents two cubic elements.

All the $I_2{}^1$ on a general C_4 gives rise to all the cubic ruled surfaces on which C_4 lies, having a bisecant for double directrix. Em. Weyr([207]).

The cuspidal C_4 is mapped on a conic by Em. Weyr([208]) by methods analogous to those used by the same author in ([201]), ([203]). Method of $I_3{}^1$ determined by trisecants does not apply if C_4 has a double point or a cusp. The method here employed is applied to C_4 with a double point in ([209]) being a direct continuation of Em. Weyr([206]). The fourth point on C_4 in a plane through three given points on it is determined, including the residual intersection of an osculating plane.

If A_1, A_2, A_3, A_4 are the points of contact of the four stationary planes of a general non-singular C_4, then the points of contact x_1, x_2, x_3, of the three osculating planes from any point y on C_4 are the three harmonic centers of y as to the tetrad $A_1A_2A_3A_4$. Em. Weyr([218]) continuing ([216]).

The $I_4{}^1$ discussed by C. Le Paige([109]) now breaks up into three I_2. The covariant T (of Le Paige) gives the three pairs of points on the axes of osculating planes.

A. Adler([2]) employs similar methods to study C_4 with one or with two linear inflexions. When C_4 has one, there exist ∞^1 axes of osculating planes, and when C_4 has two, the osculating planes on each trisecant belong to a pencil. The curve now belongs to a linear complex. A. Adler([2]). A. Adler([1]) gives new proofs of several well known theorems, including that of the existence of three concurrent axes of osculating planes. The C_2 on which C_4 is mapped and the involution conic of the $I_3{}^1$, image of the trisecants, are polar reciprocal as to a third conic. The three concurrent axes are double lines of ruled cubic surfaces on which C_4 lies; the simple directrices complete a tetrahedron. C_4 is invariant under the three axial involutions determined by opposite edges of the tetrahedron.

St. Jolles([85]) defines osculants and discusses congruences of bisecants of a given C_4.

A. Brambilla([23]) associates the lines r of a cubic surface with the conics of a quadric whose planes contain r. The pole of r as to the conic describes a cubic.

A. Brambilla [22] gives a particular case of a problem in closure, of polygons inscribed in a space nodal quartic.

In [24] A. Brambilla applies the I_3^1 to the generators of a cubic ruled surface on which the quartic curves lie.

G. Marletta [128] reproduced proofs of all known general results as to a C_4 of $p=0$, and many particular ones, by projecting the normal C_4 of S_4 into S_3. His method is synthetic.

G. Frauenfelder [72] considers a pencil of C_4 with two common linear inflexions and inflexional tangents. He shows that they are asymptotic lines on a ruled cubic surface having the line joining the inflexions for double directrix. Every generator meets C_4 in a pair of points harmonic as to the points of inflexion, which are pinch points on the surface. See L. Cremona [52], G. F. Halphen [81], V. Snyder [163].

S. Jolles [86] gives a collection of Weyr's and various related methods of generating rational space C_4. The dissertation of Schrek [158] uses Weyr's method and related methods to present a general discussion of the space C_4. No new results are given. Polygons connected with C_4, $p=0$ in S_3 were also considered by P. Locchi [119].

8. *Plane Rational Curves of Order Higher Than 4.*—For the general theory see article by G. Loria [122]. Em. Weyr [228] employs involutions in the plane to prove the theorem: Through B arbitrary points in the plane and γ further points on a rational C_n can be passed $(6-B-\gamma)(2n+B+\gamma-5)$ conics having $6-B-\gamma$ consecutive points in common.

By means of I_n of points on a line he [234] generates any C_n with an $(n-1)$-fold point linearly.

He also [232] considers (n, n) symmetric correspondence, more general than $I^1{}_{n+1}$, with numerous particular cases. He gives a proof that C_4 of S_3 has three concurrent axes of osculating planes.

J. Lüroth [123] proved the important theorem that if the points of a curve can be expressed rationally in terms of a parameter, then they can also be put into (1, 1) correspondence with a rational parameter. A rational curve cannot contain irrational involutions.

G. Garbieri [78] develops a general method of applying binary forms to express involutions on plane and space rational curves.

M. Genty [75] starts with the parametric representation $x_i=f_i(t)$ and derives all the ordinary known properties. No new results are obtained.

Em. Weyr [235] employs the principles of enumerative geometry to confirm a number of known results.

P. H. Schoute [156] gives a new derivation of quadratic inversion, by successive application of which any rational curve may be constructed.

E. de Jonquières([89]) constructs a plane C_5 with six given double points and two simple points by means of two projective pencils of cubics, with an extraneous line. Fundamental involutions on the rational plane C_5 are discussed by J. de Vries([179]). In ([178]) J. de Vries considers I_s^1 on a rational plane C_n, and considers the systems of conics through five associated points, $s>5$. In this system of conics $4(n-2)(2n-1)(s-1)$ touch the given C_n.

Another proof of Lüroth's theorem was given by E. Netto([134]).

J. E. Wright([244]) gives an alternate proof that C_n in S_2 has

$$\tfrac{1}{2}(n-1)(n-2)P_2,$$

if $x_i=f_{n,i}(t)$, where $f_{n,i}$ is an integral polynomial.

J. R. Conner([45]) constructs three linearly independent apolar binary quintics to those defining C_5 in S_2; they define a *conjugate* quintic curve. The I_5^2 is discussed. The invariant theory of rational C_5 in S_2 is identical with the covariant theory of six lines in a plane.

J. I. Tracy([168]) shows that on a quintic curve with a four-fold point, there exist three quadratic involutions determined by pairing the tangents at the multiple point. The involution conic of each is discussed.

J. R. Conner([46]) proves that a rational C_m uniquely determines and is uniquely determined by a single binary form of order $3(m-2)$, except as to projections. Any involution I_n^1 may be replaced by a single binary form of order $(r+1)(n-r)$.

A. Plamitzer([145]) studies two projective involutions, I_m^1 on a rational C_μ, I_n^1 on a rational C_ν. Lines joining associated points envelop a curve of class $m\nu+n\mu$ and of genus $(m-1)(n-1)$. Application to particular case $\mu=\nu$, $C_\mu=C_\nu$, and to ruled surfaces of order $m\nu+n\mu$.

A. B. Coble([42]) establishes relations between the ten nodes of a rational C_6 and those of the quartic surface known as Cayley's symmetroid. Two curves are equivalent under Cremona transformations only when their double points are congruent as a whole. The same property is shown to exist on the symmetroid. See also H. P. Hudson([82]) who showed that C_6 with 10 double points cannot in general be reduced to a curve of lower order by Cremona transformations. Further properties given by J. R. Conner([43]).

9. *Rational Curves C_μ, ($\mu>4$) in S_k, $k>2$.*—Em. Weyr([187]) proves the theorem: Through q points in S_3, and b points on a rational C_n in S_3, there are $k(2n-k-b+1)$ quadrics of a pencil which have k point contact with C_n, wherein $b+q+k=10$.

J. Rosanes([149]) shows by means of conjugate binary polynomials that every C_n can be expressed by the sum of $\frac{n(n-1)}{2}$ linear expressions, each raised to nth power if $n>3$, and to four if $n=3$.

G. Korndörfer([97]) gives the parametric representation of rational curves in S_3 by means of abelian integrals.

G. Castelnuovo([31]) makes a systematic study of involutions by means of hyperplanar sections of the normal C_n in S_n. All earlier theorems are derived from this point of view. Also G. A. Bordiga([20]) for $n=4$.

There are $\binom{n-m}{m+1}$ spaces S_m which meet a normal rational C_n of S_n in $m+1$ points, each meeting an S_{n-2m-1} of $S_n\left(m<\frac{n-1}{2}\right)$ in one point; through $i+1$ points of C_n pass $\binom{n-i-m}{m}$ S_{m+1} meeting C_n in $m+i+1$ points and meeting an S_{n-2m-i} of S_n in one point. The S_i having $i+1$ consecutive points on C_n envelope a variety V_{i+1} of order $(i+1)(n-i)$.

The tangents to C_n meet a fixed osculating S_{n-1} of C_n in points of C_{n-1} projective with given C_n.

The curve is self-dual. Each point referred to its own osculating S_{n-i} defines an involution on C_n.

The S_{n-1} through S_{n-k-1} as basis meet C_n in groups of I_n^k. The vertex S_{n-k-1} is called the *central space* of I_n^k. I_n^k is determined by $k+1$ groups of n elements. Each group is determined by any k points belonging to it. If two normal curves C_n of S_n, C_m of S_m are projectively equivalent, to I_m^k on C_m corresponds an I_n^k on C_n. If to each group of I_m^k there be added $n-m$ fixed points, I_n^k on C_n results and conversely. If in I_n^k ρ elements be fixed, $I_{n-\rho}^{k-\rho}$ is called the *adjoint* involution of the fixed points. In I_n^k are $(k+1)(n-k)$ elements of order $k+1$. r points constitute a *neutral* group of r points, if $k-r+1$ other points are necessary to fix a group of I_n^k.

$k-2$ arbitrary elements of I_m^k appear in $\frac{(n-k+1)(n-k)}{2}$ neutral groups of k elements.

The $n-r$ points, which with r neutral points make a group of I_n^k constitute an I_{n-r}^{k-r+1}.

If ρ points are such that $k-r+\rho$ fix a group of I_n^k, they constitute a group called the *neutral of species* ρ. If $r+s=n$, I_n^r, I_n^s have in common I_n^{r+s-n}. If $r+s+1<n$, they are both contained in I_n^{r+s+1}.

In I_n^{n-1} the R_{n-1} containing an n-fold element is called a *principal group*. If n is odd, the n groups, each consisting of an n-fold element, constitute a group of I_n^{n-1}. Let S_{n-k-1} be the central space of I_n^k; if in the points of a group the osculating S_{n-1} be constructed, they will have S_k in common, called the *axial space* of I_n^k. It is conjugate to S_{n-k-1} in the polar system defined by C_n. I_n^k, I_n^{n-k-1} are called *conjugate involutions.*

If I_n^2 contains I_n^k, the conjugate of the former is contained in the conjugate of the latter.

The $(k+1)$-fold elements of I_n^k coincide with the $(n-k)$-fold elements of its conjugate I_n^{n-k-1}.

If two conjugate involutions I_n^k, I_n^{n-k-1} $\left(k< \frac{n-1}{2}\right)$ are both in I_n^{n-k}, then the first is contained in the second.

There are $m+1$ I_{2m+1}^{2m-1} which contain four given neutral groups each of $m+1$ elements, of the first species.

G. Loria(120) develops a general parametric representation of the rational C_n in S_{n-1}, and discusses the groups of linear transformations under which they are invariant. The generalization from C_4 in S_3 is pointed out.

A. Brill(25) discusses fundamental involutions I_n^{n-p-1} on C_n of S_p, further developed by W. Stahl(164) with particular reference to rational C_4 in S_3. No further contribution to transformation theory is included.

L. Berzolari(11) gives a full historical résumé of all the results concerning rational C_n in S_3 and shows how the theory can be developed by means of combinants of four simultaneous binary quartics.

F. Derutys(61) and (59). The number of groups of $n-1$ neutral elements of I_n^k which contain an element of multiplicity $n-2k$ is $2(n-k)(n-2k)$.

An I_n^k contains $2ab(2p-3)\binom{n-k+1}{p}\binom{n-k}{p-2}$ groups of $k+p-2$ neutral elements, each consisting of two associated, multiple elements of orders a, b where

$$a+b=k-p+2.$$

If three elements, of orders a, b, c where $a+b+c=k-p+3$ are contained in I_n^k, the number is $2abc(2p-3)(2p-4)\binom{n-k+1}{p}\binom{n-k}{p-2}$.

In general, for orders $a_1, a_2, \ldots, a_\rho$, $\rho=2p-2$, where $\Sigma a_i=p+k-\rho$, the number is

$$2\prod_{i=3}^{2p-1}(2n-i)\prod_{i-1}^{2p-2}(a_i)\binom{n-k+1}{p}\binom{n-k}{p-2}.$$

J. de Vries(177). Given a C_n in S_3 the tangents of which form an I_p^1. The lines meeting two associated lines belong to a complex of order $(2n-3)(p-1)$. The lines meeting three associated tangents form a congruence of order and class $(n-2)(p-1)(p-2)$.

G. Marletta (129) shows that C_5 of S_3 with four stationary planes has one and only one quadrisecant if no two pairs of points of contact of stationary planes, or no three such points approach coincidence. G. Marletta (130) projects a normal C_n of S_n into S_3 or S_2; fundamental involutions are projectively equivalent. Particular attention is given to C_n with two tangents of $n-1$ point contact. See also G. Castelnuovo (30), L. Berzolari (13), A. Brambilla (21), V. Snyder (162).

J. R. Conner (44) gives a generalization of quadric inversion, $x_i = \dfrac{1}{x'_i}$ in S_n, and gives a discussion of I_n^{n-1} on C_n of S_n.

G. Marletta (132) studies the rational C_n in S_{2r+1}, having two S_r with $n-r$ point contact. It has ∞^1 osculating axes which form a ruled surface of order $(n-2r-1)(n-1)$. For $n=4$, $r-1$. See L. Cremona (54). Particular attention to C_n of S_3 with four points of hyperosculation. L. Berzolari (15) gives an elementary analytic discussion of C_n in S_3 having four points of hyperosculation. All the tangent lines to C_n meet the singular planes in projective groups. Further properties are found, by G. Marletta (131) by the use of the theorems given by E. Ciani (36), L. Berzolari (12) and (14), E. Ciani (37).

B. Kalicun (90) applies an I_5^1 to generate a rational quintic scroll with a four-fold directrix line.

J. F. de Vries (170) gives a general theory of normal C_4 in S_4. Polygons of $n+1$ sides inscribed in C_n of S_n are considered by H. S. White (237).

10. *Rational Involutions I_n^k on Rational Carriers.*—Em. Weyr (211). Given $k+1$ polynomials $f_1(x)=0$ in x, each of order n. The equation $\Sigma\lambda_i f_i(x)=0$ defines I_n^k. The neutral elements are those which belong to more than k groups. Plücker's equations concerning characteristics of rational curves in S_2, S_3, ... are derived. The results obtained were known before, but the method is partly new. Algebraically, a neutral group of h elements ϕ_n is a factor of more than one $f_i(x)$. Em. Weyr (222). The curves or surfaces of a linear system of order n, dimension k cut a C_1 in I_n^k, containing $\dfrac{(n-1)(n-2)}{2}$ pairs of neutral elements. Applications are made to cubic curves.

Em. Weyr (188) proves the theorem: The twelve faces of any three trihedra inscribed in a C_3 of S_3 touch the same quadric surface which contains three bisecants of C_3. This quadric is the involution surface of I_4^2 formed by such tetrahedra. Various applications were made by W. F. Meyer (133).

C. Le Paige (106) constructs an I_4^3 on C_3 of S_3 when four groups and three points on a fifth are given. Previously solved by P. Serret (159), used by P. Schlesinger([150]), and approached by C. Le Paige([114]). By the same method, a plane $C_4 p=3$, is constructed when fourteen points are given. Can be generalized to construct C_n in S_2.

C. Le Paige([102]) finds by a recurring process, the number of groups of $k+k'$ points common to $I_m{}^k$ and $I_{m'}{}^{k'}$ to be

$$\frac{(m-k)(m-k-1)\ldots(m-k-k'+1)}{\underline{|k'}} = \frac{(m'-k')(m'-k'-1)\ldots(m'-k'-k+1)}{\underline{|k}},$$

which define another involution. Application: Given two triangles and a conic. There exist two circumscribing conics, one about each, in a pencil with the given conic.

F. Deruyts([56]) represents the groups of $I_n{}^{n-1}$ by points of S_n, composite $I_n{}^1$ by normal C_n of S_n. The $|S^n{}_{n-1}|$ through the representative point meets C_n in $I_n{}^{n-1}$.

J. Rosanes([149]) proves the theorem, the groups of n points which express a binary form as the sum of n n^{th} powers constitute an $I_n{}^{n-1}$. In this, an $I_n{}^{n-m}$ has $n-k(m+1)-m$ neutral groups of $n-k$ elements $m<\dfrac{n-m}{m+1}$, which form an $I_{n-k}^{n-(1+m)k-m}$.

F. Deruyts([55]) continues the same representation. G. Castelnuovo([32]) considers C_n and an S_{n-k-1} in S_n.

The $\infty^k S_{n-1}$ having S_{n-k-1} as vertex cut C_n in groups of an $I_n{}^k$.

Lerch([118]) showed that in $I_n{}^k$ there are points of multiplicity r_1+1, $\ldots$, r^2+1, where $\Sigma r_i=k$ and $k+\rho=n$, $\rho!\binom{n-k}{\rho}\prod\limits_{i=1}^{\rho}(r_i+1)$.

F. Deruyts([55]) gives an alternate proof. C. Le Paige([103]) proves that $I_n{}^k$ may have multiply neutral groups. In $I_n{}^{n-1}$ the doubly neutral groups form an I_{n-2}^{n-5}, and the groups which leave k elements undetermined form and I_{n-k}^{n-2k-1}.

M. Genty([76]) proved that $I_n{}^k$ has $(k+1)(n-1)$ groups, each having an element of multiplicity $k+1$. If $n=2k$, there are $\dfrac{2^k(n-k)!}{k'(n-2k)!}$ groups containing k double elements. Previously proved by Em. Weyr and cited by Guccia. $I_n{}^k$ contains $(k'+1)(k-k'+1)(n-k)(n-k-1)$ groups having one element of multiplicity $k'+1$ and another of multiplicity $k-k'+1$.

M. Genty (77) considers I_n^k and a series of numbers k_i, last. $i=l$ and shows that there exist a number $\prod_{i=1}^{S}(k_i+1)\frac{(n-k)!}{(n-k-S)!}$ of groups consisting of S elements of multiplicity k_i+1. If q of the k_i are alike, the number must be divided by q^l.

F. Deruyts (57) determines the number of neutral elements common to certain given involutions: $I^3_{n_1}$, $I^3_{n_2}$ each contain an infinite number of neutral triads. The number of common neutral couples is $4\binom{n_1-2}{2}\binom{n_2-2}{2}$. The involutions $I^3_{n_1}$, $I^4_{n_2}$ have $2\binom{n_1-2}{2}\binom{n_2-2}{2}$ neutral triads in common. $I^4_{n_1}$, $I^4_{n_2}$ have $\binom{n_1-3}{2}\binom{n_2-3}{2}$ neutral tetrads in common.

For a general theory of rational involutions, see F. Deruyts (63).

Further formulas are given by F. Deruyts (58) concerning the maximum number of multiple elements and of common neutral elements. He also (59) shows that in $I_n^k k-p$ arbitrary elements can be associated with $\binom{n-k+1}{p-1}^2\cdot\frac{n-k-p+2}{p(n-k+1)}$ groups of $2p-2$ elements forming this many groups of $k+p-2$ neutral elements.

He (60) proved further that if k points of an I_n^k are in S_{n-3}, two more points are necessary to determine a group G_n having k points of this kind. Similarly for i^{th} kind, namely when i additional points are necessary to fix a group.

The number of G_k of k neutral elements of the ith kind in I_n^k is equal to the number of groups of k elements common to $i+1$ I_n^{k-1} which have in common an I_n^{k-i-1}.

He (62) showed, too, that the faces of three complete polygons of n vertices inscribed in a rational space C^m are tangent to the same surface of class $(n-2)\binom{m-1}{2}$. There exist ∞^2 similar figures.

M. Lelieuvre (117) gives a general formulation of involutions and applies his results to construction of Poncelet polygons. See (116).

J. de Vries (181) considers the various I^2_3 on C_3 of S_2. Each I_3^2 has a neutral pair, called the *associated* points. They form a symmetric (2, 2) correspondence. Application to a non-Cremona transformation of order 5. J. de Vries (175).

K. Bartels (9) applies involutions to the construction of certain ruled surfaces.

J. de Vries (182) obtains an I_3^2 from the $|C_3|$ through six points in S_2. The involution curve is a C_2. The curve of coincidences is a C_6 with six singular lines and six singular points.

BIBLIOGRAPHY

[1] Adler, A., Wien. Ber., 86: 916-36 (1882). [7.
[2] ———, Wien. Ber., 86: 1212-29, 1201-11 (1882). [7.
[3] ———, Wien. Ber., 113: 431-438 (1905). [6.
[4] Ameseder, A., Prag. Ber., 3-9 (1880). [6.
[5] Armenante, A., Giorn. di. Mat., 11: 221-232 (1873). [7.
[6] ———, Giorn. di. Mat., 12: 250-265 (1874). [7.
[7] Bartel, K., Wektor, 38 p. (1912). [3.
[8] ———, Prac. mat-fiz., 26: 65-89 (1915). [3.
[9] ———, C. R., 158: 1875-77 (1914). [10.
[10] Bertini, E., Lomb. Ist. Rend., (2), 5: 622-638 (1872). [7.
[11] Berzolari, L., Ann. di Mat., (2), 20: 101-161 (1892-3). [9.
[12] ———, Lomb. Ist. Rend., (2), 38: 446-448 (1905). [9.
[13] ———, Pal. Circ. Mat., 9: 186-197 (1895). [9.
[14] ———, Pal. Circ. Mat., 22: 214-219 (1906). [9.
[15] ———, Pal. Circ. Mat., 24: 1-16 (1907). [9.
[16] ———, Encykl. Geom. Kurve CII, (1906). [Int.
[17] ———, P. Rep., 2_1, chap. 13 (1910). [Int.
[18] ———, P. Rep., 2_2, chap. 37 (1922). [7.
[19] Bobek, C., Wien. Ber., 80: 361-396 (1879). [6.
[20] Bordiga, G. A., Ven. Ist. Atti., (6), 4: 503-525 (1885-86). [9.
[21] Brambilla, A., Nap. R. A. Atti., 24: 279-298 (1885). [9.
[22] ———, Nap. R. A., Atti., (2), 9: 33 p. (1899). [7.
[23] ———, Nap. R. A., Rend., (3), 2: 171-176 (1896). [7.
[24] ———, Ven. Ist. Atti., (6), 3: 1471-1489 (1884-85). [7.
[25] Brill, A., Math. Ann., 6: 33-65 (1873). [9.
[26] Cantone, A., Nap. R. A. Rend., (1), 25: 181-190 (1886). [5.
[27] Cardinaal, J., Amst. K. Ak. Wet. Versl., 8: 271-278 (1899-1900). [3.
[28] Caspary, F., Soc. M. F. Bull., (2), 11: 222-236 (1887). [5.
[29] Cassani, P., Ven. Ist. Atti., (3), 16: 535-554 (1870-71). [1.
[30] Castelnuovo, G., Pal. Circ. Mat., 3: 179-192 (1889). [9.
[31] ———, Ven. Ist. Atti., (6), 4: 1167-1199, 1559-1594 (1885-86). [9.
[32] ———, Ven. Ist. Atti., (6), 5: 1041-1115 (1886). [10.
[33] Cavalli, E., Ras. di Mat., 2: 73-88, 169-180 (1922). [4.
[34] Chasles, M., C. R., 41: 1097-1107 (1855). [1.
[35] ———, C. R., 72: 385-401, 577-598 (1871). [2.
[36] Ciani, E., Lomb. Ist. Rend., (2), 37: 341-353 (1904). [9.
[37] ———, Pal. Circ. Mat., 26: 315-324 (1908). [9.
[38] Ciani, E., and Wieleitner, H., Repertorium, 2: chap. 19, §§ 4, 5 (1910). [Int.
[39] Clebsch, A., J. f. Math., 64: 43-65 (1865). [4.
[42] Coble, A. B., Am. J., 41: 243-265 (1919). [8.
[43] Conner, J. R., Am. M. S. Bull., 19: 284-285 (1913). [8.
[44] ———, J. Hopk. Circ., 745-763 (1908). [9.
[45] ———, J. Hopk. Circ., 116-127 (1911). [8.
[46] ———, J. Hopk. Circ., 175-178 (1912). [8.
[47] Crelier, L., Enseign. Math., 15: 453-468 (1913). [4.
[48] Cremona, L., Ann. di Mat., (1), 4: 71-101 (1861). [7.

49 ———, Ann. di Mat. (1), 4: 22-25 (1862). [5.
50 ———, Ann. di Mat., (1), 5: 227-231 (1863). [5.
51 ———, Ann. di Mat., (1), 6: 179-190 (1864). [1.
52 ———, Ann. di Mat., (2), 1: 248-258 (1867-68). [7.
53 ———, Bologna Mem., (3), 3: 385-398 (1863). [5.
54 ———, Lomb. Ist. Rend., (2), 1: 199-202 (1868). [9.
55 Deruyts, F., Brux. Bull., (3), 14: 322-345 (1887). [10.
56 ———, Brux. Bull., (3), 14: 650-664 (1887). [10.
57 ———, Brux. Bull., (3), 26: 232-235 (1893). [10.
58 ———, Brux. Bull., (3), 35: 196-206 (1898). [10.
59 ———, Brux. Bull., (3), 35: 885-894 (1898). [9, 10.
60 ———, Brux. Bull., 36: 187-193 (1898). [10.
61 ———, Brux. Bull., 36: 194-204 (1898). [9.
62 ———, Brux. Bull., 36: 553-566 (1898). [10.
63 ———, Liège S. R. S. Mém., (2), 17: 208 p. (1892). [10.
64 Dewulf, Em., and Schoute, P. H., Soc. M. F. Bull., (2), 3: 383-400 (1879). [6.
65 Dingeldey, F., Encykl. CIII. [Int.
66 ———, P. Rep., 2: chap. 12 (1910). [Int.
67 Dixon, A. C., Quart. J., 23: 343-357 (1888-89). [5.
68 ———, Quart. J., 24: 30-54 (1890). [5.
69 Drasch, H., Wien. Ber., 85: 534-553 (1882). [4.
70 Durège, H., Prag. Ab., (6), 3: 21 p. (1869). [4.
71 Fairson, J., Liège S. R. S. Mém., (3), 5: 29 p. (1904). [5.
72 Frauenfelder, G., Monatsh f. Math., 15: 299-314 (1904). [7.
73 Gallucci, G., Nap. R. A. Rend., (3), 4: 207-215 (1898). [5.
74 Gerbaldi, F., Torino Mem., (2), 32: 309-357 (1880). [5.
75 Genty, M., Soc. M. F. Bull., 9: 115-162 (1880-81). [8.
76 ———, Soc. M. F. Bull., 20: 106-112 (1892). [10.
77 ———, Soc. M. F. Bull., 21: 52-55 (1893). [10.
78 Garbieri, G., Giorn. di Mat., 16: 1-17, 108-147 (1878). [8.
79 de la Gournerie, J., C. R., 66: 832-836 (1868). [3.
80 Gross, W., Diss. Tübingen, (1887). [6.
81 Halphen, G. F., Soc. M. F. Bull., 5: 134-136 (1877). [7.
82 Hudson, H. P., London M. S. Proc., (2), 15: 385-400 (1916). [8.
83 Janni, V., Nouv. Ann., (2), 1: 80-82 (1862). [2.
84 Joachimsthal, J., J. f. Math., 56: 44-45 (1859). [5.
85 Jolles, S., Pamphlet, Aachen, 24 p. (1886). [7.
86 ———, Arch. Math. Phys., (3), 22: 277-278 (1914). [7.
87 ———, J. f. Math., 130: 270-280 (1906). [5.
88 Jonquières de, E., Ann. di Mat., (1), 2: 86-94 (1859). [1.
89 ———, Pal. Circ. Mat., 2: 118-123 (1888). [8.
90 Kalicun, G., Wien. Ber., 120: 1299-1326 (1911). [9.
91 Kohn, G., Wien. Ber., 88: 424-431 (1883). [3.
92 ———, Wien. Ber., 89: 144-172 (1884). [3.
93 ———, Wien. Ber., 95: 318-337 (1887). [6.
94 ———, Wien. Ber., 95: 338-348 (1887). [6.
95 ———, Wien. Ber., 105: 1035-1039 (1896). [5.

96 Kohn, G., and Loria, G., Encykl. III, C5: (1906). [Int.
97 Korndörfer, G., Math. Ann., 3: 415-423 (1871). [9.
98 Laguerre, E., Nouv. Ann., (2), 17: 181-183, 337-351 (1878).° [6.
99 ———, C. R., 84: 224-227 (1877).° [6.
100 Lazarski, M., Cracow Rozpravy, 15: 224-249 (1887). [6.
101 Le Paige, C., Brux. Bull., 44: 231-237 (1877). [2.
102 ———, Brux. Bull., (3), 11: 121-127 (1886). [10.
103 ———, Brux. Bull., (3), 14: 211-214 (1887). [10.
104 ———, Brux. Mém. Sc., (2), 10. [3.
105 ———, Liège S. R. S. Mém., (2), 11: 19 p. (1885). [3.
106 ———, C. R., 98: 285-287 (1884). [6, 10.
107 ———, C. R., 98: 353-356 (1884). [6.
108 ———, Wien. Ber., 81: 159-161 (1880). [2.
109 ———, Wien. Ber., 81: 845-852 (1880). [2, 7.
110 ———, Wien. Ber., 83: 351-357 (1881). [2.
111 ———, Wien. Ber., 84: 233-236 (1881). [2, 5.
112 ———, Wien. Ber., 85: 844-847 (1882). [3.
113 ———, Wien. Ber., 87: 599-602 (1882). [5.
114 ———, Prag. Ber., (6), 11: 61-68 (1881). [10.
115 ———, Prag. Ber., (6), 14: 45-48 (1884). [6.
116 Lelieuvre, M., C. R., 132: 1172-1174 (1901). [10.
117 ———, Rev. de Math. Spec., 12: 297-299 (1901-02). [10.
118 Lerch, M., Prag. Ber., (6), 15: 597-600 (1885). [10.
119 Locchi, P., Giorn. di Mat., 59: 81-92 (1921). [7.
120 Loria, G., Pal. Circ. Mat., 2: 201-224 (1888). [9.
121 ———, Prag. Ber., 4 p. (1896). [4.
122 ———, Encykl. III C5b, (1914). [8.
123 Lüroth, J., Math. Ann., 9: 163-165 (1876). [8.
125 Majcen, G., Prag. Ber., 14 p. (1910). [6.
126 Maisano, G., Pal. Circ. Mat., 1: 7-8 (1884-87). [3.
127 ———, Pal. Circ. Mat., 1: 9 (1884-87). [3.
128 Marletta, G., Ann. di Mat., (3), 8: 97-128 (1903). [7.
129 ———, Pal. Circ. Mat., 21: 56-57 (1906). [9.
130 ———, Pal. Circ. Mat., 21: 192-210 (1906). [9.
131 ———, Pal. Circ. Mat., 24: 32-34 (1907). [9.
132 ———, Pal. Circ. Mat., 25: 376-388 (1908). [9.
133 Meyer, W. F., Apolarität und rationale Curven, Tübingen, (1883). [10.
134 Netto, E., Math. Ann., 46: 310-318 (1895). [8.
135 Osgood, W. F., Am. M. S. Bull., (2), 2: 168-173 (1896). [3.
136 d'Ovidio, E., Giorn. di Mat., 17: 310-338 (1879). [5.
137 ———, Torino Mem., (2), 32: 1-75 (1880). [5.
138 de Paolis, R., Rom. Acc. L. Rend., (4), 2_2: 335-336 (1886). [2.
139 Papin, P. Th., Rom. Pont. Acc. Atti., 32: 166-202 (1879). [1.
140 Pascal, E., P. Rep., 2_1: chap. 12 (1910). [Int.
141 ———, P. Rep., 2: chap. 29 (1922). [Int.
142 Pittarelli, G., Giorn. di Mat., 17: 260-309 (1879). [5.
143 ———, Giorn. di Mat., 21: 19-50 (1883). [2.

144 ———, Nap. R. A. Rend., (1), 24: 111-121, 216-233 (1885). [4.
145 Plamitzer, A., Wien. Ber., 125: 1343-1377 (1916). [8.
146 Reye, Th., J. f. Math., 78: 97-113 (1874). [5.
147 ———, J. f. Math., 79: 159-175 (1874). [5.
148 Rice, J. N., Diss. Catholic Univ. Amer., 32 p. (1917). [6.
149 Rosanes, J., J. f. Math., 76: 312-330 (1873). [5, 9, 10.
150 Schlesinger, O., Math. Ann., 22: 520-568; *see* 532 (1882). [10.
152 Schoute, P. H., Arch. Math. Phys., (2), 2: 113-137 (1885). [6.
153 ———, Soc. M. F. Bull., (2), 8: 278-280 (1884). [6.
154 ———, Wien. Ber., 89: 1252-1267 (1883). [6.
155 ———, Amst. K. Ak. Wet. Versl., (2), 19: 420-431 (1884). [6.
156 ———, Amst. K. Ak. Wet. Proc., (1), 20: 49-94 (1886). [8.
157 Schroeter, H., J. F. Math., 54: 31-47 (1857). [4.
158 Schrek, D. J. E., Diss. Utrecht, 87 p. (1914). [7.
159 Serret, P., C. R., 87: 643-648 (1878). [2, 10.
160 Servais, C., Brux. Bull., (3), 20: 272-280 (1890). [2, 3.
161 ———, Liège S. R. S. Mém., (3), 11: 40 p. (1922). [4.
162 Snyder, V., Am. J., 28: 237-247 (1906). [9.
163 ———, Am. M. S. Bull., 5: 343-353 (1898-99). [7.
164 Stahl, W., J. f. Math., 104: 38-61 (1889). [9.
165 Staude, O., Encykl. III C2, P. Rep., 2_2: chap. 25-28 (1922). [Int.
166 ———, P. Rep., 2_2: chap. 29 (1922). [Int.
167 Sturm, R., J. f. Math., 86: 116-145 (1879). [5.
168 Tracy, J. I., J. Hopk. Circ., 146-152 (1911). [8.
169 Treibscher, M., Wien. Ber., 81: 1080-1091 (1880). [4.
170 de Vries, J. F., Diss. Leiden, 158 p. (1922). [9.
171 de Vries, J., Mat. P. A., 1: 13-14 (1901). [4.
172 ———, Mat. P. A., 1: 81-92 (1901). [4.
173 ———, Mat. P. A., 1: 227-231 (1901). [6.
174 ———, Monatsh. f. Math., 3: 285-292 (1892). [3.
175 ———, Wien. Ber., 104: 46-59 (1895). [10.
176 ———, Amst. K. Ak. Wet. Versl., (3), 9: 696-700 (1900-01). [6.
177 ———, Amst. K. Ak. Wet. Versl., (3), 11: (1902-03), 8: 762-767. [9.
178 ———, Amst. K. Ak. Wet. Versl., 12: 740-742 (1903-04). [8.
179 ———, Amst. K. Ak. Wet. Versl., 12: 742-744 (1903-04). [8.
180 ———, Amst. K. Ak. Wet. Versl., 17: 2-6 (1908-09). [5.
181 ———, Amst. K. Ak. Wet. Versl., 18_2: 744-748 (1909-10). [10.
182 ———, Amst. K. Ak. Wet. Versl., 26: 1256-1267 (1917-8). [10.
183 Waelsch, E., Wien. Ber., 90: 160-167 (1884). [1.
184 ———, Wien. Ber., 100: 315-326, 803-821 (1891). [2, 5.
185 Wakeford, E. K., London M. S. Proc., (2), 21: 98-113 (1920-22). [5.
186 Weill, M., Nouv. Ann., (3), 1: 62-79 (1882). [2.
187 Weyr, Em., Ann. di Mat., (2), 4: 328-330 (1871). [9.
188 ———, Brux. Bull., (3), 3: 472-485 (1882). [10.
189 Weyr, Ed., J. f. Math., 73: 87-93 (1871). [2.
190 Weyr, Em., Giorn. di Mat., 10: 165-169 (1872). [2, 3.
191 ———, Math. Ann., 3: 235-237 (1871). [2, 4.

192 Weyr, Em., Lomb. Ist. Rend., (2), 4: 144-146 (1871). [2, 3, 7.
193 ———, Lomb. Ist. Rend., (2), 4: 206-210 (1871). [2, 3.
194 ———, Lomb. Ist. Rend., (2), 4: 636-643 (1871). [5.
195 ———, Lomb. Ist. Rend., (2), 5: 204-206 (1872). [3, 6.
196 ———, J. f. Math., 72: 285-292 (1870). [2.
197 ———, Soc. M. F. Bull., 1: 18-19 (1872-73). [6.
198 ———, Wien. Ber., 61: 600-606 (1870). [4.
199 ———, Wien. Ber., 63: 493-504 (1871). [7.
200 ———, Wien. Ber., 67: 286-291 (1873). [6.
201 ———, Wien. Ber., 71: 400-409 (1875). [7.
202 ———, Wien. Ber., 72: 686-702 (1875). [7.
203 ———, Wien. Ber., 73: 203-220 (1876). [7.
204 ———, Wien. Ber., 73: 654-656 (1876). [3.
205 ———, Wien. Ber., 61: 731-738, 819-827 (1870). [4.
206 ———, Wien. Ber., 75: 168-174 (1877). [7.
207 ———, Wien. Ber., 75: 458-462 (1877). [7.
208 ———, Wien. Ber., 78: 396-398 (1878). [7.
209 ———, Wien. Ber., 78: 891-895 (1878). [7.
210 ———, Wien. Ber., 79: 429-446 (1879). [2, 4.
211 ———, Wien. Ber., 79: 680-698 (1879). [10.
212 ———, Wien. Ber., 80: 1040-1046 (1879). [4.
213 ———, Wien. Ber., 81: 80-84 (1880). [2.
214 ———, Wien. Ber., 81: 162-164 (1880). [3.
215 ———, Wien. Ber., 81: 169-195 (1880). [4.
216 ———, Wien. Ber., 81: 841-844 (1880). [7.
217 ———, Wien. Ber., 81: 1007-1031 (1880). [6.
218 ———, Wien. Ber., 81: 1218-1219 (1880). [7.
219 ———, Wien. Ber., 83: 63-69 (1881). [2.
220 ———, Wien. Ber., 83: 300-321 (1881). [2.
221 ———, Wien. Ber., 83: 807-829 (1881). [2.
222 ———, Wien. Ber., 84: 884-907 (1881). [10.
223 ———, Wien. Ber., 84: 1264-1290 (1881). [5.
224 ———, Wien. Ber., 85: 513-525 (1882). [5.
225 ———, Zschr. Math. Phys., 16: 353-356 (1871). [2, 7.
226 ———, Casopis, 1: 101-104 (1872). [2.
227 ———, Prag. Ber., (6), 5: 28-30, 1-27 (1872). [2.
228 ———, Prag. Ber., (6), 5: 59-65 (1872). [8.
229 ———, Prag. Ber., (6), 6: 9-43 (1872-73). [2.
230 ———, Prag. Ber., (6), 6: 37 p. (1872). [4.
231 ———, Prag. Ber., (6), 6: 14-21 (1873). [6.
232 ———, Prag. Ber., (6), 7: 70-79 (1874). [8.
233 ———, Prag. Ber., (6), 7: 164-166 (1874). [6, 7.
234 ———, Prag. Ber., (6), 7: 198-204 (1874). [8.
235 ———, Prag. Ber., (6), 12: 158-167 (1882). [8.
236 ———, Prag. Ab., (6), 6: 9-42, 77-81 (1873). [6.
237 White, H. S., Am. M. S. Trans., 22: 80-3 (1921). [9.
238 Wieleitner, H., P. Rep., 2_1: chap. 22 (1910). [Int.
239 Winger, R. M., Am. J., 42: 129-135 (1920). [4.
240 ———, Am. M. S. Bull., 25: 27-34 (1918). [4.
241 ———, Am. M. S. Bull., 26: 75-79 (1919-20). [4.

242 van der Woude, W., Nieuw. Arch., (2), 14: 64-68 (1922). [4.
243 Wright, J. E., London M. S. Proc., (2), 6: 52-57 (1907). [4.
244 ———, Am. M. S. Bull., 13: 389-391 (1906-07). [8.
245 Zahradnik, K., Arch. Math. Phys., 56: 349-353 (1874). [4.
246 ———, Wien. Ber., 117: 1167-1190 (1908). [4.
247 ———, Prag. Ber., (6), 6: 298-305 (1873). [4.
248 ———, Prag. Ber., (6), 7: 12-15 (1874). [4.
249 ———, Prag. Ber., (6), 7: 180-189 (1874). [6.
250 ———, Prag. Ber., (6), 7: 310-315 (1874). [4.
251 ———, Prag. Ber., (6), 10: 184-191 (1877). [6.
252 ———, Prag. Ber., 19 p. (1906). [4.
253 Zecca, G., Giorn di Mat., 25: 333-355 (1887). [4.

CHAPTER VII

CORRESPONDENCES ON NON-RATIONAL CURVES

1. *Chasles's Principle of Correspondence.*—The relation established by an algebraic equation between two non-homogeneous variables λ, μ is called a *correspondence.* If $f(\lambda, \mu)=0$ is of degree m in λ, and of degree n in μ, the correspondence between λ and μ is denoted by the symbol (m, n). When λ, μ are regarded as points on straight lines, the equation $f(\lambda, \mu)=0$ defines an algebraic curve in generalized cartesian coordinates. If λ is the parameter defining a point on one rational curve and μ another, then $f(\lambda, \mu)=0$ defines an (m, n) correspondence between the points of the two curves. If λ, μ define points on the same rational curve, then the correspondence is between points on this curve. If $m=1$ or $n=1$, the correspondence is called an *involution* and if $m=1$ and $n=1$, it is called *projective.* In any case the correspondence is fixed when $mn+m+n$ pairs of corresponding elements are given.

Let λ describe C, μ describe C'. On C there are $2m(n-1)$ branch points, and on C' there are $2n(m-1)$.

When $C=C'$, there are $m+n$ points such that one of the corresponding elements coincides with the given point. Thus *an (m, n) correspondence on a rational carrier has $m+n$ coincidences.* This theorem is known as the Chasles principle of correspondence, M. Chasles([80]), but it had been used by various authors previously. See in particular E. de Jonquiéres([164]) and ([167]). L. Cremona casually mentioned it in ([97]) and ([98]). A careful history of the theorem is given by C. Segre([309]). A much earlier application was given by J. Steiner([338]), in determining the number of normals, n^2, to a general C_n from an arbitrary point. If the given curve describes a pencil, the foot of the normal from a fixed point describes a curve of order $2n$. The same principle is then applied to the evolute, the Plücker numbers of which are determined.

Later developments are given by Em. Weyr([410]), and by M. Chasles([74]), who extends the results of Plücker to cases in which the line at infinity of the plane of the given curve is a singular element. See also M. Chasles([83]). A particular case was given by M. Chasles in ([78]). A detailed application to correspondences on conics is given by M. Chasles([75]) with the remark that all the results may be extended to any rational curves. Other applications are given by Chasles in ([84]).

This theory was extended by M. Pieri[272] who showed that the locus of the foot of the normals from P to the $|C_n|$ of a pencil is of order $2n$, passes simply through P and the n^2 basis points of the pencil, through the $3(n-1)^2$ points which are nodes of particular curves of the pencil, through the $2(n-1)$ points of contact of certain curves of the pencil with the line at infinity, and through the circular points.

A (1, 3) correspondence was discussed by Em. Weyr[408].

The theorem that two general algebraic curves of order p, p' intersect in PP' points was proved by M. Chasles[73] by the principle of correspondence.

G. Fouret[125] determines the number of contacts of a system of curves of signature μ, ν (μ through a given point, ν touching a given line) with a curve of order m, class n is $n\mu+m\nu$, confirming a known result. [A. Brill[45] and Fouret in [49].] If the given curve has an r-fold point at each imaginary circular point at infinity, the number of curves (μ, ν) which meet it at a given angle is $(m+n-2r)\mu+m\nu$.

This was further developed by E. Dewulf in [107], [108] and [106]. The inclined polar of a point P as to an algebraic curve C_n is defined as the locus of a point M such that the ordinary polar of M as to the curve makes a constant angle with MP. Through P can be drawn n^2 lines meeting C_n at a given angle. The envelope of the inclined polars of C_n, as to C_n is of class nn'.

H. Schubert[296] discusses loci defined by (1, 2) correspondences on rational curves; compare also Em. Weyr[386].

G. Peano[265] obtains the enumeration of the complete system of invariant forms of (1, 1), (1, 2) and (2, 2) correspondences on rational carriers. The (2, 2) case was partially obtained by A. Capelli[52]. See also G. Peano[266]. Em. Weyr[405] applied the same principle to the determination of one-dimensional cyclic projectivities.

G. Loria[223] derives the ordinary properties of curves generated by (m, n) correspondence between rational carriers; see also R. Sturm[341]. For the application to nodal cubics, see Pittarelli[274].

A metrical application is given by K. Vahlen[364]. Two curves of orders g_0, h_0, classes g_1, h_1 have $g_0h_1+g_1h_0+g_1h_1$ common normals. The method is developed in [363]. A few new relations are developed incidentally by F. Hardcastle[141]. Plücker's numbers are re-derived by means of the coincidences of an involution by O. Zimmermann[425]. A similar application to the determination of foci is made in [426]. A historical résumé is given by C. Segre[313].

J. de Vries[376] shows that if an (n, n) correspondence between superposed carriers has $\frac{n}{2}(n+1)$ involutorial pairs, the correspondence is involutorial. If two (2, 2) correspondences have the same tetrad of branch points they are involutorial.

L. I. Neikirk([244]) showed that if an (m, n) correspondence between x and y has a single correspondence of the type $(x_i : y_i)$, $i=1, \ldots, m$; $i=1, \ldots, n$ then $f(x, y)=0$ can be reduced to the form

$$\phi(x)+\lambda\theta(x)=0, \qquad \psi(y)+\lambda\omega(y)=0,$$

in which all the functions are rational, the first two of degree m, the other two of degree n.

J. de Vries([377]) obtains further properties of (2, 2) correspondences, extending results obtained by Em. Weyr([387]) and ([393]), and by J. de Vries([374]).

2. *Inscribed and Steiner Polygons.*—For the early history of Poncelet polygons, that is, those having their vertices on one conic and their sides touching another, and the various degenerate cases, see Pascal([264]).

Application to inscribed and circumscribed polygons of a conic was made by A. Cayley([69]) especially for symmetric (2, 2) correspondences, and the result expressed more definitely in ([70]). If x, u are in symmetric (2, 2) correspondence and u, y define the same correspondence, then x, y will be a symmetric (2, 2) correspondence, but one relation must be satisfied in order that it be the same as the preceding ones.

Inscribed and circumscribed polygons of conics were discussed by E. Czuber([100]) by means of two cones of revolution with a common axis and vertex.

F. Gerbaldi([130]) discusses a cyclic symmetric (2, 2) correspondence, and applies the results to Poncelet polygons to prove the theorem:

Given two proper conics C_1, C_2 the condition in order that a polygon (and therefore ∞^1) exists, having its p vertices on C_1 and each side touching C_2 is expressed by the vanishing of an invariant of orders $2\left(\frac{p^2}{4}-1\right)$, $\frac{p^2}{4}-1$ for p even, and of orders $\frac{1}{2}(p^2-1)$, $\frac{1}{4}(p^2-1)$ when p is odd. See also G. Loria([221]) and ([219]).

E. Beltrami([23]) [also in D. Chelini([85])] obtains further conditions for inscribed and circumscribed polygons in conics, and generalizes the procedure to apply to nodal and space cubics, and outlines further generalization.

A Steiner polygon is one in which the odd sides pass through one fixed point, and the even sides pass through another fixed point.

Buchheim([50]) gives a treatment of Steiner polygons by means of abelian functions. Further properties of Steiner polygons were found by C. Küpper([202]) and ([204]). By means of parametric representation in terms of elliptic functions, H. Oppenheimer([259]) obtained various expressions for closure. See W. K. Clifford([94]) for a particular Steiner polygon.

E. Panzi([260]) derives the extended form of Plücker's formulas by direct application of Chasles's principle. The first formula for the simple case had been derived by G. Loria([222]) that giving the inflexions, by J. Bischoff([33]). All the simple cases are obtained by C. Segre([305]).

3. *Rational Involutions on Curves.*—Given two polynomials $\phi_1(x)$, $\phi_2(x)$ of degree n in x, the equation

$$\phi_1(x)+\lambda\phi_2(x)=0$$

defines an *involution of order n*. It is a $(1, n)$ correspondence between λ and x. When x is the parameter of a rational curve C, and λ is the parameter on a second curve C', then the involution on C is said to belong to C'. The lines joining the pairs of points of each group G_n envelop a curve called the involution curve J of the given involution I_n. The theory of quadratic involutions will be presupposed.

Em. Weyr([411]) develops the theory of cubic involutions, and shows that if a, b are triple elements, the equation defining it may be put in the form

$$(x-a)^3+\lambda(x-b)^3=0$$

or if $a=0$, $b=\infty$, in the form

$$x^3=\lambda.$$

When x describes a conic C_2, J_2 is another conic. The positions of the branch-points and coincidences are determined by the relative positions of C_2, J_2. Numerous special applications are given.

In ([409]) Em. Weyr studies the I_4 on a C_2; the involution curve is now of class 3. When the given involution is expressed by tangents to a conic, the involution curve is of order 3. I_4 is completely defined by two of its groups. Another form of I_4 is given by A. Ameseder([3]) by means of a pencil of planes and an I_2 of planes tangent to a quadric cone. The locus is a ruled surface of order 4, having the axis of the pencil and a conic in the involution plane for double curves. The treatment of further development is synthetic in A. Ameseder([4]). A systematic treatment of cubic involutions is given by C. Le Paige([210]), with particular reference to cross-ratios and polar groups. Particular metric involutions of order n are treated by G. Fouret([123]). The properties of I_2, I_3, ..., I_s on conics, pencils of conics, cubics and pencils of cubics are given by B. D. Fraser([126]), in a review of Kötter's synthetic geometry of algebraic curves.

An I_4 on C_4 is discussed by J. de Vries([379]), including the derivation of the 63 systems of contact C_2 with a non-singular C_4. All the I_4 on C_4

determined by pencils of curves have an involution curve of class 6 and have 12 coincidences. See A. Ameseder (7). Unless relations exist among the coefficients, the I_s on C_g have $2(g+s-1)$ coincidences. L. Berzolari (32) treats cubic involutions and their conjugates on a nodal cubic curve by means of binary forms. Problems in determinations of coincidences were solved by J. de Vries (371) for various particular cases.

A rational involution I_n is uniquely determined when three partial groups containing n_1, n_2, n_3 points, where $n_1+n_2+n_3=2n+1$, are given. An application of involutions to certain transformations of plane curves was made by M. D'Ocagne (109). O. Tognoli (354) considers two rational involutions I_{nm}, $I'_{nm'}$ on a fixed C_n by means of two pencils $|C_m|$, $|C_{m'}|$. When the latter are projective, there are $n(m+m')$ coincidences. The envelope of lines joining corresponding points and the dual are discussed.

4. *Involutions of Higher Dimensions.*—If $\phi_i(x)$, $i=1, 2, \ldots, s+1$ are $s+1$ polynomials, then $\sum_{i=1}^{S+1} \lambda_i\phi_i(x)=0$ defines an involution of order n and dimension s, designated by the symbol $I_n{}^s$. The $I_4{}^3$ on straight lines were employed by C. Le Paige (213) to find the sixth points of C_2, C_3 when 5 points are given on the latter, and to construct a C_4 through 14 points. A general discussion of the properties of several I_n on the same carrier are given by Em. Weyr (388).

It was shown by G. Castelnuovo (58) that on any elliptic curve exist $\infty^1 I_n{}^{n-1}$, each of which has n^2 n-fold points. The same author (55) determines the number N of rational involutions of order n and dimension q which exist on a curve of given genus p. If $p-1-(n-q)=Q$, then

$$N=\frac{1!\,2!\ldots q!\,1!\,2!\ldots Q!\,p!}{1!\,2!\ldots(q+Q+1)!}.$$

The number of involutions $g_n{}^q$ on a rational curve having p neutral couples is N.

Em. Weyr (389) proved that $I_n{}^{n-1}$ on a carrier of genus 1 has n^2 n-fold elements.

A number of derived $I_n{}^{n-1}$ on elliptic curves were also obtained by Em. Weyr (403) and (402), defined by the intersections of an elliptic C_3 with groups of curves. Let C_2 meet C_3 in $0_1 0_2 0_3 a_1 a_2 a_3$. The three C_2 through $0_1 0_2 0_3$ touching C_3 in a_1, a_2, a_3 respectively, meet C_3 in a'_1, a'_2, a'_3. The six points $a_1a_2a_3a'_1a'_2a'_3$ lie on a conic. If the conics having five point contact with C_3 at 0_1, 0_2, 0_3, a_1, a_2, a_3 respectively, meet C_3 again in $0'_1, \ldots, a'_3$ then these latter six points also lie on a conic. If one of the four double elements of an I_2 be considered as a triple element of an I_3, the other three form a triad. If one of the k^2 k-fold elements of an

I_k^{k-1} is chosen as a p-fold element of an I_p^{p-1} then the involution determined by this latter group, is one of the involutions by means of which I_k^{k-1} is transformed into itself.

G. Castelnuovo([60]) showed that an involution of order n and dimension $r>1$ belonging to an algebraic curve is always linear except (possibly) when each of its groups is made up of r arbitrary groups of an I_k^1, where $k=\frac{n}{r}$.

G. Humbert([155]) showed that the I_n^k on curves of genus p were of two kinds when $k>1$

(a) If $k=n$, then $p=0$

(b) If $1<k<n$, then I_n^k is rational.

For curves with arbitrary moduli, no other involutions exist. If $k=1$, the g_n^1 is either rational or, if irrational, is of genus $\omega<p$, with $2(p-1)-2n(\omega-1)$ branch points. The irrational γ_n^1 can exist only on singular curves, that is, with certain relations among their coefficients. These results were known before but new proofs were given by Humbert. Simpler proofs were given by M. De Franchis([102]) and ([103]). It was shown by G. Kohn([184]) that the groups of $n+1$ elements formed by one element of a rational carrier and the n elements corresponding to it in a symmetric (n, n) correspondence belong to an $I^n{}_{n+1}$ when n is odd.

5. *Rational Involutions on Curves of Genus* $p>0$.—H. M. Jeffery([162]) and ([163]) showed that if one group of I_4 on C_3, $p=1$ is known, the involution may be constructed. C. Küpper([203]) discusses an I_n on C_3.

A. Cayley([71]) observed that (1, 1) correspondences may become $(k, 1)$ by altering the definition. For example, the tangent to the common evolute of an ellipse and its parallel is normal to the parallel at two points when the latter is not composite, thus defining a (1, 2) correspondence.

C. Bobek([36]), ([38]) and ([35]) studied various properties of curves having an I_2, or a g_2^1, that is, hyperelliptic. These were extended to curves possessing a g_3^1 in ([39a]). The groups of g_3 on C_m of genus p are cut out by curves ϕ_{m-4} of order $m-4$, which pass through the double points of C_m. $p\geqq\frac{m+2}{2}$. If C_m has two g_3^1, $p\leqq 4$. If $p>m-2$, each group G_3 lies on a straight line. If C_m has a ϕ_{m-5} which pass through the double points, g_3^1 is impossible. If more than $m-p-2$ groups G_3 are composed of collinear points all must be. See also C. Küpper([206]). This theorem was generalized in ([199]) to G_k^1 on C_m with a point of multiplicity $m-k$. J. de Vries([372]) studied the g_2^1 on the C_5 with 4 double points. F. Amodeo([10]) and ([11]) considers g_k^1 and divides them into first and second species according as the points of each group are or are not collinear. Ways of obtaining them are obtained by C. Küpper([200]).

F. Amodeo([14]) and ([12]) defines an I_k^1 to be of species s if the lines joining points of each G_k by pairs envelop a curve of class s (involution curve). C. Küpper([194]) shows that if a C_m of genus p contains a g_k^1 it cannot have adjoints ϕ_{m-k-2}. Every adjoint through P contains the group G_k of g_k^1 through P. If p is a maximum, C_m contains a point of multiplicity $m-k$ and each G_k is composed of collinear points. F. Amodeo([15]) obtains the class of the involution curve of g_k^1.

Em. Weyr([401]) applied the same methods to rational involutions on curves of genus 1. An I_m^1 and an I_n^1 have $(m-1)(n-1)-1$ common pairs of elements. On a curve of genus p, the number is

$$(m-1)(n-1)-p.$$

The number of double elements of I_m^1 is $2(m+p-1)$. Every I_4^2 on a curve of genus 1 can be represented by the plane sections of a space quartic curve through a fixed point, and every I_4^3 by all the plane sections of C_4. These results are employed by Em. Weyr([402]) to determine Steiner polygons.

An I_n^{n-1} is uniquely fixed by one of its groups, Em. Weyr([398]). Previously proven for $n=2$; see ([395]), ([394]) and ([396]).

The two fundamental points of a Steiner polygon of $2n$ vertices on C_3, $p=1$, are n-fold points of I_n^{n-1}, ([389]).

The method was developed into a systematic symbolical calculus by Em. Weyr([404]). The same procedure may be expressed in terms of the parametric representation of points on a curve of genus 1 by means of elliptic functions. [E. Czuber([99]).]

A C_4 with one double point is hyperelliptic, $p=2$. The characteristic g_2^1 is defined by the pencil of lines through the node. The transformation of interchanging the points of each pair is a perspective quadratic inversion. The conic of invariant points passes through the points of contact of the six tangents from the double point. E. Bertini([29]); H. Picquet([271]); E. Caporali([54]) and ([53]). The conic is called the Bertini conic of the nodal C_4. If C_4 has two double points, a Bertini conic is associated with each as vertex of a perspective quadratic inversion, and passes through the other double point. E. Ciani([89]). In case the C_4 of $p=2$ has inflexional tangents at the node the Bertini conic is composite and the curve is invariant under harmonic homology. These projective inversions are all particular cases of the Jonquières perspective involution.

It was proved by H. A. Schwarz([300]) that a point on any curve of genus 2 can be expressed rationally in terms of λ and $\sqrt{f_6(\lambda)}$.

J. de Vries([378]) studies the I_2 on the hyperelliptic C_5, $p=2$, by means of the pencil of conics through the nodes, and the I_3 cut on C_5 by the lines each joining a pair of conjugates in I_2. The I_3 is cut by $|C_3|$ through the nodes and one pair in I_2.

L. Zajaczkowski([415]) applies the fundamental elliptic involution to generate certain ruled surfaces.

J. de Vries([375]) corrects a theorem of Em. Weyr on Poncelet polygons. See "Beiträge zur Curvenlehre" of the latter.

6. *Cayley-Brill Principle of Correspondence.*—See Pascal([263]). The generalization of the Chasles principle of correspondence to curves of genus greater than zero was first made by A. Cayley([66]). Particular cases were recognized by M. Chasles([82]) and by E. de Jonquières([165]). The number of coincidences of an (a, b) correspondence on a curve of genus p is $a+b+2\gamma p$. The proof was not complete, but the result was confirmed by four specific examples. See also A. Cayley([168]) and ([169]). The first proofs were given by A. Brill([40]), ([43]), ([46]) and ([47]). Other proofs were given by Junker([170]); by C. Bobek([37]); by considerations in hyperspace by C. Segre([307]); see § 12; by the enumerative method by H. Schubert([295]); by H. G. Zeuthen([419]), generalized to include the proper evaluation of multiple coincidences, in ([423]). This paper also contains a historical résumé of the preceding literature. A proof by F. Lindemann by the aid of abelian integrals is found in ([216]), and amplified in the Vorlesungen über Geometrie. Another transcendental proof is given by A. Hurwitz([159]). A proof using more elementary methods was found by F. Severi([323]). A new transcendental proof appears incidentally in connection with the study of Abel's theorem in a paper by O. Chisini([87]).

By means of the Cayley-Brill principle F. Severi, in ([322]), obtained a large number of properties of the characteristics of algebraic curves in S_r.

In the formula $u=a+b+2\gamma p$ the coefficient γ, called the *valence* (German Wertigkeit) is an integer, positive or negative, or zero. When positive or zero it indicates the number of times the point itself must be added to the group of its corresponding points, in order that the enlarged group shall belong to a linear series.

Every positive valence correspondence between the points (x) and (y) on a curve $f(x)=f(y)=0$ can be defined by one equation $\phi(x, y)=0$ in addition to the above equations. For curves with arbitrary coefficients (moduli) only valence correspondences exist, but the valence may be negative. See also G. Scorza([301]).

Two or more correspondences (a_1, b_1), (a_2, b_2) ... on the same curve are said to be dependent if integers λ_1, λ_2, ... can be found such that the group of b_1 points taken λ_1 times plus the b_2 points in the second taken λ_2 times ... move in a linear series as the point of departure describes the given curve.

A correspondence and its inverse have the same valence. On any algebraic curve correspondences exist with given valence, either positive, zero or negative.

The sum of two correspondences with valences γ_1, γ_2 is a correspondence with the valence $\gamma_1+\gamma_2$. The product of two correspondences with valences γ_1, γ_2 is a correspondence with valence $-\gamma_1\gamma_2$. For the rational determination of valence, see M. Shibayama([326]).

7. *Singular Correspondences.*—The general determination of all possible algebraic correspondence on algebraic curves was obtained by A. Hurwitz([159]), ([160]), ([161]). Earlier attempts had been made by F. Lindemann([216]). The methods of Hurwitz are treated at length in Klein-Fricke([171]) and by H. F. Baker([20]). See H. Burkhardt([51]).

The correspondences with valence are the only ones which exist on curves with general moduli, but the valence can be negative only for correspondences which are rationally composite, and can therefore be completely defined by means of two auxiliary equations. For particular moduli other correspondences, called *singular* correspondences, exist. Examples of such correspondences were pointed out by F. Klein([175]). The particular case of the harmonic cubic was found by F. Klein([174]). This with that of the equianharmonic cubic was further developed and all possible correspondences on elliptic curves were found by C. Segre([315]). See also F. Amodeo([8]).

Hurwitz showed that in every case two equations are sufficient to determine any algebraic correspondence completely. $f(x)=f(y)=0$; $\phi_1(x|y)=0$, $\phi_2(x|y)=0$. The details of a singular elliptic correspondence were given by F. Severi([323]). The theory is intimately connected with that of complex multiplication. See L. Bertini([24]). G. Tafani([350]) gives alternate proofs.

The general theory of singular correspondences is given in Chapter 15.

8. *Zeuthen's Formula.*—Given an (a, b) correspondence between two curves C, C' of genera p, p' respectively. Let η be the number of branch points on C, such that for each of them two of the b corresponding points on C' coincide, and η' the number of branch points on C'. Then

$$\eta-\eta'=2a(p'-1)-2b(p-1),$$

H. G. Zeuthen([421]). When $a=1$ or $b=1$, the groups of points on C', respectively C, describe an involution. If $a=b=1$ both η and η' are zero and we have a proof that two curves in (1, 1) correspondence must be of the same genus.

The η, η' points were obtained on the hypothesis that for each branch point only two of the corresponding points are coincident. If for a point A, ν' coincide at A', and similarly if for a point B' ν points B coincide on C, then the first member of the preceding equation is to be replaced by $\Sigma(\nu'-\nu)$. This was found by G. Halphen in ([133]) and ([139]). Other proofs are given by H. G. Zeuthen([416]); by W. Weiss([383]) with applications to ruled surfaces; C. Segre([307]); E. Bertini([25]). By other methods the theorem was found by X. Stouff([339]); R. de Paolis([262]). A corollary to the first theorem was given by H. Weber([381]). If $p=p'$ and $a=1$, then b must also be 1.

9. *Correspondence on a Cubic Curve.*—The early applications of correspondence and residuation were made to cubic curves. J. J. Sylvester([345]) considers μ points on C_3; pass a C_v through them, intersecting in $3v-\mu$ residual points. Through these pass a $C_{v'}$.... If $\mu=3i+1$ the final residual consists of a single point, the position of which is independent of the curves C_v, $C_{v'}$, ...

In ([346]) Sylvester shows how to draw a tangent to C_3 at a point P on it. Let A, B, be any two points of inflexion; let AP meet C_3 in 2, B_2 in 3, P_3 in 4, A_4 in 5, B_5 in Q. The line PQ touches C_3 at P.

In ([347]) Sylvester showed that the 27 sextatic points of C_3 may be resolved into 3 sets of 9 each; the line joining any two in the same set passes through a point of inflexion.

W. Schur([299]) gives a method of generating cubic curves by pencils of conics.

Em. Weyr([395]) generalized his treatment of correspondences on a general cubic.

Further properties of complete intersections with a cubic are given by J. J. Sylvester([348]). The generation of cubics by symmetric multiple correspondences is given by Em. Weyr([391]) and by J. de Vries([373]). Sylvester([349]) gives another illustration of residuation. See F. Morley([241]). A. S. Hart([146]) and ([145]) gives constructions for the ninth point of intersection of a pencil of cubics through 8 given ones. Other constructions were given by A. Cayley([62]) and by E. B. Elliott([111]). W. P. Milne([233]), ([234]) generates a cubic by apolar pencils of lines([231]).

F. G. Taylor([351]) applies quadratic inversion to pencils of cubics. See also W. P. Milne and D. G. Taylor([235]).

10. *Generation of Cubic Curves.*—Em. Weyr([390]) gives properties of tangentials in generating a C_3. C. Küpper([195]) considers a cubic as the envelope of a system of conics through two fixed points. The conics which osculate a cubic are discussed by J. Steiner([336]); by H. Durege([110]).

Em. Weyr([406]) and ([407]) and ([391]) generates cubic curves by means of symmetric (2, 2) correspondences of tangents to two fixed conics.

The Steiner polygons inscribed in non-singular cubics, that is, one having its odd sides all pass through one fixed point on the curve, and its even sides through another, have the property that if one such polygon exists for a given pair of points P, P', then one exists starting from any point of the curve as vertex. J. Steiner([336]); A. Clebsch([92]); C. Küpper([192]); P. H. Schoute([292]), V. Martinetti([227]).

11. *Quartic Curves.*—A. Cayley([61]) outlined the parametric representation of plane curves. If $p=0$, a point is expressed by rational functions of a parameter λ; if of genus 1, by two parameters λ, μ rationally, connected by $(\lambda^2, \mu^2)=0$; if of genus 2, by λ, μ rationally, connected by $(\lambda^3, \mu^2)=0$, and an outline of further cases. Cayley supposed that the minimum order of a curve of genus p is $p+2$. He([65]) discussed an apparent paradox in the parametric representation when radicals are involved.

M. Azzarelli([19]) obtains a particular quartic by means of reciprocal radii. Cayley proved that at a point of undulation a curve is touched by its Hessian; U. Masoni([228]) proves that the first polar of a point of undulation on C_4 consists of a tangent and of a conic and conversely. G. Humbert([156]) expressed the coordinates of a binodal quartic parametrically in terms of elliptic theta-functions.

C. Le Paige in ([211]) and ([209]) constructs C_4 through 14 points.

F. Purser and Haughton([278]) propose the question of constructing families of C_4 through restricted basis points and finding other basis points.

A. Ameseder([7]) considers involutions formed by conics through the nodes of an elliptic quartic and obtains various configurations, including conditions for inscribed polygons.

L. Gegenbauer([128]) considers the $I_4{}^3$ formed by the planes of space on an elliptic space quartic curve.

W. Wirtinger([414]) discusses the relations between the nodal quartic curve and the Kummer surface. Every such quartic can be generated by two projective pencils of conics with one common basis point.

E. Caporali([53]); C. Segre([306]); E. Ciani([90]) give various derivations of the quartic which possesses a series of apolar conics conjugate to the line meeting C_4 in 4 equianharmonic points.

The nodal quartic was discussed by H. E. Timerding([353]) as the projection of a C_5 of S_3 of type (2, 3) on a quadric surface.

M. Antscherl([16]) continuing the procedure given by C. F. Geiser([129]) shows that the general tangent cone from a point on a cubic surface contains as plane sections the general C_4. The 27 pencils of planes through the lines of the surface give twenty-seven systems of contact conics of C_4. The other thirty-six systems of contact conics arise from the pencils of space cubic curves on F_3 through the point of contact.

W. P. Milne([230]) shows that any C_4 can be generated by two projective pencils of conics. A conic of one pencil is apolar to all the conics of the second pencil. In ([232]) the same author generates a C_4 as the apolar locus of two tetrads of points on a conic.

C. M. Hebbert([148]) studies certain quartic curves by means of the transformation $z' = \frac{1}{z^2}$ in the complex plane. Isotropic coordinates $z, \bar{z}$ are employed. See also E. Beltrami([22]); G. Cesàro([72]); Perma([268]); Levy([215]).

W. V. der Woude([366]) obtains a hyperelliptic C_4 having ∞^1 configurations of Desargues $(10_3, 10_3)$.

A. Harnack([144]); Servais([316]); R. Montessus([240]) give a theory of the relation between elliptic space quartics, plane binodal quartics and plane cubic curves by means of elliptic functions.

12. *Contact Conics of Quartics.*—See ([191]). C. Küpper([201]) considers the system of conics through the nodes of an elliptic C_4, and touching it in two other points.

A. Brill([42]) studies curves having three point contact with a given curve. H. Krey([186]) applied the same method to start the study of those having four-point contact; W. Spottiswoode([335]) and ([334]), ([333]) extended the theory of four-point contact.

Härtenberger([147]) generates C_4 by means of two projective pencils of conics; A. Ameseder([5]). Every C_2 through the four points of contact of a contact C_2 of C_4 meets C_4 in another group of four points of contact. O. Hesse([150]), A. Ameseder([6]) applies the same method to the discussion of various particular cases.

C. Aronhold([18]) shows that 7 of the 28 bitangents of a non-singular C_4 can be chosen at will, and that all the others can be expressed as functions of these. The method used is the dual of the Geiser involution.

G. Kohn([183]). Given six pairs of bitangents a_i, b_i belonging to the same one-parameter system of contact conics of C_4, every C_2 touching five of them touches the other and is an involution conic of the system.

M. Noether([258]) studies the contact conics of C_4 by means of the (1, 2) correspondence between two planes which is associated with the Geiser involution. Previous studies in this direction had been made by A. Clebsch([93]); M. Noether([245]); R. de Paolis([261]).

G. Kohn([182]) obtained relations between different systems of contact conics. Two different systems have the same involution conic.

13. *The Brill-Noether Theory. The Fundamental Theorem.*—It was first shown by M. Noether([250]) that if $f=0$, $\phi=0$ are the equations of algebraic curves having P_i for r_i-fold and s_i-fold points respectively, then if $F=0$ represents a curve having P_i to multiplicity r_i+s_i-1, A, B, can always be found so that $F=0$ can be put in the form $F=Af+B\phi$ and conversely. For its application to cubic curves see J. J. Sylvester([344]) and A. Cayley([64]).

M. Noether([249]) further discussed the general case, but did not complete the proof. See also ([251]), ([256]), ([257]); G. Halphen([140]); A. Voss([370]); W. Weiss([384]); C. A. Scott([302]); F. Severi([319]). For reference to numerous other proofs from different standpoints, see ([31]), E. Löffler([218]) and ([217]) and E. Lasker([208]).

14. *Linear Series of Point Groups.*—Given $f=0$ and a linear system of curves $\Sigma\lambda_i\phi_i=0$. The latter cut from the former a linear series $g_n{}^r$ in which n is the order, or number of variable points cut from $f=0$ by each curve, and r represents the number of dimensions or of arbitrary constants (degrees of freedom) of the system $\Sigma\lambda_i\phi_i=0$, independent of $f=0$. A. Brill and M. Noether([44]) and ([41]). The same idea had been presented transcendentally by B. Riemann([282]), by A. Clebsch and P. Gordan([91]); and employed further by M. Noether([246]) and ([247]). For other methods see ([31]).

If the curves $\phi_i=0$ are *adjoints,* having each s-fold point of $f=0$ to multiplicity $s-1$, then either the curves of the system through one point of $f=0$ will not have another variable point in common on $f=0$, or the same property will hold for every point of f and the points describe an *involution,* as one basis point describes the curve.

If $f=0$ is of order n and genus p, then the adjoints $\Sigma\lambda_i\phi_i=0$ of order $n-3$ cut from $f=0$ the *canonical* series g_{2p-2}^{p-1}.

If $g_n{}^r$ satisfies the relation $r\geqq n-p+1$, the series is called a *special* series.

If the points on $f=0$ are arranged in groups of n such that r arbitrary points determine a group, the system is called an *involution* of order n and dimension r. If $p=0$ all series are involutions. If $p>0$, $g_n{}^1$ may be *irrational,* thus not contained in a one-dimensional linear series. If the canonical series $g_{2(p-1)}^{p-1}$ is compounded with an involution, the latter must be of order 2. Any curve having a linear $g_2{}^1$ is called *hyperelliptic.* G. Castelnuovo([56]); F. Amodeo([9]); R. Torelli([358]); J. D. de Porte([277]). A more geometric development of the Brill-Noether theory was given by O. Tognoli([355]) and ([356]). See also A. Brill([48]).

C. Weltzien([385]) discussed the parametric representation of hyperelliptic curves in the form

$$Px_i = (a_i t + b_i)\sqrt{E(t)} + (c_i t + d_i)\sqrt{F(t)}$$

in which a_i ... are constants and $E(t)$, $F(t)$ are fixed polynomials of degree $p+1$. The curves are of order $p+3$ and of genus p.

If $r>1$, $g_n{}^r$ is always rational except for series compounded of r groups of an irrational one-dimensional involution. G. Castelnuovo([60]); G. Humbert([155]) first proved this theorem by means of abelian integrals. Geometric proofs are given by M. De Franchis([103]) and ([102]), by F. Severi([321]) and ([323]) and for $p=1$, by G. Castelnuovo([58]). F. Enriques([113]) proved that every $\gamma_n{}^r (r>1)$ is either rational or contained in a rational series $g_n{}^s$, $s>r$. See also F. Severi([318]); R. Torelli([361]).

If the points of any group G_n cut from $f=0$ by an adjoint $\phi=0$ are divided into two parts, each is called the residual of the other.

If G_n and G'_n are residuals as to G_q, and G_n is a residual of G'_q, then G'_n is a residual of G_q. If G_n is fixed, all the coresidual groups $G_q G'_q$ are cut out by the same linear series, and are called *equivalent*. For examples of residuation see E. Study([340]).

A series $g_n{}^r$ is called complete if it is not contained in a series $g_n{}^s$, $s>r$. The complete series to which any group or any partial series belongs is unique. The order, dimension, and quality of completeness are all invariant under birational transformation.

E. Bertini([26]) showed that the canonical series g^{p-1}_{2p-2} is the only one of its order and dimension. It is simple if the curve does not possess a linear $g_2{}^1$, is complete, and has no basis points. C. Segre([307]) gives a systematic development of the Brill-Noether theory by means of projections from hyperspace. See F. S. Macaulay([225]). See also C. A. Scott([303]); F. Macaulay([226]); F. Amodeo([13]) shows that if C_m has adjoints of order $m-3-\alpha$ it defines special series $g_n{}^r$, $n \leqq 2p-2-\alpha m$, $r = p-1-\left[\alpha m - \frac{x}{2}(\alpha+3)\right]+\rho$. F. Hardcastle([142]) gives an extensive bibliography covering 1896 and previous years. F. Severi([325]) gives a new demonstration of the Brill-Noether theory.

If a group G_n on f determines a complete series $g_n{}^r$ and belongs to $r'+1$ linearly independent adjoints ϕ of order $n-3$, then $r'=p-n+r-1$. This is the *Riemann-Roch* theorem, which may also be expressed in the form:

Every special complete series $g_n{}^r$ has, as residual to the canonical series g^{p-1}_{2p-2}, another special complete series $g^{r'}_{n'}$ wherein $n+n'=2p-2$, $n-n'=2(r-r')$. See A. Brill and M. Noether([44]).

15. *Normal Curves.*—Let the canonical series g_{2p-2}^{p-1} be cut from $f=0$ by $\sum_{i=1}^{p} \lambda_i \phi_i = 0$. If we put $\rho y_i = \phi_i(x)$, and regard $y_i, \ldots, y_p$ as the homogeneous coordinates of a point in a linear space S_{p-1} of $p-1$ dimensions, than as (x) describes the curve $f=0$, (y) will describe a curve in S_{p-1} called the *canonical* curve of $f=0$. It is of order $2p-2$ and of genus p. If $f=0$ has no g_2^1, this curve is simple and has no multiple points.

The birational transformations of $f=0$ are now expressed in terms of projective transformations of the canonical curve. This concept originated with B. Riemann (281). See also H. Weber (382); L. Krauss (185); M. Noether (253); F. Klein (176), (172) and (173); and particularly C. Segre (307). By projecting this curve from an S_{p-3} determined by $p-2$ arbitrary points on it, we obtain plane curves birationally equivalent to $f=0$, of order $p+2$, with a point of multiplicity $\frac{1}{2}(p+2)$ and having $\frac{p}{4}(p-4)$ double points if p is even, or of order $p+3$, having one point of multiplicity $\frac{1}{2}(p+3)$ and $\frac{1}{4}(p-1)^2$ double points if p is odd. These were called normal curves by B. Riemann (282).

By means of a net of adjoints through $p-3$ arbitrary points on $f=0$, Clebsch-Gordan (91) obtained a curve of order $p+1$ with $\frac{p}{2}(p-3)$ double points. See also C. A. Scott (304). By projection from a proper S_{p-3} which contains the maximum number of points, the plane curve can be reduced to one of order $p-\pi+2$ having $\frac{1}{2}(p-\pi+1)(p-\pi)-P$ double points, π being the largest multiple of 3 contained in p.

The normal forms for $p=5$, 6, 7 were obtained by M. Noether (255) by means of the quadratic identities among the adjoints.

For $p=2$, 3, 4, 5, 6 see A. Wiman (413). C. E. Pengra (267). By means of the quadratic relations among the ϕ's, see A. Van Benschoten (365) for $p=4$; J. V. McKelvey (229) for $p=5$; V. Snyder (329) for $p=6$.

16. *Reduction to Forms with Double Points.*—It was proved by M. Noether (248) that any plane algebraic curve may, by Cremona transformation of the plane, be reduced to a curve having multiple points with distinct tangents at each. W. K. Clifford (95) studied the curves C_n normal in S_n or S_{n-1}. The former are rational and the latter can be projected into plane cubics. A C_n in S_{n-1} has n^2 points at which the osculating S_{n-2} has n-point contact.

C. Küpper (196), using the same method as M. Noether (251), showed that C_m in S_2 can be the projection of a C_m of S_3 only when it has at least $\frac{(m-1)^2}{4}$, $\frac{m(m-2)}{4}$ double points respectively, according as m

is odd or even. K. Bobek([39]) discussed a curve C_m, in S_3, genus p, $m-3>p>1$ having an $(m-2)$-fold secant. It is hyperelliptic and lies on a ruled surface of order $m-p$ on which the multiple secant is of multiplicity $m-p-1$. E. Bertini([28]) gave a proof that any algebraic plane curve can be transformed into one having no higher singularities than distinct double points. For further proofs, see the following:

K. Hensel and G. Landsberg([149]); G. Landsberg([207]); L. W. Thomé ([352]); G. Halphen([138]) and ([134]); A. R. Forsyth([122]); P. Appell and E. Goursat([17]); L. Kronecker([190]) and ([189]); G. Simart([327]); E. Vessiot([308]); H. Poincaré([276]); P. del Pezzo([269]); M. Pieri([273]); E. Vessiot([369]); C. Segre([308]); B. Levi([214]); G. A. Bliss([34]); F. Enriques and O. Chisini([112]).

C. Küpper([197]) showed that if a plane non-singular C_n, $n>3$ is transformed into another C_n birationally, the transformation must be a collineation. Such curves cannot be projections of space curves of the same order. E. Bertini([27]) gave another proof by means of (1, 2) plane transformations.

This theorem was generalized by V. Snyder([328]) as follows: The largest number of distinct double points which a curve of order n may have without being birationally transformable into another curve of the same order except by collineation is two less than the minimum number of double points which a space curve of the same order can have $(n>7)$. See also V. Snyder([330]).

A. Del Re([105]) proved that any plane algebraic curve can be transformed birationally into a curve C_m of S_5 having no singularities.

V. Snyder([331]) proved that plane algebraic curves of every order n and of any genus p, $0\leqq p\leqq \dfrac{(n-1)(n-2)}{2}$ exist, having distinct double points and with distinct tangents at each.

J. V. Nagy([243]) discussed the restrictions of the reducibility from the arithmetic stand-point.

F. Severi([320]) showed that every general family of curves of order n, genus p, in S_{n-p}, includes limiting curves composed of an $(n-p)$-sided polygon having $n-p-1$ nodes, together with p arbitrary chords. A family of plane C_n with not more than $\dfrac{(n-1)(n-2)}{2}$ double points includes all possible n sides, and each C_n contains composite ones composed of all possible C_{n-1} and a line. See also F. Severi [([317]) Anhang G.].

17. *Particular Curves of Order* $n>4$.—For the general theory see G. Loria([220]). Only those curves will be considered which are connected with some definite transformation or involution. The space C_5, $p=1$

was considered by Em. Weyr(396), (397) and (398). It has an I_3^1 of trisecants forming a quintic ruled surface of genus 1 on which the given curve is double. Ruled cubic surfaces through the curve have tangent planes that describe an I_3^2 which is uniquely fixed by one triad. Every I_4^3 on C_5 can be cut by quadrics through two trisecants. D. Montesano(237). C. Segre(310). Em. Weyr(399) and (400) employs similar methods to obtain involutions on the space sextic curve of genus 1.

H. Schröter(294) considered the locus of the eight points of intersection of three conjugate pairs of surfaces given by three involutions of planes. It is a C_7 having the axes for trisecants. See also P. H. Schoute(293).

D. Montesano(238) found all involutorial birational transformations in S_3 such that lines joining pairs of corresponding points belong to a linear complex. The most general one is determined by a C_{10} of genus 11 which has the property that the $\infty^3 I_{10}^3$ cut on it by the planes of space all lie on cubic curves. The curve does not lie on a cubic surface. A particular case gives C_7 of genus 5 which also appears in the study of bilinear complexes of conics. Every C_7 of genus 5 in S_3 has one point O having the property that the six residual points in which any plane through O meets it, lie on a conic; thus defining a rational I_6^2. Each trisecant meets just one other at a point not on C_7, and the plane determined by them passes through O. R. Sturm(340a) and D. Montesano(239). Other properties of particular forms of the curve are given by Th. Reye(280), G. Koenigs(181); G. Humbert(157); D. Montesano(236).

18. *Irrational Involutions.*—Given two curves C, C', in $(k, 1)$ correspondence, such that to a point on C corresponds one point on C', but to a point of C' correspond k points on C. Then C is said to contain an involution γ_k^1 of the genus p' of C'. If $p'>0$, C is said to contain an irrational involution of order k and genus p'. If p is the genus of C it follows from the Zeuthen theorem that $p' \not< p$.

A γ_4^1 on C_6 of genus 5 was given by A. Cayley(63); it is of index 5.

If $p'=0$, the notation g_k^1 will be used. In this case the k points of each group G_k of g_k^1 belong to a linear series. Any curve containing a g_2^1 is hyperelliptic.

L. Fuchs(127) published the theorem that the necessary and sufficient condition that an algebraic curve $f(x, y)=0$ is invariant under an involutorial birational transformation is that it can be transformed birationally into the form $\psi(x)y^2=\phi(x)$, or that it is hyperelliptic, but it was pointed out by A. Hurwitz, that this is incorrect. It is true if the curve does not have a γ_2^1.

C. Küpper and C. Bobek(205) determine the forms of hyperelliptic curves of order $3n$ which belong to the Geiser or the Bertini involutions.

C. Segre (315) determines all possible involutions on general and singular elliptic curves. On the former the only involutorial birational transformations are the central g_2^1 associated with every point of the curve, and the three γ_2^1 which can be expressed as the product of two central projections at any points having a common first tangential. For harmonic curves, cyclic involutions I_4^1 also appear, and for equianharmonic curves cyclic involutions of order 6. This is intimately connected with the theory of complex multiplication of elliptic functions. See G. Greenhill (132); G. Halphen (135); A. Hurwitz (159) and (160); (158) and (161); F. Klein (174); C. Segre (310).

The involutions can be expressed in terms of the parameter u of an elliptic function by $u' = \pm u + C$ in the general case, and by $u' = \pm \omega u + C$, $u' = \pm \omega^2 u + C$, where $\omega^2 + \omega + 1 = 0$, in the equianharmonic case, and by $u' = \pm iu + C$ in the harmonic case. The earlier literature contained theorems which were stated incorrectly. See A. Harnack (143) and (144); Em. Weyr (393) and (394); C. Küpper (192).

Associated with every (n, n') correspondence between two curves of genera p, p' is an intermediary curve of genus

$$P = 2(p+1) + n' = 2(p'-1) + n,$$

n' and n being the number of branch points on C', C, on which exist two involutions, one of order n and genus p', the other of order n', genus p. By means of this theorem V. Snyder and F. R. Sharpe (332) showed how (n, n') correspondences could be constructed, which can be expressed by two auxiliary equations, but not by one.

G. Fontene (121) considered a composite (m, n) correspondence on a given curve by means of a connex.

G. Castelnuovo (56) showed that if a curve of genus p contained an irrational involution γ_ν^1 of genus π and also a rational series $g_n^{\nu-1}$, the second not containing the first, then there exist $y = (n-\nu+1) - p + \nu\pi$ groups of the first which belong to the second. If $n - \nu < p - \nu\pi$, then a group of the former series imposes at most $\nu - 1$ conditions on a group of the second in order to be contained in it (generalization of the Riemann-Roch theorem). If on the canonical C_{2p-2} of S_{p-1} exists a γ_ν^1 of genus π, through $\eta = 2(p-1) - 2\nu(\pi-1)$ coincidences passes an $S_{p-\pi+1}$ from which the curve can be projected into $C_{2\pi-2}$ of $S_{\pi-1}$ counted ν-fold. F. Amodeo (9); C. Segre (314); C. Küpper (198), determines a number of conditions on point-groups in the plane such that an C_m through them has an excess on the number of degrees of freedom left. R. Torelli (358) applied these results to hyperelliptic curves.

The theorem of Lüroth (224) that rational curves cannot contain irrational involutions was extended by E. Goursat (131) to apply to curves of genus p as follows: if the coordinates of a point on an algebraic curve C are rational functions of those of C', the genus of C is at most equal to that of C'.

R. Torelli (362) showed that if on C of genus p there exists $\gamma_\mu{}^1$ of index ν, then the number z of groups containing $r+1$ points of a $g_n{}^r$ is given by $z=\binom{\mu-1}{r}\nu n-\frac{1}{2}\binom{\mu-2}{r-1}d$, in which d is the number of double points of $\gamma_\mu{}^1$. In (357) R. Torelli showed that if a curve of genus $p=2$ contains a $\gamma_\nu{}^1$, it also contains a second irrational involution. If C is of genus 3 and contains a $\gamma_3{}^1$ of $\pi=3$, it also contains an elliptic involution.

G. Castelnuovo (57) obtained the condition that a $\gamma_n{}^1$ of index ν is contained totally in a linear series. The index means the number of groups of $\gamma_n{}^1$ through a given point. Such a series can have at most $2\nu(n+p-1)$ coincidences. The necessary and sufficient condition that the series is contained in a linear series is that this number is actually attained. This was generalized by R. Torelli (360). If a $\gamma_n{}^1$ of index ν is so constituted that the totality of the ν groups G_n of the series through a point P (including P itself counted ν times) describes a linear series as P describes the curve, then the given series is linear. This was first proved by transcendental means by F. Severi (318), see also (324). Also proved by G. Castelnuovo by an application of the theorem just cited.

If a curve C of genus p contains an involution $\gamma_2{}^1$ of genus π C depends on $2p-\pi-1$ moduli. C. Segre (307).

R. Torelli (359) obtains further relations among curves with singular moduli which contain irrational $\gamma_n{}^1$. Some of these are extended to series of higher dimensions by E. S. Allen (2), and (1); J. V. de Porte (277).

Every curve of genus 3 representable on a double curve of genus 2 is hyperelliptic. F. Enriques (114).

O. Chisini (86). Given two plane curves $f_1(x, y)=0$, $f_2(x, y)=0$ in (2, 2) point correspondence. Three cases are possible:

(a) f_1, f_2 are birationally equivalent,
(b) f_1, f_2 have the same $\gamma_2{}^1$,
(c) f_1, f_2 are conjugate of each other.

If the correspondence is not periodic, both f_1 and f_2 are rational. A curve ϕ can be constructed in (1, 2) correspondence with each f_1, f_2. ϕ may be of genus 0, 1, 2, ..., and (2, 2) may be of period 2^r.

19. *Relations to Enumerative Geometry.*—J. Steiner (337) treated the problem of generating plane algebraic curves of order $m+n$ by means

of projective pencils of orders m and n, by a method previously employed to generate cubics and quartics by M. Chasles([76]) and ([77]). That an involution is associated with a (1, 2) correspondence was shown in ([78]), generalized in ([79]) by the same author. The concept of an involution of higher order was given by E. de Jonquières([164.1]). The number of coincidences of an (a, b) correspondence on a straight line was given by E. de Jonquières([167]) and used repeatedly by L. Cremona([97]). M. Chasles explicitly gave the formula $u=a+b$ ([80]) and applied it to numerous problems on conics([81]).

E. de Jonquières([166]) showed that in a system ∞^1 of curves of order m, index μ, the number ν which touch a given line is $\nu=2(m-1)$. Problems in multiplicity of elements and in recognition of extraneous factors were considered by H. G. Zeuthen([419]) and applied to the derivation of Plücker's formulas([424]). The later derivation of these formulas by O. Zimmermann([425]) is an extension of the same method. The formulas were extended to space curves by H. G. Zeuthen([418]); to systems of rational curves([420]); to algebraic surfaces in ([422]); and by R. Sturm([342]). Application to the surface of centers of curvature by G. Darboux([101]); G. Fouret([124]). Other applications by H. Schubert([297]) and ([298]); by H. Krey([187]) and ([188]); by L. Saltel([288]) and ([289]) and ([286]). The extensive applications made by Em. Weyr were collected and reproduced in his book([386]). G. Halphen([136]) showed that in a system ∞^1 of conics the characteristic is always of the form $a\mu+b\nu$, μ being the number through a point, ν the number touching a given line.

H. G. Zeuthen([421]) proved that in an (a, b) correspondence between two curves of genera p_1, p_2 with η, η^1 coincidences, then

$$\eta-\eta^1=2b(p_1-1)-2a(p_2-1).$$

The first member was replaced by $\Sigma(\eta-\eta^1)$ by G. Halphen([139]); H. G. Zeuthen([417]), in which higher coincidences are replaced by their equivalents in terms of elementary ones. See also A. Cayley([67]); M. Noether([252]); G. Halphen([137]).

The Cayley-Brill formula has already been given. E. Bertini([30]); G. Castelnuovo([59]); M. Noether([246]) and ([254]) gave proofs of the theorem that if C of genus p contains $g_n{}^r$, then

$$p\leqq(k+1)\left(n-r-\frac{k(r-1)}{2}\right)$$

wherein k is the maximum integer in $\frac{n-r}{r-1}$. For the application of the theorem to possible projections of C_n in S_r, see G. Veronese([367]); C. Segre([311]) and ([312]).

If C_m is the projection of another C_m of S_r, the order of the minimum adjoint of C_m is in the interval μ, where

$$m-3>\mu>m-3-\frac{m-r}{r-1}.$$

See C. Küpper([193]). G. Schaake([291]) determined the effect of actual multiple points on the surface of trisecants of a space curve.

20. *Applications of Binary Forms.*—For cubic curves inscribed in a hexagon, see A. Cayley([68]). For concyclic points, see E. Beltrami([21]). For problems in elimination see L. Saltel([287]) who obtains the order of the eliminant of three ternary polynomials in the general case. F. Morley([241.1]) obtained the resultant of three ternary forms of the same order. The general problem in any number of variables is discussed by Morley and A. B. Coble([242]). V. Retali([279]) employs quadratic inversion and elimination to obtain various theorems regarding conics having multiple contact with the given curve. E. Ciani([88]) studies the $(16_6, 16_6)$ configuration when the six coplanar points are in k-fold involution. J. Fairon([115]) illustrates properties of binary forms by means of points on a conic. P. Savio([290]) obtains the complete system of 18 forms derived from the classification of (2, 2) correspondences. The method is purely algebraic; binary in two sets.

G. Roch([283]) showed that if $F(s, z)=0$ defines an algebraic function z of s, then any s', rational in s and z has $m-p+1+q$ arbitrary constants, in which m is the order of s', p is the genus of $F(s, z)=0$ and q is the number of poles of $\dfrac{\phi}{F'}$.

The logarithmic integrals when $p=2$ and $p=3$ were determined by the same author in ([284]). An application to the determination of the bitangents of the non-singular quartic was given by the same author([285]). The theory of automorphic functions was applied to algebraic curves by G. Humbert([151]), ([152]), ([153]) and ([154]). Two new covariants were added by H. S. White([412]).

A parametric representation for curves of genus 4 was determined by H. Poincaré([275]). The canonical representation of the abelian integrals of all three kinds for the case in which the function is represented by a curve having only linear branches at multiple points was obtained by J. C. Fields([117]).

The general problem of rationalization of the branches of an algebraic curve has been exhaustively discussed by P. Koebe([177]), ([178]), ([179]) and ([180]). Criteria for adjointness were given by J. C. Fields([118]) and relations between branch-points and double points of an algebraic curve were obtained by the same author([119]), who develops the whole general theory of algebraic functions of one variable([120]). The periodicity of abelian integrals of the second kind was determined in a different form by E. Picard([270]).

J. Fairon([116]) gives a representation of a binary quartic on a ruled cubic surface and on a cubic curve.

M. Stuyvaert([343]) studies various geometric configurations which arise from the condition that two simultaneous binary forms have more than one common root. A. B. Coble([96]) generalizes the idea of closure to apply to (k, m) correspondences. See also ([96.1]). Moreover, using the notation and interpretation of W. F. Meyer([229.1]), he obtains a considerable number of new results in the applications of binary forms to involutions. See ([96.2]), ([96.3]), ([96.4]).

21. *Polar Properties.*—E. Waelsch([380]), Em. Weyr([392]); also C. Le Paige([212]), and A. Del Re([104]) gave a number of theorems concerning polar properties which are applied to adjoints of a given curve.

BIBLIOGRAPHY

1 Allen, E. S., Pal. Circ. Mat., 37: 345-370 (1914). [18.

2 ———, Rom. Acc. L. Rend., (5), 22_2: 424-428 (1913). [18.

3 Ameseder, A., Wien. Ber., 81: 271-299 (1880). [3.

4 ———, Wien. Ber., 81: 615-647 (1880). [3.

5 ———, Wien. Ber., 86: 396-423 (1882). [12.

6 ———, Wien. Ber., 87: 15-81 (1883). [12.

7 ———, Wien. Ber., 93: 357-379 (1886). [3, 11.

8 Amodeo, F., Ann. di. Mat., (2), 19: 1-27, 145-157 (1891). [7.

9 ———, Ann. di Mat., (2), 20: 227-235 (1892). [14, 18.

10 ———, Ann. di Mat., (2), 21: 221-236 (1893). [5.

11 ———, Ann. di. Mat., (2), 22: 1-22 (1896). [5.

12 ———, Nap. R. A. Atti., 9: 23 p. (1889). [5.

13 ———, Nap. R. A. Rend., (3), 2: 316-333 (1896). [14.

14 ———, Nap. R. A. Rend., (3), 3: 207-223 (1897). [5.

15 ———, Nap. R. A. Rend., (3), 6: 174-191 (1900). [5.

16 Antscherl, M., Wien. Ber., 123: 1189-1203 (1914). [11.

17 Appell, P., and Goursat, E., Theorie d. funct. alg., 276 (1895). [16.

18 Aronhold, C., Berlin, Monatsber., 499-523 (1864). [12.

19 Azzarelli, M., Rome Pont. Acc. Atti., 23: 234-253 (1875). [11.

20 Baker, H. F., Abel's Theorem, p. 639 (1897). [7.

21 Beltrami, E., Bologna Mem., (2), 9: 607-658 (1869). [20.

22 ———, Bologna Mem., (3), 10: 626 (1870). [11.

23 ———, Bologna Mem., (3), 10: 233-312 (1870). [2.

24 Bertini, E., Le fonzioni ellittiche, 545 (1903). [7.

25 ———, Introd. alla geom. proi. degli iperspazii, 399 (1907). [8.

26 ———, Ann. di. Mat., (2), 22: 1-40 (1894). [14.

27 ———, Math. Ann., 44: 158-160 (1894). [16.

28 ———, Riv. F. M., 1: 22-24 (1891). [16.

29 ———, Rom. Acc. L. Rend., (3), 1: 92-96 (1877). [5.

30 Bertini, E., Torino Atti., 26: 118-130 (1890-1901). [19.
31 Berzolari, L., Encykl., III C4 no. 24, 406. [14, 13.
32 ———, Nap. R. A. Rend., (2), 5: 35-40 (1891). [3.
33 Bischoff, J., Ann. di. Mat., (2), 6: 144-147 (1873-75). [2.
34 Bliss, G. A., Am. M. S. Bull., 29: 161-183 (1923). [16.
35 Bobek, C., Mat. Ann., 29: 386-412 (1887). [5.
36 ———, Wien. Ber., 93: 601-617 (1886). [5.
37 ———, Wien. Ber., 93: 899-911 (1886). [6.
38 ———, Wien. Ber., 94: 861-873 (1886). [5.
39 ———, Wien. Ber., 95: 349-354 (1887). [16.
39a ———, Wien. Ber., 98: 141-172 (1889). [5.
40 Brill, A., Gött. Nachr., 507-518 (1871). [6.
41 Brill, A., and Noether, M., Gött. Nachr., 116-132 (1873). [14.
42 Brill, A., Math. Ann., 3: 456-458 (1870). [12.
43 ———, Math. Ann., 6: 33-65 (1873). [6.
44 Brill, A., and Noether, M., Math. Ann., 7: 269-316 (1873). [14.
45 Brill, A., Math. Ann., 8: 534-538 (1873). [1.
46 ———, Math. Ann., 7: 607-622 (1874). [6.
47 ———, Math. Ann., 31: 374-409 (1888). [6.
48 ———, Math. Ann., 36: 321-360 (1890). [14.
49 ———, C. R., 82: 1328-1330 (1876) (Fouret). [1.
50 Buchheim, A., London, M. S. Proc, (1), 13: 180-185 (1882). [2.
51 Burkhardt, H., C. R., 26: 1854-1856 (1898). [7.
52 Capelli, A., Giorn. di Mat., 69-148 (1879). [1.
53 Caporali, E., Mem. di Geom., nos. 20-22 (1888). [5, 11.
54 ———, Nap. R. A. Rend., 20: 143-147 (1881). [5.
55 Castelnuovo, G., Rom. Acc. L. Rend., (4), 5_2: 130-133 (1889). [4.
56 ———, Rom. Acc. L. Rend., (4), 7_2: 294-299 (1891). [14, 18.
57 ———, Rom. Acc. L. Rend., (5), 15_1: 337-344 (1906). [18.
58 ———, Torino Atti., 24: 4-22 (1888-9). [4, 14.
59 ———, Torino Atti., 24: 346-373 (1888). [19.
60 ———, Torino Atti., 28. 727-738 (1892-3). [4, 14.
61 Cayley, A., London M. S. Proc., 1: 8 p. (1869-71). [11.
62 ———, Math. Qu., 61: 122 (1894). [9.
63 ———, Math. Ann., 8: 359-362 (1875).
64 ———, Mess. of Math., 3: 62-74 (1874). [13.
65 ———, Mess. of Math., 14: 12-14 (1885). [11.
66 ———, C. R., 62: 586-590 (1866). [6.
67 ———, Quart. J., 7: 212-223 (1866). [19.
68 ———, Quart. J., 9: 210-221 (1868). [20.
69 ———, Quart. J., 11: 83-91 (1871). [2.
70 ———, Quart. J., 12: 197-198 (1873). [2.
71 ———, Quart. J., 15: 32-33 (1877-78). [5.
72 Cesàro, G., Nouv. Ann., (4), 1: 1-9 (1901). [11.
73 Chasles, M., J. de Math., (2), 18: 202-219 (1873). [1.
74 ———, Nouv. Ann., (2), 10: 97-104 (1871). [1.
75 ———, Nouv. Ann., (2), 10: 385-401 (1871). [1.
76 ———, C. R., 36: 943-953 (1853). [19.

77 ———, C. R., 37: 272-277 (1853). [19.
78 ———, C. R., 41: 1097-1107 (1855). [1, 19.
79 ———, C. R., 53: 985-996 (1861). [19.
80 ———, C. R., 58: 1167-1176 (1864). [1, 19.
81 ———, C. R., 59: 209-218 (1864). [19.
82 ———, C. R., 62: 579-586 (1866). [6.
83 ———, C. R., 72: 577-588 (1871). [1.
84 ———, C. R., 73: 927-932, 970-971 (1871). [1.
85 Chelini, D., Mem. D. Chelini, (1881). [2.
86 Chisini, O., Ann. di Mat., (3), 24: 121-156 (1915). [18.
87 ———, Lomb. Ist. Rend., (2), 54: 552-569 (1921). [6.
88 Ciani, E., Giorn. di Mat., 36: 68-76 (1898). [20.
89 ———, Giorn. di Mat., 53: 192-197 (1915). [5.
90 ———, Nap. R. A. Rend., (3), 2: 126-144 (1896). [11.
91 Clebsch, A., and Gordan, P., Theorie der Abel'schen Funk., 15 (1866). [14, 15.
92 Clebsch, A., J. f. Math., 63: 94-121 (1864). [10.
93 ———, Math. Ann., 3: 45-75 (1871). [12.
94 Clifford, W. K., Math. Quart., 27: 49-50 (1892). [2.
95 ———, Lond. R. S. Trans., 169: 663-681 (1878). [16.
96 Coble, A. B., Am. J., 43: 1-19 (1921). [20.
96.1 ———, Am. M. S. Trans., 28: 357-383 (1926). [21.
96.2 ———, Am. J., 31: 183-212 (1909). [21.
96.3 ———, Am. J., 31: 355-364 (1909). [21.
96.4 ———, Am. J., 32: 333-364 (1910). [21.
97 Cremona, L., Introd., §§ 83, 87, 98, 106, 116, 117 (1863). [1, 19.
98 ———, Nouv. Ann., (2), 3: 25-30 (1864). [1.
99 Czuber, E., Jahresbericht, 4: 100-107 (1894-95). [5.
100 ———, Prag. Ber., 156-162 (1875). [2.
101 Darboux, G., C. R., 70: 1328-1333 (1870). [19.
102 De Franchis, M., Pal. Circ. Mat., 17: 104-121 (1903). [4, 14.
103 ———, Rom. Acc. L. Rend., (5), 12: 303-310 (1903). [4, 14.
104 Del Re, A., Giorn. di Mat., 26: 348-351 (1888). [21.
105 ———, Nap. R. A. Rend., (3), 7: 202-203 (1901). [16.
106 Dewulf, E., Soc. M. F. Bull., (2), 2: 41-48, 372-392 (1878). [1.
107 ———, Nouv. Ann., (1), 18: 314-319 (1859) (Salmon). [1.
108 ———, Nouv. Ann., (2), 11: 297-305 (1872). [1.
109 D'Ocagne, M., Ann. Ac. Polyt. Porto, 9: 5-9 (1914). [3.
110 Durege, H., Prag. Ber., 47-63 (1871). [10.
111 Elliott, E. B., Math. Quart., 71: 90-91 (1899). [9.
112 Enriques, F., and Chisini, O., Theoria Equaz., 2: 550. [16.
113 Enriques, F., Pal. Circ. Mat., 10: 30-35 (1895). [14.
114 ———, Rom. Acc. L. Rend., (5), 7_2: 281-286 (1898). [18.
115 Fairon, J., Liège S. R. S. Mém., (3), 3: 73 p. (1901). [20.
116 ———, Liège S. R. S. Mém., (3), 6: 23 p. (1906). [20.
117 Fields, J. C., J. f. Math., 127: 277-308 (1904). [20.
118 ———, London M. S. Proc., (2), 11: 127-132 (1912-13). [20.
119 ———, Math. Ann., 73: 560-570 (1912-13). [20.

120 Fields, J. C., Lond. R. S. Trans., 212: 339-373 (1913). [20.
121 Fontene, G., Soc. M. F. Bull., 25: 247-267 (1896). [18.
122 Forsyth, A. R., Mess. of Math., 30: 1-7 (1900). [16.
123 Fouret, G., Pal. Circ. Mat., 3: 42-48 (1889). [3.
124 ———, Soc. M. F. Bull., 1: 122-124, 258-259 (1873). [19.
125 ———, Soc. M. F. Bull., 5: 19-24 (1876-77). [1.
126 Fraser, B. D., Edinb. M. S. Proc., 7: 46-61 (1889). [3.
127 Fuchs, L., Berlin Sitzungsber, 797-804 (1886). [18.
128 Gegenbauer, L., Wien. Ber., 93: 790-797 (1886). [11.
129 Geiser, F., Math. Ann., 1: 129-138 (1868). [11.
130 Gerbaldi, F., Pal. Circ. Mat., 43: 78-104 (1919). [2.
131 Goursat, E., Am. J., 16: 291-298 (1894). [18.
132 Greenhill, G., London M. S. Proc., (1), 19: 301-364 (1888). [18.
133 Halphen, G. H., Étude (Salmon), 626 (1884). [8.
134 ———, J. de Math., (3), 2: 87-144 (1876). [16.
135 ———, J. de Math., (4), 5: 5-52 (1889). [18.
136 ———, London M. S. Proc., (1), 9: 149-170 (1878). [19.
137 ———, C. R., 78: 1105-1108 (1874). [19.
138 ———, C. R., 80: 638-642 (1875). [16.
139 ———, Soc. M. F. Bull., 5: 9-18 (1876). [8, 19.
140 ———, London M. S. Proc., 5: 160-163 (1877). [13.
141 Hardcastle, F., London M. S. Proc., (1), 29: 132-139 (1897-98). [1.
142 ———, Am. M. S. Bull., 4: 390-402 (1897-98). [14.
143 Harnack, A., Math. Ann., 9: 1-54 (1876). [18.
144 ———, Math. Ann., 12: 47-86 (1877). [11, 18.
145 Hart, A. S., Camb. Dub. Math. J., 6: 181-182 (1851). [9.
146 ———, Quart. J., 22: 199 (1886-87). [9.
147 Härtenberger, J. f. Math., 58: 54-79 (1861). [12.
148 Hebbert, C. M., Tôhoku M. J., 13: 1-14 (1918). [11.
149 Hensel, K., and Landsberg, G., Theorie der alge. Funk., chap. 24, 365 (1902). [16.
150 Hesse, O., Sven. Vet. Ak. Bih., 49: 263-79 (1855). [2.
151 Humbert, G., J. de Math., (4), 2: 239-329 (1886). [20.
152 ———, J. de Math., (4), 3: 327-404 (1887). [20.
153 ———, J. de Math., (4), 5: 81-134 (1889). [20.
154 ———, J. de Math., (4), 6: 233-292 (1890). [20.
155 ———, J. de Math., (4), 10: 169-201 (1894). [4, 14.
156 ———, C. R., 97: 989-991, 1042-1045, 1136-1137, 1287-1290 (1883). [11.
157 ———, J. Éc. Polyt., 64: 123-149 (1894). [17.
158 Hurwitz, A., Gött. Nachr., 85-107 (1887). [18.
159 ———, Leipz. Ber., 38: 10-38 (1885-86). [6, 7, 18.
160 ———, Math. Ann., 28: 561-585 (1887). [7, 18.
161 ———, Math. Ann., 32: 290-308 (1888). [7, 18.
162 Jeffery, H. M., Brit. Assn. Rep., 47: 28-30 (1877). [5.
163 ———, Quart. J., 15: 198-224 (1877-78). [5.
164 de Jonquières, E., Mél. de géom., chap. 4 (1856). [1.
164.1 ———, Ann. di Mat., 2: 86-94 (1859). [19.
165 ———, J. f. Math., 66: 289-321 (1866). [6.
166 ———, Giorn. di. Mat., 4: 45-53, 212-213 (1866). [19.
167 ———, J. de Math., (2), 6: 113-134 (1861). [1, 19.

168 ——— (Cayley), London M. S. Proc., 1: 1-6 (1866-69). [6.
169 ——— (Cayley), London R. S. Trans., 158: 75-172 (1868). [6.
170 Junker, Diss. Tübingen, (1889). [6.
171 Klein-Fricke, Theorie d. ellip. Modulfunk., 2: 518 (1892). [7.
172 Klein, F., Riemannsche Flächen, 1: 152-170, 184 (1892). [15.
173 ———, Riemannsche Flächen, 2: 88 (1892-93). [15.
174 ———, Math. Ann., 15: 252-282 (1879). [7, 18.
175 ———, Math. Ann., 17: 62-70 (1880). [7.
176 ———, Math. Ann., 36: 1-83 (1889). [15.
177 Koebe, P., Math. Ann., 67: 145-224 (1909). [20.
178 ———, Math. Ann., 69: 1-81 (1910). [20.
179 ———, Math. Ann., 72: 437-516 (1912). [20.
180 ———, Math. Ann., 75: 42-129 (1914). [20.
181 Koenigs, G., Nouv. Ann., (3), 2: 301-306 (1883). [17.
182 Kohn, G., Monatsh. f. Math., 1: 71-91, 129-158 (1890). [12.
183 ———, Wien. Ber., 97: 325-328, 1381-84 (1887). [12.
184 ———, Wien. Ber., 106: 488-489 (1897). [4.
185 Kraus, L., Math. Ann., 16: 245-259 (1879). [15.
186 Krey, H., Math. Ann., 10: 221-227 (1876). [12.
187 ———, Math. Ann., 15: 211-238 (1879). [19.
188 ———, Math. Ann., 19: 497-517 (1882). [19.
189 Kronecker, L., Werke, 2: 193-225 (1897). [16.
190 ———, J. f. Math., 91: 301-334 (1881). [16.
191 Küpper, C., Encykl., III C5: 527-532 (1908). [12.
192 ———, Math. Ann., 24: 1-42 (1884). [10, 18.
193 ———, Math. Ann., 31: 291-301 (1888). [19.
194 ———, Monatsh. f. Math., 8: 57-78 (1897). [5.
195 ———, Prag. Ber., 43: 63-68 (1871). [10.
196 ———, Prag. Ber., 477-485 (1887). [16.
197 ———, Prag. Ber., 264-272 (1892). [16.
198 ———, Prag. Ber., 403-412 (1892). [18.
199 ———, Prag. Ber., 16 p. (1895). [5.
200 ———, Prag. Ber., 11+9 p. (1896). [5.
201 ———, Prag. Ab., (6), 5: 20 p. (1872). [12.
202 ———, Prag. Ab., (6), 6: 10 p. (1873). [2.
203 ———, Prag. Ab., (3), 12: 18 p. (1883-84). [5.
204 ———, Prag. Ab., (6), 12: 10 p. (1884). [2.
205 Küpper, C., and Bobek, C., Prag. Ab., (7), 1: 47 p. (1886). [18.
206 Küpper, C., Prag. Ab., (7), 3: 18 p. (1889). [5.
207 Landsberg, G., J. f. Math., 132: 1-20 (1906). [16.
208 Lasker, E., Math. Ann., 60: 20-116 (1905). [13.
209 Le Paige, C., Acta. Math., 3: 181-200 (1883). [11.
210 ———, Liège S. R. S. Mém., (2), 10: 103 p. (1883). [3.
211 ———, C. R., 98: 353-355 (1884). [11.
212 ———, Casopis, 10: 212-216 (1881). [21.
213 ———, Prag. Ber., 20: 61-68 (1881). [4.
214 Levi, B., Rom. Acc. L. Rend., (5), 7: 111-113 (1898). [16.
215 Levy, G., Ann. Éc. Norm., 32: 49-135 (1915). [11.
216 Lindemann, F., J. f. Math., 84: 294-305 (1878). [6, 7.
217 Löffler, E., Diss. Tübingen, 49 (1907). [13.

218 Löffer, E., Math. Ann., 65: 400-412 (1908). [13.
219 Loria, G., I polig. di Poncelet, Torino, 50 (1889). [2.
220 ———, Encykl., III C5b: 614-627 (1914). [16.
221 ———, Bibl. Math., 3: 67-74 (1889). [2.
222 ———, Bull. Sci. M., (2), 17: 257-289 (1893). [2.
223 ———, Giorn. di Mat., 34: 354-374 (1896). [1.
224 Lüroth, J., Math. Ann., 9: 163-165 (1876). [18.
225 Macaulay, F. S., London M. S. Proc., 26: 495-544 (1895). [14.
226 ———, Am. M. S. Bull., 4: 540-544 (1897-98). [14.
227 Martinetti, V., Pal. Circ. Mat., 5: 109-120 (1891). [10.
228 Masoni, U., Nap. R. A. Rend., (1), 21: 45-74 (1882). [11.
229 McKelvey, J. V., Am. J., 34: 115-146 (1902). [15.
229.1 Meyer, W. F., Apolarität und rationale Curven, (1883). [20.
230 Milne, W. P., Edinb. R. S. Proc., 35: 10-23 (1916-17). [11.
231 ———, Edinb. R. S. Proc., 35: 24-31 (1916-17). [9.
232 ———, Edinb. R. S. Proc., 36: 84-90 (1917-18). [11.
233 ———, London. M. S. Proc., (2), 5: 235-243 (1910-11). [9.
234 ———, London M. S. Proc., (2), 10: 207-210 (1911-12). [9.
235 Milne, W. P., and Taylor, D. G., London M. S. Proc., (2), 20: 101-105 (1922). [9.
236 Montesano, D., Nap. R. A. Atti., (2), 15: 12 p. (1913). [17.
237 ———, Nap. R. A. Rend., (2), 27: 181-188 (1888). [17.
238 ———, Rom. Acc. L. Rend., (4), 4_1: 207-211, 277-282 (1888). [17.
239 ———, Torino Atti., 27: 660-690 (1892). [17.
240 Montessus, R., Bull. Sci. M., (2), 44: 31-36, 42-50, 58-64 (1920). [11.
241 Morley, F., Math. Qu., 53: 125 (1890). [9.
241.1 ———, Am. J., 47: 91-97 (1925). [20.
242 Morley, F., and Coble, A. B., Am. J., 49: 463-488 (1927). [20.
243 Nagy, J. V., Math. Ann., 73: 230-242 (1913). [16.
244 Neikirk, L. I., Ann. of Math., (2), 13: 52-54 (1911). [1.
245 Noether, M., Erl. Ber., 10: 80-86 (1878). [12.
246 ———, Berlin Ab., 1-120 (1882). [14, 19.
247 ———, J. f. Math., 93: 271-318 (1882). [14.
248 ———, Gött. Nachr., 267-278 (1871). [16.
249 ———, Gött. Nachr., 490-498 (1872). [13.
250 ———, Math. Ann., 2: 292-316, 314 (1869). [13.
251 ———, Math. Ann., 6: 351-359 (1873). [13, 16.
252 ———, Math. Ann., 8: 533-578 (1874). [19.
253 ———, Math. Ann., 17: 263-284 (1880). [15.
254 ———, Math. Ann., 23: 311-358 (1884). [19.
255 ———, Math. Ann., 26: 143-150 (1886). [15.
256 ———, Math. Ann., 30: 410-417 (1887). [13.
257 ———, Math. Ann., 40: 140-144 (1892). [13.
258 ———, Münch. Ab., 17: 105-150 (1889-92). [12.
259 Oppenheimer, H., Monatsh. f. Math., 20: 141-170 (1909). [2.
260 Panzi, E., Giorn. di Mat., 53: 349-354 (1915). [2.
261 de Paolis, R., Rom. Acc. L. Mem., (3), 2: 31-50 (1878). [12.
262 ———, Rom. Soc. It. Sc. Mem., (3), 7: 165 p. (1890). [8.
263 Pascal, P., P. Rep., 2_1: 345-347 (1910). [6.
264 ———, P. Rep., 2_2: 238-240 (1922). [2.

265 Peano, G., Giorn. di Mat., 20: 79-101 (1882). [1.
266 ———, Torino Atti., 17: 73-79 (1881). [1.
267 Pengra, C. E., Univ. Wis. Trans., 14: 655-669 (1903). [15.
268 Perna, Pal. Circ. Mat., 17: 65-72 (1903). [11.
269 del Pezzo, P., Nap. R. A. Rend., (2), 7: 15-21, 45 (1893). [16.
270 Picard, E., Ann. Éc. Norm., (3), 33: 1-15 (1916). [20.
271 Picquet, H., Assoc. Fr. Sc., 7: 95-132 (1878). [5.
272 Pieri, M., Giorn. di Mat., 24: 13-22 (1896). [1.
273 ———, Riv. F. M., 4: 40-42 (1894). [16.
274 Pittarelli, G., Rom. Acc. L. Mem., (4), 3: 375-400, 401-416 (1886). [1.
275 Poincaré, H., J. de Math., (5), 1: 219-314 (1895). [20.
276 ———, C. R., 117: 18-23 (1893). [16.
277 de Porte, J. D., Am. J., 40: 47-68 (1918). [14, 18.
278 Purser, F., and Haughton, S., Math. Qu., 43: 101-102 (1885). [11.
279 Retali, V., Liège S. R. S. Mém., (2), 18: 35 p. (1895). [20.
280 Reye, Th., Zschr. f. Math. Phys., 13: 521-526 (1868). [17.
281 Riemann, B., Werke, 2d ed., 487 (1892). [15.
282 ———, J. f. Math., 54: 115-155 (1857). [14, 15.
283 Roch, G., J. f. Math., 64: 372-376 (1865). [20.
284 ———, J. f. Math., 65: 42-51 (1866). [20.
285 ———, J. f. Math., 66: 97-120 (1866). [20.
286 Saltel, L., Brux. Mém., 39: 15 (1875). [19.
287 ———, Nouv. Ann., (2), 12: 565-570 (1873). [20.
288 ———, C. R., 80: 1064-1067 (1875). [19.
289 ———, C. R., 82: 63-86 (1876). [19.
290 Savio, P., Giorn. di Mat., 40: 192-222 (1902). [20.
291 Schaake, G., Amst. K. Ak. Wet. Versl., 33: 160-168 (1924). [19.
292 Schoute, P. H., J. f. Math., 95: 105-119, 375 (1883). [10.
293 ———, Nieuw. Arch., 10: 54-65 (1884). [17.
294 Schröter, H., Math. Ann., 5: 50-82 (1872). [17.
295 Schubert, H., Kalkül der abz. Geom., § 18 (1879). [6.
296 ———, J. f. Math., 88: 311-342 (1880). [1.
297 ———, Math. Ann., 11: 347-378 (1877). [19.
298 ———, Math. Ann., 12: 202-221 (1877). [19.
299 Schur, W., Math. Ann., 18: 1-32 (1881). [9.
300 Schwarz, H. A., J. de Math., (3), 6: 111-114 (1880). [5.
301 Scorza, G., Torino Atti., 35: 285-301 (1900). [6.
302 Scott, C. A., Math. Ann., 52: 593-597 (1899). [13.
303 ———, Am. M. S. Bull., 4: 261-273 (1897-98). [14.
304 ———, Quart. J., 28: 377-381 (1896). [15.
305 Segre, C., Lezioni di geom. enum., lith. (1900). [2.
306 ———, Ann. di Mat., (2), 20: 237-242 (1892). [11.
307 ———, Ann. di Mat., (2), 22: 41-142 (1894). [6, 8, 14, 15, 18.
308 ———, Ann. di Mat., (2), 25: 1-53 (1896). [16.
309 ———, Bibl. Math., (2), 6: 33-40 (1892). [1.
310 ———, Math. Ann., 27: 296-314 (1886). [17, 18.
311 ———, Camb. Ph. Soc. Trans., 30: 203-226 (1887). [19.
312 ———, Math. Ann., 34: 1-25 (1889). [19.
313 ———, Am. M. S. Bull., (2), 10: 442-468 (1903-04). [1.

314 Segre, C., Rom. Acc. L. Rend., (4), 3_2: 149-153 (1889). [18.
315 ———, Torino Atti., 24: 734-758 (1889). [7, 18, 14.
316 Servais, C., Ann. Ac. Polyt. Porto, 8: 146-187 (1913). [11.
317 Severi, F., Alg. Geom., 134-135 (1921). [16.
318 ———, Ann. di Mat., (3), 12: 44-79 (1906). [14, 18.
319 ———, Ann. Éc. Norm., (3), 24: 449-468 (1908). [13.
320 ———, Rom. Acc. L. Rend., (5), 24_1: 877-888, 1011-1020 (1915). [16.
321 ———, Torino Atti., 38: 119-134 (1903). [14.
322 ———, Torino Mem., (2), 51: 87-114 (1901). [6.
323 ———, Torino Mem., (2), 54: 1-49 (1903). [4, 6, 7.
324 ———, Ven. Ist. Atti., 65_2: 625-643 (1906). [18.
325 ———, Ven. Ist. Atti., (9), 4: 929-938 (1919-20). [14.
326 Shibayama, M., Tôhoku M. J., 16: 89-91 (1919). [6.
327 Simart, G., C. R., 116: 1047-1050 (1893). [16.
328 Snyder, V., Am. J., 30: 10-18 (1908). [16.
329 ———, Am. J., 30: 325-336 (1908). [15.
330 ———, Am. J., 30: 337-346 (1908). [16.
331 ———, Am. M. S. Bull., (2), 15: 1-4 (1908-09). [16.
332 Snyder, V., and Sharpe, F. R., Am. M. S. Trans., 22: 31-40 (1921). [18.
333 Spottiswoode, W., London M. S. Proc., (1), 8: 29-34 (1877). [12.
334 ———, C. R., 83: 627-630 (1876). [12.
335 ———, Lond. R. S. Trans., 155: 653-669 (1865). [12.
336 Steiner, J., J. f. Math., 32: 182-184 (1845). [10.
337 ———, J. f. Math., 47: 1-108 (1854). [19.
338 ———, J. f. Math., 49: 333-348 (1855). [1.
339 Stouff, X., Ann. Éc. Norm., (3), 5: 212-326 (1888). [8.
340 Study, E., Math. Ann., 36: 216-229 (1890). [14.
340.1 Sturm, R., Flächen 3 Ord., 388 p. (1867). [17.
341 ———, Liniengeom., 1: 16 (1892). [1.
342 ———, J. f. Math., 72: 350-359 (1870). [19.
343 Stuyvaert, M., Liège S. R. S. Mém., (3), 7: 230 p. (1907). [20.
344 Sylvester, J. J., Salmon's Curves, 154 (1873). [13.
345 ———, Math. Misc., 8: 14 (1868). [9.
346 ———, Math. Misc., 25: 27 (1876). [9.
347 ———, Math. Misc., 35: 134 (1880). [9.
348 ———, Math. Misc., 38: 37 (1883). [9.
349 ———, Math. Misc., 49: 21 (1888). [9.
350 Tafani, G., Giorn. di. Mat., 52: 1-10 (1914). [7.
351 Taylor, F. G., Edinb. M. S. Proc., 34: 163-175 (1915-16). [9.
352 Thomé, L. W., J. f. Math., 126: 52-72 (1903). [16.
353 Timerding, H. E., J. f. Math., 123: 284-311 (1901). [11.
354 Tognoli, O., Giorn. di. Mat., 20: 270-286 (1882). [3.
355 ———, Giorn. di. Mat., 22: 308-332 (1884). [14.
356 ———, Giorn. di. Mat., 23: 247-262, 345-365 (1885). [14.
357 Torelli, R., Nap. R. A. Rend., (3), 17: 412-420 (1911). [18.
358 ———, Pal. Circ. Mat., 19: 297-304 (1905). [14, 18.
359 ———, Rom. Acc. L. Rend., (5), 22_1: 772-775 (1913). [18.
360 ———, Torino Atti., 42: 86-99 (1907). [18.
361 ———, Ven. Ist. Atti., (8), 10: 831-835 (1907-08). [14.
362 ———, Ven. Ist. Atti., (8), 10: 1323-1336 (1907-08). [18.

363 Vahlen, K., J. f. Math., 113: 348-352 (1894). [1.
364 ———, J. f. Math., 118: 251-256 (1897). [1.
365 Van Benschoten, A., Am. J., 31: 213-252 (1909). [15.
366 Van der Woude, W., Amst. K. Ak. Wet. Versl., 27II: 962-966 (1918-19). [11.
367 Veronese, G., Math. Ann., 19: 161-234 (1882). [19.
368 Vessiot, E., Toulouse Ann., 10: 1-26 (1896). [16.
370 Voss, A., Math. Ann., 27: 527-536 (1886). [13.
371 de Vries, J., Monatsh. f. Math., 3: 285-292 (1892). [3.
372 ———, Wien. Ber., 101: 46-59 (1895). [5.
373 ———, Nieuw. Arch., 12: 82-93 (1886). [9.
374 ———, Nieuw. Arch., 14: 193-200 (1888). [1.
375 ———, Nieuw. Arch., (2), 5: 68-85 (1902). [5.
376 ———, Nieuw. Arch., (2), 7: 469-470 (1907). [1.
377 ———, Nieuw. Arch., (2), 9: 1-5 (1911). [1.
378 ———, Amst. K. Ak. Wet. Versl., 3: 115-117 (1895). [5.
379 ———, Arch. Néerl., 23: 93-114 (1889). [3.
380 Waelsch, E., Wien. Ber., 88: 418-423, 1039-1048 (1883). [21.
381 Weber, H., J. f. Math., 76: 345-348 (1873). [8.
382 ———, Math. Ann., 13: 35-48 (1877). [15.
383 Weiss, W., Math. Ann., 29: 382-385 (1887). [8.
384 ———, Monatsh. f. Math., 7: 321-324 (1896). [13.
385 Weltzien, C., Math. Ann., 30: 535-545 (1887). [14.
386 Weyr, Em., Beiträge zur Curvenl., 64 p. (1880). [1, 19.
387 ———, Ann. di Math., (2), 4: 272-280 (1871). [1.
388 ———, Liège S. R. S. Mém., (2), 10: 11 p. (1883). [4.
389 ———, Monatsh. f. Math., 2: 458 (1891). [4, 5.
390 ———, Wien. Ber., 61: 633-644 (1868). [10.
391 ———, Wien. Ber., 69: 784-794 (1874). [9, 10.
392 ———, Wien. Ber., 79: 429-446 (1879). [21.
393 ———, Wien. Ber., 87: 592-598 (1883). [1, 18.
394 ———, Wien. Ber., 87: 837-872 (1883). [5, 8.
395 ———, Wien. Ber., 88: 459-482 (1883). [5, 9.
396 ———, Wien. Ber., 90: 206-225 (1884). [5, 17.
397 ———, Wien. Ber., 92: 498-523 (1885). [17.
398 ———, Wien. Ber., 97: 592-617 (1888). [5, 17.
399 ———, Wien. Ber., 99: 932-951 (1890). [17.
400 ———, Wien. Ber., 100: 457-466 (1891). [17.
401 ———, Wien. Ber., 100: 589-606 (1891). [5.
402 ———, Wien. Ber., 101: 1457-1483, 1695-1741 (1892). [4, 5.
403 ———, Wien. Ber., 101: 1506-1519 (1892). [4.
404 ———, Wien. Ber., 103: 365-441 (1894). [5.
405 ———, Casopis, 11: 191-212 (1882). [1.
406 ———, Prag. Ber., 1-18 (1869). [10.
407 ———, Prag. Ber., 1-10 (1870). [10.
408 ———, Prag. Ber., 8 p. (1874). [1.
409 ———, Prag. Ber., 131-133 (1877). [3.
410 ———, Prag. Ab., (6), 4: 35 p. (1870). [1.
411 ———, Prag. Ab., (6), 7: 56 p. (1874). [3.
412 White, H. S., Math. Ann., 36: 597-601 (1890). [20.
413 Wiman, A., Oslo. Selsk. For., 21: 21 p. (1895). [15.

414 Wirtinger, W., Jahresbericht, 4: 97-99 (1894). [11.
415 Zajaczowski, L., Crac. Bull., 111-112 (1896). [5.
416 Zeuthen, H. G., Acta Math., 1: 171-188 (1882). [8.
417 ———, Math. Ann., 40: 99-124 (1892). [19.
418 ———, Ann. di Mat., (2), 3: 175-217 (1869). [19.
419 ———, Soc. M. F. Bull., 5: 186-190 (1873). [6, 19.
420 ———, Soc. M. F. Bull., 7: 97-106 (1874). [19.
421 ———, Math. Ann., 3: 150-156 (1871). [8, 19.
422 ———, Math. Ann., 10: 446-546 (1876). [19.
423 ———, Math. Ann., 40: 99-124 (1892). [6.
424 ———, Nouv. Ann., (2), 6: 200-206 (1867). [19.
425 Zimmermann, O., J. f. Math., 123: 1-32, 175-209 (1901). [1, 19.
426 ———, J. f. Math., 126: 171-193 (1903). [1.

CHAPTER VIII

CREMONA TRANSFORMATIONS IN SPACE AND HYPERSPACE

After preliminary papers by Cayley([21]) and Noether([125]) in 1870 the main lines of this theory were developed for space S_3 by Cremona and Noether in 1871. The account given here follows in a general way the same outline as Chapter IV. However, the involutorial transformations which have received much attention are discussed in the following chapter. At the end of each section such extensions to hyperspace as have been made are indicated. General accounts of space transformations are given by Sturm[([166]); Chapter XI], by Doehlemann[([47]); Part II], and by Salmon[([149]); Chapter 17]. Loria[([97.1]), 250-254] gives a historical account, and Pascal[([138.1]), 963-967, 978-985] a short résumé of the subject. Hudson([78.4]) gives an admirable treatment of the matters discussed in sections 1, 2, 4, 5.

1. *Definitions and General Properties.*—We set the planes of an $S_3(x)$ into projective correspondence with the surfaces, ϕ, of a homaloidal web of order n in $S_3(y)$ and obtain a Cremona transformation T from a point x of $S_3(x)$, the base point x of a net of planes, to point y of $S_3(y)$, the base-point variable with x of the corresponding net of surfaces ϕ in the homaloidal web. Then planes in $S_3(y)$ are mapped by means of their sections with the web ϕ upon surfaces ϕ' of order n' in $S_3(x)$ which necessarily constitute the homaloidal web in $S_3(x)$ determined by T^{-1}. A line l' in $S_3(x)$ passes by T into a curve λ of order n' in $S_3(y)$; and a line l in $S_3(y)$ passes by T^{-1} into a curve λ' in $S_3(x)$ of order n. We use unprimed letters for occurrences in $S_3(y)$ and the corresponding primed letters for corresponding occurrences in $S_3(x)$. Thus the use of a primed letter in $S_3(y)$ indicates a theorem; e. g., if above the orders of T, T^{-1} for planes are n, n', then the orders of T, T^{-1} for lines are n', n, respectively.

If, in S_r, T similarly defined transforms the S_{r-1}'s, S_{r-2}'s, $\ldots$, S_1's of $S_r(x)$ into manifolds of orders $n, n', n'', \ldots, n^{(r-2)}$ in $S_r(y)$, then T^{-1} transforms the S_{r-1}'s, $\ldots$, S_1's of $S_r(y)$ into manifolds of orders $n^{(r-2)}, n^{(r-3)}, \ldots, n', n$, respectively in $S_r(x)$.

The first paper which brings out clearly the distinctions between the various kinds of *fundamental* or *F-points* and the various kinds of *principal* or *P-surfaces* and *P-curves* is that of Cremona[([41]) (1871)]. The

n^3-1 fixed intersections of three surfaces of the web ϕ are absorbed by a *fundamental basis* or F-basis of the web which consists of certain F-curves $C_{q_1}{}^{r_1}, \ldots, C_{q_i}{}^{r_i}, \ldots, C_{q_\rho}{}^{r_\rho}$ of orders $q_1, \ldots, q_i, \ldots, q_\rho$ and of multiplicity $r_1, \ldots, r_i, \ldots, r_\rho$ for the general surface ϕ; and of certain F-points $f_1, \ldots, f_j, \ldots, f_\sigma$ of multiplicity $k_1, \ldots, k_j, \ldots, k_\sigma$ for the general surface ϕ. At a point of the F-basis not all surfaces ϕ contain a general direction. There is then a net of surfaces ϕ on such a direction and therefore an image x of this direction in $S_3(x)$. At an *isolated* F-point f_j, the directions correspond to points x on a rational P-surface in $S_3(x)$ whose order is the number of times a curve λ passes through f_j [Cremona(38), Noether(126).] This P-surface is a double factor of the jacobian $J(\phi')$ [Cremona(38)]. At an F-point on the F-curve $C_{q_i}{}^{r_i}$ the directions correspond in general to points x on a rational curve of order r_i called the P-curve of the F-point. In general, also, as the F-point varies on $C_{q_i}{}^{r_i}$ the P-curve in $S_3(x)$ varies over a P-surface, of genus equal to that of $C_{q_i}{}^{r_i}$ and of order equal to the number of variable intersections of a curve λ with $C_{q_i}{}^{r_i}$ [Cremona(38), (41); Noether(126)]. Such a P-surface is a simple factor of $J(\phi')$ [Cremona (38)]. It may happen, however, that the P-curve of an F-point on $C_q{}^r$ does not vary as the F-point runs over $C_q{}^r$. The P-curve is then an F-curve $C'^{r'}{}_{q'}$ of the web ϕ' and directions at an F-point on either curve correspond to the points of the other F-curve. Such corresponding F-curves in inverse webs are said to be of the *second kind*. Cremona(38), (41) had inferred from particular cases that for two such curves $r'=q$ and $q'=r$ a view which persisted for a long time [cf. Sturm(166), p. 342]. Nevertheless Montesano[(114) (1918); cf. also (115)] by constructing examples shows that transformations in S_3 exist for which two corresponding F-curves of the second kind of orders ν', ν have multiplicities on the webs ϕ', ϕ equal to $k\nu$, $k\nu'$, respectively, where k, ν, ν' are arbitrary positive integers. Marletta(98.1) extends this theorem to transformations in S_n.

The isolated F-point is characterized by the fact that its consecutive points determine a P-surface. It may be a particular point on one or more F-curves. An F-curve is of the first or the second kind according as it is met by a general curve λ in variable points or in fixed points only.

The jacobian $J(\phi)$ has an isolated F-point f_j of multiplicity k_j on the web ϕ as a point of multiplicity $(4k_j-2)$; it has an F-curve of the first kind $C_{q_i}{}^{r_i}$ as a curve of multiplicity $(4r_i-1)$ [Cremona(38), (41); Noether(126), for special cases (125)]; and it has an F-curve of the second kind $C_q{}^r$ as a curve of multiplicity $4r$ [Cremona(38)]. Doehlemann(48), Levi(94) and Gerbaldi(65) discuss the jacobian of four surfaces by algebraic methods and reach the same conclusions.

A plane α in $S_3(y)$ is mapped by the system of curves cut out by the web ϕ—a system of order n with q_i base-points of order r_i—upon a surface of the web ϕ'; similarly, a plane α in $S_3(x)$ is mapped upon a surface of the web ϕ. Hence

$$n'=n^2-\sum_{i=1}^{\rho} q_i r_i^2, \qquad n=n'^2-\sum_{i=1}^{\rho'} q'_i r'^2_i$$

[Cayley(21); Cremona(41)]. The general surface ψ', the transform of α by T^{-1}, meets the P-surface of an isolated F-point only in F-curves; and meets the P-surface of an F-curve $C_{q_i}^{r_i}$ in q_i P-curves [Cremona(41)].

Loria(97) observes that a plane α and a surface ϕ in $S_3(y)$ correspond respectively to a surface ϕ' and a plane α in $S_3(x)$ whence the genus of plane sections of surfaces of either web is called the *genus* of the transformation (cf. § 2). The equal genus of the two sets of plane sections had also been noted by Noether(127) who combined the resulting relation

$$(n-1)(n-2)-\sum_{i=1}^{\rho} q_i r_i(r_i-1)=(n'-1)(n'-2)-\sum_{i=1}^{\rho'} q'_i r'_i(r'_i-1)$$

with the above equations to get

$$4(n-n')=\sum_{i=1}^{\rho} q_i r_i-\sum_{i=1}^{\rho'} q'_i r'_i.$$

Pannelli in seeking an extension of the Clebsch-Cremona theorem that in the plane the number of F-points of inverse transformations is the same finds relations among the characteristic numbers of the F-bases(133), (134), (135). Of these the most noteworthy is the theorem that if as above the web ϕ has σ isolated F-points and ρ F-curves $C_{q_i}^{r_i}$ of genus p_i respectively, then

$$\sigma+\rho+\sum_{i=1}^{\rho} p_i=\sigma'+\rho'+\sum_{i=1}^{\rho'} p'_i.$$

Tummarello(171.4) states that if p, p' are the genera of the F-curves in the two spaces then $\sum p=\sum p'$ and for each p there is an equal p'. This is proved by Hudson[(78.3) and (78.4) p. 277].

Miss Hudson(77) shows that a transformation can be constructed with an F-curve of any genus however great; and that a product $T=T_1T_2$ has F-curves of genera equal to those of the factors T_1, T_2; and that consequently the general space transformation cannot be expressed as a product of a finite number of types of elementary transformations. She proves also [(78.4), p. 358] that a transformation can be constructed which has an arbitrarily given curve as an F-curve of the first kind.

Autonne(7.1) shows that to an isolated F-point which is a higher singularity of the web there corresponds a degenerate P-surface, or merely

F-curves, or both. He introduces(7.2) a very artificial nomenclature for the F-system and P-system.

For Cremona transformation in space $S_k (k>3)$ there may exist F-loci of dimensions 0, 1, ..., $k-2$ but no exhaustive analysis of the possible behavior of such F-loci with respect to their P-loci has been made.

2. *Determination of the Types.*—Cayley(21) remarked that a type of transformation T would be determined if an F-basis could be selected whose *postulation* (i. e., the number of linear conditions imposed by the prescribed basis) for surfaces of order n is $\frac{1}{6}(n+1)(n+2)(n+3)-4$ and whose *equivalence* (i. e., the number of intersections of three surfaces absorbed by the prescribed basis) for surfaces of order n is n^3-1. The surfaces of such a web would necessarily be rational. He gave for a curve of order m with h apparent nodes the postulation $m(n+1)-\frac{1}{2}(m-1)m+h$ and the equivalence $m(3n-2)-m(m-1)+2h$.

Noether(127) first gives reasonably complete formulas for (a) postulation: the ϕ constitute a web; (b) equivalence: three ϕ's have a single variable intersection; and (c) genus: the general ϕ has arithmetic genus $p_a=0$. Let as before the web ϕ of order n have ρ F-curves $C_{q_i}^{r_i}$ of order q_i and rank (order of developable of tangents) a_i which are r_i-fold on the ϕ's; $C_{q_i}^{r_i}$ have c_{ii} actual double points and meet $C_{q_j}^{r_j}$ in c_{ij} points $(i\neq j)$; let the ϕ's have σ isolated F-points f_l of multiplicity k_l and let the curve $C_{q_i}^{r_i}$ pass through f_l with d_{il} branches, the tangent cones of the ϕ's at f_l having no part fixed for all the ϕ's; and finally let the ϕ have certain points of contact of order $s-1$, i. e., a common $(s-1)$-fold point with common tangent cone of order $(s-1)$. Then the conditions (a), (b), (c) on the ϕ's lead to the following relations in which

$$\Sigma_1=\sum_{i=1}^{\rho}, \qquad \Sigma_2=\sum_{i,j=1}^{\rho}(i\leqq j), \qquad \Sigma_3=\sum_{l=1}^{\sigma}, \qquad \Sigma_4=\sum_{i=1}^{\rho}\sum_{l=1}^{\sigma}:$$

(a) $$\tfrac{1}{6}(n+1)(n+2)(n+3)-4=\Sigma_1\tfrac{1}{6}r_i(r_i+1)\{(3n-2r_i+5)q_i-\tfrac{1}{2}(2r_i+1)a_i\}-\Sigma_2\tfrac{1}{6}r_i(r_i+1)(3r_j-r_i+1)c_{ij}+\Sigma_3\tfrac{1}{6}k_l(k_l+1)(k_l+2)-\Sigma_4\tfrac{1}{6}r_i(r_i+1)(3k_l-2r_i+2)d_{il}+\Sigma\tfrac{1}{2}s(s+1);$$

(b) $$n^3-1=\Sigma_1 r_i^2\{(3n-2r_i)q_i-r_ia_i\}-\Sigma_2 r_i^2(3r_j-r_i)c_{ij}+\Sigma_3 k_l^3-\Sigma_4 r_i^2(3k_l-2r_i)d_{il}+\Sigma s^2;$$

(c) $$\tfrac{1}{6}(n-1)(n-2)(n-3)=\Sigma_1\tfrac{1}{6}r_i(r_i-1)\{(3n-2r_i-5)q_i-\tfrac{1}{2}(2r_i-1)a_i\}-\Sigma_2\tfrac{1}{6}r_i(r_i-1)(3r_j-r_i-1)c_{ij}+\Sigma_3\tfrac{1}{6}k_l(k_l-1)(k_l-2)-\Sigma_4\tfrac{1}{6}r_i(r_i-1)(3k_l-2r_i-2)d_{il}.$$

From these by forming proper linear combinations there follow:

$$
\begin{aligned}
(1)\quad & \tfrac{1}{6}(4n-1)(4n-2)(4n-3)-1=\textstyle\sum_1 \tfrac{1}{6}4r_i(4r_i-1)\{(12n-8r_i-5)q_i\\
&\quad -\tfrac{1}{2}(8r_i-1)a_i\}-\textstyle\sum_2 \tfrac{1}{6}4r_i(4r_i-1)(12r_j-4r_i-1)c_{ij}\\
&\quad +\textstyle\sum_3 \tfrac{1}{6}4k_l(4k_l-1)(4k_l-2)\\
&\quad -\textstyle\sum_4 \tfrac{1}{6}4r_i(4r_i-1)(12k_l-8r_i-2)d_{il}+\sum s(7s-3);\\
(2)\quad & 2(n^2-1)=\textstyle\sum_1 r_i\{(n+r_i)q_i-\tfrac{1}{2}r_ia_i\}-\sum_2 r_ir_jc_{ij}+\sum_3 k_l^2\\
&\quad -\textstyle\sum_4 r_ik_ld_{il}+\sum \tfrac{1}{2}s(s+1);\\
(3)\quad & 11(n-1)=\textstyle\sum_1 r_i(5q_i-\tfrac{1}{2}a_i)-\sum_2 r_ic_{ij}+\sum_3 2k_l\\
&\quad -2\textstyle\sum_4 r_id_{il}+\sum s(s+3).
\end{aligned}
$$

Equation (1) expresses the fact that the jacobian $J(\phi)$ is completely determined by its $(4k_l-2)$-fold points f_l and its $(4r_i-1)$-fold curves $C_{q_i}^{r_i}$. Equation (2) in connection with an equation of Section 1 leads to

$$
\begin{aligned}
(4)\qquad 4n'-4&=\textstyle\sum_1 2r_i\{(n-r_i)q_i-\tfrac{1}{2}r_ia_i\}-\sum_2 2r_ir_jc_{ij}+\sum_3 2k_l^2\\
&\quad -\textstyle\sum_4 2r_ik_ld_{il}+\sum s(s+1)\\
&=\textstyle\sum_1 Q_i+\sum_3 2R_l+S;
\end{aligned}
$$

where

$$
\begin{aligned}
(5)\quad Q_i&=2r_i\{(n-r_i)q_i-\tfrac{1}{2}r_ia_i\}-r^2c_{ii}-\sum_{j=1}^{\rho} r_ir_jc_{ij}-\textstyle\sum_3(2r_ik_l-2r_i^2)d_{il}\\
R_l&=l^2-\textstyle\sum_1 r_i^2d_{il}\\
S&=\textstyle\sum s(s+1)
\end{aligned}
$$

[Noether gives for Q_i the value

$$
\begin{aligned}
Q_i=2r_i\{(n-r_i)q_i-\tfrac{1}{2}r_ia_i\}&-\sum_h r_h^2c_{hi}\\
&-\sum_j(2r_ir_j-r_i^2)c_{ij}-\sum_l(2r_ik_l-2r_i^2)d_{il}
\end{aligned}
$$

where $h<i$ and $j\geqq i$, a value which does not have the requisite symmetry.]

Since $4n'-4$ is the number of variable intersections with the $J(\phi')$ of a line in $S_3(x)$ the term Q_i is the number of variable intersections of a curve λ in $S_3(y)$ with the F-curve $C_{q_i}^{r_i}$ and is the order of its P-surface in $S_3(x)$. The term R_l is the order of the P-surface in $S_3(x)$, which occurs twice in $J(\phi')$, of the isolated F-point f_l. Also, the order of the P-surface in $S_3(x)$ which corresponds to the contact of order $s-1$ in $S_3(y)$, and which occurs $(s+1)$ times in the $J(\phi')$, is s [cf. Cremona(38) Nota 2ª].

Miss Hudson(75) remarks that Noether pointed out that the foregoing formulas for the postulation and equivalence of an isolated F-point A of a homaloidal web require modification when the tangent cone at A degenerates and contains a fixed part due to the passage through A of certain branches of F-curves. She then studies the equivalence and postulation of the condition that the tangent cone at A contain a fixed part irrespective of the passage through A of F-curves. She considers also the condition that certain sheets of the surfaces have contact of higher order at A [cf. also ([74]), ([73])]. Applications are made to *monoidal* transformations for which A is a point of multiplicity $n-1$.

The determination of types by means of the above formulas for postulation and equivalence occurs only incidentally. The first general method for obtaining a wide variety of transformations in space is due to Cremona([38]). A surface ϕ of a homaloidal web is cut by the other surfaces of the web in a homaloidal net of curves on ϕ. Since ϕ is rational it can be mapped on a plane π by a net $\sum$ of curves of order r on a certain base B. All surfaces of order n cut ϕ in curves which map from a system $n\sum$ in π of curves of order nr on the base nB. To determine the space transformations in which ϕ can figure as a member of the homaloidal web we select from the system $n\sum$ a net which contains a fixed part L and a variable part which constitutes a plane homaloidal net H on a base B'. In space the surfaces which cut out on ϕ the maps of the homaloidal net LH form with ϕ a homaloidal web whose F-curve is the map of L. The order n' of the inverse transformation is the order of the map on ϕ of the variable curve H on π. Since the types of nets H presumably are known, those types which can be augmented by L to lie in the system $n\sum$ may be counted out. Cremona applies (Nota 1[a]) this process to the cases where ϕ is a quadric, a general cubic surface, a nodal cubic surface, and a ruled cubic surface; also (Nota 2[a]) where ϕ is a cubic surface with two or three nodes or with a uniplanar point, a Steiner quartic surface, and a quintic surface with a double cubic curve, for all of which the mapping on the plane π is known in advance. De Paolis([136]) uses this method to determine monoidal transformations of order n, particularly those for which $n'=n$ [cf. also Bianchi([15])]. This is the first determination of a type with arbitrary n. Loria([96]) followed Cremona in the discussion of the case where ϕ is a general cubic surface. Sturm([170]) reestablishes the types $n=3$ by a discussion of the homaloidal nets on the cubic surface itself. Aroldi([4]) applies Cremona's method to the general quartic surface with a double conic. Tummarello([171]) applies the method to the determination of monoidal transformations for which $n'=n^2$. They are divided into two types according as the ϕ's have not or have the same tangent cone of order $n-1$ at O; of the first type there are three classes, of the second one, for each type of plane homaloidal net.

Montesano([104]) considers transformations for which the ϕ's are surfaces of order n with an $(n-1)$-fold line. These are of genus zero and are included among the types given by Loria([97]) of genus zero: (a) the ϕ's and ϕ''s are Steiner quartics; (b) the ϕ are Steiner quartics, the ϕ''s quadrics; (c) the ϕ's have an $(n-1)$-fold line, the ϕ''s an $(n'-1)$-fold line. Dr. Grazia Nobile([124]) refers to lectures at Naples (1918-19) in which Montesano constructed all webs $\phi(n=3)$ which are monoidal or pluri-monoidal; and to unpublished theses of Drs. De Cesare and Berardi in which all webs ϕ consisting of quartics with a double conic, and quintics with a triple line and two skew double lines incident to the triple line, are determined. All these types are of genus one. Dr. Nobile completes the construction of webs of genus one by adding 32 new types.

Segre([154.1]) studies an (n, n) transformation with an $(n-1)$-fold F-line obtained by two projections from higher space. Perazzo([138.2]) obtains transformations of low degree in S_3 from incidence theorems in S_5. Marletta([98.1]) also obtains by projection a new homaloidal web with an $(n-2)$-fold F-line. Tummarello([171.1]) obtains new webs of order n with an $(n-3)$-fold line and $(n-3)$ triple points. He([171.2]) also gives an $(n, 2n-3)$ transformation for which the web of order n is monoidal and the other consists of surfaces due to DeVries; and later([171.3]) discusses monoidal (n, n) transformations with osculation at a vertex.

Loria([96]) discusses webs ϕ of general cubic surfaces with an F-curve, certain simple F-points and a number of common contacts of order r with a given plane at a given point. The latter conditions give rise to diophantine equations similar to those in the plane and the geometric solutions are obtained.

Noether([126]) first makes systematic use of products to obtain a variety of types of space transformations. Miss Hudson([76]) adds to the seven types $(n=3)$ for which ϕ is a non-singular cubic surface those types for which $\phi(n=3)$ has one, two, three, or four nodes or a nodal line. Of the 75 types enumerated, 26 can be obtained as products of two $(n, n'=2, 2)$ transformations; 16 as products of a $(2, 2)$ transformation and others of the list; and 33 are not expressible as products of simpler types. She further([78]) lists those types which are products of any two $(2, 2)$ transformations; and also([72]) lists 127 types of $(3, 3)$ transformations for which the sextic F-curve degenerates into 6 lines which meet or coincide in various ways and gives for each type the triad of bilinear forms which determine it. Cech[([21.1]) cf. also ([78.1])] adds to these one type. Montesano([117]) takes as elements of a product the cubic transformation determined by three pairs of projectively related pencils of planes whose F-basis in either space consists of a space cubic curve and three bisecants, say C'_3, b'_1, b'_2, b'_3, and C_3, b_1, b_2, b_3. If a number of such transformations $K_1, \ldots, K_t$ are carried out in succession in such a way that K_l, K_{l+1} have $C_3^{(l)}$, $C'^{(l+1)}_3$ in coincidence and perhaps also some of the bisecants there arises a transformation of order $2n-1$ which has the final $C_3^{(t)}$ as an $(n-1)$-fold F-curve and a number ρ of bisecants of $C_3^{(t)}$ as F-curves of multiplicity $s_1, \ldots, s_\rho$ where $n, s_1, \ldots, s_\rho$ form a ternary "Cremona group of numbers." In this way an aggregate of types of space transformation simply isomorphic with the aggregate of types of plane transformations is obtained. Tinto([170.2]), ([170.3]) also studies these transformations. In an earlier paper Montesano([116]) had considered the case where the C_3's degenerate into a conic and a secant line.

Miss Beloch([13]) proves the two following theorems stated by Castelnuovo: (I) For a homaloidal web the adjoint surfaces of all indices h (i. e., the surfaces of order $n-4h$ with multiplicity $i-h$ (or zero) for every i-fold F-curve and with multiplicity $\alpha-2h$ (or zero) at every α-fold isolated F-point of the given web) are lacking. (II) Every web of order $n=4k$ possesses either (a) an F-curve of multiplicity $i \geqq k+1$ and of order $\leqq 15$ or (b) an isolated F-point of multiplicity $\alpha \geqq 2k+1$; with similar results for $n \equiv 1, 2, 3$ (mod. 4). She then considers transformations of type (a) for which the F-curve of greatest multiplicity is (1) a plane curve; (2) a curve on a quadric; (3) a curve on a cubic surface, and determines for them some of the other F-curves and F-points. She shows that there are no transformations of type (b) with only F-points without contacts; that every transformation with only F-points with contacts is monoidal and of type: (1), (2, 4), quadrics on a tetrahedron with a fixed tangent plane at one vertex; or (2), (3, 9), cubics with a double point and simple points with a contact of third order at one; or (3), (4, 16), quartics with a triple point and a contact of the fifth order at a simple point. Of these three types the first two were noted by Cremona while the third is new. Miss Beloch also proves that if in type (a) the F-curve of maximum multiplicity is of genus 3 and order 6 the transformation is the product of a (3, 3) type and of one of order less than n.

Montesano([108]) obtains a type for which the ϕ's contain a space quartic of the first kind C_4 as an m-fold F-curve and $2m$ bisecants of C_4 as simple F-curves. The inverse web is of the same kind, $n=n'=2m+1$, and the genus is m. In superposed spaces it may be involutorial. Del Pezzo([141]) studies transformations for which the ϕ's are cones. The simplest is a quadric type T and higher types of order n are obtained as products of types T. Berardi([13.2]) gives fourteen types of quintic transformations with a triple and two double lines.

Many types such as that of Montesano([108]) above have been discussed in connection with involutorial transformations. These are reported more fully in the next chapter. We mention also the determination by Montesano([106]) of involutorial types ($n'=n$) for which the ϕ's have an $(n-2)$-fold line. Also, many types are mentioned in connection with the complex Γ of Section 5 and in connection with the regular transformations of Section 3.

In higher space relatively few homaloidal systems have been exhibited. Cayley([21]) remarks that as two bilinear relations in S_2, and three in S_3, determine respectively transformations of orders 2 and 3, so r bilinear relation in S_r determine a transformation of order r. The properties of this transformation known in S_3 have been generalized by Godeaux([67]) to S_r. The involutorial case in S_4 determined by four polarities is studied by Miss Alderton([1]); cf. also Wong([177]). Del Pezzo([139]) determines 16 homaloidal systems of quadrics in S_4; and also([140]) exhibits a homaloidal system of quadric cones in S_4 which is then generalized to S_r. He([142]) considers also a quadratic transformation in S_4 whose inverse is of order 8, as well as a number of its special cases; also a quadratic transformation whose inverse is of order 4. Carrone([19]) follows the method of Cremona explained above in using the mapping system which arises from stereographic projection of a quadric in S_r upon an S_{r-1}. He obtains thereby various homaloidal systems $(n=2)$ which for $r=4$ are nearly all given by del Pezzo. Carrone also obtains a type from the fact that in S_{r+1} r independent S_{r-1}'s, $\sigma_1, \ldots, \sigma_r$ have the property that on a point P of S_{r+1} there is a single line which meets $\sigma_1, \ldots, \sigma_r$. If then in S_{r+1} two S_r's are given, a point P of the one determines a line which meets the other in a point P'. This transformation from P to P' is discussed. This symmetrical case is extended to the case of fixed spaces $S_{r_1}, S_{r_2}, \ldots, S_{r_{p-r+1}}$ in $S_p (p>r)$ where $r_1+r_2+\ldots+r_{p-r+1}=n$. Then the S_{p-r}'s of S_p which meet the fixed spaces form a system ∞^r, one on each point of S_p and again points P, P' of two given spaces S_r, S'_r are images in a Cremona transformation. Veneroni([174]) observes that the manifolds of order r in S_r which contain $r+1$ arbitrary S_{r-2}'s form a homaloidal system. They contain also the variety $V_{r-2}^{(r-1)(r-2)}$ formed by all the lines which meet the $r+1$ given S_{r-2}'s. The inverse system is of the same type and in superposed spaces the transformation has 2^r fixed points. Marletta([98.1]) obtains new homaloidal systems in S_r.

3. *Integer Linear Transformations Associated with Regular Groups.*— Coble([27]) has defined the *regular* Cremona group in S_r to be that group which is generated by a transformation of the form $x_j x'_j=1$ $(j=1, \ldots, r+1)$ and collineations; but the discussion is confined to those elements for which the isolated F-points are placed in general position. In S_3 the transformations of this group have been described by Kantor([88]) as having no F-curves of the first kind. He gives their more striking properties in S_3 and notes the possibility of generalization to S_r. We take up at once the general case. The regular transformations have properties

quite analogous to those in the plane since they are determined to within collineations by their finite number ρ of isolated F-points. This number ρ is the same for the direct and inverse transformations. In S_r they have F-spaces of dimension $r-2, \ldots, 1$ but the existence of these is a necessary consequence of the existence of the isolated F-points. There is but one type of principal *variety*—that whose points correspond to the aggregate of directions at an isolated F-point. Such a P-variety is determined uniquely by its behavior at the isolated F-points.

The order n of such a regular transformation T from $S_r(x)$ to $S_r(y)$ has the form $(r-1)\mu+1$. It transforms an $S_{r-1}(x)$ into a variety in $S_r(y)$ of order n with multiple points of orders $(r-1)_{s1}, \ldots, (r-1)_{s\rho}$ at the isolated F-points in $S_r(y)$. The P-variety which corresponds to the F-points, $F_i(i=1, \ldots, \rho)$, in $S_r(x)$ has the order r_i with multiplicity a_{ij} at the F-point F_j in $S_r(y)$. If a variety $V_{r-1}(x)$ has the order x_0 and the multiplicity x_i at the point F_i this variety is transformed by T into one in $S_r(y)$ with the order y_0 and the multiplicity y_j at F_j where

$$
G:\quad
\begin{aligned}
y_0 &= [(r-1)\mu+1]x_0 - r_1x_1 - \ldots - r_ix_i - \ldots - r_\rho x_\rho,\\
y_1 &= (r-1)s_1x_0 - a_{11}x_1 - \ldots - a_{i1}x_i - \ldots - a_{\rho 1}x_\rho,\\
&\ldots\ldots\ldots\ldots\ldots\ldots\ldots\ldots\ldots\ldots\ldots\ldots\\
y_j &= (r-1)s_jx_0 - a_{1j}x_1 - \ldots - a_{ij}x_i - \ldots - a_{\rho_i}x_\rho,\\
&\ldots\ldots\ldots\ldots\ldots\ldots\ldots\ldots\ldots\ldots\ldots\ldots\\
y_\rho &= (r-1)s_\rho x_0 - a_{1\rho}x_1 - \ldots - a_{i\rho}x_i - \ldots - a_{\rho\rho}x_\rho.
\end{aligned}
$$

The behavior of $V_{r-1}(x)$ and its transform at pairs of ordinary corresponding points may be indicated by further equations

$$
\begin{aligned}
y_{\rho+1} &= -(-1)x_{\ +1},\\
&\ldots\ldots\ldots\ldots\\
y_m &= -(-1)x_m.
\end{aligned}
$$

Since T is generated by the particular transformation above, the linear transformation G is generated by the particular element,

$$
A:\quad
\begin{aligned}
y_0 &= rx_0 - (x_1 + \ldots + x_{r+1}),\\
y_j &= (r-1)x_0 - (x_1 + \ldots + x_{r+1}) + x_j, \qquad (j=1, \ldots, r+1),
\end{aligned}
$$

and by permutations of the variables $x_1, \ldots, x_m$. This particular element, and therefore G itself has an invariant linear form, and an invariant quadratic form,

$$
\begin{aligned}
&(r+1)x_0 - x_1 - x_2 \ldots - x_m,\\
&(r-1)x_0^2 - x_1^2 - x_2^2 \ldots - x_m^2.
\end{aligned}
$$

The existence of these forms imposes a system of relations on the coefficients μ, r_i, s_j, a_{ij} of the transformation G entirely analogous to the system [Chapter IV (2)] which prevails in the plane [([27]) p. 366]. To the inverse Cremona transformation T^{-1} there corresponds the inverse linear transformation G^{-1} which is obtained from G by interchanging the r_i, s_j and the a_{ij}, a_{ji}.

To every type of regular transformation T with prescribed coordination of the two sets of isolated F-points there corresponds a linear transformation G; and the aggregate of such types with $\rho \lesseqgtr m$ F-points gives rise to a collineation group $g_{m,r}$ which is infinite and discontinuous for $m > r+3$ except in the cases $(m, r) = (6, 2)(7, 2)(7, 3)(8, 2)(8, 4)$ [([27]) p. 362]. For $r=3$ the symmetric types

$$(n, \rho, r_i) = (3, 4, 1)(7, 6, 2)(15, 7, 4)$$

have been given by Kantor[([88]) p. 27]; for higher spaces Coble[([27]) p. 368-69] adds the types: in $S_4(49, 8, 10)$; in $S_r(r, r+1, 1)$; in $S_r(r \text{ odd}), (\frac{1}{2}(r-1)(r+3)+1, r+3, \frac{1}{2}(r+1))$. By considering the particular case for which the $\rho \lesseqgtr m$ F-points are placed on a normal elliptic curve, Coble[([27]) § 6] derives a collineation group $e_{m,r}$ with rational coefficients which is a transform of $g_{m,r}$ for which the invariant linear space becomes a fixed coordinate space.

Kantor pursues the study of the transformation G with particular reference to the determination([87]) of periodic cubic transformations in S_3, and([88]) of finite groups of regular transformations in S_3 (cf. § 6). Coble pursues a similar study for *congruent sets of points* in S_r. Two ordered sets of m points, $P_m{}^r$ in $S_r(x)$ with points $p_1, \ldots, p_m$ and $Q_m{}^r$ with points $q_1, \ldots, q_m{}^r$ in $S_r(y)$ are congruent under the regular transformation T if $\rho \lesseqgtr m$ of the pairs p_i, q_i are the F-points of T, T^{-1} and if the remaining $m-\rho$ pairs are ordinary corresponding pairs. By mapping such a set $P_m{}^r$ by means of its absolute projective invariants upon a point P of a linear space $\Sigma_{r(m-r-2)}$, the aggregate of sets congruent to it under regular Cremona transformation is mapped upon an aggregate of points P in $\Sigma_{r(m-r-2)}$ which form an aggregate of conjugate points under a Cremona group $G_{m,r}$ in $\Sigma_{r(m-r-2)}$ which is isomorphic with the collineation group $g_{m,r}$. This group is discussed in Sections 7 and 10.

With any set $P_m{}^r$ there is "associated" projectively a set $P_m{}^{m-r-2}$ in S_{m-r-2}. For such associated sets the groups $G_{m,r}$ and $G_{m,m-r-2}$ in $\Sigma_{r(m-r-2)}$ coincide. This association of sets of discrete points is studied by Coble([26]), ([27]), ([31]), and, in a particular case, by Musselman([123]). The groups $g_{m,r}$ and $g_{m,m-r-2}$ are simply isomorphic.

If in forming products with the generator $x_j x'_j = 1$ one F-point is held fixed, the types thus generated have a monoidal character and the coefficients of the transformation G for T in S_r may be inferred from those of a transformation G' for a regular transformation T' in S_{r-1} [cf. Coble(30)]. This procedure has been called the dilation of T'. If two points are fixed, the types obtained in S_3 are called *dyoidal* by Kantor(88). When the spaces are superposed and one F-point is held fixed, all lines on this point are transformed among themselves; if two F-points are held fixed, all planes on the line of the two are transformed among themselves, etc.

The set of points P_m^r defines a discontinuous aggregate of P-varieties which consists of the directions at any one of the m points, the S_{r-1}'s on any r of the m points, etc.; in general, of any variety V^k_{r-1} defined by its behavior at the points of P_m^r and of a type which can be obtained by regular transformation from the directions at a point. Under regular transformation the P-varieties of a set P_m^r are transformed into the P-varieties of a congruent set Q_m^r. There is also attached to a set P_m^r an aggregate of discriminant conditions of which the simpler types are the condition that two points of P_m^r coincide in some direction, and the condition that $r+1$ points of P_m^r are on an S_{r-1}. The other types are such as arise from these simpler types by regular transformation. Properties of P_m^r which persist under regular transformation to congruent sets Q_m^r, i. e., invariant properties of P_m^r under regular Cremona transformation are expressed in terms of these discriminant conditions [cf. Coble(27) §8; (28) §3; (30); (32)].

Kantor has developed a theory similar to the above for transformations in S_r which can be generated by collineations and a quadratic transformation with an isolated simple F-point and an F-variety which is a quadric in an S_{r-1}. His main objective has been the periodic types and the finite groups [cf. § 6].

Kantor(90) considers the behavior of the group of all Cremona transformations in S_r from the standpoint of general group-theory. He concludes that it is a group whose elements contain certain continuous parameters $a_1, a_2, \ldots$; and certain discontinuous parameters $\alpha_1, \alpha_2, \ldots$; each set of parameters infinite in number. The α's vary with the type of transformation; for given type the a's vary continuously with the choice of the F-varieties. Subgroups may be characterized by invariant relations on either type of parameter.

4. *Projective Descriptions of Particular Types.*—The simplest Cremona transformation in space, the quadratic ($n=2$, $n'=2$) transformation, is studied by Schiaparelli(152) and Geiser(62) as a quadric inversion; see also Cayley(21) and Cremona(38), (41). Aschieri(5) constructs this type by means of a correlation between two S_3's and two projectively related bundles of rays, and later(6) studies it in connection with a certain linear complex. The transformation ($n=2$, $n'=3$) is studied by

Cayley([21]) and Cremona([38]), ([41]). The transformation ($n=2$, $n'=4$) is discussed in detail by Cremona([40]) and by Medugno([99]).

The most interesting space transformation ($n=n'=3$) is that with a sextic F-curve of genus three (or its special cases) in either space for which the P-surfaces are the octavic loci of trisecants of the F-sextics. It is first obtained in 1837 from three correlations by Magnus([98]) who also recognizes the existence of the sextic F-curve, C^6. It is discussed from the same point of view by Cremona([37]); by Cayley([21]) who notes the special case where C^6 consists of four lines and their two transversals; and by Sturm([169]) who observes that it is determined by 13 pairs of corresponding points and that it has in superposed spaces 8 fixed points [cf. also Terquem([170.1]), Kantor([82]), Machovec([97.2]), and Baker([12.1])]. The properties of the transformation are developed by Reye[([148]); Vol. III].

Of particular cases of the above transformation, von Krieg([92]) first gives that which arises by extending to space the Sedydewitz construction for a planar quadratic transformation. It is generated by three pairs of projective pencils of planes and has for F-curve the three base-lines of the pencils and a cubic curve bisecant to each. It is applied by von Krieg to various construction problems. This case is more fully investigated by Ascione([7]) particularly for phenomena which appear in superposed spaces; and later is discussed by Doehlemann([49]) who observes a case which has a cubic space curve as a locus of fixed points. The particular case where the three base lines of the pencils form a triangle and the transformation is of the form $x_i x'_i=1$, ($i=1, \ldots, 4$) was given earlier by Silldorf([159]) who observes that it is determined by the two F-tetrahedra and one ordinary corresponding pair; and that the lines on two related F-points are in a quadratic correspondence.

Miss Hudson([71]) gives 28 types of the cubic transformation (3, 3) for which the one sextic F-curve consists of six lines with 8 simple intersections or their equivalent in multiple intersections. The cases for which the six lines are the edges of a tetrahedron is discussed by Beltrami([13.1]), Amigues([2.1]) and Obeslo([128.1]); and, in metrical aspects, by Hunyadi([78.5]) and Meyer([99.1]), ([99.2]). Eckhardt([49.1]) compounds such transformations with two fixed F-points to produce regular transformations.

If the three correlations which define the (3, 3) transformation become polarities, the transformation is an involution with pairs x, x' apolar to a net of quadrics. The geometry associated with this involution is developed by Sturm([167]), ([168]) without particular reference to the general theory. The case in which the net of quadrics is on a cubic space curve is discussed by Cantone([18]), by Reye[([148]) Vol. 2]; and by Schoute([154]); also by Haskell([70]) who employs the comitants of a binary cubic.

An extensive account of the quadratic and cubic transformations is given in Sturm's treatise[(166) p. 347-401]; a more elementary account by Doehlemann[(47) Part 2]. The cubic reciprocity defined by three collineations is discussed by Sturm(169); and a metric case, in which the plane determined by a point P is on the feet of the perpendiculars dropped from P to three skew lines, by Guradze(69).

Cremona(38) gives a great variety of particular cases for each of which the bases and the jacobians of the inverse homaloidal webs are described. If the general surface of the web ϕ is a general quadric, the order of the inverse web ψ is 2, 3, or 4; for ϕ a general cubic surface, the ψ are of orders 3, 4, 5 (two types), 6 (two types) to which Sturm(170) adds another); for ϕ a nodal cubic surface, the ψ are of orders 5, 6, 7 (three types), 8, 9; for ϕ a ruled cubic surface, the ψ are of orders 2, 3, 4, 5; for ϕ a binodal cubic surface, the ψ are of orders 3 (two types), 4, 5, 6, 7; for ϕ a trinodal cubic surface, the ψ are of orders 3, 4, 5, 6; for ϕ a cubic surface with a uniplanar point, the ψ are of orders 4, 5, 6, 7, 8, 9; for ϕ a Steiner quartic surface, the ψ are of orders 2, 4; and for ϕ a quintic surface with double space cubic, the ψ are of orders 4, 7. Cremona returns later(41) to the cases ($n=2, n'=2, 3, 4; n=3, n'=3, 4, 5$). Bonicelli(16) investigates the (3, 6) case defined by cubic surfaces on three skew lines, a point, and a point at which there is a fixed tangent plane. Cremona(44) discusses a (6, 4) transformation; as well as(45) a (6, 5) transformation. Aschieri(5.1) treats a (4, 4) transformation, and Godeaux(67.2) discusses transformations which carry a pencil of planes into a pencil of planes, or a linear congruence into a linear congruence.

Saltel(150) defines the Aguesienne transformation in space to be the involution whose pairs x, x' are on a line with a fixed point P and in the involution cut out on this line by a pencil of quadrics [cf. also Deruyts(46)]. The web has a node at P and is on the base curve C^4 of the pencil of quadrics; the P-surface of P is the quadric on C^4 and P, the P-surface of C^4 is the quartic cone on C^4 with vertex at P. Godeaux(66) replaces the bundle of lines on P by the rays of a linear congruence with directrices a_1, a_2; the web consists of sextics with double lines a_1, a_2 and simple C^4. There is a quintic locus of fixed points on a_1, a_2, C^4. Schoute(153) extends the transformations described in Chap. IV, Section 4 to space and obtains for the first involution a cubic transformation with F-tetrahedron and for the second the transform of this by an inversion.

De Paolis(136) gives a geometric and analytic account of transformations whose homaloidal web of order n has an $(n-1)$-fold point; and in particular of those for which $n'=n$.

As noted above, interesting types arise and have been discussed in connection with their involutorial cases. Thus Geiser(63) describes the involution of order 7 defined by pairs x, x' which form with six given points the base points of a net of quadrics. Montesano(108), (110) constructs involutions for which planes pass into surfaces of order $2n+1$ with a space quartic of the first kind as an n-fold curve, and with $2n$ chords of the curve as simple lines. The genus of such a transformation is n. Montesano(113) also discusses involutions determined by bilinear complexes of conics in space.

Conner(65) gives an account of the Kantor involution determined by 7 points in space; and Coble(31) describes certain involutions which leave a symmetroid unaltered.

Snyder(163) describes two types of involutions in S_4 which transform a V_3^3 in S_4 into itself. The first is the projection of the variety upon itself from a point of the V_3^3. In the second, each plane on a line π of V_3^3 meets V_3^3 in a conic, and the projection of this conic into itself from the pole of π is part of a transformation in S_4.

Further types are mentioned in Section 5 in connection with the complex Γ, and in Chapter IX on Involutions.

5. *Fixed Points and Cyclic Sets. The Complex* Γ.—When the spaces $S_3(x)$ and $S_3(y)$ are superposed, the transformation T carries x into a point $y=x'$ of the same space and, as in the plane, fixed points, or invariant cycles of points, may arise. T then may be periodic or belong to a group in $S_3(x)$. In this and the three following sections the theory thus arising is reported. The line which joins x to its correspondent x' will in general occupy ∞^3 positions and thus will run over a complex Γ. For particular transformations Γ may reduce to a congruence. The complex Γ has been widely used as a means of classifying transformations S_3.

Aschieri(5) finds that the general quadratic transformation has 6 fixed points and finds also the fixed points and loci of fixed points for the involutorial cases. Kantor(81) shows that the number N_k of cyclic sets of k points for the transformation $y_i=1/x_i (i=1, \ldots, 4)$ where $k=e_1^{m_1}e_2^{m_2}\ldots e_\nu^{m_\nu}$ ($e_j=$a prime) is given by

$$N_k=2[3^k-\sum 3^{k/e_j}+\sum 3^{k/e_ie_j}\ldots(-1)^\nu 3^{k/e_1\cdots e_\nu}]/k.$$

He also(81) finds that the number N_k for the transformation $y_i=1/x_i$ $(i=1, \ldots, r+1)$ in S_r is given by

$$N_k=(r-1)![r^k-\sum r^{k/e_j}+\sum r^{k/e_ie_j}\ldots(-1)^\nu r^{k/e_1\cdots e_\nu}]/k.$$

James(80) proves for a Cremona transformation in S_r whose orders are $n, n', \ldots, n^{(j)}, \ldots, n^{(r-2)}$ (the respective orders of the transforms of an $S_{r-1}, S_{r-2}, \ldots, S_{r-j-1}, \ldots, S_1$) that the number of fixed points is $(\sum_{j=0}^{r-2} n^{(j)})+2$, the number of involutorial pairs is $\sum_{j-0}^{r-2}\frac{1}{2}n^{(j)}(n^j-1)$; and gives also a formula for the number N_k of cycles of k points. Hudson and Wren([78.2]) discuss the two involutorial pairs of the (2, 2) transformation in S_3.

Pannelli([130]) uses the formulas of Noether given in Section 2 to make a detailed study of the complex Γ. De Paolis([137]) studies the complex Γ for the involution associated with a (1, 2) transformation in space; also([138]) involutorial transformations for which Γ degenerates into a congruence consisting of (a) lines on a point; or (b) bisecants of a cubic space curve; or (c) transversals of a space curve of order μ and a $(\mu-1)$-fold secant. Vogt([175]) studies the transformations for which Γ degenerates into a linear congruence. T has orders (n, n) and an F^{n-1} of fixed points. Montesano([101]) finds that if Γ is a non-special linear complex and T involutorial an upper limit for n is 11; a special case is treated later([103]). Pieri([143]) finds that for Γ a special linear complex the order of n is unlimited. Aschieri([6]) considers the case where Γ is linear and special and T quadratic. Montesano([105]) discusses the case where Γ is tetrahedral for which n in general is 19 though reductions occur in particular cases. Del Re([147.1]) constructs transformations of the third and higher orders for which Γ is linear or is a congruence and gives extensions to hyperspace. Pieri([144]) considers the case when Γ is a Hirst complex determined by two correlative planes in space. Montesano([107]) discusses the transformations T for which the complex Γ is quadratic; and Pieri([145]) the particular case where Γ consists of the secant lines of a conic. Montesano([109]) defines an involution, for which Γ consists of the tangents of a surface, in terms of a linear (one conic on a point) congruence of conics in space whose planes envelop a surface F of class >1. For given x the plane of the conic on x has a contact with F and the line through x and this contact meets the conic again at x'. Pieri([146]) lists cases where Γ consists of the tangents of a surface; and Montesano([112]) adds to this list. Montesano([111]) studies the type (a) of Loria([97]) in which the webs are Steiner quartics and finds that Γ is either a special linear complex or a cubic complex. Miss Caldarera([17]) investigates the cases for which Γ consists of the secants of a space cubic curve.

6. *Periodic Transformations and Finite Groups.*—Montesano([100]) defines an abelian G_{16} of quadratic involutions by means of a plane π and five points $O_i (i=1, \ldots, 5)$ in space. The quadrics Q_i with O_i, π as pole and polar and the other four points as self-polar tetrahedron meet π in the same conic K. The involution I_i has F-point O_i and F-conic K and $I_iI_j=I_jI_i=I_kI_lI_m$. The G_{16} possesses invariant cubic surfaces on which the group is the map of the plane Cremona group G_{16} determined by five points. Montesano also finds a Cremona G_{32} in space by mapping

the lines of a quadratic complex upon an S_3 and taking the map of the group of collineations and correlations which is defined by the complex. Del Re([147]) considers the transformation T, $x_i x'_i = a_i$ $(i=1, \ldots, r+1)$ in S_r with 2^r fixed points. These fixed points and T itself are invariant under the linear group G, $x'_i = \pm x_i$ of order 2^r. Then G and T generate a Cremona group of order 2^{n+1}. An analogous group is obtained by combining $x'_i = \mu_i x_i (\mu_i^m = 1)$ with T.

S. Kantor in a series of papers([83]), ([84]), ([85]), ([86]) derives the types of periodic transformations and of finite groups for that particular subgroup of the Cremona group in S_r which is generated by collineations and a quadratic transformation defined by an isolated F-point P, and a general quadric in an S_{r-2} not on P as an $F-M^2_{r-2}$. In forming products of these generators the F-manifolds of the factors are so chosen that an isolated F-point does not lie on the S_{r-2} of an $F-M^2_{r-2}$. This sub-group has an arithmetic character analogous to that of ternary transformations. A cyclic group generated by a periodic transformation, or a finite group has an invariant linear system of varieties V_{r-1} whose order and dimension may be reduced by a properly defined "adjoint" process to the point where the resulting linear system can be transformed into one of a finite number of types, and the associated group into one of a similar set of types.

Kantor carries out a similar process for regular periodic substitutions in S_3([87]) and for regular finite groups in S_3([88]). A regular transformation is one generated by $x_i x'_i = 1 (i=1, \ldots, 4)$ and collineations. The list obtained in ([88]) contains (a) a collineation group in S_3; (b) 34 types with an invariant bundle of lines; (c) 10 types with an invariant pencil of planes on a line a, 5 of which leave the line bundles on points a_1, a_2 of a each invariant while the other 5 interchange these bundles; (d) 5 types of groups of cubic transformations; (e) a group G_{720} determined by 5 points; (f) 7 types of groups determined by 6 points; and (g) 14 types of groups determined by 7 points. The groups (f) have an invariant Weddle surface determined by the 6 points and arise from collineation groups on the Kummer surface which is birationally equivalent to the Weddle. The groups (g) have an invariant locus of nodes of quadrics on the 7 points and arise from collineation groups which the plane quartic curve, birationally equivalent to the nodal locus, admits.

E. H. Moore([118]) observes that all of the cross-ratios of a binary $(r+3)$-ic are rationally expressible in terms of r of the cross-ratios which he takes for the coordinates of a point P in S_r. When the roots are permuted, P is transformed by a Cremona transformation which in general is of order r. The $G_{(r+3)!}$ thus generated has a collineation subgroup

$G_{(r+2)!}$ and is determined by $r+2$ points. The F-spaces of the elements are linear spaces of various dimensions determined by these points. For $r=3$ this is Kantor's group (e) just above. For $r=2, 3, 2p-1$ Coble([24]), ([25]), ([33]) has discussed this group in connection with the solution of the equation of degree 5, 6, $2p+2$ (cf. § 10). In ([26]) he generalized this group in connection with a set of m points in S_r, P_m^r. If the first $r+2$ of the m points be placed at the reference basis and the unit point in S_r and if the last coordinate of the other points be taken as equal then the $(m-r-2)r+1$ coordinates of these points are taken as coordinates of a point P in a linear space $\Sigma_{r(m-r-2)}$. As the points of P_m^r are permuted the point P is transformed under a Cremona $G_{m!}$ in $\Sigma_{r(m-r-2)}$. The set P_m^r in S_r is "associated" with a set P_m^{m-r-2} in S_{m-r-2} whose $G_{m!}$ coincides with that of P_m^r. For $r=1$ or $r=m-3$ this group is Moore's cross-ratio group. One case $m=6$, $r=2$ had been given earlier by Kasner([91]) who obtained it from four independent irrational invariants of a double six on the cubic surface mapped from the plane by cubic curves on P_6^2.

Coble([27]) obtains a group $G_{m,r}$ of transformations in $\Sigma_{r(m-r-2)}$ by adding to the permutations of the points P_m^r the operation of passing from P_m^r to a set Q_m^r *congruent* to P_m^r under regular Cremona transformations in S_r (cf. § 3 Chap. IV). Again the group $G_{m,r}$ attached to P_m^r coincides with the group $G_{m,m-r-2}$ attached to the associated set P_m^{m-r-2}. This group is in general infinite (cf. §7) but three finite cases $G_{6,2}$, $G_{7,2}=G_{7,3}$ and $G_{8,2}=G_{8,4}$ are of particular interest being isomorphic with the group of respectively, the 27 lines of a cubic surface, the 28 double tangents of the plane quartic, and the 120 tritangent planes of a space sextic of genus four on a quadric cone. Coble([28]) uses the $G_{6,2}$ in Σ_4 for the determination of the lines on a cubic surface (cf. § 10).

Coble([33]) defines by means of $2p+2$ points in S_{2p-1} an abelian $G_{2^{2p+1}}$. It is generated by transformations of the type $xx'=1$ with isolated F-points at $2p$ of the points which interchange the other two. The group has an invariant variety of dimension p, a generalization of the Weddle surface, which is mapped by the simplest linear system invariant under the group upon the hyperelliptic Kummer manifold of genus p.

7. *Infinite Discontinuous Groups.*—Certain infinite discontinuous groups in S_r which are subgroups of the general Cremona group in S_r are themselves sufficiently general to have the constitution ascribed by Kantor[([90]), cf. § 4] to the Cremona group itself. This is true in S_3 of the monoidal group of Cremona transformations with an invariant bundle of lines, and of the dyoidal [cf. Kantor([88])] group with an invariant pencil of planes. The group of regular transformations and the group

generated by a quadratic transformation (cf. § 6) have a similar character. Other examples may be given such as the group in S_3 which has an invariant quadric or the group in S_r which has an invariant elliptic curve C^{r+1}. Such groups are characterized by the fact that their elements contain continuous parameters in addition to the discontinuous parameters. Other infinite groups are properly discontinuous in the sense that an element contains no continuous parameters.

Hutchinson(79) observes that the Kummer quartic surface admits 60 cubic involutions ($x_i x'_i = 1$) with F-points at a Göpel tetrad of nodes, and that the Weddle surface admits 15 similar involutions, which on either surface generate an infinite discontinuous group. Baker(12) notes that an element to which Hutchinson ascribes an infinite period is actually of period two. He adds relations among various products of involutions. Sharpe and Craig(36) use Severi's basis theory to confirm Hutchinson's general result and to add others. Snyder(162) shows that the infinite discontinuous group of birational transformations which occur on certain surfaces of Rosenblatt, which contain a pencil of elliptic curves, can be replaced by a similar group of Cremona transformations. Snyder and Sharpe(157) obtain an infinite discontinuous group of non-involutorial Cremona transformations which leaves invariant a quartic surface containing a space sextic of genus three. They prove also that the known birational group of a quartic surface which contains a space sextic of genus two can be replaced by a group of Cremona transformations; and earlier(156.1) prove a similar theorem for the space sextic of genus three. Snyder(163) combines the two types of involutions which carry a V_3^3 in S_4 into itself [cf. § 4] to form a discontinuous group with continuous parameters.

Coble has defined a doubly infinite array of infinite discontinuous groups, $G_{m,r}$ (cf. § 6), in $\Sigma_{r(m-r-2)}$. To an aggregate of conjugate points P under $G_{m,r}$ in $\Sigma_{r(m-r-2)}$ there corresponds in S_r the aggregate of sets of points P_m^r congruent to one of the sets under regular Cremona transformation in S_r or the associated aggregate of sets of points P_m^{m-r-2} in S_{m-r-2}. However, the transformations of $G_{m,r}$ in Σ are not regular. The group $G_{m,r}$ contains subgroups simply isomorphic with $G_{m'r'}$ if $m' \leqq m$ and $r' \leqq r$ [(30), p. 258]. The infinite case occurs for $m > r+3$ with the exceptions noted in Section 6. For particular points P in Σ the aggregate of conjugate points under $G_{m,r}$ may be only finite. Then the set of points P_m^r in S_r is *special* and is self-congruent under an infinite discontinuous Cremona group in S_r. Coble[(27), pp. 377-8; (30)] gives as examples 9 points on an elliptic plane cubic, or 8 points on an elliptic space quartic, the sum of whose canonical parameters is a sub-multiple of a period;

also the ten nodes of a rational plane sextic, and nine or ten nodes of a symmetroid. Another method of defining infinite discontinuous subgroups of $G_{m,r}$ arises from its isomorphism with the collineation group $g_{m,r}$ (cf. § 3) with integer coefficients. Those elements of $G_{m,r}$ whose corresponding elements in $g_{m,r}$ are congruent to the identity modulo j form a discontinuous infinite subgroup of $G_{m,r}$ whose factor group is finite. For $j=2$ these finite factor groups have been determined by Coble(**29**) to be known groups which occur in connection with period transformations of theta functions in p variables. Coble(**31**) also notes that the normal variety V_2^{r-1} in S_r admits an infinite discontinuous group of regular transformations, a property shared by the Veronese V_2^4 in S_5.

8. *Continuous Groups.*—Noether(**128**) shows that every continuous group of quadratic transformations in S_3 is one of the following 5 types or a subgroup of one: (a) the group (∞^{10}) which transforms the system of quadrics on a proper conic into itself; (b) the group (∞^{11}) of a system of quadrics on two intersecting lines; (c) the group (∞^{13}) of a system of quadric cones which touch along a common generator; (d) the group (∞^{11}) of a system of quadrics on a line and an outside point; and (e) the group (∞^{13}) of a system of quadrics which are on a line and touch at a point of the line.

Enriques and Fano(**53**) show that every continuous Cremona group in space can be transformed into one of the following 6 types or a subgroup of one: (a) the projective group; (b) the group (∞^{10}) of conform transformations which transform spheres into spheres; (c) a generalized Jonquières group with an invariant bundle of lines; (d) a generalized Jonquières group with an invariant pencil of planes; (e) a group (∞^3) of cubic transformations with an invariant linear system of cubic surfaces on a space quartic of the second kind; and (f) a group (∞^3) of transformations of order 7 with an invariant linear system of dimension 12 of surfaces of order 7, with a common double conic, with a simple line l tangent to the conic at p, with a fixed tangent plane along l, and with a triple point at p.

Fano(**59**) divides the primitive types (a), (b) above into 10 sub-types, 9 of which had been given by Lie, and a tenth—the ∞^6 conform transformations of a sphere into itself. Fano(**56**) divides the primitive types (c), (d) into 12 sub-types. Fano(**57**) shows that the types of continuous groups of Cremona transformations given by Lie are reducible to the types of Enriques-Fano. Fano(**55**) gives a résumé of the above and states the theorem that any continuous Cremona group in S_r has an invariant linear system of dimension n and can be mapped upon a projective group in S_n which has an invariant rational V_r. Fano(**58**) gives a complete enumeration of continuous groups not contained in a larger group as well as the infinitesimal transformations of such groups. Fano([60]) gives a general report on continuous groups.

Young and Morgan([120]) study the continuous group defined by three invariant pencils of planes in space in each of which projectivities are simultaneously defined, a type given by Fano([56]). The authors indicate the generalizations to S_r and study particularly the geometry under this group of the system of space cubic curves with the base lines of the three pencils for bisecants.

9. *Geometric Applications.*—One of the first applications of space transformations was to the study of surfaces of higher order as the transforms of known surfaces of lower order. Geiser([64]) considers the quartic surface with a double conic as the transform of a cubic surface by a quadric inversion. He([62.1]) uses metric cases both of the general and of the tetrahedral (3, 3) transformation in a study of the quadric and cubic surface. Noether([126]) employs the quadratic transformation and the (3, 3) transformation to obtain from planes or other rational surfaces a variety of different types of rational surfaces. Cremona([39]) uses a similar method to extend this variety. He([37.1]) uses the special (2, 2) transformation with F-point on the F-conic to get the quartic surface with double conic from a cubic surface and a special case from a quadric surface. Cremona([42]) uses a quadratic transformation to transform a certain cubic surface into a quartic surface with a cuspidal conic, and a cubic transformation to transform a quadric into a quintic surface with a cuspidal space quartic. Eckhardt([49.1]) uses the tetrahedral transformation to study the Steiner quartic surface. Saltel([150]) proceeds from surfaces already constructible to others by the Arguesienne transformation; and([151]) studies the behavior of transforms at acquired singular points. Cremona([43]) transforms a cubic surface into a quartic surface with a self-contact by a quadratic transformation. Caporali([18.1]) employs a (3, 5) transformation in connection with the quintic surface with double quintic curve. Montesano([111.2]), ([111.3]) obtains by transformation a large variety of quintic surfaces. Pieri([142.1]) uses an involution whose complex Γ is quadratic and special to study the sextic surface with a four-fold line and 10 nodes. Nobile([124.1]) uses monoidal transformations in the study of rational surfaces of order n with an $(n-\nu)$-fold line and ν-fold points. Wren([177.1]) uses Cayley's (2, 3) transformation to study curves on a surface of order n with an $(n-1)$- or $(n-2)$-fold line. Von Krieg([92]) applies the (3, 3) transformation determined by three pairs of projective pencils of planes to the construction of a conic in space, a cubic space curve, a rational space quartic, a quadric surface, and a cubic surface. Chizzoni([22]), ([23]) uses a particular involution of order 3 in space to study certain families of conics and rational surfaces (including

cyclides) connected with them; and also gives new constructions for the space quartic on 8 points and the quadric on 9 points. Morley and Conner([34]) define the Weddle surface to be the aggregate of pairs x, x' apolar to the net of quadrics on a cubic space curve and also apolar to another quadric. Snyder([161]) uses the involution determined by a web of quadrics on 6 points in the study of the Kummer and Weddle surfaces. Van Veen([172]) employs an involutorial transformation from plane to plane to establish theorems concerning space quartic curves of the first kind which have 3 or 4 intersections with a fixed space cubic curve. Miss Castelli([20]) applies a particular (3, 3) transformation noted by Cremona (defined by pairs x, y whose line is in a linear congruence and which satisfy a correlation) to the study of quartic and quintic surfaces with particular singularities.

The study of Cremona null-systems was begun by Ameseder([2]) who developed synthetically the properties of the general quadratic null-reciprocity. Lazzeri([93]) first discusses the Cremona reciprocity and finds that in a transformation from point to plane of orders (n, n') for incident point and corresponding plane, the locus of the point is a surface of order $n'+1$, and the envelope of the plane is a surface of class $n+1$ whose relations to the F-bases are described. The reciprocities are polar only in the case of the polar system of a quadric $(n, n'=1, 1)$; of pole and polar plane as to a tetrahedron $(n, n'=3, 3)$; and one case $(n, n'=2, 2)$. He finds only two null-systems for which $n, n'=1$, 1 and 3, 3. However, Montesano([102]) obtains infinitely many types by combining an involution whose pairs are on lines of a linear complex Γ with the linear null-system defined by Γ and proves that all Cremona null-systems arise in this way. In connection with other transformations Francois([61]) employs the reciprocity which relates a point P to the plane of the feet of the perpendiculars from P on three coordinate axes.

On linear systems of surfaces in S_3 which relate to Cremona transformations and on the reduction of such systems by Cremona transformation we find papers by Guccia and Enriques. Guccia([68]) defines p_0 to be the dimension of a linear system of surfaces determined by an ordinary base L consisting of curves and points; p_1 to be the geometric genus of (F); p_2 to be the genus of the variable intersection M of F, F'; and p_3 to be the number of variable intersections of F, F', F''. He proves that if (F) is determined by L and $p_1=0$ and if on a generic F the linear system M contains a smaller system of genus 0 or a smaller system of genus 1 and dimension >1 then $p_0+p_2-p_3=2$. This is extended to the case where $p_1>0$ and F contains a smaller system F' with $p'_1=0$ and properties like the above, in which case $p_0-p_1+p_2-p_3=2$. From these he deduces that any linear system with $p_3=1$ is determined by its base; also that $p_3=1$ is the necessary and sufficient condition for a homaloidal web.

Enriques([52]) considers linear systems of surfaces which are *simple* (i. e., surfaces on x do not necessarily pass through an x') and such that the variable intersection of two surfaces is a curve which is (a) rational; (b) elliptic; or (c) hyperelliptic. He shows that such systems can be transformed by Cremona transformation into the following types: (a) into (1) quadrics on a conic; or (2) quadrics tangent at a point to a plane; or (3) surfaces of order n with an $(n-1)$-fold base line and perhaps other base elements; (b) [cf. also Enriques([51])] if the grade (number of variable intersections of three surfaces) >3 into one of the following systems of grade n and dimension $n+1$ or into a system contained in one of them; (1) for $n=4$, the system of cubic surfaces determined by a quintic partial intersection with a quadric (which may degenerate); (2) for $n=5$, 6, 7, 8, 9 the system ∞^{n+1} of cubic surfaces with a double base point P, $9-n$ base lines on P, and the same tangent quadric cone at P; *as well as* (3) for $n=5$ the system ∞^6 of cubic surfaces determined by a quartic of the second kind (which may degenerate); (4) for $n=6$ the system ∞^7 of cubic surfaces on three skew lines; (5) the system ∞^7 of cubic surfaces with a double point P and a space cubic curve (which may degenerate) on P; (6) the system ∞^7 of cubic surfaces with a biplanar base point P with a fixed osculating plane at P, and a plane cubic curve with a node at P; (7) for $n=7$, 8 the system ∞^8 or ∞^9 of quadrics with 1 or no base points; and (8) for $n=8$ the system of quartics with a triple point P, two double base lines on P and the same tangent cone at P; (c) [cf. also Enriques([50])] into a system of surfaces of order n with an $(n-2)$-fold base line, and a base curve which cuts a plane on the base line in two points outside the line, and perhaps other base elements.

Pannelli([132]) shows that a twisted curve in space with any singularities whatever can be transformed by Cremona transformations of the type $x_ix'_i=1$ into a curve with only ordinary multiple points. Levi([95]) shows that any space curve can be transformed by Cremona transformation into a space curve without multiple points.

Of isolated geometric applications we note the following. Hesse([70.1]) uses the involution defined by a net of quadrics in the study of the plane quartic curve. Montesano([106.1]) applies an (8, 13) transformation in a study of the analytic dependence of two complex variables. He also([111.1]) uses a space involution of order 7 in connection with the problem of projectivity for a group of linear complexes or congruences. Montesano([106.2]) employs a (4, 5) transformation to convert a linear system of conics on a point into a line congruence; and Godeaux([67.1]) a similar transformation to convert a congruence of twisted cubics into one of lines. Montesano([110.1]) uses many types of transformation in the study of linear congruences of conics. Doehlemann([46.1]) uses a monoidal transformation in discussing the surface of incidence of corresponding rays of two sheafs.

Pannelli([131]) and Wimmer([176]) put the points of a space Σ' in projective correspondence in two ways with the web of planes in a space Σ and also with a web of quadrics in Σ. Thus P' in Σ' determines a pair x, x' of points in Σ which describe a quartic surface as P' describes a plane. The properties of this involution and of the quartic surface are developed. Autonne([8]), ([9]), ([10]), ([11]) extends to space and to $N-1$ dimensions his theory of birational transformations (cremoniennes) of forms in two series of contragredient variables, which leave invariant the condition of contact. Kantor([89]) remarks that the Plücker formulas which express the class, the number of double tangents and the number of flexes $(=x, y, z)$ in terms of the order, the number of double points, and the number of cusps $(=x', y', z')$ constitute a Cremona involution in S_3 which he discusses. This idea is extended to the Cayley formulas for space curves. Berry([14]) uses quadratic and cubic transformations in the discussion of integrals of the first kind of total differentials on certain quintic surfaces. Aprile([3]) discusses systems of twisted quartics of the second kind. Ota([129]) develops the transformation $x_i x'_i = a_i$ $(i=1, \ldots, 4)$ from the geometric side; [cf. also Vegas([173])].

Snyder([160]) employs the involution determined by a net of quadrics and other transformations in the investigation of line congruences contained in the complex of generators of the net. Sharpe and Morgan([119]) find Cremona transformations which leave a quartic surface with two nodes invariant and which are periodic on the surface. Snyder and Sharpe([158]) discuss transformations of the Jacobian of a web of quartic surfaces into itself which may be effected by Cremona transformations. They express the transformations in terms of a Severi base on the surface which consists of a plane section, a sextic of genus three, and nine of the ten lines.

10. *Algebraic and Other Applications.*—The solution of the quintic equation described in Chap. IV § 12 in which the algebraic part is simplified by the use of an auxiliary form-problem defined by a finite Cremona group has been extended by Coble to equations of higher degree. For the sextic case the cross-ratio group of Moore, a g_6 in S_3, has for simplest invariant linear system the system of quadrics on five points which suggests a natural resolvent sextic developed earlier by Brioschi and Maschke([25]). Related to these is the discussion of the form-problems of the groups $G_{5!}$, $G_{6!}$ attached to the sets P_5^2, P_6^2 (cf. § 6) of 5, 6 points in S_2 given in [([26]) §§ 8, 10]. Coble([33]) also outlines a program for a

development, along similar lines, of a solution of the general equation of even degree.

A similar procedure may be applied to the equation for the 27 lines on a cubic surface([28]). The group is the extended group $G_{6,\,2}$ (cf. § 7) defined by P_6^2, a group in space Σ_4. The simplest invariant linear system divides under the operations of the invariant subgroup g_{25920} of $G_{6,\,2}$ of index two into two linear systems of dimension four contragredient to each other under a collineation group of Maschke which is in immediate relation with hyperelliptic modular functions of grade 3 and genus 2.

The complete system of invariants of the $G_{6\,!}$ of P_6^2 is given in [([26]) § 5]; systems of irrational invariants of $G_{6,\,2}$ are given in [([25]) § 3]; and of $G_{7,\,2}$ in [([27]) § 8]. These are connected respectively with the cubic surface and the planar quartic curve. The latter system is used by Morley([122]) in the determination of the condition that the curve be of the Lüroth type. Other examples of invariants of the groups $G_{m,\,r}$ appear in [([27]), pp. 380-381] and [([31]), § 4].

Applications of this theory to abelian modular functions are given by Coble[([32]) § 14].

BIBLIOGRAPHY

1 Alderton, N., Cal. Pub., 1: 345-358, no. 15 (1920). [2.

2 Ameseder, A., J. f. Math., 97: 62-92 (1884). [9.

3 Aprile, G., Note e Mem., 1: 245-262 (1921). [9.
(1880). [4.

3 Aprile, G., Note e Mem., 1: 245-262 (1921). [9.

4 Aroldi, G., Giorn. di Mat., (3), 11: 175-192 (1920). [2.

5 Aschieri, F., Lomb. Ist. Rend., (2), 14: 21-26 (1881). [4, 5.

5.1 ———, Lomb. Ist. Rend., (2), 14: 251-267 (1881). [4.

6 ———, Lomb. Ist. Rend., (2), 21: 216-226, 285-293, 446-449 (1888). [4, 5.

7 Ascione, E., Giorn. di Mat., 31: 55-94 (1893). [4.

7.1 Autonne, L., C. R., 121: 881-883, 1129-1130 (1895). [1.

7.2 ———, C. R., 122: 1043-1045 (1896). [1.

8 ———, Brux. Mém. Sav. Etr., 59: 254 p. (1901). [9.

9 ———, C. R., 135: 776-778 (1902). [9.

10 ———, J. Éc. Polyt., (2), 8: 17-73 (1903). [9.

11 ———, J. Éc. Polyt., (2), 17: 147-206 (1913). [9.

12 Baker, H. F., London. M. S. Proc., (2), 11: 302-312 (1912-13). [7.

12.1 ———, London. M. S. Proc., (2), 21: 114-133 (1923). [4.

13 Beloch, Margherita, Ann. di Mat., (3), 16: 27-68 (1909). [2.

13.1 Beltrami, E., Giorn. di Mat., 1: 208-217, 354-360 (1863); Op. Mat. 1, Milan., 73-91 (1902). [4.

13.2 Berardi, N., Giorn. di Mat., (3), 14: 109-122 (1923). [2.

14 Berry, A., Camb. Ph. Soc. Trans., 18: 333-347 (1900); 19: 249-296 (1904); 20: 74-112 (1908). [9.

15 Bianchi, L., Giorn. di Mat., 16: 263-266 (1878). [2.

16 Bonicelli, Maria, Giorn. di Mat., (2), 9: 184-191 (1902). [4.
17 Caldarera, Grazia M., Pal. Circ. Mat., 18: 205-217 (1904). [5.
18 Cantone, A., Nap. R. A. Rend., 25: 181-190 (1886). [4.
18.1 Caporali, E., Ann. di Mat., (2), 7: 149-188 (1875). [9.
19 Carrone, C., Cat. Acc. Gio. Atti., (4), 11: 1-62 (1898). [2.
20 Castelli, Teresita, Giorn. di Mat., (3), 7: 10-26 (1916). [9.
21 Cayley, A., London M. S. Proc., 3: 127-180 (1870). [1, 2, 4.
21.1 Cech, Casopis, 50: 219-249 (1921). [2.
22 Chizzoni, F., Rom. Acc. L. Rend., (4), 2_1: 470-476 (1886). [9.
23 ———, Rom. Acc. L. Rend., (4), 2_1: 476-482 (1886). [9.
24 Coble, A. B., Am. M. S. Trans., 9: 396-424 (1908). [6.
25 ———, Am. M. S. Trans., 12: 311-325 (1911). [6, 10.
26 ———, Am. M. S. Trans., 16: 155-198 (1915). [3, 6, 10.
27 ———, Am. M. S. Trans., 17: 345-385 (1916). [3, 6, 7, 10.
28 ———, Am. M. S. Trans., 18: 331-372 (1917). [3, 6, 10.
29 ———, Am. J., 40: 317-340 (1918). [7.
30 ———, Am. J., 41: 243-265 (1919). [3, 7.
31 ———, Am. M. S. Trans., 24: 1-20 (1922). [3, 4, 7, 10.
32 ———, Am. M. S. Bull., 28: 329-364 (1922). [3, 10.
33 ———, Am. M. S. Bull., 30: 301-313 (1924). [6, 10.
34 Conner, J. R., with Morley, F., (*see* Morley). [9.
35 Conner, J. R., Am. J., 38: 155-176 (1916). [4.
36 Craig, C. F., with Sharpe, F. R., (*see* Sharpe). [7.
37 Cremona, L., J. f. Math., 68: 1-133 (1868). [4.
37.1 ———, Lomb. Ist. Rend., (2), 4: 140-144, 159-162 (1871). [9.
38 ———, Lomb. Ist. Rend., (2), 4: Nota 1ª: 269-279; Nota 2ª: 315-324 (1871). [1, 2, 4.
39 ———, Math. Ann., 4: 213-230 (1871). [9.
40 ———, Bologna Mem., (3), 1: 365-386 (1871). [4.
41 ———, Ann. di Mat., (2), 5: 131-162 (1871). [1, 4.
42 ———, Bologna Mem., (3), 2: 117-127 (1872). [9.
43 ———, In Mem. Chelini, Coll. Math., Milan, 413-424 (1881). [9.
44 ———, Dublin Trans., 28: 279-284 (1884). [4.
45 ———, London M. S. Proc., 15: 242-246 (1884). [4.
46 Deruyts, F., Liège S. R. S. Mém., (2), 14: Mém. 5, 14 p. (1888). [4.
46.1 Doehlemann, K., Diss. Muenchen, 40 p. (1889). [9.
47 ———, Geom. Transformationen, Leipzig, v. 2: (1908). [1, 4.
48 ———, Math. Ann., 41: 545-570 (1893). [1.
49 ———, Münch. Ber., 24: 41-50 (1894). [4.
49.1 Eckhardt, F. E., Math. Ann., 5: 30-49 (1872). [4, 9.
50 Enriques, F., Rom. Acc. L. Rend., (5), 2_2: 281-287 (1893). [9.
51 ———, Rom. Acc. L. Rend., (5), 3_1: 481-487, 536-543 (1894). [9.
52 ———, Math. Ann., 46: 179-199 (1895). [9.
53 Enriques, F., and Fano, G., Ann. di Mat., (2), 26: 59-98 (1897). [8.
54 Fano, G., with Enriques, F., (*see* Enriques). [8.
55 Fano, G., Monatsh. f. Math., 9: 17-29 (1898). [8.
56 ———, Torino Mem., (2), 48: 221-278 (1898). [8.
57 ———, Rom. Acc. L. Rend., (5), 7_1: 302-308 (1898). [8.
58 ———, Rom. Acc. L. Rend., (5), 7_1: 332-340 (1898). [8.
59 ———, Torino Atti, 33: 480-504 (1898). [8.

60 ———, Encyklopädie III AB 4b, 339-343. [8.
61 Francois, C., Liège S. R. S. Mém., (3), 9: Mém. 4, 19 p. (1912). [9.
62 Geiser, C. F., Bern. Mitt., 97-107 (1865). [4.
62.1 ———, J. f. Math., 69: 197-221 (1868). [9.
63 ———, J. f. Math., 67: 78-89 (1867). [4.
64 ———, J. f. Math., 70: 249-257 (1869). [9.
65 Gerbaldi, F., Pal. Circ. Mat., 10: 158-160 (1896). [1.
66 Godeaux, L., Nouv. Ann., (4), 8: 69-72 (1908). [4.
67 ———, Lomb. Ist. Rend., (2), 43: 116-119 (1910). [2.
67.1 ———, Nouv. Ann., (4), 11: 1-17 (1911). [9.
67.2 ———, Brux. Mém., (2), 6: 1-75 (1922). [4.
68 Guccia, G. B., Pal. Circ. Mat., 1: 338-349 (1884-87). [9.
69 Guradze, H., Arch. Math. Phys., (3), 4: 288-292 (1903). [4.
70 Haskell, M. W., Am. Math. Month., 10: 1-3 (1903). [4.
70.1 Hesse, O., J. f. Math., 49: 279-332 (1855). [9.
71 Hudson, H. P., London M. S. Proc., (2), 9: 51-66 (1910-11). [4.
72 ———, London M. S. Proc., (2), 10: 15-47 (1911-12). [2.
73 ———, London M. S. Proc., (2), 11: 398-410 (1912-13). [2.
74 ———, Math. Ann., 73: 73-85 (1912-13). [2.
75 ———, Ann. di Mat., (3), 19: 45-56 (1912). [2.
76 ———, Am. J., 34: 203-210 (1912). [2.
77 ———, Pal. Circ. Mat., 35: 286-288 (1913). [1.
78 ———, Am. J., 35: 183-188 (1913). [2.
78.1 ———, London M. S. Proc., (2), 23: 27-29 (1925). [2.
78.2 Hudson, H. P., and Wren, T. L., (*see* Wren).
78.3 Hudson, H. P., London M. S. Proc., (2), 25: (1927). [1.
78.4 ———, Cremona Trans. in Plane and Space, Cambridge Univ. Press, (1927). [1.
78.5 Hunyadi, Zschr. Math. Phys., 11: 356-359 (1866). [4.
79 Hutchinson, J. I., Am. M. S. Bull., (2), 7: 211-217 (1900-01). [7.
80 James, C. G. F., London M. S. J., 1: 3-7 (1926). [5.
81 Kantor, S., C. R., 90: 1156-1158 (1880). [5.
82 ———, Math. Ann., 20: 297-298 (1882). [4.
83 ———, Lomb. Ist. Rend., (2), 27: 477-485 (1894). [6.
84 ———, Lomb. Ist. Rend., (2), 27: 712-722, 749-759 (1894). [6.
85 ———, Lomb. Ist. Rend., (2), 28: 249-263, 298-312 (1895). [6.
86 ———, Am. J., 18: 219-263 (1896). [6.
87 ———, Am. J., 19: 1-59, 382 (1897). [3, 6.
88 ———, Acta Math., 21: 1-78 (1897). [3, 6, 7.
89 ———, Ven. Ist. Atti., (8), 3: 769-773 (1901). [9.
90 ———, Wien. Ber., 112_{2a}: 667-754 (1903). [3, 7.
91 Kasner, E., Am. J., 25: 107-122 (1903). [6.
92 von Krieg, F., Zschr. Math. Phys., 29 Supp.: 38-72 (1884). [4, 9.
93 Lazzeri, G., Rom. Acc. L. Rend., (4), 2_2: 73-79 (1886). [9.
94 Levi, A., Giorn. di Mat., (2), 3: 215-249 (1896). [1.
95 Levi, B., Rom. Acc. L. Rend., (5), 7_1: 111-113 (1898). [9.
96 Loria, G., Torino Atti., 26: 275-299 (1890-91). [2.
97 ———, Lomb. Ist. Rend., (2), 23: 824-834 (1890). [1, 2, 5.
97.1 ———, Il passato ed il presente ... Torino, (1896). [1.

97.2 Machovec, Prag. Ber., 104-137 (1888). [4.
98 Magnus, L., Sammlung von Aufgaben, Berlin, (1837). [4.
98.1 Marletta, G., Pal. Circ. Mat., 49: 252-262 (1925). [1, 2.
99 Medugno, M., Nap. Acc. delle Sci., 18 p. 8vo. (1910). [4.
99.1 Meyer, W. Fr., Arch. Math. Phys., (3), 5: 168-175 (1903). [4.
99.2 ———, Heidelberg Kong., Verh. 3: 322-346 (1905). [4.
100 Montesano, D., Ven. Ist. Atti., (6), 6: 1425-1444 (1887-88). [6.
101 ———, Rom. Acc. L. Rend., (4), 4_1: 207-215, 277-285 (1888). [5.
102 ———, Rom. Acc. L. Rend., (4), 4_1: 583-590 (1888). [9.
103 ———, Nap. R. A. Rend., (2), 2: 181-188 (1888). [5.
104 ———, Lomb. Ist. Rend., (2), 21: 688-690 (1888). [2.
105 ———, Rom. Acc. L. Rend., (4), 5_1: 497-501 (1889). [5.
106 ———, Rom. Acc. L. Rend., (4), 5_2: 123-130 (1889). [2.
106.1 ———, Lomb. Ist. Rend., (2), 24: 889-907 (1891). [9.
106.2 ———, Torino Atti, 27: 660-690 (1892). [9.
107 ———, Lomb. Ist. Rend., (2), 25: 795-821 (1892). [5.
108 ———, Giorn. di Mat., 31: 36-50 (1893). [2, 4.
109 ———, Lomb. Ist. Rend., (2), 26: 589-604 (1893). [5.
110 ———, Giorn. di Mat., 31: 36-50 (1893). [4.
110.1 ———, Nap. R. A. Rend., (3), 1: 93-110, 155-181 (1895). [9.
111 ———, Lomb. Ist. Rend., (2), 30: 563-571 (1897). [5.
111.1 ———, Ann. di Mat., (3), 1: 313-357 (1898). [9.
111.2 ———, Nap. R. A. Rend., (3), 7: 67-106 (1901). [9.
111.3 ———, Nap. R. A. Rend., (3), 13: 66-88 (1907). [9.
112 ———, Nap. R. A. Rend., (3), 13: 256-262 (1907). [5.
113 ———, Nap. R. A. Atti, (2), 15: no. 8, 55 p. (1914). [4.
114 ———, Rom. Acc. L. Rend., (5), 27_1: 396-400, 438-441 (1918). [1.
115 ———, Rom. Acc. L. Rend., (5), 30_2: 447-451 (1921). [1.
116 ———, Nap. R. A. Rend., (3), 27: 116-127 (1921). [2.
117 ———, Nap. R. A. Rend., (3), 27: 164-175 (1921). [2.
118 Moore, E. H., Am. J., 22: 279-291 (1900). [6.
119 Morgan, F. M., with Sharpe, F. R., (*see* Sharpe). [9.
120 Morgan, F. M., with Young, J. W., (*see* Young). [8.
121 Morley, F., and Conner, J. R., Am. J., 31: 263-270 (1909). [9.
122 Morley, F., Am. J., 41: 279-282 (1919). [10.
123 Musselman, J. R., Am. J., 40: 69-86 (1918). [3.
124 Nobile, Grazia, Giorn. di Mat., (3), 12: 147-174 (1921). [2.
124.1 ———, Giorn. di Mat., (3), 13: 23-32 (1922). [9.
125 Noether, M., Math. Ann., 2: 293-316 (1870). [1.
126 ———, Math. Ann., 3: 547-580 (1871). [1, 2, 9.
127 ———, Ann. di Mat., (2), 5: 163-177 (1871-3). [1, 2.
128 ———, Deutsche Math. Ver., 5: Heft. 1, p. 68 (1896). [8.
128.1 Obeslo, V., Casopis, 44: 208-219 (1915). [4.
129 Ota, T., Tôhoku Sc. R., 10: 233-272 (1921). [9.
130 Pannelli, M., Giorn. di Mat., 28: 245-256 (1890). [5.
131 ———, Giorn. di Mat., 29: 133-153 (1891). [9.
132 ———, Lomb. Ist. Rend., (2), 26: 216-222 (1893). [9.
133 ———, Rom. Acc. L. Rend., (5), 19_1: 449-455 (1910). [1.
134 ———, Rom. Acc. L. Rend., (5), 20_1: 404-409 (1911). [1.

135 ———, Lomb. Ist. Rend., (2), 47: 1041-1052 (1914). [1.

136 de Paolis, R., Giorn. di Mat., 13: 226-248, 282-297 (1875). [2, 4.

137 ———, Rom. Acc. L. Mem., (4), 1: 576-608 (1885). [5.

138 ———, Rom. Acc. L. Rend., (4), 1: 735-742, 754-758 (1885). [5.

138.1 Pascal, E., Repertorium, Leipzig, Teubner, (1910). [1.

138.2 Perazzo, U., Torino Mem., (2), 54: 149-182 (1904). [2.

139 del Pezzo, P., Nap. R. A. Rend., (3), 1: 133-139 (1895). [2.

140 ———, Nap. R. A. Rend., (3), 2: 336-344 (1896). [2.

141 ———, Nap. R. A. Rend., (3), 2: 288-296 (1896). [2.

142 ———, Nap. R. A. Rend., (3), 3: 33-50 (1897). [2.

142.1 Pieri, M., Torino Atti, 24: 514-526 (1889). [9.

143 ———, Pal. Circ. Mat., 6: 234-244 (1892). [5.

144 ———, Lomb. Ist. Rend., (2), 25: 1037-1060 (1892). [5.

145 ———, Pal. Circ. Mat., 7: 296-306 (1893). [5.

146 ———, Giorn. di Mat., (2), 2: 167-178 (1895). [5.

147 Del Re, Rom. Acc. L. Rend., (4), 6_2: 271-276 (1890). [6.

147.1 ———, Giorn. di Mat., 28: 257-282 (1890). [5.

148 Reye, T., Geometrie der Lage, v. 2, Stuttgart (1907); v. 3, Leipzig (1892). [4.

149 Salmon, G., Anal. Geom. of Three Dimen., 5th ed. (Rogers) London (1915). [1.

150 Saltel, L., Brux. Mém. Cour., 22: Mém. 5, 53 p. (1872). [4, 9.

151 ———, Brux. Mém. Cour., 27: Mém. 2, 66 p. (1876). [9.

152 Schiaparelli, Torino Mem., (2), 21_a: 227-319 (1862). [4.

153 Schoute, P. H., Bull. Sci. M., (2), 6: 152-168, 174-188 (1882). [4.

154 ———, Nieuw. Arch., (2), 4: 90-100 (1899). [4.

154.1 Segre, C., Torino Atti, 21: 95-115 (1886). [2.

155 Sharpe, F. R., and Morgan, F. M., Ann. of Math., (2), 15: 84-92 (1913-14). [9.

156 Sharpe, F. R., and Craig, C. F., Am. M. S. Trans., 15: 236-244 (1914). [7.

156.1 Sharpe, F. R., and Snyder, V., Am. M. S. Trans., 15: 266-276 (1914). [7.

157 ———, (*see* Snyder(**164**)). [7.

158 ———, (*see* Snyder(**165**)). [9.

159 Silldorf, Zschr. Math. Phys., 18: 523-542 (1873). [4.

160 Snyder, V., Am. M. S. Trans., 11: 371-387 (1910). [9.

161 ———, Am. M. S. Trans., 12: 354-366 (1911). [9.

162 ———, Am. M. S. Trans., 14: 105-108 (1913). [7.

162.1 Snyder, V., with Sharpe, F. R., (*see* Sharpe(**156.1**)).

163 Snyder, V., Pal. Circ. Mat., 38: 344-352 (1914). [4. 7.

164 Snyder, V., and Sharpe, F. R., Am. M. S. Trans., 16: 62-70 (1915). [7.

165 ———, Am. M. S. Trans., 19: 275-290 (1918). [9.

166 Sturm, R., Geometrische Verwandtschaften, v. 4, Leipzig (1909). [1, 4.

167 ———, Math. Ann., 1: 533-574 (1869). [4.

168 ———, J. f. Math., 70: 212-240 (1869). [4.

169 ———, Math. Ann., 19: 461-486 (1882). [4.

170 ———, Deutsche Math. Ver., 14: 18-24 (1905). [2, 4.

170.1 Terquem, Nouv. Ann., (1), 5: 497-508 (1846). [4.

170.2 Tinto, J. F., Edinb. M. S. Proc., 34: 133-145 (1915-16). [2.

170.3 ———, Edinb. M. S. Proc., 36: 2-16 (1917-18). [2.

171 Tummarello, A., Nap. R. A. Rend., (3), 17: 427-439 (1911). [2.
171.1 ———, Giorn. di Mat., (3), 11: 60-66 (1920). [2.
171.2 ———, Note e Mem., 1: 282-288 (1921). [2.
171.3 ———, Note e Mem., 1: 289-298 (1921). [2.
171.4 ———, Soc. Ital. Prog. Sci. Atti, 13: 326-327 (1924). [1.
172 van Veen, H. J., Nieuw. Arch., (2), 11: 127-131 (1914). [9.
173 Vegas, M., R. de Ci. Madrid Rev., 1: 90-123 (1904). [9.
174 Veneroni, E., Lomb. Ist. Rend., (2), 34: 640-644 (1901). [2.
175 Vogt, W., Jahresb. Schles. Ges. f. Vaterl. Kultur, 8 p. (1906). [5.
176 Wimmer, B., Zschr. Math. Phys., 36: 214-230 (1891). [9.
177 Wong, B. C., Am. J., 49: 383-388 (1927). [2.
177.1 Wren, T. L., London M. S. Proc., (2), 15: 144-165 (1916). [9.
177.2 Wren, T. L., with Hudson, H., London M. S. Proc., (2), 24: 28-29 (1926) [5.
178 Young, J. W., and Morgan, F. M., Am. M. S. Trans., 17: 233-244 (1916). [8.

CHAPTER IX

INVOLUTIONS AND (1, 2) CORRESPONDENCES IN S_3, S'_3

1. *Reciprocal Radii.*—The early history and the properties of reciprocal radii will be found in Chapter I. B. Tortolini([124]) showed that the element of surface and of volume, when transformed by reciprocal radii, become

$$dS' = \left(\frac{k}{r}\right)^4 dS, \qquad dV' = \left(\frac{k}{r}\right)^6 dV,$$

wherein

$$\frac{x'}{x} = \frac{y'}{y} = \frac{z'}{z} = \frac{k^2}{r^2}.$$

Envelopes of spheres touching three given spheres were studied by O. Bonnet([9]), M. Chasles([15]), J. A. Serret([101]). The latter showed that a sphere cutting a cyclide along a line of curvature meets it always at the same angle. Joachimstahl([43]) proved that any sphere having double contact with a cyclide meets it in two circles. A. Mannheim([49]) and ([50]) showed that the three fixed spheres could be transformed by reciprocal radii into others having their centers on a straight line. The Dupin cyclide can be generated in two ways as the envelope of spheres touching three given ones. A. Mannheim([51]) defined as principal pole the center of a sphere of inversion under which a given surface remains invariant. It can always be transformed into a surface symmetric about a plane. If it has a curve (circle or line) of principal poles, the surface can be transformed into one of revolution. J. Moutard([71]) considered a surface of order m, having the pole O to multiplicity p, and containing the circle at infinity to multiplicity q. By reciprocal radii it is transformed into another of order m', etc., such that

$$m' = 2m - p - 2q; \qquad p' = m - 2q; \qquad q' = m - p - q.$$

It may be invariant if $p + 2q = m$. Such a surface is called anallagmatic; it is the envelope of a sphere orthogonal to the sphere of inversion and having its center describe a fixed surface. Quartic surfaces for which $q = 2$ are anallagmatic as to five poles. The five spheres of inversion are mutually orthogonal and the surfaces of centers are confocal quadrics. The line joining any two poles is perpendicular to the plane of the other three. The tetrahedron having any four poles for vertices has the fifth for orthocenter. Through any point of the surface pass ten circular sections. See also J. C. Maxwell([55]). G. Fouret([29]) showed that the Dupin cyclide is the only surface invariant under a continuous system of inversions. The lines of curvature of the cyclide are used as parameters by E. Colot([16]).

S. L. Ravier([88]) associated transformation by reciprocal radii with stereographic projection to generate quadrics. Given a circle C with center O and in plane π. Draw through O any line not in π, and on it take any two points A, A'. Join A to any point M in π and A' to M', inverse of M as to C. Let AM, $A'M'$, intersect in μ. As M describes π, μ will describe a quadric surface.

A somewhat similar transformation was devised by d'Ocagne([72]) and ([73]). Two points, P, P' are called conjugate if they are diametrical opposites as to a sphere passing through a fixed circle. By linear transformation it can be reduced to reciprocal radii. R. Bricard([10]) gave an alternate proof of the theorem that lines of curvatures are preserved under transformation by reciprocal radii.

E. Meyer([56]) enumerated more minutely the possible conformal point transformations. They are perspective reciprocal radii with a real fixed sphere, or this transformation and (a) reflection as to the center, or (b) as to a plane through the center or (c) as to a line through the center. Also proved by A. Leveque([46]). Earlier proofs had been given by E. Goursat([34]) and by G. Darboux([18]).

G. Valiron([125]) gives a geometric proof that lines of curvature are invariant under transformation by reciprocal radii.

For the literature and general theory of transformation by reciprocal radii in space see J. K. Doehlemann([20]), R. Sturm([121]), V. Snyder and C. H. Sisam([108]), F. S. Woods([139]).

2. *Involutorial Transformations of Space.*—The first generalization of involutorial transformations to those defined by surfaces of order n was made by R. de Paolis([77]); the method was similar to that employed by E. de Jonquières([44]), for the plane involutions which bear his name, but which was not then published. Given a surface of order n with an $(n-2)$-fold point O. The transform of a point P is on OP and is the harmonic conjugate of P as to the residual points of intersection with the surface. The conjugate of a plane is a surface of order n, having O to multiplicity $n-1$, and passing through the curve of contact of the tangent cone to the fixed surface from O. This is known as the perspective monoidal transformation. If $O=(0, 0, 0, 1)$, and the fixed surface has the equation

$$u_{n-2}x_4^2+2u_{n-1}x_4+u_n=0,$$

u_i being ternary in x_1, x_2, x_3 of degree i, the equations of the involution have the form

$$x'_i = x_i(u_{n-2}x_4 + u_{n-1}), \qquad x'_4 = -(u_{n-1}x_4 + u_n).$$
$$i = 1, 2, 3.$$

If u_{n-2} is identically zero the fixed surface is itself a monoid, and P' is defined as the harmonic conjugate of P as to O and the residual intersection of OP with the fixed monoid. In this case all the surfaces of the web conjugate to the planes of space have the same tangent cone at O, and the basis curve consists of $n(n-1)$ of its generators.

V. Martinetti([54]) obtained all the monoidal involutions compounded of the perspective monoidal form and any harmonic homology which leaves the bundle of lines through the vertex invariant. In the plane all the non-perspective involutions of type JH can be reduced to homology. The corresponding question in S_3 has not been answered. Eleven types were obtained by Martinetti as follows, in terms of the invariant elements:

1. n odd. C_n point by point and of C_1 not in its plane.
2. n even. C_n point by point, and two points on C_1.
3. n arbitrary. C_n point by point. No other point.
4. n arbitrary. C_{n-r} point by point and C_1 (n-odd), two points (n-even).
5. n arbitrary. C_{n-r} point by point only.
6. n even. Plane π point by point and C_1.
7. n odd. Plane π point by point and two points on C_1.
8. n arbitrary. Plane π point by point only.
9. n odd. π fundamental. C_1 point by point.
10. n even. Two fixed points on C_1.
11. n arbitrary. No fixed elements.

It was shown by D. Montesano([60]) that any birational involution which leaves a bundle of lines invariant is either the perspective monoidal, M, or this compounded with a ternary involution H, G, B or J which leaves the bundle at the vertex invariant. Thus in addition to the types found by Martinetti, the MG gives $s_1 \sim s_{3n} : \sum h_i{}^n C_{n(2n-3)}$ and a curve of invariant points of order $4n$, in which h_i is a generator of the cone. The invariant cone K_6 of G may or may not be composed of invariant points. In MB, $s_1 \sim s_{3n} : \sum_{i=1}^{\infty} h_i C_{n(n-3)}$ and a curve of invariant points of order $3n$. The invariant cone K_9 of B may or may not be composed of invariant points.

In MJ, $s_1 \sim s_n O^{n-1} k^{n-2m} (l_1 \ldots l_{2(p+1)})^m C\gamma$,

$$\gamma = (4m-1)n - 2(p+3)m^2, \qquad n = p + v + 2,$$
$$m = \beta + 1, \qquad b + p + 1 = n - 2m.$$

Not all the forms of possible invariant configurations have been determined.

A number of other types have been considered by this same method. Thus D. Montesano(61) considers the involutions of genus 0 in which $s_1 \sim s_n : c_1^{n-1}$; the pencil of planes through c_1 remains invariant. The two double planes of the involutorial pencil are point by point invariant.

Systems of involutions forming finite groups are obtained by D. Montesano(69). The first, of order 8, can be defined as the plane projection of the harmonic homologies which leave each quadric of a web with common self-polar tetrahedron invariant. All are quadratic, four perspective and three non-perspective. Similarly, from quadratic forms in S_4 an analogous group of order 16 in S_3, a second one of order 16, and one of order 32 in S_2. The last is the plane representation of Klein's G_{32} of collineations and correlations which leave the Kummer surface invariant.

Another type found by D. Montesano(66)

$$s_1 \sim s_n : c_1^{n-2} c_{3n-4}, \qquad [c_1, c_{3n-4}] = 3n-7,$$

wherein $[c_i, c_k]$ denotes the number of intersections of c_i, c_k.

Planes of the pencil c_1 are either invariant or form an involution. For $n=3$ this type is included in the preceding one. Both of these types probably include irrational involutions. D. Montesano(68) considers two non-monoidal transformations which leave every conic of a linear congruence invariant. The congruence is defined by a bundle of planes and a projective net of quadrics. A monoidal transformation of order 4, vertex at that of the bundle of planes, leaves each conic of the congruence fixed. A rational plane curve of order n, having a point of multiplicity $n-1$ at the vertex meets the plane of each conic in one point, center of an involution which interchanges the points of the conic. A third involutorial transformation results when the locus of the center in each plane is a surface in (1, 1) correspondence with the planes of the bundle. Lines joining conjugate points form a complex of order μ.

$$s_1 \sim s_{6\mu+4} : C_7^{2\mu+1} C_\nu^2 \gamma_1^{2s_1} \dots \gamma_h^{2s_h} t_1 \dots t_{5(2\mu+1)},$$

Another type for n general is $s_1 \sim s_{2n+1} : c_4^n 2n g_i$, c_4 of genus 1, $[g_i, c_4] = 2$. It leaves the congruence of lines formed by the bisecants of C_4 invariant. D. Montesano(59). Two particular involutions of order 4 and genus 0 were given by D. Montesano(63) which were not included in the earlier types. One can be reduced to central harmonic homology and the other to axial harmonic homology.

F. London(47) provides a linear construction for involutions determined by conjugate points as to one or more quadric surfaces.

O. Tognoli([122]) includes certain restricted I_2 of S_3 as particular cases of much more general correspondences discussed in Chapter X.

F. Romano([93]) gives a detailed treatment of the involution leaving a web of cubic surfaces through three skew lines and four simple points invariant. This and seven other forms were defined by C. Moffa([57]). Of these, four types had been found previously, and four were new.

J. R. Conner([17]), following S. Kantor([44.1]), constructed an I_2 of order 15 by means of a group $[a]$ of associated points. Let $|C_4|$ be the linear system of quartic surfaces having 7 of these points as double points, and $|Q_2|$ the net of quadrics through all eight. A space quartic curve through $[a]$ is uniquely determined by one point P_1, and it meets those C_4 through P_1 in a second point P_2.

$$s_1 \sim s_{15} : 7a_i{}^8 l_{ij} r^3{}_{ij}$$

in which l_{ij} is the (parasitic) line through a_i, a_j and r_{ij} is the (parasitic) cubic curve through all but a_i, a_j. The surface of invariant points is $K_6 : \sum a_i{}^3 l_{ij} r_{ij}$.

A metrical interpretation of oriented pairs of points was found by H. Rothe([97]).

V. Snyder([111]) supplements D. Montesano([60]) by giving the equations of all nonperspective monoidal types and more details concerning properties of perspective monoidal types which satisfy restricted conditions. An example of the latter type was later developed by L. Godeaux([33a]); it had been already employed by M. M. Torrey([123]), but derived in a different way. Every monoidal involution of the form MJ belongs to a special linear line complex, the axis of which passes through the vertex of the monoid.

F. R. Morris([70]) classifies the involutions determined by conjugate points as to a net of quadrics according to the number of composite quadrics contained in the net. The conjugate points as to one or more linear complexes is not included. Fourteen types are enumerated and their equations derived. Other methods of derivation of the general cubic involution are found in the following papers: Castelli([14.1]), Pelzner([82.1]), Schoute([99.1]), ([99.2]), Cantone([12.2]), Chizzoni([15.1]), Del Re([19.1]), Fiedler ([27.2]). For the books and bibliography of this section, see R. Sturm([121]), J. K. Doehlemann([20]), V. Snyder([114]).

For groups generated by this transformation and collineations, see Chapter VIII.

3. (1, 2) *Correspondences Between Two Spaces.*—Given a web of surfaces $|\phi|$, all of the same order, such that any three surfaces of the web intersect in two variable points. If we put $x'_1 = \phi_i$, we obtain a

(1, 2) correspondence between the points of (x') and of (x). Let P' be a point of (x'), P_1, P_2 its two images in (x). The lines joining P_1P_2 are called principal lines. The locus of $P_1=P_2$ is the surface K of coincidences and its image in (x') is the surface of branch points L'. The points P_1, P_2 are conjugate in an involution I_2, associated with the correspondence.

Particular systems were found by Em. Weyr(137).

The web consisting of quadrics through 6 points was considered by T. Reye(91). K_4 is the Weddle surface, locus of the vertex of the cones of the web.

$$s_1 \sim s'_4 : 3c_1{}^2 \cdot P'^3$$
$$c_1 \sim c'_2$$
$$s'_1 \sim s_2 : 6P$$
$$c'_1 \sim c_4 : 6P$$
$$K_4 \sim L'_4 : 16P'^2$$

The 15 lines joining the basis points by pairs are transformed into 15 double points of L'_4, and the cubic curve through all six has for image the other double point of L'_4.

The surface of branch points is the Kummer quartic surface with 16 double points. The six bundles of lines with vertex at a basis point are transformed into bitangents of L'_4. All the principal lines are bisecants of the cubic curve γ_3 through the six basis points. Their images in (x') are lines of a bundle, vertex a double point of L'.

P. H. Schoute(99) studied properties of particular associated I_2, without developing systematically the idea of the double space.

A particular (1, 2) correspondence, in which to the planes of (x') correspond quadric surfaces was obtained by C. F. Aschieri(6), following F. Schur(100). Given a regulus a^2, designate by (a^2) the net of linear line congruences through it, and let Σ be the system of ∞^3 linear complexes through two given skew lines. If (a^2) is a net in Σ, the polar planes of a point A_1 as to (a^2) form a bundle (A_1) of planes and of lines with vertex at A_1, projective with (a^2). Let (b^2) be another net in Σ, with a bundle (B_1). Make these two bundles respectively projective with bundles (A) and (B). A point M determines a regulus M^2 of Σ. When M describes a line g, m^2 describes a congruence. Given any complex θ not in Σ, any regulus of Σ has two lines in common with θ, hence to any point of S_3 correspond two lines of θ; to a line of S_3, a regulus, and to a plane, a linear congruence. To a plane correspond the two directrices of its image congruence; to a pencil of planes, a regulus, and to a bundle, a linear congruence. The pairs of directrices, images of the planes of S_3, generate a linear complex θ_1 in involution with θ. If θ is now mapped birationally on S'_3, a (2, 1) correspondence is obtained.

The surfaces K, L' are both quadrics, and I_2 is quadratic. A sphere may be associated with each linear complex in involution with a given one, furnishing a (1, 2) correspondence between a system of non-euclidean spheres and ordinary ones.

By means of the properties developed in ([91]) Th. Reye([92]) showed that the $(16_6, 16_6)$ Kummer configuration can be constructed linearly when six points are given arbitrarily.

Every involution in S_r, the points of each group of which can be mapped rationally on the points of another space S'_r is called rational. For $r=1$ and $r=2$ all involutions I_n are rational. R. de Paolis([82]) first presents the general problem of (1, 2) correspondences between S'_3 and S_3 and raises the question whether all I_2 in S_3 are rational or not, but does not attempt to answer it.

When P_1 describes a locus, P_2 may describe the same or a different locus. In the former case the locus is invariant under I_2, and in the latter the loci are conjugate. Every curve invariant under I_2, image of a rational curve in (x'), is hyperelliptic, elliptic, or rational.

If any curve C meets K at P, but is not tangent to P_1P_2, its conjugate passes through the same point with a tangent different from either. The image C' in (x') of both the given C curve and of its conjugate touches L' at the corresponding point. If C touches P_1P_2 at P, so does its conjugate, and C' has a cusp at P' on L'. A self-conjugate C meets K along P_1P_2 and C' cuts L' at C'. Th. Reye([91]). If C' touches L' at every common point, C has a double point at every intersection with K, and is composite. A surface and its conjugate pass through the same curve on K. Their residual intersection is invariant under I_2. The image surface touches L' at every common point, and has the image of the residual intersection of two conjugate surfaces for double curve.

Let the lines c'_1 of (x') be transformed into $|R_{n'}|$ of (x).

A fundamental point of (x) is called a fundamental point of the first class if all the $R_{n'}$ of the ∞^4 system pass through it; otherwise of the second class. Let ρ be the number of fundamental points f_i of the first class, of multiplicity r_i on $|\phi_n|$ and of multiplicity θ_i on $|R_{n'}|$. Let there be r fundamental curves γ_i of order m_i, multiple to order i_i on $|\phi|$, and having s_i points on $|R_{n'}|$. Then

$$n'=n^2-\sum i_r{}^2 m_r, \qquad nn'-\sum_{i=1}^{\rho} r_i\theta_i-\sum_{i=1}^{r} i_i s_i=2.$$

The genus of a general $R_{n'}$ is called the genus of the transformation. Let K be of order v, and denote by N the order of the surface conjugate to a general plane in I_2.

4. *Fundamental Elements in the Two Spaces.*—A fundamental point of (x) of the first class may also be fundamental in I_2. The $R_{n'}$ through it may have any direction, in which case its conjugate in I_2 is a surface, or it may be fixed in direction. In this case the point lies on a fundamental curve of I_2. If not fundamental in I_2, its conjugate has the same property in the (1, 2) correspondence, or finally it may be self-conjugate in I_2. These four types of possible fundamental points in the (1, 2) correspondence are known as the first, second, third and fourth kinds respectively.

If a point f in (x) is fundamental of the first kind, its image surface in (x') is rational. Such a surface, of order θ, is a fundamental surface of the first kind. If f is of the second kind, the image surface is met by the images of the planes of (x) in variable curves, but not all the R'_n, images of the lines of (x) meet it in variable points. If the point is of the third or fourth kind, its image surface in (x') is not met in variable points, by the images of the planes of (x).

To a fundamental point of the second class in (x), of multiplicity r on $|\phi|$, corresponds a curve of order r in (x').

A fundamental curve γ of (x) is of the first class when every $R_{n'}$ meets it in variable points; otherwise it is of the second class. A multiple curve γ_m of multiplicity i on $|\phi|$ may be fundamental in I_2. If the ∞^4 curves R_I, conjugates of lines in I_2, meet it in σ variable points, each of its points is conjugate to a curve and these generate a surface of order σ, image of γ in I_2. Such a curve is of the first class and the first kind. If the $|R_I|$ do not meet γ, each point of γ is conjugate to the same fixed curve. If the conjugate is different from γ, then γ is of the second kind, if it coincides with γ, of the third kind. If γ is fundamental in the (1, 2) correspondence but not in I_2, but is conjugate to another curve, also fundamental in (1, 2), it is of the fourth kind. If conjugate to itself, but not point for point, it is of the fifth kind, and if each point is invariant in I_2, of the sixth kind. A similar diversity of fundamental elements exists in (x').

A curve a in (x) which does not intersect $|\phi|$ in variable points is called parasitic. To any point of a corresponds the whole curve in I_2. A parasitic curve of order r corresponds to a fundamental point of (x') of second class and second kind; it is of multiplicity r on all the surfaces of the quadratic system, images of the planes of (x). It is fundamental in I_2 and lies on K.

The jacobian of the web $|\phi|$ consists of K, of the fundamental surfaces, images of fundamental curves of the first class of (x'), and of the fundamental surfaces, images of the fundamental points of the first class in (x'), the latter each counted twice.

$$\gamma = \Sigma\sigma + 2\Sigma\theta' = 4(n-1).$$

L' is of order $2(p+1)$, p being the genus of $|R_{n'}|$. An i-fold fundamental curve of the first class on $|\phi|$ is of multiplicity $4i-1$ on $J(\phi)$; an i-fold fundamental curve of the second class is of multiplicity $4i$ on $J(\phi)$.

5. *Various Particular Cases.*—A case in which the web $|\phi|$ is composed of cubics with basis elements in restricted position was given by M. Pieri([87]). Given an elliptic space C_4 and four coplanar points A_i not on C_4. These and the points of C_4 in their plane determine a pencil of cubic curves with a ninth basis point A_5. Through C_4, $5A_i$ pass a web of cubic surfaces $|\phi|$. Two meet in a space quintic curve of genus 2 and any three meet in two variable points. The C_4 may be composite. The surface L'_6 has a four-fold line and 10 double points. $K_7: C_4{}^3 5A_i$. The associated involution is of order 11. Through each point A_i pass two parasitic lines. They lie on K_7, and their images are the double points of L'_6.

Another particular case was given by G. de Longchamps([48]) and by Elgé([22]). Given a fixed point O. As M describes a given locus, find M', M'' on OM such

$$OM.MM' = OM \cdot MM'' = k^2, \qquad k \text{ being given.}$$

The (1, 2) correspondence associated with the web of quadrics through six points was treated exhaustively by R. de Paolis([80]).

M. Pieri([85]) mapped the involutions belonging to a special linear complex with axis p on the points of S'.

$$s'_1 \sim s_n : p^{n-3} 0_{5n-11}, \qquad [p, 0_{5n-11}] = 5n-18.$$
$$C'_1 \sim C_{n+2}, \text{ genus } n-1, \qquad [C_{n+2}, p] = n, \qquad [C_{n+2}, 0_{5n-11}] = 5n-4.$$
$$L'_{2n} \sim K_{3n-4}.$$

Within the web $|s_n|$, is a linear system ∞^1 consisting of a fixed plane through p, and a pencil s_{n-1}. The image in (x') of the plane is a line d'; any point of d' is the image of a cubic of the pencil PP' and the seven basis points in the plane. The s_{n-1} having p, PP' to multiplicity $n-3$ is also singular; its image in (x') is another line e'. The residual s_1 is any plane through p. The points of e' have for images the plane sections of s_{n-1} by the planes through PP'. d', e' intersect—their point has for images in (x) the exceptional conjugate points P, P'.

The involutorial correspondences found by M. Pieri([84]) give rise to a Hirst quadratic complex. T. A. Hirst([35]) and ([36]). Given two planes in (1, 1) point-line correspondence. The lines joining the points of each to all the points of the associated line in the other generate a Hirst complex.

A (1, 2) correspondence defined by a web of quadrics through a fixed conic was investigated in great detail by J. Cardinaal([14]) with application to quartic surfaces with a double conic, image of quadrics generated by two projective pencils of planes. The transformation has $p=0$. L is a quadric and K is a quadric. The same method was applied by J. Cardinaal([13]) to determine and classify the quartic surfaces having a double line. The correspondence is now (1, 4) but is made $(1, 4-k)$ by the appearance of k isolated double points.

The (1, 2) correspondence between the Kummer and Weddle surfaces was developed from a different standpoint by R. W. H. T. Hudson([37]) in his book, and still more in detail by V. Snyder([117]), wherein a number of infinite discontinuous groups are derived, all generated by involutions.

R. Bricard([11]) again proved that the lines P_1P_2 in (x) are bisecants of a space C_3.

C. Rosati([94]) obtains the asymptotic lines of the Kummer surface.

Another (1, 2) correspondence defined by ∞^1 quadric surfaces was given by E. Cairo([12]).

A systematic treatment of (1, 2) correspondences which includes all rational involutions was begun by F. R. Sharpe and V. Snyder([105]) by means of three equations in two sets of coordinates (x) and (x'), together with an enumeration of the cases associated with webs of non-singular surfaces of orders 2-5, having independent basis elements. For a regular system, the genus of the transformation is one greater than the arithmetic genus of a general surface of the web $|\phi|$ of invariant surfaces. The types there enumerated include the rational types previously known, some twenty new isolated types, and seven families of types in terms of an arbitrary n. The problem of reducibility is not considered, except in a few cases. The same process was applied to obtain those types which leave a web $|\phi|$ of sextic surfaces each invariant, and having certain basis elements, by J. O. Osborn([74]). It is not complete as restrictions were put on the basis elements. That connected with the non-monoidal cubic involution was found by F. R. Sharpe and V. Snyder([106]). It is expressed by the table:

$$s'_1 \sim s_{12} : \gamma_6{}^3 + 7l + 21m; \qquad c'_1 \sim c_{27} : P_8{}^6 + 7P_i{}^4.$$
$$s_1 \sim s'_{27} : \gamma 6'^{16} + 7l^{8'}; \qquad c_1 \sim c'_{12} : [c'_{12}, l'] = 4.$$
$$\gamma_6 \sim \Gamma'_{54} : \gamma'_6{}^{12} + 7l'^{16}; \qquad \gamma_6' \sim \Gamma_{16} : \gamma_6{}^4 + P_8{}^8 + 7P_i{}^6 + 21m^2.$$

The particular case $x_i = \dfrac{1}{x_i}$ is considered in detail. A further study of restricted basis elements for a web of cubics $|\phi|$, mentioned by M.

Pieri([87]) was made by V. Snyder([110]). All three of the types can be reduced to those discussed by M. Pieri. The basis elements may be

(a) c_2, c_1 skew to it, 5 pts. $A_i : A_5$ on $F_2 : c_2c_12A_i$, then A_6 also on F_2.
(b) c_1, c_1 skew P^2 on c_1.
(c) plane c_3, c_1 $[c_1, c_3]=1$.

In all these cases $p=2$, $p_a=0$.

Generalizing the scheme of C. Moffa([57]), V. Snyder([112]) obtained all the types associated with a web of invariant monoids with a non-composite basis curve.

If there are seven isolated basis points, the involution is monoidal; that with six does not exist. Those with 5, ..., 0 all exist. For every value of n, L' is of order 4. If there are no basis points on $|\phi|$, L' has no singular plane, and has eleven double points.

Two of these forms for $n=3$ had been found by C. Moffa. Another category of monoidal involutions, namely those in which the monoidal conjugates of the planes of space all have a common tangent cone at the vertex have been determined by the preceding methods by M. M. Torrey([123]).

A. Emch([22.1]) considers the I in S_n defined by $x'_i=\phi_i(x)$ and employs $p_{ik}=x_i\phi_k-x_k\phi_i$ which changes sign under I, and $g_{ik}=x_i\phi_k+x_k\phi_i$ which is invariant. Functions of the p, g are used to investigate contact loci. The lines joining pairs of conjugate points are mapped on certain varieties, illustrated by known cases. G. M. Caldarera([12.1]) discusses an I of order 27 in which lines PP' meet a fixed cubic curve. The correspondence on the bisecants of C_3 is isomorphic with the Geiser G in the plane. L. Godeaux([33.2]) generates an I of order 5 by means of two skew directrix lines and a pencil of quadrics. $s_1 \sim s_5 : d^2d'^212l$. F. Aschieri([6.1]) constructs an I_4 by means of a quadric polarity, a null system, and a projectivity. $s_1 \sim s_4 : d^2c_3c_22c_l$. In ([33.3]) an I is discussed which leaves a congruence of lines invariant. Montesano([64.2]) defines an I_7 by means of 5 arbitrary linear complexes, and the null polarity as to the complex k in involution with each of them. Lines PP' belong to k and each such line contains three pairs of conjugate points. In ([64.1]) the same author develops the theory of an I_6 contained in a linear complex. The cubic surfaces of a web are invariant.

6. (1, 2) *Correspondence between S_4 and S'_4.*—A (1, 2) correspondence between the points of two linear S_4 was obtained by G. Aprile([2]) by means of ∞^4 quadric varieties having for basis elements a c_3 of S_3 and three simple points, the plane of which does not meet c_3. These quadrics are made projective with the planes of (x').

The lines of (x') are transformed into c_5, $p=1$, and the lines of (x) go into conics of (x').

Each basis point $T_i \sim S'_3$. The basis $c_5 \sim \Gamma'_5$, enveloped by S'_3. The complete image in (x) of a tangent S'_3 of Γ'_5 is a point on c_3 and a conical variety of order 4 with vertex on c_3. The variety of branch points is rational, of order 4.

In the associated I_2 in (x), linear spaces of three dimensions $\sim V_8 : 3T^5c^4$. Any S_3 through the isolated basis points meets c_3 in three points P_i. The k_3 through the six points T_i, P_i is parasitic. $\infty^1 k_3 = K_3$ on the quartic variety of coincidences, and triple on the conjugate of S_3.

H. C. Shaub([107.1]) derived the equations of the $(1, 2)$ correspondences which leave a ∞^4 system of quadric varieties in S_4 invariant.

7. *Involutions of Lines in S_3.*—The problems associated with lines in S_3 can be interpreted as point problems in S_4, but they are sufficiently numerous and have such direct applications to ruled surfaces, congruences, and complexes, that they shall be treated in terms of lines.

F. Aschieri([4]) determined an I_2 of lines of a particular cubic complex as follows: Given a pencil of quadric surfaces. The polar planes of a point as to the quadrics of the pencil form a pencil, the axis of which is associated with the given point. The lines form a tetrahedral complex, the fundamental tetrahedron being the self-polar tetrahedron of the pencil of quadrics. Lines meeting pairs of quadrics of the pencil harmonically generate a quadratic complex of Battaglini.

When p is a line of T, its polar complex as to all possible complexes B from pairs of quadrics having the fundamental tetrahedron of T for self-conjugate tetrahedron have a line p' in common also in T. The $p \sim p'$ becomes $P \sim P'$ expressible by $x_i x'_i = 1$. The lines PP' generate a cubic complex.

Another I_2 was given by F. Aschieri([3]). Given a linear line complex θ_1 and a fixed cone γ_2. Given any point P. Its polar plane as to θ_1 is π, and as to γ_2 is π_1. The lines π, $\pi_1 = l$ generate the quadratic complex θ_2. Similar dual construction. The line l is the polar line of the given point P as to θ_2.

To every line l there is a pole X as to θ_2, and a conjugate l' as to θ_1, which has a pole X'. The points X, X' are conjugate in a cubic I_2 of points, image of that defined by the lines l, l'.

An I_p of tangents on a rational C_n of S_3 gives rise to a complex made up of the (l_2) pairs of associated tangents and the linear congruences of which they are the directrices. If $n=3$, $p=2$, a cubic complex results.

A further study of Cremona transformations among lines of S_3 was made by F. Aschieri([5]).

A particular involution of lines derived from a generalization of Pascal's theorem to S_3 was given by F. Klein([45]).

J. de Vries([130]) generates a line complex by means of an I_p on a rational space curve C_n. Draw the tangents to C_n at all the p points of a group of I_p and construct the linear congruences having each pair of tangents of a group for directrices. As I_p describes C_n, the congruences describe a complex of order $(2n-3)(p-1)$. The complex curve in each plane has $(n-2)(2n-5)(p-1)^2$ double tangents, which form a congruence of this order and class. Detailed study of $n=3$, $p=2$.

A. Del Re([19]) gives a representation of line complexes and line congruences in descriptive geometry, and applies it to graphical statics. J. de Vries([130]) constructs a particular line complex as follows: Given two space cubic curves C_3, K_3. From a point P can be drawn one bisecant of each, determining a plane, but in each plane are 9 such points. Construct R, conjugate of P as to cubic involutorial transformation defined by C_3. Similarly for the point Q on K_3. Consider the complex of lines R, Q. It is of order 18.

Similarly, any line t determines the C_3 of the system of which it is a bisecant, meeting C_3 in P and Q. Find P', Q' in the point involution. $t'=P'Q'$ is the conjugate of t. A line through a basis point $\sim$ quadric cone; a line in plane $A_i a_k \sim$ plane pencil. The complex of coincident lines is the cubic PP'. A bundle of lines $\sim$ congruence (4, 6) and a plane field $\sim$ congruence (6, 6).

Another example is given by two harmonic homologies, A, a in plane α and B, b in plane β. A line t of S_3 meets α in P, β in Q. $t'=P'Q'$.

The congruence having a, b for directrices is composed of invariant lines. The planes α, β are singular.

A plane pencil $\sim$ quadric regulus.

A bundle $\sim$ linear congruence.

A special linear complex $\sim$ a quadratic complex.

See J. de Vries([134]) and ([135]). The same author constructs a different I_2 by means of a plane homology A, a in α and a congruence of space cubic curves $|C_3|$ through 5 basis points. A line t fixes the C_3 of the congruence of which it is a bisecant and meets α in P.

Through P' draw the bisecant t' to the same C_3. The bundle A and the special linear complex with axis a are all invariant lines. Basis points are all singular, the bundles $\sim$ complexes of order 4.

A plane pencil $\sim$ ruled surface of order 9.

Plane field $\sim$ congruence (8, 5).

Bundle $\sim$ congruence (7, 8).

Linear congruence $\sim$ congruence (15, 13).

Special linear complex $\sim$ complex of order 9.

A slight variation is furnished by that I_2 defined by the $\infty^2|C_4|$ through 8 associated basis points. [J. de Vries(136).] A line t fixes the C_4 of the system, of which it is bisecant. t meets a fixed plane in P, through which can be drawn the other bisecant lines. This I_2 belongs to the complex of generators of the quadrics of the net through the 8 points.

G. Schaake(98.1) considers the I_2 in line space in which a pencil is conjugate to a regulus. Given a linear complex Ψ, a line m not in Ψ, and a line n in Ψ. Construct the quadric determined by l, m, and n. It has a second line 0 in Ψ. The line l', harmonic conjugate of l as to m and 0 is a line of the regulus to which l, m, n, 0 belong. This defines I_2. When l describes a pencil, l' describes a regulus through n, and if l describes a plane, l' describes a linear congruence containing n. A linear complex is transformed into quadratic complex having n as double line.

Given r, q any two space quartic curves. There are ∞^4 space quartics of the first kind which meet r in eight points and q in eight points. Any line t of S_3 is bisecant of a single curve of the system. Now by means of a plane x the I_2 is extended to all the lines of S_3.

Pencil $\sim$ ruled surface of order 9.

Bundle $\sim$ congruence (22, 24).

Plane field $\sim$ congruence (24, 23).

An involution I_{10} of lines of S_3 may be constructed as follows: Given the bilinear congruence of $|C_3|$ through 5 fixed points in S_3. A curve is uniquely fixed by having a given line for bisecant. Given two such congruences A and B. A line t determines a curve of each, and these have 10 common bisecants, hence the lines of S_3 are arranged in an involution of order 10. All the basis points are singular; a line through one is transformed into a ruled surface of order six. A general pencil of lines $\sim$ ruled surface of order 36, on which all 10 basis points are 12-fold. [J. de Vries(132).]

An I_4^4 of points determined by four points on each line of S_3 can be constructed by a general pencil of cubic surfaces; the surface of trisecants of the basis C_9 is composed of singular lines. It contains the 27 lines of each surface of the pencil. C. H. van Os(126).

Given four pencils of lines in four different planes. A line p of space pierces each plane, hence fixes a line of each pencil. These four lines have a second transversal p'. The relation between p, p' was obtained by J. de Vries(131). A plane pencil is transformed into a ruled surface of order 7. In (133), the two pencils are replaced by a quadric surface. Similarly, given a bilinear congruence of space cubic curves through two fixed points A_1, A_2 and having three fixed bisecants a_i. A point P determines the C_3 of the system passing through it. On C_3 find the harmonic conjugate P' of P as to A_1, A_2. This defines a cubic point involution. A line p is bisecant of a single C_3 of the system, meeting it in P_1, P_2. Construct P'_1, P'_2 by the point involution. $P'_1P'_2=p'$ is the conjugate of the given line in an involution of lines. C. H. van Os(127).

G. Schaake(98) shows that the ∞^3 systems of quadrics through a line of each of three of the defining pencils in the above scheme is not linear. Through three arbitrary points pass 5 quadrics of the system. The involution I_2 found previously contains 6 congruences (1, 1) of singular lines. A general bundle of lines goes into a congruence (5, 6).

8. *Line Complexes and Congruences Associated with* I_2.—Lines joining pairs of conjugate points in I_2 generate either a complex or a congruence of lines. When the line joining one pair P_1P_2 of conjugate points contains no further pairs, the involution is said to belong to the complex.

R. de Paolis(81) obtained the order of the complex. The bundles having vertices in fundamental points are singular. The lines determined by two conjugate fundamental points are simple lines of the complex. Moreover, every bisecant of a parasitic curve is a line of the complex. Now follow detailed analyses of number and kind of fundamental and singular lines in terms of the elements of the (1, 2) point transformation. A scheme is outlined by means of which all rational (1, 2) point transformations can be obtained. In a second memoir R. de Paolis(79) determines those involutions in which the lines joining conjugate points belong to a congruence instead of a complex. There are three types:

(a) The lines form a bundle;
(b) The lines are bisecants of a space cubic curve;
(c) The lines are secants of a rational C_m and of a line which meets it in $m-1$ points.

A limiting case determined by an $(1, m)$ correspondence between the points of a line and planes through the line is not considered there. The first is the monoidal involution; the second is a particular case of the cubic involution expressed by three bilinear equations—a polar system as to a net of quadric surfaces through the cubic curve. The third case becomes axial harmonic homology when $m=1$. This case was considered further by W. Vogt(128) in connection with a wider problem of a Cremona transformation.

V. Eberhard(21) showed that the involutorial transformation of order 7 associated with the web of quadric surfaces through 6 basis points belongs to type (b), but not in the particular way defined by the bilinear equations. Also proved by G. Fontene(28).

D. Montesano([60]) showed that cases (b) and (c) of R. de Paolis can be reduced to monoidal type, hence if the lines P_1P_2 of any I_2 belong to a congruence of rational curves of odd order the involution can be reduced to a monoidal form. The I_2 for which the lines P_1P_2 generate a linear complex were obtained by D. Montesano([65]). Given any line r of S_3. In any plane through it draw the line P_1P_2 belonging to the complex Γ. Since the line meets r, it meets r', its conjugate as to Γ. On lines of Γ meeting r, r' points P_1P_2 describe a quartic surface F_4. A pencil of linear congruences in Γ have F_2 in common, through r_1, r_2. $[F_2, F_4]=C$, $p=3$ that is hyperelliptic. Construct another F_4 with second pair of lines r, r' but same C_6. Residual is C_{10} of $p=11$, basis curve of $\infty^4 F_4$. Any F_4 of system through P also passes through P'. I_2 is of order 11, $s_1 \sim s_{11}: C^3{}_{10}a_1 \ldots a_{20}$—quadrisecants of C_{10}. Any plane meets C_{10} in 10 points of C_3, but C_{10} does not lie on F_3. C_{10} may be composite. $s_1 \sim s_{11-\mu}: L_0{}^3 \ldots L_3{}^0 a_1 \ldots a_{20}$.

If the complex is special, I_2 for every $n \geqq 3$ exist. [M. Pieri([85]).] It is determined and generated by a pencil of invariant surfaces F_{n-1}, each having a variable $(n-3)$-fold point on the axis p, containing the line to multiplicity $n-4$ all having the same tangents through p'. There is a basis curve of order $5n-11$, having $5n-18$ points on p. Every plane through p is invariant, and in it the involution is the Geiser plane involution.

$$s_1 \sim s_{4n-5}: p^{4n-3}c_{5n-11}: c_{5n-11} \sim \Omega_{12n-18}: c^8{}_{5n-11}p^{12n-39}.$$
$$p \sim \pi_{4n-6}: p^{4n-14}c^3{}_{5n-11}: K_{3n-4}: p^{3n-10}c^2{}_{5n-11}.$$

The I_2 associated with the web $|\phi_3|: C_5 2P$ is of this type. It is a generalization of the procedure used by C. Moffa([57]), including the case in which C_5 has a triple point at O, and all the surfaces of the web have a double point there. V. Snyder([112]).

The types contained in a tetrahedral complex were also determined by D. Montesano([67]).

$$s_1 \sim s_{19}: c^5{}_{11}4p^2.35u_iK_{12}: c^3{}_{11}4p35u_i.$$

The same form was studied by J. Wimmer([138]) and by F. R. Sharpe and V. Snyder([105]), the last named by means of two bilinear equations and one quadratic in (x), linear in (x'). The type defined by a web $|\phi_3|: C_4 3P$ belongs in this category.

Montesano([68]) considered the involutions which leave every conic of ∞^2 system invariant. Lines joining corresponding points form a complex Γ_μ with a singular bundle O. On each conic the conjugate points form an involution. Three cases arise. The center of the involution (a) is always fixed at O, or (b) describes a curve $C_\mu : O^{\mu-1}$, in which case every bisecant OG is associated with all the conics of the system through OG, or (c) the center describes a surface F_m.

In (a) it is the perspective monoidal of R. de Paolis([78]) and is of order 4. In (b), C_μ is fundamental, and also the C_7, $p=5$ residual of a conic, intersection of two cubic surfaces. The $5(2\mu+1)$ trisecants of C_7 which meet C_μ not at O are all parasitic, t.

$$s_1 \sim s_{6\mu+4} : C_7^{2\mu+1} C_\mu^2 C_2^{2\mu} \Sigma t_i.$$

The jacobian is composed of

$$\Gamma_{3(2\mu+1)} : C_7^{2\mu+1} C_\mu C_2^{2\mu} \Sigma t_i,$$
$$\Gamma_{9(2\mu+1)} C_7^{6\mu+2} C_\mu^6 C_2^{6\mu} \Sigma t_i,$$

conjugates of C_7, C_μ. The surface of invariant points is

$$K_{3\mu+4} : C_7^{\mu+1} C_\mu C_2^\mu \Sigma t_i.$$

In (c), the complex Γ is composed of ∞^2 pencils, point on the surface F_m as vertex and containing a line of the bundle O.

$$s_1 \sim s_{6\mu+4} : C_7^{2\mu+1} C^2{}_{2m+\mu} \Sigma \gamma_s^{2s} \Sigma t.$$
$$K : C_7^{\mu+4} C_{2m+\mu} \Sigma \gamma_s^{2s} \Sigma t.$$

No I_2 can belong to the general quadratic complex. [D. Montesano([62]).] If it contains a bundle of lines, it is included in those just mentioned. If I_2 contains a system of linear congruences, then

$$s_1 \sim s_{13} : C_4^5 C^2{}_{10} t_1 \ldots t_{16}$$

in which $K_8 : C_4^3 C_{10} t_1 \ldots t_{16}$.

That contained in the tetrahedral complex has been considered.

A particular quadratic complex was considered by M. Pieri([84]), namely that of characteristic [(11) (11) 2] generated by two correlative planes. [T. A. Hirst([35]).] The associated involutions are:

$s_1 \sim s_{10} : C_2^2 C'{}_2^2 0_2^5 C_4^2 V h_1 \ldots h_8,$ $\quad V \sim 0_2,\ C_2 \sim \Gamma_8,\ C'_2 \sim \Gamma'_8.$

$s_1 \sim s_{13} : C'{}_3^3 C_5^3 0_2^3 0_4^5 h_1 \ldots h_{16},$ $\quad C'_3 \sim \Gamma_{12},\ C_5 \sim \Gamma_{12}.$

$s_1 \sim s_{16} : 0_2^2 0'{}_2^2 C_6^2 C_7^5 h_1 \ldots h_{25},$ $\quad C_6 \sim \Gamma_{15},\ 0_7 \sim \Gamma_{15} : 0_2^6 0'{}_2^6 C_6^7 h_1^3 \ldots.$

All the lines h_i are parasitic.

They can be generated by webs of invariant surfaces of orders 2, 3, 4 respectively. None is contained in a tetrahedral complex. A second particular quadratic complex is that composed of the secants of a fixed conic. M. Pieri([86]) determines the involution belonging to this complex. The lines P_1P_2 from each point of C_2 generate a surface F_5 with C_2 and the vertex of the bundle for triple point. These ∞^1 F_5 have C_{13}, $p=14$ in common, $[C_{13}, C_2]=13$.

$$\begin{aligned} &s_1 \sim s_{28}: C_2^{14}C^5{}_{13}h_1 \ldots h_{39} \qquad \text{all } h_i \text{ parasitic,} \\ &C_2 \sim P_{27}: C_2^{13}C^5{}_{13}h_1 \ldots h_{39}, \\ &C_{13} \sim \Omega_{78}: C_2^{13}C_{13}^{39}h_1 \ldots h_{39}, \\ &K_{18}: C_2^9C^3{}_{13}h_1 \ldots h_{39}. \end{aligned}$$

P_{27} and K_{18} touch along all 9 sheets of the latter at every point of C_2; Ω_{78}, K_{18} touch along all 3 sheets of the latter at every point of C_{13}. Those involutions contained in a special complex, i. e., one composed of the tangents to a surface were derived in part by M. Pieri ([83]) and completed by D. Montesano ([64]). The ∞^3 tangents can be transformed into the points of ∞^2 conics forming a linear system. Each conic is invariant in the involution and lies in a tangent plane to the given surface. Instead of conics, there may be a linear congruence of invariant lines, reducible to monoidal types. The former may be formulated thus: Given in S_3 a linear congruence of $|C_2|$ situated in the tangent planes of a surface. If the two points of each conic collinear with the point of contact of its plane are regarded as conjugate, the correspondence thus established is a birational and involutorial transformation, belonging to the complex of tangents to the surface. The latter are contained in the two categories which follow: Given two linear line congruences γ, γ', with d as common directrix, and d', d'', as second respective directrix. If two lines are homologous which meet d at the same point, and d', d'' in projective ranges, this homology must be involutorial, or if a linear system Σ of homologies be established between d', d'', this system may be referred projectively to the points of d, homologous lines meeting on d, and meeting d', d'' in the pair of points associated with the homology belonging to the point on d. This system Σ must be made up of involutions.

The complexes (or congruence) to which those involutions belong which leave a web of cubic surfaces invariant were determined by C. Moffa ([57]). The cubic complex associated with the cubic involution was found by Th. Reye ([90]) and its equation derived by F. R. Sharpe and V. Snyder ([106]). The I_2 of order 15 found by S. Kantor ([44.1]) and J. R. Conner ([17]) belongs to this same complex.

Another involution belonging to Γ was obtained by D. Montesano ([58]). Let $B_1 \ldots B_8$ be the basis points. Consider the fixed quadric cone $B_6^2B_1 \ldots B_5$. A point P fixes the C_4 of the net which passes through it. This C_4 meets K_2 in one variable point Q. The line PQ is in Γ. Conversely, given any line p of Γ. It meets K_2 in two points Q_1, Q_2, each of which determines a C_4 passing through it. Each curve meets p in one pair of points P_1, P_2. Thus between the lines of Γ and the points of S_3 there is a $(1, 2)$ correspondence, and the associated involution is that of P_1, P_2

$$s_1 \sim s_9 : d^8{}_{78}c^5{}_{78}c_{16}\ldots c_{56}g_1\ldots g_5,$$
$$d_{78} \sim D_7 : d^4{}_{78}c^2{}_{78}c_{16}\ldots c_{56}g_1\ldots g_5,$$
$$c_{78} \sim c_{78},$$
$$c_{18} \sim s_1 : B_i d_{78}.$$

The lines g_i are also parasitic. Each g_i lies on the quadric through $c_{78}c_{6i}d_{78}d_{6i}$, and is a generator of the other system through B_i.

9. *Infinite Discontinuous Birational Groups.*—Compare Section 7 of Chapter VIII. As nearly all the infinite discontinuous groups of Cremona transformations thus far known can be generated by involutorial transformations, they should be treated in this connection, although the transformations themselves are not involutorial. G. Humbert([38]), ([39]) and ([40]) gave the first illustration in connection with a particular form of the Kummer surface. The method employed is transcendental, but it can be expressed as the product of two projections of the surface upon itself from two nodes—thus two monoidal involutions. Another illustration was furnished by P. Painlevé([76]), which involved those elliptic functions which admit complex multiplication. J. I. Hutchinson([41]) and ([42]) showed that the methods previously employed could be applied to the general Kummer surface, and to certain other surfaces birationally equivalent to it. G. Fano([25]) showed that any quartic surface through a hyperelliptic sextic curve of genus 2 has two infinite series of such curves, is invariant under the characteristic $g_2{}^1$ of each, and that their product is not periodic. This quartic surface does not contain a system of elliptic curves; the cases previously mentioned all depend essentially on the non-central involutions on a non-singular cubic curve. F. Enriques([23]) proved that every surface invariant under an infinite discontinuous group of birational transformations has a pencil of elliptic curves, except when $p_a = P_1 = 1$.

In these papers the authors were concerned only with points on the surfaces, hence did not raise the question whether the transformations were contained in a Cremona group or not.

Numerous examples of the first category are given by V. Snyder([115]). Quartic surfaces with $m(1 < m \leqq 16)$ nodes exist under such a group. Another form was given by Snyder([109]) by means of congruence of lines which can be arranged in conjugates on quadric surfaces. Interchanging the points of contact with the focal surface on each generator is an involution, and the two conjugate systems generate a non-periodic transformation. Further development by H. B. Owens([75]).

A. Rosenblatt([95]) gave examples of irregular surfaces with a system of elliptic curves; in ([96]) these results were considerably extended. Particular examples of quartic surfaces were given by Garnier([30]) and by Remy([89]). Further developed by V. Snyder([116]).

F. Severi([102]) expressed the transformations of Fano by means of the base, obtaining all the linear transformations which leave a certain quadratic form invariant. V. Snyder([117]) gave a number of further infinite groups belonging to the Kummer surface. The question raised by H. F. Baker([7]) is answered and the result expressed in terms of the Severi base by F. R. Sharpe and C. F. Craig([103]). Later Snyder([118]) generalized the result of Rosenblatt to surfaces of any given index of irregularity, and showed that all the transformations employed are Cremonian.

L. Godeaux([31]) proved that the most general quartic through a conic is not invariant under any birational transformation. His proof is not complete. For certain other rational curves on quartics, see Godeaux([32]). F. R. Sharpe and V. Snyder([105]) proved a similar theorem for quartics through a straight line, a conic. For a single space cubic one involution exists; for two space cubics the product of these involutions is not periodic.

A résumé is given in the presidential address of H. F. Baker([8]).

L. Godeaux([33]) showed that the quartic through a general space sextic curve of genus 3 is invariant under an infinite discontinuous group of birational transformations. V. Snyder and F. R. Sharpe([119]) showed that this and all other known transformations of infinite order under which a surface remains invariant are contained in a Cremona transformation. This one does not contain involutions nor does the surface contain infinite systems of hyperelliptic curves, thus being more general than the Fano surface. The quartic through an unrestricted C_8, $p=2$ is invariant under two Cremona involutions, which generate an infinite group. F. R. Sharpe and V. Snyder([107]). This and two other forms were considered by G. Fano([27]) and ([26]). Thus, a quartic surface with two skew lines has three operations which generate an infinite group. The first is the involution of points on the transversals of the two lines, and the second and third are expressed by an even projection of each cubic curve through one line or the other into itself. The symbolic form of every transformation of the group is expressed in terms of the Severi base.

This is more general than the particular case $n=1$ of C_1 and a rational C_m meeting it in $m-1$ points treated by Sharpe and Snyder([104]), since the parts played by the two skew lines may be interchanged. The second case is that of a quartic with a double point and a simple line through it, and the third is an F_4 with C_m, $p=2$. If $m=6$, the Fano surface results; $m=8$ gives that discussed by F. R. Sharpe and V. Snyder. An infinite group is associated with every even value of $m>4$, and for certain odd values.

10. *Involutions Belonging to M_3 in S_4.*—Although all involutions I_n of order n in S_1 or S_2 are rational, it was proved by F. Enriques([24]), that this is not true in S_3. Groups of n points exist such that any point uniquely determines the group to which it belongs, that cannot be mapped rationally on the points of S'_3. Examples with smaller value of n were given by G. Aprile([1]). I_2 belonging to varieties M_3 in S_4 were obtained by G. Marletta([52]) and ([53]). One new theorem is: Every M_n of S_4 having an S_2 to multiplicity $n-3$ and in it an S_1 to multiplicity $n-2$ is representable by the pairs of an I_2 in S_3.

It is not known whether this M_n is rational or not. It has been shown by G. Fano([27a]) that irrational varieties in S_4 exist which have all invariants zero.

It was shown by V. Snyder([120]) that the projection of the cubic variety $V_3{}^3$ of S_4 into itself from points on it cannot form a continuous group. Nor can the transformation defined in any plane through a line p on $V_3{}^3$ by means of the harmonic homology having p for axis and its pole as to the residual conic in the plane as center. Nor can the combination of the two generate a continuous group.

The details of mapping the general cubic variety of S_4 on a double S_3 were given by V. Snyder([113]). The lines joining pairs of conjugate points of I_2 in S_3 form a special linear complex, but each line contains two pairs of points. The I_2 is of order 6 and is monoidal of type MJ.

It is not known whether certain of the involutions I_2 in S_3 are rational or not, in particular

$$s_1 \sim s_n : c_1{}^{n-2} c_{3n-4},$$

given by D. Montesano([66]) and the types given by him in ([69]).

Numerous I_2 belonging to the cubic V_3 of S_4 having 10 double points were given by E. Ascione([6a]), but these are all rational.

A more general form was given by G. Marletta([53]). Given in S_3 a point A and a C_8 of $p=7$, the residual intersection of two cubic surfaces through a common line. An S_2 through A meets C_8 in 8 points, through which can be drawn C_4 having A^3. The $\infty^2|C_4|$ describe a rational congruence of order 2. The $|F_5|: A^4C_8$ is a pencil.

Each C_4 is on one F_5; residual is S_1 through A, not meeting C_8. Any S_2 through this S_1 meets F_5 in a C_4 of the congruence. Each F_5 contains two such S_1 through A, forming I_2. Their S_2 meets $|F_5|$ in basis C_3. In S_4 construct $M_n: S_2{}^{n-2}$ with $S_1{}^{n-2}$ in the multiple S_2, with curve a meeting any S_3 through S_2 in 1 point. Take also $M'_3: S_1$ and Σ_3. Take a point P on M_n, and construct $\Sigma'_3 = PS_2$. It meets M_n in F_3 and meets M' in F'_3, which have S_1 in common, hence residual is C_8. The point $A = (\Sigma', a)$ is triple on C_4 meeting C_8 in 8 points. Residual is 1_1, 1_2 meeting Σ, S_2. Hence there is a (1, 2) correspondence between M_n and Σ_3.

BIBLIOGRAPHY

1 Aprile, G., Giorn. di Mat., 56: 90-112 (1918). [10.
2 ———, Giorn. di Mat., 57: 129-150 (1919). [6.
3 Aschieri, F., Bologna Mem., (3), 10: 549-574 (1879). [7.
4 ———, Giorn. di Mat., 13: 328-337 (1875). [7.
5 ———, Lomb. Ist. Rend., (2), 14: 21-26, 123-131, 219-227 (1881). [7.
6 ———, Lomb. Ist. Rend., (2), 14: 673-683 (1881). [3.
6.1 ———, Lomb. Ist. Rend., (3), 5: 251-267 (1881). [5.
6.2 Ascione, E., Nap. R. A. Rend., (3), 2: 13-29 (1896). [10.
7 Baker, H. F., London M. S. Proc., (2), 11: 302-312 (1912-13). [9.
8 ———, London M. S. Proc., (2), 12: 1-40 (1913). [9.
9 Bonnet, O., J. Éc. Polyt., 35: 119-308, 137 (1853). [1.
10 Bricard, R., Nouv. Ann., (4), 3: 16-17 (1903). [1.
11 ———, Nouv. Ann., (4), 16: 171-173 (1916). [5.
12 Cairo, E., Per. di Mat., (3), 10: 155-164 (1913). [5.
12.1 Caldarera, G. M., Pal. Circ. Mat., 18: 205-207 (1904). [5.
12.2 Cantone, L., Nap. R. A. Rend., 25: 181-190 (1886). [5.
13 Cardinaal, J., Amst. K. Ak. Wet. Verh., 1: 1-63 (1892-93). [5.
14 ———, Amst. K. Ak. Wet. Versl., (3), 8: 88-146 (1891). [5.
14.1 Castelli, S. T., Giorn. di Mat., 54: 10-25 (1916). [2.
15 Chasles, M., J. Éc. Polyt., 25: 266-316, 314 (1832). [1.
15.1 Chizzoni, F., Rom. Acc. L. Rend., (4), 2_1: 470-476 (1886). [2.
16 Colot, E., Nouv. Ann., (2), 4: 169-177 (1865). [1.
17 Conner, J. R., Am. J., 38: 155-176 (1916). [2, 8.
18 Darboux, G., Arch. Math. Phys., (3), 1: 34-37 (1900). [1.
19 Del Re, A., Nap. R. A. Rend., (3), 10: 278-295 (1904). [7.
19.1 ———, Giorn. di Mat., 28: 257-282 (1890). [2.
20 Doehlemann, J. K., Geom. Trans., 2: Ch. 6-7 (1908). [1, 2.
21 Eberhard, V., Diss. Breslau, 61 p. (1885). [8.
22 Elgé, M., J. de M. E., (2), 21: 61 p. (1897). [5.
22.1 Emch, A., Am. J., 48: 21-44 (1926). [5.
23 Enriques, F., Rom. Acc. L. Rend., (5), 15: 665-669 (1906). [9.
24 ———, Rom. Acc. L. Rend., (5), 21_1: 81-83 (1912). [10.
25 Fano, G., Lomb. Ist. Rend., (2), 39: 1071-1086 (1906). [9.
26 ———, Rom. Acc. L. Rend., 29_2: 113-118, 175-182, 231-236 (1920). [9.
27 ———, Rom. Acc. L. Rend., (5), 29_1: 408-415, 458-491 (1920). [9.
27.1 ———, Torino Atti., 43: 541-552 (1903). [10.
27.2 Fiedler, W., Z. Vierteljschr., 21: 369-383 (1876). [2.
28 Fontene, G., Nouv. Ann., (4), 15: 515-516 (1915). [8.
29 Fouret, G., Nouv. Ann., (3), 7: 113-117 (1888). [1.
30 Garnier, R., C. R., 149: 1054-1057 (1909). [9.

31 Godeaux, L., Bull. Soc. Sc. Hain., 62: 1-8 (1912). [9.
32 ———, Nieuw. Arch., (2), 11: 88-97 (1915). [9.
33 ———, Crac. Bull., 16: 529-547 (1913). [9.
33.1 ———, Instituto, 73: 16 p. (1925). [5.
33.2 ———, Bull. Soc. Sc. Hain., (7), 1: 1-5 (1910). [5.
33.3 ———, Brux. Bull., 359-364 (1907). [5.
34 Goursat, E., Am. Éc. Norm., (3), 6: 9-102 (1889). [1.
35 Hirst, T. A., Mem. D. Chelini, 50-73 (1881). [5, 8.
36 ———, London M. S. Proc., 10: 131-143 (1879). [5.
37 Hudson, R. W. H. T., Kummer Quartic Surface, 160-172 (1905). [5.
38 Humbert, G., J. de Math., (5), 6: 279-386 (1900). [9.
39 ———, C. R., 126: 394-397 (1898). [9.
40 ———, C. R., 126: 508-510 (1898). [10.
41 Hutchinson, J. I., Am. M. S. Bull., 6: 328-337 (1900). [9.
42 ———, Am. M. S. Bull., 7: 211-217 (1901). [9.
43 Joachimstahl, J., J. f. Math., 30: 347-350 (1846). [1.
44 de Jonquieres, E., Giorn. di Mat., 23: 48-75 (1885). [2.
44.1 Kantor, S., Acta Math., 21: 1-78 (1897).
45 Klein, F., Math. Ann., 22: 246-249 (1883). [7.
46 Leveque, A., Nouv. Ann., (4), 20: 356-358 (1920). [1.
47 London, F., Math. Ann., 38: 334-368 (1891). [2.
48 de Longchamps, G., J. de M. E., (2), 10: 153-158 (1886). [5.
49 Mannheim, A., Nouv. Ann., (1), 17: 89 (1856). [1.
50 ———, Nouv. Ann., (1), 19: 67-79 (1860). [1.
51 ———, Nouv. Ann. (1), 20: 218-222 (1861). [1.
52 Marletta, G., Mem. Acc. Acircale, 10: 1-2 (1917-18). [10.
53 ———, Rom. Acc. L. Rend., (5), 27_1: 371-374 (1918). [10.
54 Martinetti, V., Lomb. Ist. Rend., (2), 18: 132-142 (1885). [2.
55 Maxwell, J. C., Nouv. Ann., (2), 10: 162-181 (1871). [1.
56 Meyer, E., Arch. Math. Phys., (3), 13: 149-152 (1908). [1.
57 Moffa, C., Diss. Naples, 45 p. (1923). [2, 5, 8.
58 Montesano, D., Bologna Mem., (4), 3: 549-577 (1893). [8.
59 ———, Giorn. di Mat., 31: 36-50 (1893). [2.
60 ———, Lomb. Ist. Rend., (2), 21: 579-594, 684-688 (1888). [2, 8.
61 ———, Lomb. Ist. Rend., (2), 21: 688-690 (1888). [2.
62 ———, Lomb. Ist. Rend., (2), 25: 795-821 (1892). [8.
63 ———, Lomb. Ist. Rend., (2), 30: 563-571 (1897). [2.
64 ———, Nap. R. A. Rend., (3), 13: 256-262 (1907). [8.
64.1 ———, Nap. R. A. Rend., (2), 2: 181-188 (1888). [5.
64.2 ———, Ann. di Mat., (3), 1: 313-357 (1898). [5.
65 ———, Rom. Acc. L. Rend., (4), 4: 207-215, 277-285 (1888). [8.
66 ———, Rom. Acc. L. Rend., (4), 5: 123-130 (1889). [2, 10.
67 ———, Rom. Acc. L. Rend., (4), 5_1: 497-501 (1889). [8.
68 ———, Torino. Atti., 27_1: 660-690 (1891-92). [2, 8.
69 ———, Ven. Ist. Atti., (6), 6: 1425-1444 (1888). [2, 10.
70 Morris, F. R., Univ. Calif. Pub., 1: 223-240 (1920). [2.
71 Moutard, J., Nouv. Ann., (2), 3: 306-309 (1864). [1.
72 d'Ocagne, M., Liège S. R. S. Mém., 5: 1-10 (1889). [1.
73 ———, Soc. M. F. Bull., 25: 8-9 (1896). [1.
74 Osborn, J. O., Am. J., 46: 17-36 (1923). [5.

75 Owens, H. B., Am. J., 35: 323-356 (1913). [9.
76 Painlevé, P., C. R., 126: 512-515 (1898). [9.
77 de Paolis, R., Giorn. di Mat., 13: 226-248, 282-298 (1875). [2.
78 ———, Rom. Acc. L. Rend., (4), 1: 735-742 (1885). [8.
79 ———, Rom. Acc. L. Rend., (4), 1: 754-758 (1885). [8.
80 ———, Rom. Acc. L. Rend., (4), 6_2: 3-11 (1890). [5.
81 ———, Rom. Acc. L. Mem., (4), 1: 84-107 (1885). [8.
82 ———, Rom. Acc. L. Mem., (4), 1: 576-608 (1885). [3.
82.1 Pelzner, O., Münch. Diss., (1913). [2.
83 Pieri, M., Giorn. di Mat., 33: 167-178 (1895). [8.
84 ———, Lomb. Ist. Rend., (2), 25: 1037-60 (1892). [8, 5.
85 ———, Pal. Circ. Mat., 6: 234-244 (1892). [8, 5.
86 ———, Pal. Circ. Mat., 7: 296-306 (1893). [8.
87 ———, Torino Atti., 24: 514-526 (1889). [5.
88 Ravier, S. L., Nouv. Ann., (3), 10: 371-373 (1891). [1.
89 Remy, L., C. R., 149: 1057-1060 (1909). [9.
90 Reye, T., Geom. der Lage, (1892). [8.
91 ———, J. f. Math., 86: 84-107 (1879). [3.
92 ———, J. f. Math., 86: 209-212 (1879). [3.
93 Romano, F., Diss. Catania, 47 p. (1906). [2.
94 Rosati, C., Pal. Circ. Mat., 35: 324-344 (1913). [5.
95 Rosenblatt, A., Pal. Circ. Mat., 33: 212-216 (1912). [9.
96 ———, Crac. Bull., 15: 761-810 (1912). [9.
97 Rothe, H., Wien. Ber., 126: 831-855 (1917). [2.
98 Schaake, G., Amst. K. Ak. Wet. Versl., 27: 957-961 (1918-19). [7.
98.1 ———, Amst. K. Ak. Wet. Versl., 34: 478-485 (1924-25). [7.
99 Schoute, P. H., Assoc. Fr. Sc., 156-179 (1880). [3.
99.1 ———, Bull. Sci. Math., (2), 6: 152-168, 174-188 (1882). [2.
99.2 ———, Nieuw Arch., (2), 4: 90-100 (1899). [2.
100 Schur, F., Math. Ann., 15: 432-464 (1879). [3.
101 Serret, J. A., J. de Math., (1), 18: 123-162; 128 (1853). [1.
102 Severi, F., Pal. Circ. Mat., 30: 265-288 (1911). [9.
103 Sharpe, F. R., and Craig, C. F., Am. M. S. Trans., 15: 236-244 (1914). [9.
104 Sharpe, F. R., and Snyder, V., Am. M. S. Trans., 15: 266-276 (1914). [9.
105 ———, Am. M. S. Trans., 20: 185-202 (1919). [5, 8, 9.
106 ———, Am. M. S. Trans., 21: 52-78 (1920). [5, 8.
107 ———, Am. M. S. Trans., 25: 1-12 (1923). [9.
107.1 Shaub, H. C., Am. J., 49: 367-382 (1927). [6.
108 Snyder, V., and Sisam, C. H., Anal. Geom. Space, Chap. 14. (1914). [1.
109 Snyder, V., Am. J., 38: 177-185 (1910). [9.
110 ———, Am. J., 46: 131-141 (1924). [5.
111 ———, Ann. of Math., (2), 25: 279-284 (1924). [2.
112 ———, Ann. of Math., (2), 26: 165-172 (1925). [5, 8.
113 ———, Giorn. di Mat., 61: 125-128 (1924). [10.
114 ———, Am. M. S. Bull., 30: 101-124 (1924). [2.
115 ———, Am. M. S. Trans., 11: 15-24 (1910). [9.
116 ———, Am. M. S. Trans., 11: 371-387 (1910). [9.
117 ———, Am. M. S. Trans., 12: 354-366 (1911). [5, 9.
118 ———, Am. M. S. Trans., 14: 105-108 (1913). [9.
119 Snyder, V., and Sharpe, F. R., Am. M. S. Trans., 16: 62-70 (1915). [9.

120 Snyder, V., Pal. Circ. Mat., 38: 344-352 (1914). [10.
121 Sturm, R., Geom. Verwandt., 4, Chap. 11: 339-360 (1909). [1, 2.
122 Tognoli, O., Giorn. di Mat., 10: 1-16; 141-147 (1872). [2.
123 Torrey, M. M., Am. J., 47: 181-206 (1925). [5.
124 Tortolini, B., Ann. di Mat., (1), 2: 189-195 (1859). [1.
125 Valiron, G., Rev. de Math. Spec., 21: 238-239 (1911-12). [1.
126 Van Os, C. H., Amst. K. Ak. Wet. Versl., 25_2: 963-971 (1917-18). [7.
127 ———, Amst. K. Ak. Wet. Versl., 27_1: 337-351 (1918-19). [7.
128 Vogy, W., Schl. Ges. f v. Kultur, 8 p (1006). [8.
129 de Vries, J., Amst. K. Ak. Wet. Versl., 11: 762-767 (1902-03). [7.
130 ———, Amst. K. Ak. Wet. Versl., 21_2: 812-815 (1912-13). [7.
131 ———, Amst. K. Ak. Wet. Versl., 27_1: 256-259 (1918-19). [7.
132 ———, Amst. K. Ak. Wet. Versl., 27_1: 260-265 (1918-19). [7.
133 ———, Amst. K. Ak. Wet. Versl., 27_1: 842-848 (1918-19). [7.
134 ———, Amst. K. Ak. Wet. Versl., 27_2: 1070-1073 (1918-19). [7.
135 ———, Amst. K. Ak. Wet. Versl., 27_2: 1074-1082 (1918-1919). [7.
136 ———, Amst. K. Ak. Wet. Versl., 27_2: 1197-1200 (1918-19). [7.
137 Weyr, Em., Prag. Ab., (6), 5: 24 p (1871). [3.
138 Wimmer, J., Zschr. Math. Phys., 36: 214-230 (1891). [8.
139 Woods, F. S., Higher Geom. Chap., 15: 423 p. (1922). [1.

CHAPTER X

REDUCTION OF SINGULARITIES OF SPACE CURVES AND SURFACES

1. In this very short chapter we shall not attempt to do anything further than to bring up to date the two existing and readily accessible accounts by Castelnuovo-Enriques[(5) p. 639], and by Berzolari[(3) Ch. 31 § 5, 36 § 4] which cover the literature down to about 1915. A critical account of singularities of surfaces still remains to be written. Due to the doubts surrounding many of the papers on the subject, to the difficulties inherent in the question, the writing of a report of that nature would be an arduous task indeed. It is one which the committee hopes may be accomplished in a not too distant future.

2. There is relatively little to say concerning singularities of space curves. The only recent writings are Enriques(9), Enriques-Chisini[(11) p. 551], Albanese(2). Enriques-Chisini give a very complete treatment of consecutive singularities. They analyze those of C of S_3 by considering those of its plane projection, or else directly by means of quadratic transformations of the first kind (with fundamental point and conic). They show that the two methods give the same result. They also show that the transformations in question enable one to reduce the singularities of C to multiple points with distinct tangents. They prove that the reduction to a curve without any singularities whatever can be obtained by means of quadratic transformations of the second kind (with three fundamental points and a fundamental straight line). The same authors have also given a good historical account [p. 575].

Albanese(2) gave a new and very rapid reduction birational on C but not on its carrying space.

3. We pass now to singularities of surfaces. Enriques(9), (11), Enriques-Chisini[(12) p. 577], Chisini(7), (8) have investigated at length the behavior of a surface in the vicinity of a singular point. Enriques-Chisini examine also many special cases and give valuable historical data(12) [p. 654]. They consider the relations between multiple points and multiple curves, consecutive singularities and the behavior of the first polars at the singular points. They also examine the behavior of $z(xy)$ and the permutation of its values when (x, y) is in the vicinity of (x_0, y_0) where (x_0, y_0, z_0) is a singular point of F. Of particular interest is the case when the point is an intersection of two or more branches of the

total multiple curve, especially in regard to the local parametric representations by means of power series, Chisini(7), (8). Enriques in his papers, also Enriques-Chisini[(12), p. 634] examine the character of the singularity in regard to the number of distinct locally irreducible branches of F through it; the approximation of F by means of distinct rational surfaces to an assigned infinitesimal degree; the relation of all this to the local parametric representation of F by sets of holomorphic or meromorphic functions of two variables.

Related investigations: Schmeidler(16), H. W. E. Jung(13) first five chapters. Jung in particular gives in Ch. 5 valuable contributions to the theory of the consecutive curves to an assigned point on F. His book is a compendium on his investigations on algebraic surfaces. Its cardinal feature is the extension to surfaces of the notion of *divisor* of the so-called arithmetical theory of functions of one variable, associated with the names of Kronecker, Hensel, Landsberg and presented in a recent book(14) by Jung himself, also by Bliss(4). In the case of surfaces, complications arise due to the fact that *divisors* naturally ascribed to curves, must sometimes be assigned to certain point-neighborhoods or consecutive curves. It is this question which Jung studies in very complete fashion in (13).

4. Chisini(6) has reduced the singularities of a surface assumed in S_3, by birational transformations of S_3 itself. He removes the isolated singularities and exceptional singularities of the multiple curve by a sequence of quadratic transformations whose fundamental lines pass through the points. That the process is finite he shows by means of a certain number, the *rank* ρ of a singular point P which is a non-increasing integer under the operations considered. Let L be the complete intersection of the first polar surface of an arbitrary point O of S_3 with F, each component of L being counted only once in the curve. L is what Chisini calls the *simplified* polar curve. On the projection L' of L from O on an arbitrary plane the projection of P produces a reduction of the class of L' by an amount say v. Let also r be the multiplicity of L at P and c the number of distinct branches of the multiple curve issuing from P. Then

$$\rho = 4v - 5c - r(r-1).$$

Thus at nodes of the multiple curve of F, $\rho = -4$, at cusps of that curve $\rho = 1$. The rank ρ is invariant under a birational transformation regular at P. Under the transformations that he uses Chisini shows that the rank is never increasing. Utilizing this fact he reduces F to a surface whose singular points of the type considered are of rank 1, -4, and are merely double points with distinct tangents for the simplified polar L. F will then have no other exceptional singularities than points where two distinct multiple curves cross one another without contact, or base points of the variable part of L where it is not tangent to the multiple curve. The elimination of sets of consecutive multiple curves is accomplished by Chisini by means of monoidal transformations, his results in (6), (7) playing an important part in this last step. The reduced surface has no other singularities than multiple curves with distinct tangent planes at arbitrary points and with no other exceptional points than those considered above.

5. Albanese gives in (1) a reduction birational on F but not on its carrying space. He first constructs on F a linear system of curves $\sum_k^h$, where $h>6$ is the dimension and k the number of variable intersections of any two of its curves at finite distance from the base points. Furthermore the system is simple, i. e., its curves that are made to pass through an arbitrary point of F, do not necessarily pass through any other. Finally it has the property that any similar subsystem $\sum_{k-j}^{h-i}$, $i\leqq 4$, is irreducible and such that $j<2i+1$. By means of the system F is reducible to F' having only the following singularities: A finite number of isolated double points that may be consecutive to one another; double lines that do not intersect and carry a finite number of simple cuspidal points.

In regard to the singularities of surfaces see also Severi (17). For those of algebraic varieties see Schmeidler (16). Lefschetz [(15) p. 368] established that any algebraic variety of rank one is birationally equivalent to a non singular variety in some space. See also Ch. 17, No. 8.

BIBLIOGRAPHY

1 Albanese, G., Pal. Circ. Mat., 48: 321-332 (1924).

2 ———, Rom. Acc. L. Rend., (5), 33_1: 13-14 (1924).

3 Berzolari, L., P. Rep., II_2 (1922).

4 Bliss, G. A., Ann. of Math., (2), 26: 95-124 (1924).

5 Castelnuovo-Enriques, Encykl. III C 6a, (1915).

6 Chisini, O., Bologna Mem., (7), 8: 3-42 (1921).

7 ———, Bologna Rend., 3-9 (1919).

8 ———, Rom. Acc. L. Rend., (5), 26_2: 8-12 (1917); 29_1: 127-130, 170-174, 241-247 (1920).

9 Enriques, F., Bologna Rend., (1916); (1917).

10 ———, Rom. Acc. L. Rend., (5), 26_1: 415-420 (1917).

11 ———, Rom. Acc. L. Rend., 26_2: 35-43 (1917).

12 Enriques-Chisini, Lezioni sulla teoria geometrica delle equazioni e delle funzioni algebriche, vol. 2, Bologna (1918).

13 Jung, H. W. E., Algebraische Flaechen, Hannover (1925).

14 ———, Einfuehrung in die Theorie der algebraischen Functionen einer Veraenderlichen, Berlin: 1 vol. (1923).

15 Lefschetz, S., Trans. Am. M. S., 22: 317-482 (1921).
16 Schmeidler, W., Math. Ann., 81: 223-234 (1920); 84: 303-320 (1921).
17 Severi, F., Rom. Acc. L. Rend., (5), 23_2: 527-539 (1914).

The following is a list of references to earlier works not considered in this chapter. As further assistance to the reader the explicit titles have been added throughout.

Autonne, L., Sur l'équation différentielle du premier ordre et sur les singularités de ses intégrales algébriques, J. Éc. Polyt., (2), 2: 51-109 (1897); (2), 3: 1-74 (1897).

Bertini, E., Einfuehrung in die projektive Geometrie mehrdimensionaler Raeume. 318-332, 411-428 (1924). Germ. Transl., by A. Duschek (Vienna).

Black, C. W. M., The parametric representation of the neighborhood of a singular point of an analytic surface, Boston Am. Ac. Proc., 37: 281-330 (1901-1902).

Cayley, A., On a singularity of surfaces, Math. Papers, 6, 123-128. *Also* in Quart. J., 9: 332-338 (1868).

Del Pezzo, P., Estensione di un teorema di Noether, Pal. Circ. Mat., 2: 139-144 (1888).

———, Sui sistemi di curve e di superficie, Pal. Circ. Mat., 3: 236-240 (1889).

———, Intorno ai punti singolari delle superficie algebriche, Pal. Circ. Mat., 6: 139-154 (1892).

———, Osservazioni su un memoria del Prof. Corrado Segre e risposta ad alcuni suoi appunti; Replica ad una nota del Prof. Segre in risposta ad alcuni mie osservazioni, Rom. Pont. Acc. Atti 27, (2), 2: 4, 13 p., 10, 7 p. (1897).

De Paolis, R., Sopra un sistema omaloidico formato da superficie d'ordine n con un punto $(n-1)$ — plo., Giorn. di Mat., 13: 226-248, 282-297 (1875).

Dumas, G., Sur la résolution des singularités des surfaces, C. R., 152: 682-684 (1911); 154: 1495-1497 (1912).

Geck, E., Ueber uniplanare Knotenpunkte, Dissertation (Tuebingen 1909); Math. naturw. Mitt. Wuertemb, (2), 6: 65-78 (1904); 7: 1-8 (1905).

Halphen, G., Sur les lignes singulières des surfaces algébriques, Ann. di Mat., (2), 9: 68-105 (1878).

Hensel, K., Ueber eine neue Theorie der algebraischen Functionen zweier Variabeln, Acta Math., 23: 339-416 (1900); Jahresbericht, 8: 221-231 (1900).

Hudson, Hilda, On pinch points, Quart. J., 44: 161-166 (1913).

Jung, H. W. E., Darstellung der Funktionen eines algebraischen Koerpers zweier unabhaengigen Veraenderlichen x, y in der Umgebung $x=a$, $y=b$. J. f. Math., 142: 289-314 (1912).

———, Ueber die ausgezeichneten Kurven algebraischer Flaechen, J. f. Math., 142: 61-117 (1912).

Kobb, G., Sur la théorie des fonctions algébriques de deux variables, J. de Math (4), 8: 385-419 (1892); Soc. M. F. Bull., 21: 76-80 (1893).

Levi, B., Risoluzione delle singolarità puntuali delli superficie algebriche, Torino Atti, 33: 66-86 (1897); C. R., 134: 222-225 (1902).

———, Sulla riduzione delle singolarità puntuali delle superficie algebriche dello spazio ordinario per trasformazioni quadratiche, Ann. di Mat., (2), 26: 219-253 (1897).

———, Sulla trasformazione di una curva algebrica in un'altra piiva di punti multipli, Rom. Acc. L. Rend., (5), 7: 111-113 (1898).

——, Intorno alla composizione dei punti generici delle linee singolari delle superficie algebriche, Ann. di Mat., (3), 2: 127-138 (1899).

Noether, M., Zur Theorie des eindeutigen Entsprechens algebraischer Gebilde von beliebig vielen Dimensionen, Math. Ann., 2: 293-316 (1870).

——, Ueber die algebraischen Funktionen einer und zweier Variabeln, Gött. Nachr., 267-278 (1871).

——, Zur Theorie des eindeutigen Entsprechens algebraischer Gebilde, Math. Ann., 8: 495-533 (1875).

——, Anzahl der Moduln einer Klasse algebraischer Flaechen, Berlin. Ber., 123-127 (1888).

Pannelli, M., Sulla riduzione delle singolarità di una curva gobba, Lomb. Ist. Rend., (2), 26: 216-222 (1893).

——, Sulla riduzione delle singolarità di una superficie algebrica per mezzo di trasformazioni birazionali dello spazio, Ann. di Mat, (2), 25: 67-138 (1897).

Pfeiffer, G. F., Series of papers on singularities of surfaces and related topics, Kiew. Univ. Nachr. (Russian), II and 134 p (1910); Bull. de l'Univ. de Kiew, 1-31 (1909); Recueil math. de Moscou, 27: 228-231 (1909); Atti IV. Congr. Mat. Roma (1908), 2: 309-315 (1909).

Pieri, M., Trasformazione d'ogni curva algebrica in altra priva di punti multipli, Riv. di Mat., 4: 40-42 (1894).

Pilgrim, L., Binomische und trinomische Naeherungsflaechen algebraischer Flaechen, Math. naturw. Mitt. Wuerttemb., (2), 7: 19-46 (1905).

Poincaré, H., Sur les transformations birationnelles des courbes algébriques, C. R., 117: 18-23 (1893).

Rohn, K., Ein Beitrag zur Theorie der biplanaren und uniplanaren Knotenpunkte, Math. Ann., 22: 124-144 (1883); Leipz. Ber., 36: 1-8 (1884).

Salmon, G., On the degree of a surface reciprocal to a given one, Cam. Dub. Math. J., 2: 65-74 (1847).

——, On the cone circumscribing a surface of the *m*th order, Camb. Dub. Math. J., 4: 188-191 (1849).

Segre, C., Sulla scomposizione dei punti singolari delle superficie algebriche, Ann. di Mat., (2), 25: 54 (1896).

——, Intorno ad un carattere delle superficie e delle varieta superiori algebriche; intorno ad una mia Memoria "Sulla scomposizione," etc., Torino Atti., 31: 485-501 (1895); 32: 521-529, 781-789 (1896).

——, Su un problema relativo alle intersezioni di curve e superficie, Torino Atti., 33: 19-23 (1897).

Voss, A., Beitraege zur Theorie der algebraischen Flaechen, Math. Ann., 27: 527-536 (1886).

Woelffing, E., Die singulaeren Punkte der Flaechen, Zschr. Math. Phys., 42: 14-36 (1897).

Zeuthen, H. G., Recherche des singularités qui ont rapport à une droite multiple d'une surface, Math. Ann., 4: 1-20 (1871).

——, Révision et extension des formules numériques de la théorie des surfaces réciproques, Math. Ann., (10), 446-546 (1876).

CHAPTER XI

MULTIPLE CORRESPONDENCE AND MAPPING IN SPACE AND HYPERSPACE

1. (1, k) *Correspondence between* S_3, S'_3, $K>2$.—A (1, 8) correspondence between the points of (x') and (x) is obtained by associating the quadric surfaces of a web $|\phi|$ with the planes of (x'), $x'_i=\phi_i(x)$. A line in (x') has an elliptic c_4 for image, and on it lie all eight associated points, image of a variable point on the given line. The line joining a pair of associated points P_1P_2 is a component of $\infty^1|c_4|$ of the system through them. It is called a principal line. Th. Reye([114]). The surface of coincidences is K_4, jacobian of the web. Its image in (x'), with which it is in (1, 1) correspondence is L'_{16} of class 4. The images of the principal lines are straight lines in (x'), bitangents of L'_{16}, the surface is the complete focal surface of the congruence. An arbitrary line of (x) is transformed into a conic c'_2 which touches L'_{16} in four points.

A plane of (x) has for image a Steiner surface of order 4, having three concurrent double lines. The web of quadrics contains 10 composite ones—pairs of planes; the 10 lines lie on K_4.

If $|\phi|$ has r basis points, there are r bundles of lines composed of principal lines, making r systems of bitangents to L'_{16-2r}. Each congruence is of class 2.

The case of $r=6$ has been considered in Art. 8 of Chapter X. If $|\phi|$ represents a web of surfaces of order n without basis elements, $x'_i=\phi_i(x)$ defines a (1, n^3) correspondence between (x') and (x). The jacobian $J(\phi)$ is the locus of coincidences: L' is of order $4n^2(n-1)$, and of class $4(n-1)^3$. A locus in (x) which is not generated by all n^3 associated points has for image in (x) a locus which touches L' in every image point of those on K. Th. Reye([115]).

$$c_1(x)\sim c'_n(x), \quad \text{touching } L' \text{ in } 4n-4 \text{ points.}$$
$$s_1(x)\sim s'_{n^2}, \quad \text{touching } L' \text{ along a curve of order } 4n(n-1).$$

The case $n\geqq 2$ was further employed by Th. Reye([116]) to obtain systems of surfaces containing series of conics. The (1, 8) transformation defined by $x'_i=x_i^2$ which sends the lines of a Battaglini quadratic complex into the ∞^3 conics which touch the faces of a tetrahedron was considered by C. Segre([133]).

Many of the properties derived by D. Montesano([89]) as applying to I_8 are at once obtainable from the results of Segre.

H. E. Timerding([159]) treated a particular case of $n=2$ by putting $x'_i=x_i^2$. The web of quadrics has a common self-polar tetrahedron. Both K and L' are composite, each consisting of the faces of the tetrahedron of reference. A quadric inscribed in L' is transformed into a tetrahedroid in (x)—a special Kummer surface in which the six nodes in any singular plane are in involution. The image in (x') of a general plane of (x) is the general Steiner surface, as in any web $|\phi|$ of quadrics. J. de Vries([167]) discusses a (1, 8) correspondence between the Kummer quartic surface and a plane. V. Martinetti([82]) gives further properties of the $(16_6, 16_6)$ configuration of Kummer.

J. de Vries([169]) obtains further properties of the (1, 8) correspondence. K. W. Rutgers([119]) gives alternate proofs for properties of the surface L'.

R. Scaccionoce([121]) gave an analytical representation of a surface generated by two planes and a bundle of surfaces of class p having a $(p-1)$ fold line, all birationally related.

Further properties of 8 associated points were found by H. Newmann([91]).

The $(1, 8-r)$ correspondence determined by a web of quadrics with $r=0\ldots5$ was considered by V. Snyder and F. R. Sharpe([149]). L'_{16} has a double curve of order 60 and a cuspidal curve of order 36. When $P_1=P_2$ describes K_4, the other six points describe a surface R_{24} of order 24, having a cuspidal curve of order 90 and a double curve of order 120. Similar properties when $|\phi|$ has 1, ..., 5 basis points.

J. R. Musselman([90]) showed that K_4 contains two distinct ∞^3 systems of sextics of $p=3$. See also J. R. Conner([29.1]). It can be transformed birationally into the symmetroid. The ∞^3 self-polar triangles as to a given conic were mapped on the points of S_3 by G. Schaake([122]) by means of a (1, 1) correspondence between the points of the plane of the conic and the bisecants of a rational space quartic curve. The vertices of self-polar triangles correspond to three concurrent bisecants. In the plane of the conic is a c_5 of $p=6$ containing ∞^1 configurations $(10_3, 10_3)$ of Desargues.

A. Del Re([32]) obtains various multiple correspondences by means of linear systems of monoids with a common vertex.

H. de Vries([166]) defines a (1, 9) correspondence between the planes and points of S_3 by means of two general space C_3.

J. de Vries([168]) considers an involution of tangents I_p of a rational space curve R_n. The lines in each group are divided into couples, and the linear congruences constructed having each pair for directrices. As the group describes I_p this congruence describes a complex, the order and properties of which are determined. A particular case is developed in detail, leading to a quadratic system of quadrics and its dual which defines a Kummer configuration $(16_6, 16_6)$.

2. *Lines of Complexes Mapped on Points of S_3.*—The lines of a linear complex can be mapped birationally on the points of space. F. Klein([60]); M. Noether([92]); F. Klein and S. Lie([59]); F. Klein([61]); S. Lie([73]).

The quadratic identity $p_{12}p_{34}+p_{13}p_{42}+p_{14}p_{23}=0$ and the equation of the complex define a three-dimensional quadratic variety M_2 in S_4, which can be projected stereographically upon the points of space. (Noether, Klein). F. Klein([62]). S. Lie considered two equations $f_1(x_1, x_2, x_3, x_4; x'_1, x'_2, x'_3, x'_4)=0$, $f_2=0$ homogeneous in two sets of coordinates, making the points of either space correspond to an ∞^3 system of curves in the other. When both equations are bilinear, the curves are lines of a complex.

Given two bilinear equations $\Sigma a_{ik}x_ix'_k=0$, $\Sigma b_{ik}x_ix'_k=0$ between the point coordinates of S_3 and S'_3, if the points (x) of one of them are images of the lines of a linear complex of the other (x), then the points of (x) are images of the lines of a special quadratic complex of the other, consisting of the lines which meet a fixed conic, or the tangents of a quadric surface. S. Lie([72]).

For the early history see S. Lie([68]). See also S. Lie([74]) and the statement made by S. Lie and F. Engel([69]).

In particular

$$x_1x'_4-(x'+ix'_2)x_4+x_3x'_3=0,$$
$$(x'_1-ix'_2)x_3-x_2x'_4+x'_3x_4=0,$$

map the lines of the complex $p_{12}-p_{34}=0$ on the points of (x'), and the points of (x) on the complex of secants of $x_1^2+x_2'^2+x_3'^2=0$, $x'_4=0$. When the point describes a line in (x), the image line in (x') describes a regulus through the conic. The lines of the conjugate regulus are images of the points of the conjugate of the given line as to the complex $p_{12}-p_{34}=0$. If the conic is the imaginary circle at infinity, the line-sphere transformation results. S. Lie and G. Scheffers([70]). C. M. Jessop([55]). V. Snyder([148]) has given the details, nothing new.

R. Bricard([20]) mapped the points of S_3 on the tangent lines to a quadric surface birationally. When the point describes a line the image line describes a quadric surface circumscribed about the given one. The line-sphere transformation of Lie appears as a particular case. Further cases have been given by E. Study([150]), H. Beck([10]), M. Fouche([46]), L. Berwald([13]). See the criticism by S. Kantor([58]). E. O. Lovett([78]), H. Liebmann([75]), have generalized the method to apply to much wider cases.

P. F. Smith([147]) mapped the points of S_3 on the spheres of a linear spherical complex, employing a direct generalization of the transformation by reciprocal radii.

P. del Pezzo([105]) maps the complex on the points of S_3 by means of the quadrics through two skew lines.

Special linear complexes were mapped on S_3 by A. Wiman([181]) by means of the bilinear equations

$$x'_1x_1+x'_2x_2=0, \qquad \Sigma a_{ii}x_ix'_i=0,$$

and applied to the determination of ruled surfaces having a directrix line. The same method is also employed to obtain ruled surfaces in a general linear complex, and those which contain a conic. Further details are given by R. Sturm([152]), R. Schuhmacher([127]), T. Reye([111]), F. Schur([128]), L. Cremona([31]).

The tetrahedral complex—consisting of the ∞^3 lines which meet the faces of a fixed tetrahedron in points having a fixed cross-ratio—was mapped on the points of S_3 by A. Weiler([178]) and ([179]).

If the complex is $p_2p_5+kp_3p_6=0$, we may put

$$\begin{aligned} p_1&=x_1x_2, & p_4&=(k-1)x_3x_4,\\ p_2&=x_1x_3, & p_5&=-kx_4x_2,\\ p_3&=x_1x_4, & p_6&=x_2x_3. \end{aligned}$$

See also E. Bertini([11]). The mapping also follows directly from its method of generation by means of two projective S_3. T. Reye([110]). For further details and consideration of special cases, see J. Cardinaal([23]); C. Servais([141]); St. Jolles([57]).

Linear complexes are mapped on the points and circles of a plane by L. Eckhart([34]). The complex $\Sigma a_ip_i=0$ is associated with the circle $a_3(x^2+y^2)+(a_2+a_5)x-(a_1+a_4)y-a_6=0$ and the point $(-a_5, a_4, a_3)$.

The general quadratic complex was mapped on the points of S_3 by F. Klein; see M. Noether([92]). Using $x_1, \ldots, x_6$ for line coordinates, the equation may be put in the form

$$\phi(x_1, x_2, x_3, x_4)+Ax_5+Bx_6=0,$$

in which A, B contain only $x_1, \ldots, x_4$.

$$\begin{aligned} \rho x_i&=\xi_i(A\xi_3-B\xi_2), & i&=1, \ldots, 4.\\ \rho x_5&=\xi_1\xi_4B-\xi_3\phi, & \rho x_6&=-\xi_1\xi_4A+\xi_2\phi. \end{aligned}$$

G. Darboux([33]) shows that the system of quadric surfaces through 4 points gives a map of the general quadratic line complex on the points of S_3. A fundamental curve of order 5 appears in (ξ). See also S. Lie([71]), A. Voss([165]), and F. Aschieri([3]).

The tetrahedral complex is also the projection in S_3 of the lines which meet three planes in S_4, C. Segre(135) and (138); and also the intersection with S_3 of the ∞^3 planes which meet a normal C_4 in S_4 in three points. C. Carrone(24).

Another method was employed by J. de Vries(171). Given a pencil of quadric surfaces $|F_2|$. The polar planes of a point P form a pencil, axis p. As P describes S_3, p generates the tetrahedral complex having the self-polar tetrahedron of the pencil of quadrics for fundamental tetrahedron.

The quadrics circumscribing the tetrahedron go into (2, 2) congruences contained in the complex. A general surface F_n of order n is transformed into a $(n, 3n)$ congruence. G. Castelnuovo(27).

K. Ogura(93) and (94) gives a different method of mapping the tetrahedral complex on the points of S_3, and applies a similar method to map various cubic complexes on S_3.

E. Caporali(21) and (22) investigated in detail the (2, 2) congruences contained in a given quadratic complex by the method of Klein. The same results were obtained by E. Bertini by projection from an S_1 in S_5, see C. Rosati(118). A (1, 2) representation was given by R. Sturm(155) previously proposed and partially carried out by F. Aschieri(6) and (7). A particular form of quadratic complex, generated by two projective pencils of linear complexes was given by F. Aschieri(5). More generally D. Montesano(84) mapped the same quadratic complex on the points of S_3. R. Sturm(156). The method was proposed by C. Segre(139).

Special cases of the complex of Battaglini were mapped on S_3 by D. Montesano(87).

H. Beck(10) maps the points of space on the tangents to a quadric, but each tangent corresponds to two points, conjugate in an axial harmonic homology. The points of the two directrices are transformed into the two reguli of the quadric. The Lie line-sphere transformation appears as one interpretation of a limiting cases when a parameter vanishes.

The cubic complex Σ_3 of the generators of the quadrics in a net has the 8 basis points as vertices of singular bundles. R. Sturm(157) and (158). See also (151). F. Aschieri(2). T. Reye(113) and (112). See also J. C. Kluyver(64). D. Montesano(85) mapped Σ_3 birationally on the points of S_3 as follows. Let π be the plane of any three basis points. A point P determines a pencil in the net, basis curve C_4, which meets π in one variable point P'. PP' is a line of Σ_3, unique when P is given. Conversely, given a line p of Σ_3. It meets π in P' and fixes the C_4 passing through it; p meets C_4 in a second point P. There are 56 such mappings, and the product of any two defines a (1, 1) point transformation in S_3.

These are of three different kinds, according to the number of common basis points in π, π'.

Montesano also defines a (1, 2) correspondence between the lines of Σ_3 and the points of S_3. Let K_2 be the quadric cone through the basis points $B_1 \ldots B_5$ and having its vertex at B_6. A point P fixes the C_4 of the net passing through it, and C_4 meets K_2 in Q. PQ is a line of Σ_3. Conversely, given a line p of Σ_3. It meets K_2 in two points Q_1, Q_2, each of which determines a C_4, meeting p in P_1, P_2. In the associated involution of P_1, P_2 each line of Σ_3 joins one pair of conjugate points. If C_{ik} is the cubic through all basis points except B_i, B_k and d_{ik} is the line joining B_iB_k and g_i is a generator of the quadric through d_{78} and d_{6i}, through B_i, we may write

$$s_1 \sim s_9 : C_{78}{}^3 d_{78}{}^5 \cdot C_{16} \ldots C_{56} \cdot g_1 \ldots g_5,$$
$$C_{78} \sim \gamma_7 : C_{78}{}^2 d_{78}{}^4 \cdot C_{16} \ldots C_{56} \cdot g_1 \ldots g_5,$$
$$C_{i6} \sim s_1 : B_i d_{78};$$

each g_i is parasitic. The lines Σ_3 contain conjugate points of 168 such involutions.

Another method of mapping Σ_3 is obtained by associating each line p of Σ_3 with the point P on it such that the tangent plane to the quadric of the net through p at P shall pass through a given basis point B_8. Conversely, given P; the line PB_8 determines a quadric of the net, and the other generator through P is p. There are 8 such mappings, and the product of any two defines a birational transformation, in which the lines PP' belong to Σ_3. F. R. Sharpe and V. Snyder([145]).

It was shown by G. Fano([43]) that the general complex of order higher than 2 cannot be mapped birationally on the points of S_3. The ∞^3 circles of a plane were mapped on the points of S_3 by P. H. Schoute([124]). Given a triangle ABC, $U=0$ its circumscribed circle, and P any point. Project A, B, C from P on opposite sides, a is length of side $B'C'$, etc., any circle then has the form

$$U+\lambda a^2+\mu b^2+\nu c^2=0.$$

Let λ, μ, ν be rectangular coordinates of a point in S_3. The condition for orthogonality is then expressed as polarity as to a fixed quadric surface. A variation was employed by K. W. Walstra([177]) the center being the x_1, x_2 of the image point, and the third coordinate of the point being the power of the origin as to the given circle. Further developed by J. de Vries([170]). Another form was employed by G. Bordiga([16]) who considered a complex of circles in S_4 and mapped them on the points of S_3.

M. Pieri([108]), D. Montesano([86]) and ([88]) mapped the complex of tangents to an algebraic surface on the points of S_3. The necessary and sufficient condition that the correspondence is (1, 1) is that the surface is rational. A summary of complexes associated with (1, 1) correspondences is given by G. Loria([76]).

The lines of S_3 were mapped on the pairs of lines and points of a plane by B. Mayor([83]). Given two projective bundles of planes; map the planes of one on the lines of an auxiliary plane, and those of the other on the points of the same plane. A line not through the vertex of either bundle uniquely determines a line and a point, and conversely. The locus of united elements is the image of a linear complex. The exceptional elements are a special linear complex, and the image of its axis. Further applications were made by G. Marletta([79]) including the theory of algebraic congruences. A more extensive study of defining lines by an ordered pair of points in the plane was made by J. Grünwald([51]) with applications to kinematics. Two different methods are developed, one linear, and the other by means of the Plücker quadratic identity. The Lie line-sphere transformation is included in the latter.

L. Eckhart([34.1]) generalized the various methods employed in descriptive geometry and gives in analytic form the principal schemes of mapping the lines of space or of hyperspace on groups of points in the plane.

A similar study of pairs of points of S_2 mapped on a quartic variety of S_5 was made by G. Schaake([122.1]). The pairs of conjugate points as to a fixed conic of the plane are mapped on the points of S_3. This was carried further by the same author in ([122.2]), by means of a fixed space cubic curve.

3. *Lines of S_3 Mapped on Points of S_4.*—The quadratic identity $\pi=0$ among the six homogeneous line coordinates defines a four-dimensional variety in S_5 which can be projected stereographically into S_4. F. Klein([62]). The linear transformations under which π remains invariant can be interpreted in S_4 in terms of motion, similarity and transformation by reciprocal radii. A similar interpretation can be made concerning an $n-1$ quadratic variety in S_n. A. Giacomini([47]).

If $\pi=0$ be written in the form $x_1^2+x_2^2+x_3^2+x_4^2-x_5x_6=0$, then

$$\rho x_i=x'_ix'_5, \qquad i=1, \ldots, 4.$$
$$\rho x_5=x_5'^2.$$
$$\rho x_6=x_1'^2+x_2'^2+x_3'^2+x_4'^2.$$

The exceptional elements in S_4 are the hyperplane $x'_5=0$ and the quadric hypercone $x_1'^2+x_2'^2+x_3'^2+x_4'^2=0$, and in line space, the directrix of the special complex $x_5=0$.

Other methods of mapping the lines of S_3 on the points of S_4 were given by F. Aschieri([8]); F. Chizzoni([29]); G. Loria([77]). Let Σ_3, S_4 be two linear spaces. In S_4 given two points, vertices of projective systems of S_1, S_2, S_3. To a line r of Σ_3 associate a perspective S_2 of the first hyper-bundle, and construct its projective image S'_2. Let S_2, S'_2 intersect in P, image of r, and conversely. This is a direct generalization of F. Aschieri. The method of Chizzoni was also generalized by G. Loria. Given Σ_3 and in it two planes α, β. In S_4 given two lines a, b. Make points of α projective with planes of S_4 through a; points of β projective with the planes of S_4 through b. A line r of Σ_3 meets α, β each in a point. The image planes of these points meet in P, image of r, and conversely.

G. Lazzeri (65) mapped the lines of S_3 on a plane point line connex.

G. Scorza (129) gave an interpretation of the complete ternary field in terms of the lines of space.

The lines of S_n can be mapped birationally on the points of a rational variety of dimensions $2(n-1)$ in $S_{\binom{n+1}{2}-1}$ by the method of H. Grassmann (48) and (49). Let (x), (y) be any two points of S_n. The coordinates of the line joining them are the two rowed determinants of the matrix

$$\begin{vmatrix} x_1 & x_2 & \cdots & x_{n+1} \\ y_1 & y_2 & \cdots & y_{n+1} \end{vmatrix}$$

These determinants are regarded as point coordinates of $S_{\binom{n+1}{2}-1}$, and the identities among them define the variety $M_{2(n-1)}$. E. d'Ovidio (95) and (96). H. Grassmann (50). F. Engel (38). The order of $M_{2(n-1)}$ was found by H. Schubert (126) to be $\dfrac{1}{2n-1}\dbinom{2n-1}{n-1}$. The case of $n=4$ was derived again by G. Castelnuovo (28) and $n=3$ by C. Segre (140). This method was employed by B. Levi (67) to determine the system of bisecants of an algebraic curve in S_n. The resulting forms will not be multiple, since not all bisecants of C_m in S_n can be trisecants. See E. Bertini (12) and the note by G. Castelnuovo.

A more general proposition is given by P. del Pezzo (103). For reduction of singularities by projection and by Cremona transformations, see P. del Pezzo (106); M. Pannelli (99); C. Segre (132); H. G. Zeuthen (183); and B. Levi (66).

R. Schumacher (127) considers two fixed skew lines g_1, g_2 in S_4 and two fixed planes π_1, π_2 in line space (77). A point P of S_4 determines the planes $g_1{}^P$, $g_2{}^P$ and a line g determines two points P_1 in π_1, P_2 in π_2. Now establish a projectivity between the ∞^2 planes through g_1, and the ∞^2 points of π_1. Similarly for g_2, π_2. This establishes a projectivity between g and P. M. Pannelli (101); M. Pieri (107); E. Veneroni (164). Compare R. Sturm (153); C. M. Jessop (50).

4. *Linear Line Complexes Mapped on Other Elements.*—The system of ∞^5 linear complexes can be mapped on a part of the plane line-point connexes. G. Lazzeri([65]). Let π, π', π_1, π_2 be any four planes of S_3, Associate the points of π and of π_1 projectively; and the points of π_2 with the lines of π' projectively.

A line meets π in P, and π_2 in P_2, hence is uniquely associated with the point P_1 of π_1 and the line p' of π'. In this way a linear line complex is represented by a lineo-linear connex with a fixed pencil. Or on the conics of a plane [C. Segre([137]); see also R. Sturm([154])] or on a point and a circle of a plane, L. Eckhart([34]), or the totality of quadric surfaces through four fixed basis points, J. Kloboncek([63]). In the last memoir the special linear complexes are mapped on the quadratic system of conics previously considered by L. Cremona([30]) and by F. Aschieri([4]). In the latter the lines of space (axes of special complexes) which are images of composite conics form a cubic complex.

Another method was given by J. de Vries([172]). Given a pencil of lines (A, a) not in the complex, and a point M not in a. Let the bundle A' be made projective with the bundle A. Given a line r of the complex. Call the plane $(A, r)=\rho$, and let ρ' be its associated plane of A'. If r meets a in R, then the point ρ', $MR=R'$ is the image of r.

Conversely, if R' is given, draw $A'R'$; it is the axis of a pencil of planes of A', one plane of which passes through the point a, $MR'=R$. The image line r is the intersection of this plane and the polar plane of R. A pencil in the complex $\sim$ conic through M.

A linear congruence $\sim$ cubic surface with node at M.

A regulus $\sim$ space curve of order 4 with node at M.

The singular elements are all derived.

Another method is given by the same author([176]), by means of an axial involution, the axes of which are not in the complex nor conjugate with regard to it, and an arbitrary point C. Given a line r of the complex. Let its image in the involution be r'. The transversal of r and r' through C meets r in its image point R. In this representation each line of the complex passes through its image point. Conversely, given any point R, let ρ be its polar plane, and L, λ its conjugate in the involution. Then R is the image of that line of ρ which meets CR and its conjugate in L, λ.*

A pencil $\sim$ conic meeting the invariant lines a, b of the involution.

A linear congruence $\sim$ cubic surface through a, b and the directrices of the congruence.

A [2, 2] congruence $\sim$ plane.

A cubic ruled surface $\sim$ line.

* The author is indebted to Dr. Marian M. Torrey for detailed abstracts of a number of recent papers to which he did not have access.

M. N. van der Bÿl([20a]) systematizes the latter method by obtaining all the mappings in which each line of the complex passes through its image point; in particular, those in which a pencil of the complex is mapped by a conic in its plane and passing through its vertex, and a linear congruence is mapped on a cubic surface. The singular elements are obtained in the general case and in various particular cases.

J. de Vries([173]) generalizes the method employed in ([172]) to map a quadratic complex on the points of S_3.

A quadric cone in the complex $\sim$ space C_4 with a fixed double point M.

A pencil cone in the complex $\sim$ conic through M.

A conic cone in the complex $\sim$ plane C_4 with singularity at M.

A congruence (4, 4) in the complex $\sim$ plane.

5. *Various Correspondences in Hyperspace.*—The theorem of Chasles concerning the number N_1 of coincidences in (α_0, α'_0) correspondences on two superposed S_1's, $N=\alpha_0+\alpha'_0$ was generalized by G. Salmon([120]). If there is no locus of ∞^1 invariant points, then if α_1 is the order of the locus in (x') image of a straight line in (x), then

$$N_2=\alpha_0+\alpha'_0+\alpha_1(\alpha_1=\alpha'_1).$$

This was extended by H. G. Zeuthen([184]).

$$N_3=\alpha_0+\alpha'_0+\alpha_1+\alpha'_1(\alpha_1\neq\alpha'_1).$$

See also H. Schubert([125]).

M. Pieri([109]) generalized this by induction, with the following results: If S_k, S'_k are in (α_0, α'_0) point correspondence such that an S_i in S_k has a locus of order α_i in S_k and there is no ∞^1 locus of invariant points, then the number of coincidences N_k has the two forms

$$N_{2m}=\sum_{i-1}^{m}\alpha_i+\sum_{i-1}^{m}\alpha'_i \qquad \text{with } \alpha_m=\alpha'_m$$

$$N_{2m+1}=\sum_{i-1}^{m}\alpha_i+\sum_{i-1}^{m}\alpha'_i, \qquad \alpha_m\neq\alpha'_m \text{ in general.}$$

F. Palatini([97]) established a correspondence between the points (x, y) of a plane, and the real points (a, b, c, d) of S_4, wherein $x=a+ib$, $y=c+id$. Another representation was given by A. Henschel([52]). [C. Segre([134]). G. Scheffers([123])].

G. Castelnuovo([26]), ([27]), and ([28]), constructs a (3, 6) congruence of lines in S_4 by means of three projective ∞^2 systems of S_3 through three skew lines as vertices. The congruences are projected into S_3 and mapped on the points of a plane. A configuration of 15 planes and 10 points in S_4 is found; through each point pass 6 planes and in each plane are 4 points.

G. Bordiga([14]) and ([15]) proposes a generalization of the linear line complex of S_n, and the generation of normal curves by its means; also surfaces by projective systems of S_4 with S_3 as basis in S_6, the resulting surface can be mapped on a plane. A complex is defined as the totality of ∞^3 lines in S_4 which satisfy three given conditions, in particular, those meeting three given surfaces, or the bisecants of one which meet another or finally the trisecants of one surface. A birational transformation between the points of two S_3 is established, in which the planes of each are transformed into the web of quintic surfaces through a space cubic doubly and a plane quartic simply. Numerous errors are contained in the papers. E. Ascione([9]) considers the complex of trisecants of certain F_6, F_5, F_4 in S_4 and cuts the latter by S_3, the section being the general Steiner quartic surface with three concurrent double lines. F. Palatini([98]) determined the characteristics of certain linear systems of linear line complexes in space of 5 dimensions by means of the methods of enumerative geometry. By means of projection from a singular element on a surface or a variety of hyperspace, P. del Pezzo([102]), ([103]), and ([104]) obtained various theorems concerning tangent hypercones, and the condition under which spatial sections are normal curves.

J. Fairson([44]) obtained a correspondence between S_n and S_{2n-1} by means of invariants of systems of binary forms.

C. H. van Os([163]) considers q linear systems of order m_i, ∞^p in S_n, projectively related. The locus generated by them is of dimension $n+p=q$ and of order Σm_i.

F. Klein([62]) proved that every general line complex could be represented as the complete intersection of M_2 and M_n in S_5.

C. H. Sisam([146]) obtained a complete representation of lines of hyperspace, S_1 in S_r by means of connexes; and this was generalized to complete representation of the totality of S_k in S_r by F. Severi([142]) by means of Grassmann matrix coordinates of order $k+1$.

A. Brambilla([17]) employs a (1, 16) correspondence in S_4 by means of $x'_i=x_i^2$ to study surfaces of order 8, sections by a fixed S_3 of the images of $|S_3|$. This is a generalization to S_4 of the method employed by C. Segre([133]) and by H. E. Timerding([159]). Other generalizations had previously been made by A. Brambilla([18]) and ([19]).

G. Marletta([80]) and ([81]) showed that the quartic variety with a double plane in S_4 is rational and mapped it on S_3.

C. Segre([136]) studied the varieties which represent pairs of points of two S_2 and of two S_3; more generally of S_p and S_q. The resulting variety is of dimension $n=p+q$ in space of $(pq+p+q)$ dimensions, and of order $\dfrac{n!}{p!q!}$. It can be mapped on the points of S_n.

G. Aprile([1]) showed that the variety generated by the tangent lines to a normal ruled cubic variety of S_4 which meet a generating plane of it is rational, and mapped it on the points of S_3.

J. Eiesland([35]), ([36]) and ([37]) generalized the line-sphere transformation of Lie to apply to hyperplanes and hyperspheres.

F. Severi([143]) employs the methods of enumerative geometry to obtain the number of coincidences in a $\infty^{(k+1)(r-k)}$ series of pairs of S_k in S_r. The bibliography of the known particular cases is included.

F. Severi([144]) obtains for surfaces and varieties in S_r, $p_q=\dfrac{\pi(\pi-1)}{2}$, $p_a=p_q-\pi$, $p^{(1)}=(\pi-2)(4\pi-5)$ of the surface in (1, 1) correspondence with the non-ordered pairs of points of a curve of genus π. The points are in (1, 1) correspondence with the congruence of bisecants of the given curve.

G. Scorza([130]) and ([131]) considers those varieties whose curve sections by S_k are elliptic, extending to varieties the theorems of G. Castelnuovo([25]) and F. Enriques([40]) and ([41]). The variety consists of an ∞^1 elliptic system of kS_{k-1} or, if $n>8$, is rational and is representable on S_k by means of a linear system of cubic forms. If $n=4$, $k>3$, it is the quartic base of a pencil of quadric varieties in S_{k+2}. If $n=5$ it can be reduced either to $V_4{}^5$ of S_7, or $V_5{}^5$ of S_8 or $V_6{}^5$ of S_9. If $n=6$, $V_6{}^6$ of S_8. The case of $V_4{}^4$ had been done previously by C. Rosati([118]).

The $V_6{}^5$ is the $\infty^6|C_1|$ in S_4, and the other cases of $n=5$ can be represented by the lines of one, or two, or three linear complexes of lines in S_4, respectively.

F. Enriques([39]) considered the parametric representation of a variety of order n in S_4 with an $n-2$-fold line. It can be represented by an involution in S_3 if it contains a rational non-composite surface of conics through C_1. If V_n of S_r has a linear congruence of rational curves it can be represented by an involution of S_3 if it contains a surface not made up of curves of the congruence.

This idea was extended by G. Fano([42]) to apply to three dimensional varieties with rational surface sections. Apart from $V_3{}^3$ of S_4, all such varieties are rational.

J. de Vries([174]) maps the lines of S_3 on the conics of a ∞^4 system in the plane. Given a space cubic r_3 and its plane image conic r_2. A pencil of planes through an axis I cut r_3 in the groups of an $I_3(A_1A_2A_3)$. Let (B) be its image.

Draw the tangents to r_2 at the points B_k, forming a triangle with vertices C_x. As the plane describes a pencil, the points C_k describe a conic k_2, involution conic of r_2. If l is a secant of r_3, I_3 has one fixed point and an I_2. The congruence of bisecants of r_3 is mapped by the pairs of tangents of r_2.

When l describes a pencil, the conic k_2 describes a pencil.

When l describes a plane the conic k_2 describes a net with 3 basis points, vertices of a circumscribed triangle to r_2.

A regulus is mapped by a system of index two.

A bundle becomes a net with a fixed basis point, pole of the chord, image of the bisecant through its vertex.

If r_2 lies in an osculating plane of r_3, k_2 and r_2 touch each other.

The congruence (3, 1) of axes of osculation is mapped by the k_2 having double contact with r_2.

The pencils of lines in S_3 can be mapped birationally on the conics of a plane; J. Wolff([182]). Given four non-coplanar points A_i and S_3, and a plane α which contains none of them. Let the line A_iA_k meet α in B_{ik}, and the plane $A_iA_jA_k=a_i$ meet α in the line b_i. Take a general point A in α. Denote the ∞^5 quadric surfaces through the points A_i by 0. Let (P, π) designate a general pencil. Associate with it that quadric of 0 for which P is pole of α and π polar plane of A. This quadric is cut by α in a conic which is image of the pencil (P, π). Conversely, given a conic in α, there is one quadric of the system which contains it. The other steps can then be retraced. The singular elements are obtained in both systems.

The lines of S_3 are mapped on the pairs of points of a plane by J. de Vries([175]) as follows:

Given two fixed planes α, α' and two fixed points A, A' not in either. An arbitrary line l meets α in L, α' in L'. Let AL meet a fixed plane π in P, $A'L'$ meet π in P'. The P, P' are the images of l. The steps taken in inverse order will define a line l. A pencil of lines is mapped by two projective point ranges; a regulus by two projective conics; a bundle by two projective plane fields. A linear congruence determines a quadratic correspondence between pairs of points in the plane.

6. *Compound Involutions.*—Given two multiple correspondences $(1, m)$, $(1, n)$, the first between (x') and (x), the second between (ξ', ξ), and a (1, 1) correspondence between (x') and (ξ'). By this means an (m, n) correspondence is established between the points of (x) and of (ξ), with the property that m associated points in (x) have the same n points for images in (ξ), and conversely. Such a multiple correspondence is called a compound involution.

O. Tognoli([160]) applies to S_3 the method employed by C. Wiener([180]) for S_2. Given three pencils of surfaces of orders p, q, r, with m common basis points. They determine an involution of order $n=pqr-m$. In (x') given p', p', r', m', n' projective with respective pencils in (x). Particular attention is given to the case $n=n'$.

Similar constructions for involutions were given by J. S. Vanecek([161]). Given a fixed quadric F_2, a space curve C_m, and a surface F_p. Let π be the polar plane of a variable point P as to F_2, meeting C_m in m points. The polar plane π_m of one of them, M, meets π in a line which pierces F_p in p points X. The polar plane π_p of any of them meets the line π, π_m in a point P'. The four points P, M, X, P' are vertices of a self-polar tetrahedron as to F_2. When P describes a curve or surface of order μ, P describes a curve or surface of order $4\mu mp$. The dual construction was given by J. S. Vanecek([162]). The method of O. Tognoli([160]) was applied by M. Pannelli([100]) to determine fundamental elements, K, K', L, L' and includes a discussion of systems having basis curves.

A particular I_8 and the configuration 8_4 was given by G. Fontené([45]).

A particular (2, 2) involution was obtained by H. Rothe([117]). Given a fixed plane and a number $m \neq -1$. Let P_1, P_2 be a pair of ordered points in S_3. Construct P_0 such that $P_1P_0 : P_0P_2 = m$. Pass through P_0 a plane parallel to the fixed one. The points P'_1, P'_2 symmetric as to this plane are the images of P_1, P_2. The pairs can be mapped birationally on the points of S_6.

A different (2, 2) compound involution is that obtained from two arbitrary triangles ABC, $A'B'C'$ in S_3. Given any point P, find P' such that

$$PA = P'A', \qquad PB = P'B', \qquad PC \neq P'C'.$$

The images of planes are quadric surfaces. The transformation was suggested by A. Jacobi([53]) and ([54]).

BIBLIOGRAPHY

1 Aprile, G., Giorn. di. Mat., 54: 294-313 (1916). [5.

2 Aschieri, F., Giorn. di Mat., 13: 328-337 (1875). [2.

3 ———, Giorn. di Mat., 41: 261-284 (1903). [2.

4 ———, Lomb. Ist. Rend., (2), 12: 265-269, 341-347 (1879). [4.

5 ———, Lomb. Ist. Rend., (2), 16: 531-539 (1880). [2.

6 ———, Lomb. Ist. Rend., (2), 14: 500-509 (1881). [2.

7 ———, Lomb. Ist. Rend., (2), 21: 216-226, 285-293 (1888). [2.

8 ———, Rom. Acc. L. Mem., (4), 4: 172-180 (1887). [3.

9 Ascione, E., Rom. Acc. L. Rend., (5), 6_1: 162-169 (1897). [5.

10 Beck, H., Münch. Ab., 51-72 (1917). [2.

11 Bertini, E., Rom. Acc. L. Rend., (3), 4: 30-33 (1879). [2.

12 ———, Torino Atti., 26: 118-130 (1890). [3.

13 Berwald, L., Wien. Ber., 128: 1403-51 (1919). [2.

14 Bordiga, G., Ven. Ist. Atti., (6), 4: 163-190, 1461-1501 (1885-86). [5.

15 ———, Ven. Ist. Atti., (6), 6: 919-966 (1887-88). [5.

16 ———, Ven. Ist. Atti., (8), 6: 733-747. [2.

17 Brambilla, A., Giorn. di Mat., 35: 1-21 (1897). [5.
18 ———, Lomb. Ist. Rend., (2), 21: 269-270, 511-520 (1888). [5.
19 ———, Torino Atti., 20: 784-790 (1885). [5.
20 Bricard, R., Nouv. Ann., (4), 5: 221-226 (1905). [2.
20a van der Bÿl, M. N., Amst. K. Ak. Wet. Versl., 34: 421-427 (1924-25). [4.
21 Caporali, E., Memorie, Naples, 54-85 (1888). [2.
22 ———, Rom. Acc. L. Mem., (3), 2: 749-769 (1878). [2.
23 Cardinaal, J., J. f. Math., 101: 142-153 (1887). [2.
24 Carrone, C., Nap. R. A. Rend., (3), 7: 57-66 (1901). [2.
25 Castelnuovo, G., Rom. Acc. L. Rend., (5), 3_1: 59-61 (1894). [5.
26 ———, Ven. Ist. Atti., (6), 5: 1249-81 (1886-87). [5.
27 ———, Ven. Ist. Atti., (6), 6: 527-579 (1888). [2, 5.
28 ———, Ven. Ist. Atti., (7), 2: 855-901 (1890-1). [3, 5.
29 Chizzoni, F., Cat. Acc. Gio. Atti., (3), 20: 273-308 (1888). [3.
29.1 Conner, J. R., Am. J., 37: 29-42 (1915). [1.
30 Cremona, L., Giorn. di Mat., 10: 47 (1872). [4.
31 ———, Lomb. Ist. Rend., (2), 3: 209-219 (1870). [2.
32 Del Re, A., Nap. R. A. Rend., (3), 18: 296-308 (1912). [1.
33 Darboux, G., Soc. M. F. Bull., (1), 2: 155-158 (1871). [2.
34 Eckhart, L., Wien. Ber., 127: 91-118 (1918). [2, 4.
34.1 ———, Wien. Ber., 132: 177-192 (1922). [2.
35 Eiesland, J., Am. J., 35: 201-228 (1913). [5.
36 ———, Am. J., 40: 1-46 (1918). [5.
37 ———, Tôhoku M. J., 16: 185-235 (1919). [5.
38 Engel, F., Grassmann's Werke, 1-2: 402 (1903). [3.
39 Enriques, F., Ann. di Mat. (3), 20: 109-111 (1913). [5.
40 ———, Math. Ann., 46: 179-199 (1895). [5.
41 ———, Rom. Acc. L. Rend., (5), 3_1: 481-487, 536-543 (1894). [5.
42 Fano, G., Ann. di Mat., (3), 24: 49-88 (1915). [5.
43 ———, Torino Atti., 43: 541-552 (1908). [2.
44 Fairson, J., Liège S. R. S. Mém., 9: 27 p. (1912). [5.
45 Fontené, G., Soc. M. F. Bull., 34: 3-16 (1906). [6.
46 Fouche, M., Soc. M. F. Bull., 45: 37-53 (1917). [2.
47 Giacomini, Ann. de la scuo. norm. Pisa, 8: 33 p. (1899). [3.
48 Grassmann, H., Ausdehnl., 3: 65-70 (1865). [3.
49 ———, Werke, 1-2: 44 (1902). [3.
50 ———, J. f. Math., 83: 57-65 (1877-78). [3.
51 Grünwald, J., Wien. Ber., 120: 677-741 (1911). [2.
52 Henschel, A., Diss. Jena, 15 p. (1892). [5.
53 Jacobi, A., Werke, 3: 42-68 (1886). [6.
54 ———, J. f. Math., 73: 179-206 (1871). [6.
55 Jessop, C. M., Line Complex, chap. 12 (1903). [2.
56 ———, Line Complex, 244-257 (1903). [3.
57 Jolles, St., J. f. Math., 149: 19-56 (1919). [2.
58 Kantor, S., Giorn. di Mat., 40: 278-280 (1902). [2.
59 Klein, F., and S. Lie., Berlin Sitzungsber, 891-899 (1870). [2.
60 Klein, F., Gött. Nachr., 258-276 (1869). [2.
61 ———, Gött. Nachr., 164-175 (1872). [2.
62 ———, Math. Ann., 5: 257-278 (1872). [2, 3, 5.
63 Kloboncek, J., Prag. Ber., 23 (1914). [4.

64 Kluyver, J. C., Nieuw. Arch., 19: 1-34 (1892). [2.
65 Lazzeri, G., Ven. Ist. Atti., (6), 3: 247-268, 437-474 (1884-85). [3, 4.
66 Levi, B., Ann. di Mat., (2), 26: 218-253 (1897). [3.
67 ———, Torino Mem., (2), 48: 83-142 (1898-9). [3.
68 Lie, S., Chris. Verhl., 22: 16-38, 107-146 (1869). [2.
69 Lie, S., and Engel, F., Trans. Gruppen., 3: 137-139 (1893). [2.
70 Lie, S., and Scheffers, G., Ber. Trans., 411-475 (1896). [2.
71 Lie, S., Gött. Nachr., 53-86 (1870). [2.
72 ———, Leipz. Ab., 49: 687-740 (1897). [2.
73 ———, Math. Ann., 5: 145-256 (1872). [2.
74 ———, Math. Ann., 5: 251 (1872). [2.
75 Liebmann, H., Münch. Ber., 189-198 (1915). [2.
76 Loria, G., Giorn. di. Mat., 25: 97-108 (1887). [2.
77 ———, Giorn. di. Math., 27: 224-225 (1889). [3.
78 Lovett, E. O., Cong. Math. Paris, 291-298 (1900). [2.
79 Marletta, G., Sulla. rappres. Catania, 22 p. (1906). [2.
80 ———, Giorn. di. Mat., 40: 265-274 (1902). [5.
81 ———, Giorn. di. Mat., 41: 46-61, 113-128 (1903). [5.
82 Martinetti, V., Pal. Circ. Mat., 16: 196-203 (1902). [1.
83 Mayor, B., C. R., 135: 1318-1321 (1902). [2.
84 Montesano, D., Diss. Rome, 50 p. (1886). [2.
85 ———, Bologna Mem. (5), 3: 549-577 (1893). [2.
86 ———, Lomb. Ist. Rend., (2), 26: 589-604 (1893). [2.
87 ———, Nap. R. A. Rend., (1), 25: 192-203 (1886). [2.
88 ———, Nap. R. A. Rend., (3), 13: 66-88, 167, 256-262 (1907). [2.
89 ———, Ven. Ist. Atti., (6), 6: 1425-1444 (1888). [1.
90 Musselman, J. R., Am. J., 40: 69-86 (1918). [1.
91 Newmann, H., Math. Ann., 73: 86-89 (1912-13). [1.
92 Noether, M., Gött. Nachr., 298-306 (1869). [2.
93 Ogura, K., Tôhoku M. J., 13: 172-204 (1918). [2.
94 ———, Tôhoku M. J., 14: 28-63 (1918). [2.
95 d'Ovidio, E., Rom. Acc. L. Mem., (3), 1: 929-986 (1877). [3.
96 ———, Torino Atti., 12: 334-349 (1877). [3.
97 Palatini, F., Palmi, 20 p. (1891). [5.
98 ———, Ven. Ist. Atti., (8), 3_3: 371-383 (1900-01). [5.
99 Pannelli, M., Lomb. Ist. Rend., (2): 26: 216-222 (1893). [3.
100 ———, Nap. R. A. Rend., (2), 1: 153-161 (1887). [6.
101 ———, Rom. Acc. L. Rend., (4), 6: 479-487 (1890). [3.
102 del Pezzo, P., Nap. R. A. Rend., (1), 24: 212-216 (1885). [5.
103 ———, Nap. R. A. Rend., (1), 25: 176-180; 205-213 (1886). [3, 5.
104 ———, Nap. R. A. Rend., (2), 1: 40-42 (1887). [5.
105 ———, Pal. Circ. Mat., 1: 241-271 (1889-7). [2.
106 ———, Pal. Circ. Mat., 6: 139-154 (1894). [3.
107 Pieri, M., Giorn. di Mat., 28: 209-218 (1890). [3.
108 ———, Giorn. di Mat., 33: 167-178 (1895). [2.
109 ———, Rom. Acc. L. Rend., (4), 3_1: 196-199 (1887). [5.
110 Reye, T., Geom. der Lage, 3: 1-11, 175-182 (1892). [2.
111 ———, Geom. der Lage, 2: 177-187 (1892). [2.
112 ———, Geom. der Lage, 3: Lesson 17 (1892). [2.
113 ———, J. f. Math., 82: 154-183 (1877). [2.

114 ———, J. f. Math., 86: 84-107 (1879). [1.
115 ———, J. f. Math., 94: 312-319 (1883). [1.
116 ———, Math. Ann., 48: 113-141 (1896-97). [1.
117 Rothe, H., Wien. Ber., 126_2: 831-855 (1917). [6.
118 Rosati, C., Ann. di. Mat., (3), 1: 25-38 (1898). [2, 5.
119 Rutgers, R. W., Amst. K. Ak. Wet. Versl., 18_2: 1636-1640 (1916). [1.
120 Salmon, G., Geom. three dimen., 511 (1874). [5.
121 Scaccionoce, R., Per. di Mat., (3), 5: 35-38 (1908). [1.
122 Schaake, G., Amst. K. Ak. Wet. Versl., 33: 432-449 (1924). [1.
123 Scheffers, G., Leipz. Ab., 47: 201-208 (1895). [5.
124 Schoute, P. H., Wien. Ber., 94: 786-793 (1886). [2.
125 Schubert, H., Math. Ann., 10: 1-116 (1876). [5.
126 ———, Math. Ann., 26: 26-73 (1886). [3.
127 Schumacher, R., Math. Ann., 37: 100-140 (1890). [2, 3.
128 Schur, F., Math. Ann., 15: 432-464 (1879). [2.
129 Scorza, G., Reg. Emilio, 90 p. (1885). [3.
130 ———, Ann. di Mat., (3), 15: 217-273 (1908). [5.
131 ——, Rom. Acc. L. Rend., (5), 17: 10-11 (1908). [5.
132 Segre, C., Ann. di Mat., (2), 25: 2-54 (1896). [3.
133 ———, Giorn. di Mat., 21: 355-378 (1883). [1, 5.
134 ———, Math. Ann., 40: 413-467 (1892). [5.
135 ———, Pal. Circ. Mat., 2: 45-52 (1888). [2.
136 ———, Pal. Circ. Mat., 5: 192-204 (1891). [5.
137 ———, Torino Atti., 20: 487-504 (1885). [4.
138 ———, Torino Atti., 22: 791-801 (1887). [2.
139 ———, Torino Mem., (2), 36: 3-84 (1884). [2.
140 ———, Torino Mem., (2), 36: 87-158 (1885). [3.
141 Servais, C., Brux. Mém., (2), 1: 53 p. (1905). [2.
142 Severi, F., Ann. di Mat., (3), 24: 89-120 (1915). [5.
143 ———, Rom. Acc. L. Rend., (5), 9_2: 321-326 (1900). [5.
144 ———, Torino Atti., 38: 119-134 (1902-03). [5.
145 Sharpe, F. R., and V. Snyder, Am. M. S. Trans., 25: 1-12 (1923). [2.
146 Sisam, C. H., Torino Atti., 46: 481-487 (1911). [5.
147 Smith, P. F., Am. M. S. Trans., 1: 371-390 (1900). [2.
148 Snyder, V., Diss. Göttingen, 47 p. (1895). [2.
149 Snyder, V., and F. R. Sharpe, Am. M. S. Trans., 19: 275-290 (1918). [1.
150 Study, E., Jahresbericht, 25: 96-113 (1916). [2.
151 Sturm, R., Liniengeom., 1: 9, 68 (1892). [2.
152 ———, Liniengeom., 1: 257-272 (1892). [2.
153 ———, Liniengeom., 1: 272-291 (1892). [3.
154 ———, Liniengeom., 1: 291-308 (1892). [4.
155 ———, Liniengeom., 3: 261-382 (1896). [2.
156 ———, Liniengeom., 3: 389-410 (1896). [2.
157 ———, J. f. Math., 70: 212-242 (1869). [1, 2.
158 ———, Math. Ann., 6: 513-550 (1873). [2.
159 Timerding, H. E., Ann. di Mat., (3), 1: 95-135 (1898). [1, 5.
160 Tognoli, O., Giorn. di Mat., 10: 1-15, 141-146 (1872). [6.
161 Vanecek, J. S., C. R., 94: 1463-1464, 1583-1586 (1882). [6.
162 ———, Prag. Ber., (1888). [6.
163 Van Os, C. H., Nieuw. Arch., (2), 11: 249-256 (1915). [5.

164 Veneroni, E., Giorn. di Mat., 36: 300-305 (1898). [3.
165 Voss, A., Math. Ann., 8: 54-135, 62 (1875). [2.
166 de Vries, H., Nieuw. Arch., (2), 10: 258-268 (1913). [1.
167 de Vries, J., Monatsh. f. Math., 12: 45-48 (1901). [1.
168 ———, Amst. K. Ak. Wet. Versl., 11: 762-767 (1902-03). [1.
169 ———, Amst. K. Ak. Wet. Versl., 15_2: 1263-70 (1913). [1.
170 ———, Amst. K. Ak. Wet. Versl., 28_1: 270-273 (1919-20). [2.
171 ———, Amst. K. Ak. Wet. Versl., 32: 478-482 (1923). [2.
172 ———, Amst. K. Ak. Wet. Versl., 34: 5-7 (1924-25). [4.
173 ———, Amst. K. Ak. Wet. Versl., 34: 8-12 (1924-25). [4.
174 ———, Amst. K. Ak. Wet. Versl., 34: 13-15 (1924-25). [6.
175 ———, Amst. K. Ak. Wet. Versl., 34: 208-213 (1924-25). [6.
176 ———, Amst. K. Ak. Wet. Versl., 34: 322-324 (1924-25). [4.
177 Walstra, K. W., Amst. K. Ak. Wet. Versl., 25_2: 960-963 (1917-18). [2.
178 Weiler, A., Zschr. f. Math. Unt., 22: 261-267 (1877). [2.
179 ———, Zschr. f. Math. Unt., 28: 188-192 (1883). [2.
180 Wiener, C., Math. Ann., 3: 11-33 (1871). [6.
181 Wiman, A., Diss. Lund., 111 p. (1892). [2.
182 Wolff, J., Amst. K. Ak. Wet. Versl., 34: 325-330 (1924-25). [5.
183 Zeuthen, H. G., Ann. di Mat., (2), 3: 175-218 (1867-70). [3.
184 ———, C. R., 78: 1553-1557 (1874). [5.

CHAPTER XII

THE MAPPING OF A RATIONAL SURFACE ON A PLANE

1. *The Quadric Surface.*—The stereographic projection of the sphere was known in ancient times; see E. Kötter (104). For the quadric of revolution see A. J. Fresnel and Deviel (78); J. N. P. Hachette (86). For the mapping of the elliptic paraboloid see C. Ciamberlini (40). For the mapping of the general quadric on a plane see M. Chasles (33), (34), (35), (36), (37) and (38). J. Steiner (132). G. P. Dandelin (64). J. Plücker (126).

A. Cayley (30). M. Chasles (38) and (39). If the quadric is projected from a point O of the quadric upon a plane which meets the generators through O in A and B, the image of O is the line AB and the images of the generators OA, OB are A and B respectively. A plane section goes into a conic through A and B; in general a curve of order $p+q$ meeting OA in p points and OB in q points goes into a curve of order $p+q$ having A for p-fold and B for q-fold point. The generators of the systems OB, OA go into lines through A and B respectively. The papers of Plücker and Cayley express the projection analytically. In the case of a cone the points A and B coincide and the images of the plane sections are conics having a common tangent at a point A.

2. *The Cubic Surface.*—The mapping of a cubic surface F_3 on a plane was first made by A. Clebsch (41) by means of the generation of F_3 by three projective bundles of planes, used by H. Schröter (128) in discussing the 27 lines on the surface and previously given by J. Steiner (134) and H. Grassmann (81) and (82). If (x) is a point on a plane and (y) on the surface there are three relations of the form $\Sigma a_{ij}x_iy_j=0$. These were first given by L. J. Magnus (107). The particular case in which $a_{ij}=a_{ji}$ was further developed by O. Hesse (87). Solving for (y) it is seen that a plane section goes into a cubic curve through 6 fixed points. Each of these points corresponds to a line a_i on F_3. Six other lines b_i correspond to the conics through five of the six points and the remaining 15 lines c_{ij} to the lines joining two of the six points, L. Cremona (57). A line in (x) goes into a space cubic on F_3 meeting each b line in 2 points, each c line in one point and not meeting each a line. A curve of order n in (x) goes into a curve of order $3n-\Sigma m_i$ where m_i is the multiplicity of the curve at the point a_i. A cubic surface with a double point is depicted by the special case in which 6 points lie on a conic or 3 points lie on a line or two are consecutive.

A mapping which is geometrically simpler was given by A. Clebsch([51]) by means of the transversals of 2 skew lines on F_3, the skew projection of J. Steiner([131]) and ([133]). The plane sections go into quartic curves with 2 common double points and 5 simple basis points. This system of quartics is transformable into cubics through 6 points so that the two mappings are equivalent. Similarly the mappings given by F. Dimont([73]) by curves or orders 4, 5, or 6 are all transformable into cubics through six points. The case of sextics with 3 triple points and 6 simple points corresponds to the mapping by means of the bisecants of a space cubic on F_3. The same remark applies to other mappings given by L. Zängli([145]). The mapping given by G. Majcen([108]) obtained by defining the cubic surface by the intersection of the transversals of 2 skew lines and a projective bundle of planes is included in the skew projection of Clebsch.

The modification in the mapping of a cubic surface which has singular points was treated by J. Diekmann([72]). The 6 basis points of the cubics which represent the plane sections have in each case some special position. The cubic surface with 4 double points is discussed by E. Laguerre([106]). The plane sections are depicted as cubics through the 6 vertices of a complete quadrilateral. W. P. Milne([109]) applies the mapping of a cubic surface to the consideration of triads of points apolar to a pencil of plane cubics.

3. *Quartic Surface with a Double Line.*—A. Clebsch([46]) showed that the quartic surface F_4 with a double line contains 64 plane sections composed of two conics each meeting the double line in the same point. The depiction on a plane is then obtained analytically by forming the equation of a quartic surface with the double line and passing through one of these conics. Geometrically the surface is mapped by drawing the lines which meet the double line and conic. The plane sections are represented by quartic curves with one common double point and 8 simple basis points. The cubic through the 9 points is the image of the double line. W. Frahm([77]) considers the case of additional double points. The plane sections are depicted as quintics with one triple point, one double point, and 8 simple points, transformable into quartics with one double point and 8 simple points. J. Cardinaal([19]) considers the surface as the locus of the intersection of 2 projective pencils of quadrics through the line.

4. *F_4 with a Double Conic.*—In the first part of a paper by A. Clebsch([43]) the author starts with the parametric representation of a quartic surface $y_i=f_i(x_1, x_2, x_3)$ $i=1$ to 4, where $\Sigma a_i f_i=0$, $i=1$ to 5, is a system of cubics through 5 fixed points. He shows that there are 16 lines on the surface represented by the 5 points, the 10 connecting lines and the conic through the points, also the systems of conics repre-

sented by the lines through one point and the conics through 4 points. He also shows that there the surface has a double conic represented by a cubic of the system $\Sigma a_i f_i = 0$. In the second part commencing with the equation of a quartic surface with a double conic $\phi=0$, $p=0$, namely, $\phi^2 - 4p^2q = 0$, he shows how to find the depiction of the surface with which he began in the first part.

This analytical method was continued by G. Korndörfer (100) and (101) when the surface had additional double points and for the case of 2 intersecting double lines in (102) and of 2 intersecting coincident lines in (103). The case of a cuspidal conic was discussed by L. Cremona (59), the image of a cubic surface with two double points by a special (2, 2) quadric transformation. See also Tötössy Béla (137).

C. F. Geiser (79) and L. Cremona (63) showed that an F_4 with a double conic can be found by applying a (2, 2) quadratic transformation to a cubic surface through a conic and a point and also to a quadric. See also J. Cardinaal (19) and (20), and G. Darboux (65.1).

A. Del Re (66) considers the surface $(x^2+y^2+z^2)z^2 - ht^2w^2 = 0$ which has two double lines in the plane $z=0$. The plane sections are depicted as cubics through 3 fixed points and a 4th point with a fixed tangent.

5. *The Steiner Surface.*—The Steiner surface which has 3 double lines meeting in a point which is a triple point on the surface was mapped on a plane by K. Weierstrass (143), the plane sections going into conics, so that $y_i = f_i(x_1, x_2, x_3)$, $i=1$ to 4, f_i being a quadratic ternary form.

The equation of the surface was found by eliminating x_1, x_2, x_3 by A. Brill (16). A. Clebsch derived the normal form of mapping $y_1 = 2x_2x_3$, $y_2 = 2x_3x_1$, $y_3 = 2x_1x_2$, $y_4 = x_1^2 + x_3^2$ from the equation of the surface in the form $y_2^2y_3^2 + y_3^2y_1^2 + y_1^2y_2^2 = 2y_1y_2y_3y_4$. He also showed that the general form of Weierstrass could be reduced to the normal form by a suitable choice of coordinates, the fundamental triangle in (x) representing the 3 double lines and the triple point. See also L. Cremona (62), where the cases of 2 or 3 coincident double lines are discussed. For various parametric representations see G. Koenigs (99).

6. *Rational Quartics with No Double Curve.*—The quartic with a triple point is mappable by projection from the triple point. The rational quartics with any other type of singular point can be mapped on a double plane by projection from the double point. Since this double plane is rational, the sextic curve of branch points must belong to one of three types: (1) quartic; (2) a sextic with a 4-fold point; (3) a sextic with two consecutive triple points. The first type of quartic surface was found by L. Cremona (53). The plane sections can be mapped on a plane by sextics with $7P^2$ and $4P^1$ all lying on a cubic which is the image of the

singular point. The other two types were found by M. Noether([120]). The second type is mapped on a plane by $C_7(a^3b_1{}^2b_2{}^2\ldots b_9{}^2)$ and the third by $C_9(a_1{}^3, a_2{}^3, \ldots, a_8{}^3b_1{}^2b)$ the 10 points each lying on a cubic as in the first case. See G. Jung([91]) and D. Montesano([111]).

7. *Quintics with 2 Double Skew Lines.*—A. Clebsch([46]) determined analytically the parametric equations of the surface, the plane sections being mapped as sextics with 2 triple points A, B and 13 simple basis points. The 2 skew lines are depicted as quintics with a triple point at A (or B) and a double point at B (or A) and through the 13 simple points. The 13 points cannot be taken arbitrarily but are intersections of two such quintics. A pencil of lines through B (or A) and a quintic A^3B^2 (or A^2B^3) give the web of sextics. There are 13 lines on F_5 images of the 13 basis points. The line AB is a fundamental line. The system of sextics can be transformed into quintics A^2B^2 with 12 simple basis points, the line AB is now the image of the 13th line on F_5 and the 12 points are the intersection of 2 quartics A^2B and AB^2. See A. Cayley([32]). See also T. Castelli([20.1]).

8. *Quintics with a Double Cubic.*—A. Clebsch([46]) found analytically the equation and the parametric equations of the quintic with a double cubic. The plane sections are depicted as quartics through 11 fixed points, the double curve as C_7 with 11 double points.

F_5 with 3 double lines meeting in a triple point. E. Caporali([17]) gives the depiction of this surface by sextics with $7P^2$ and $3P^1$.

9. *Quintics with Double C_4, $p=1$.*—The equation of the surface is of the form

$$y_1H_1{}^2+y_2H_1H_2+y_3H_2{}^2=0,$$

where $H_1=0$, $H_2=0$ are quadrics. The surface contains the conics

$$H_2=\lambda H_1, \qquad y_1+\lambda y_2+\lambda^2 y_3=0.$$

A straight line through $O\equiv(0, 0, 1)$ meets the surface in 2 pairs of points PQ, $P'Q'$, corresponding to the two values of λ given by $y_1+\lambda y_2+\lambda^2y_3=0$. The locus of the harmonic conjugate of OPQ, $OP'Q'$ is a ruled cubic R, in (1, 2) correspondence with F_5. If R_3 is depicted on a plane, then F_5 can be depicted on a double plane, with a sextic curve of branch points with a P^4. The plane sections correspond to quartics with a P^2 and 7 simple basis points, see A. Clebsch([49]) and ([44]). M. Noether([118]) mapped the quintic surface on a second quartic with a triple line and hence upon a plane.

L. Cremona([55]) discusses a quintic surface with a quartic cuspidal curve of genus one and a double line which is the reciprocal of a cubic surface with a uniplanar point. A. Cayley([31]) gives a summary of Clebsch's work on F_4 and F_5.

10. *Quintics with Double Quintic with a Triple Point.*—A. Clebsch([49]), footnote, mentions this surface as being mappable on a double plane with a quartic curve of branch points, the depiction of plane sections on a simple plane being cubics through 4 points. There are 10 lines corresponding to the 4 points and the 6 connecting lines. For a special case see G. Darboux([65]). L. Cremona([61]) obtained the surface as the image of a quadric under a (3, 3) birational transformation in which the sextic basis curve was composed of a space cubic curve, a line, and conic meeting the cubic in 3 points and the line in 2. The other sextic curve consists of the quintic with a triple point and a line meeting it in 2 points. E. Caporali([18]) considers a (3, 5) transformation in which the cubic surfaces have in common a rational quartic and a point with a common tangent plane. The quintic surfaces have in common a double quintic curve with a triple point and a conic meeting it in 4 points. The plane depiction of the quintic surface follows immediately.

11. *Quintics with Singular Points.*—D. Montesano([111]), ([112]), and ([113]) has investigated various types of quintics having either no multiple curve or multiple curves insufficient for the rationality of the surfaces. The method is that of Cremona, namely, to apply a birational transformation to a known rational surface to obtain a new quintic surface. E. Pensa([122]) has developed a general process of determining the multiple elements of a surface so that $p_a=P_2=0$, (G. Castelnuovo([28])) which ensures the rationality of the surface and has applied the method to find all the combinations which give rationality to the quintic surface.

12. *Rational F_n $n>5$.*—For $n>5$ the rational surfaces have not been completely enumerated. L. Cremona([60]) and ([61]) and J. E. Hill([88]) and ([89]) have found various rational surfaces by considering (3, 3) birational transformations in which the sextic basis curves are composite. M. Noether([119]) used a repetition of (2, 3) and (3, 3) transformations. E. Caporali([17]) gives 8 kinds of sextic surfaces obtained from the theory of triply infinite linear systems of plane curves. L. Cremona([58]) discusses an F_{12} whose plane sections are depicted as sextics with six double points.

M. Noether([118]) discusses an F_6 with a double cubic and a non-intersecting double line; a plane section is mapped by a C_7 with $P^3 6P^2$ and 10 simple points.

G. Bordiga([8]) applied the known cubic involution of order 2 in which a point P and its conjugate P' are conjugate with respect to a quadric F and lie on the same secant of 2 skew lines a, b conjugate to the quadric. The image of a quadric surface is of the 6th order and has four-fold points where a, b meet F and has for double lines the edges of the tetrahedron determined by these 4 points.

M. Bonicelli(3) discusses sextic surfaces with 3 double lines, a triple conic and a fundamental cubic, the inverse of a cubic web with fundamental elements, 3 lines, a point with a fixed tangent plane, mappable by cubics through 6 points. P. del Pezzo(124) discusses a sextic surface with 9 double lines and a 4-fold point. A. Brambilla(12), (11), (13), (14), and (15) considers $x_i = \theta_i^n$ $(i=1 \text{ to } 4)$, $\Sigma\theta_i = 0$ and its generalization.

D. Montesano(114) considers a rational surface of order n such that to any point P of the surface corresponds a plane of the bundle through a fixed point O which passes through P.

13. *F_n with an $n-2$-fold Line.*—M. Noether(118) showed that there exists on the surface a rational curve of order $n-2$ meeting the multiple line $x_3 = x_4 = 0$ in $n-3$ points and therefore each conic on the plane of the pencil $\frac{x_3}{x_4} = \lambda$ in one point. The equation of the surface can be written in the form $AB = CD$ when $ABCD$ are linear in x_1 and x_2. The equations $x_3 = \lambda x_4$, $A = \mu C$, $D = \mu B$ map the surface on the plane λ, μ. The plane sections are depicted as curves of order n with a $n-2$-fold point and $3n-4$ simple basis points. The multiple line is depicted as a curve of order $n-1$ with a $n-3$-fold point and passing through the $3n-4$ basis points.

14. *F_n with an $n-3$-fold Line and $n-3$ Triple Points.*—A. Tummarello(138) obtains these surfaces as the images of cubic, or known quartic and quintic surfaces by a transformation of order n with an $n-1$-fold line and $3(n-1)$ simple basis points.

15. *Rational Ruled Surfaces.*—L. Cremona(54) depicted a rational ruled surface on a plane so that the system of generators is depicted as a pencil of lines and therefore the plane sections as curves of order μ with a $\mu-1$-fold point at the vertex of the pencil. He considers the case of a R_{m+n} with an m-fold and an n-fold directrix $m \geqq n$ and $(m-1)(n-1)$ double generators. The plane sections are depicted as curves of order m with an $m-1$-fold point having $m-n-1$ fixed tangents. If $m=n$ the least value of μ is $m+1$.

A. Armenante(1) found the analytic form of the equations of C_μ by the method of H. A. Schwarz(129) and gave a partial discussion of the representation of the double curve of R_n; A. Clebsch(42) discusses the mapping of the cubic ruled surface on a plane by means of conics through a fixed point which is the image of the double line, the simple directrix being represented by a line. See also A. Clebsch(52).

16. *R_4 with a Double Cubic.*—A. Clebsch(48) depicts the plane sections of the ruled quartic with a double cubic by means of cubics with a double

point and a simple basis point. He finds the equation of the surface and the depiction of the double curve as a quartic with double points at the two basis points. This depiction follows from the definition of the generators as the lines joining corresponding points on 2 conics (in different planes) whose points are in (1, 1) correspondence. See L. Cremona([56]).

A. Clebsch([50]) showed that by considering 2 plane sections of R_n and the generators joining corresponding points, the plane sections are depicted as curves of order $n+1$ with an n-fold point. By quadratic transformations it is possible to transform these curves to curves of lower order having either (1) one other simple basis point, (2) no simple basis point, or (3) basis points, consecutive to the multiple point so that there are $1, 2, \ldots, n-1$ fixed tangents.

17. *R_n with an $n-1$-fold Line.*—M. Noether([118]) discussed the mapping of R_n with an $n-1$-fold line. By projection from a point on the multiple line the plane sections are depicted as curves of order n with an $n-1$-fold point and $n-1$ simple basis points corresponding to the $n-1$ generators through the center of projection. The curves can be transformed into curves of lower order by successive quadratic transformations, the number of simple basis points being reduced by 2 each time until there is either one or no simple basis point.

The case $n=5$ is discussed by B. Kalicum([95]) and ([96]).

18. *General Theory of Mapping a Rational F_N on a Plane.*—A. Clebsch([46]). If the coordinates of a point (y) on the surface are expressible in terms of the coordinates of a point (x) on a plane by equations of the form

$$y_i=f_i(x_1, x_2, x_3) \qquad i=1 \text{ to } 4$$

where f_i is a homogeneous polynomial of order n, the section of F_N by the plane $\Sigma a_i y_i=0$ is represented by the curve $\Sigma a_i f_i=0$. If this system of curves has a_i i-fold basis points the order N of the surface is given by the equation

$$N=n^2-\Sigma i^2 a_i.$$

The genus of the plane sections and the image curves is the same. The image of an i-fold basis point is a rational curve, of order i and that of a curve of order m having a j-fold point at an i-fold basis point is a curve of order $mn-\Sigma ij$.

The image of the double curve of F_N is a curve of order $(N-4)n+3$ having a $(N-4)i+1$-fold point at an i-fold basis point which is in (2, 1) correspondence with the double curve. See also M. Noether([117]).

E. Caporali([17]) considers a triply infinite system of plane curves ϕ of order n genus p and grade N with s basis points, images of the plane sections of genus p of a surface F_N of order N. The linear system of jacobian curves of nets of curves of the system ϕ corresponds to the curves of contact ψ' of the tangent cones of F_N from points outside and are of order $(N-4)n+3$ and multiplicity $(N-4)r+1$ at an r-fold basis point. Two nets of curves of ϕ have in common a pencil. The number of curves of this pencil with an extra double point is

$$3(n-1)^2-\Sigma(r-1)(3r+1)=N+4p+s-1.$$

Hence the class of F_N is $N+4p+s-1$. Two jacobian curves meet in $6(n-1)^2-2\Sigma(3r^2-2r+1)=2(N+4p-s-1)=k$ other points which are common to all the jacobian curves and correspond to pinch points on the double curve Δ' of F_N and to coincidences in the (1, 2) correspondence between the points of Δ' and its image Δ. The curves Δ and ψ and therefore Δ' and ψ' meet in $\alpha=(N-1)(2N-7)+2p(N-11)+3s$ other points. By considering the intersection of Δ' and the second polar of a point not on F_N we have

$$3t+\alpha=(N-2)(\tfrac{1}{2}(N-1)(N-2)-p).$$

The curve Δ has 3 double points for each triple point of Δ'. Hence its genus P can be found.

The genus P' of Δ' can then be found by applying Zeuthen's formula namely from

$$K=2(P-1)-4(P'-1).$$

G. Humbert([90]) shows by considering the double integrals belonging to the rational surface F_n that the adjoint surfaces of order $n+q-4$ meet the surface in curves whose images on a plane are curves of order $hq-3$, the images of the plane sections of genus p being of order h, and are adjoint to the curves of order hq, images of the curves of intersection of F_n with surfaces of order q. The dimension of the system of curves is $\frac{1}{2}q(q-1)n+q(p-1)+\frac{1}{6}(q-1)(q-2)(q-3)$.

19. *Surfaces in S_3 Derived by Projection from a Higher Space.*— P. del Pezzo([123]) showed that surfaces of order n in S_n are ruled for $n>9$ and for $3\leq n\leq 9$ are mappable on a plane by means of cubics, $p=1$. Their projection on S_3 is a F_n with a double curve of order $\dfrac{n(n-3)}{2}$ with $\dfrac{(n-2)(n-3)(n-4)}{6}$ triple points and genus $\dfrac{(n-4)(n-5)}{2}$.

G. Veronese([139]) found F_4 with a double conic as the projection of the intersection of two quadric varieties in S_4. He also([140]) found the Steiner surface as the projection of F_4 in S_5, which can be mapped on the plane by conics.

G. Bordiga([4]) obtained an F_6 with a double C_7, $p=1$ and one P_3 by projecting an F_6 in S_4 which is the locus of the intersection of 4 projective forms of the second kind and is mapped on a plane by quartics through 10 points. In ([7]) he considered 4 forms of the 2d kind in S_5 in which the S_3 of one corresponds to the S_4 of the others. The locus of the intersection of corresponding elements is a rational F_7. G. Bordiga([5]) discusses the plane representation of ruled normal surfaces F_{n-1} in S_n defined as the locus of the intersection of 2 systems of planes with straight line axes, projectively related. The surface F_{n-1} contains ∞^1 lines and a S_{n-1} meets it in a C_{n-1} depicted on a plane by a S_{n-2} through $n-2$ points on F_{n-1} as a C_{n-1} with $n-2$ fixed simple points and a P_{n-2}.

G. Bordiga([6]) derives F_5 and F_6 from S_6, the surfaces being generated by 3 collinear bundles out of S_4 with the centers of 3 dimensions "im Auge Gefast." The sections by S_5 are sextic curves $p=1$ represented on a plane by cubics.

20. *Miscellaneous Methods.*—G. Jung([93]) describes a general method of considering a rational surface as the locus of the intersection of a plane of a bundle and the corresponding line of a congruence when the planes and lines are connected by a Cremona transformation, and applies the method to the case of a monoid.

P. Visalli([141]) applies a similar method using the lines connecting corresponding points on two planes between which there is a Cremona correspondence. See also G. Jung([94]) and A. Del Re([70]).

J. de Vries([142]) considers a surface derived from the lines of a rational congruence and having the directrices of the congruence for multiple lines so that a line of the congruence meets the surface in only one point not on a directrix. See also R. Sturm([135]), G. B. Guccia([85]), C. Minea([110]).

T. Reye([127]) derives rational surfaces from the correspondence between 2 spaces (x') and (x) in which to a plane (x') corresponds a quadric in (x). A plane in (x) goes into a Steiner surface in (x'). A quadric in (x) goes into a rational surface of order 8. If the quadrics which define the correspondence have basis points, a quadric in (x) through r basis points goes into a rational surface of order $8-r$.

S. Kantor([98]). The lines joining a point P on a quadric to 3 fixed points meet the quadric again in 3 points which determine a plane whose envelope is a surface of the sixth class. Its image in a certain point-plane transformation is the quadric. Hence the surface is rational. The image of a line in (x) is a conic in (x'). Hence the rational surfaces contain 2 systems of conics corresponding to the generators of the quadric in (x).

A. Del Re([67]), ([68]), ([69]) and ([71]) discusses various quintic surfaces and the F_4 with a double conic obtained as the fundamental surface of a (1, 2) connex.

21. *Rational Surfaces and Linear Systems of Curves.*—G. Jung(**92**) states that if all the linear systems of least order with genus p are supposed known then all the rational surfaces with plane sections of genus p may be deduced and conversely, if all the rational surfaces with plane sections of genus p are supposed known then the linear systems which are the images of the plane sections may be deduced. See also C. Segre(**130**) and G. Jung(**91**). If we study rational surfaces and determine the general types we may omit particular cases and so fail to determine all linear systems and we are certain to omit the linear system not representable by rational surfaces. Two linear systems of least order which are not birationally equivalent may correspond to two surfaces one of which is a special case of the other. Any linear system of least order with consecutive basis points may be considered a special case of a system with non-consecutive basis points but not of least order.

22. *Rational Surfaces Mappable on a Multiple Plane.*—A. Clebsch(**49**) considered various rational surfaces which can be mapped on a double plane, namely the quadric by projection from an external point, the cubic surface by projection from a point on the surface, the quartic with a double conic by projection from a point on the conic and the quintic with a double quartic of $p=1$. The last surface contains a system of conics lying on planes through a point π. The locus of the polar lines of π with respect to the conics is a cubic surface in (1, 2) correspondence with the quintic surface. By mapping the cubic surface on a plane the quintic surface can be mapped on a double plane.

M. Noether(**116**) shows that a double plane with a curve of branch points of order 2_n and a_i $2i$ or $2i+1$-fold points is rational if

$$\tfrac{1}{2}i(i-1)a_i=\tfrac{1}{2}(m-1)(m-2).$$

M. Noether(**121**) stated that rational double planes can be transformed birationally into one of 3 types with a curve of branch points either,

(1) C_{2n} with P_{2n-2}
(2) General C_4
(3) C_6 with 2 consecutive P_3

and refers to his paper(**116**) in which he finds the condition that a double plane shall be rational; he also refers to E. Bertini(**2**), where the latter determined the 4 types of plane involutions.

G. Castelnuovo and F. Enriques(**24**) showed that these are the types of curves for which the first adjoints of order $2n-3$ exist but the second and higher adjoints of order $2n-6$, $2n-9$—do not exist; and from this showed that the bigenus P_2, P_3, etc., are all 0 for a surface mappable on such a double plane. When a rational surface mappable on a double plane is mapped on a simple plane, then between the 2 planes there is a (1, 2) correspondence, and the image of line on the double plane is a curve on the simple plane belonging to the involution which interchanges points associated in the correspondence.

A. Bottari([9]) has considered the cases of multiple planes of order $n>2$. The corresponding involutions are now cyclic of order n and generated by Cremona transformations. The types of periodic Cremona transformations have been determined by S. Kantor([97]) and A. Wiman([144]) and the former has also found the simple curves which belong to the involution and the curves of invariant points. By mapping these curves on the multiple plane Bottari found the following types of multiple planes, namely with curve of branch points:

(1) For arbitrary n, $\mu n+2$ lines, μn of which pass through a point O.
(2) $n=3$, an arbitrary cubic
(3) $n=3$, $C_6 : P^4$ with 2 consecutive double points
(4) $n=5$, and arbitrary C_3 and an inflexional tangent.

A. Bottari([10]) extends the results of the previous paper to cases in which the cyclic involution is the product of 2 or more of the simple involutions already considered.

23. *General Conditions for the Rationality of a Surface.*—M. Noether([115]) and ([118]) showed that a surface containing a rational pencil of rational curves is rational. The method used was to first map the rational curves on the planes of a pencil and then transform them either into lines or conics so as to get a surface with a $n-1$ or $n-2$ fold line. The former is mappable upon a plane by projection. The latter contains rational curves meeting each conic lying on the planes through the $n-2$ fold line in one point P. The surface can be mapped on a plane by projecting each conic by lines through the corresponding point P.

G. Castelnuovo([23]). From Noether's theorem a surface whose plane sections are rational is rational. E. Picard([125]) and G. B. Guccia([84]) showed that such a surface is either ruled or a Steiner surface. G. Castelnuovo([26]) found that any surface whose plane sections are elliptic curves is either ruled or is rational and in the latter case is of order $n\leq 9$ and the plane sections can be mapped on a plane as cubics through $9-n$ points.

G. Castelnuovo([27]). A surface containing a net of elliptic curves is either rational or transformable to a ruled surface. A surface containing a net of hyperelliptic curves of genus p whose characteristic series is not special or composed of pairs of points of the $g_2{}^1$ of the generic curve is either rational or transformable into a ruled hyperelliptic surface of genus p. If the characteristic series is composed of pairs of points of the g_2^1 the surface can be transformed into a rational surface, a ruled hyperelliptic surface or a ruled elliptic surface. F. Enriques([75]) showed that a surface whose plane sections are hyperelliptic curves of genus $p>1$ is either ruled and of genus p or a rational surface containing a pencil of conics.

F. Enriques([74]) and ([76]) investigates the irrationalities which may occur in representing a rational surface on a plane and shows that a rational surface F_n may be reduced by the aid of the adjoints of order $n-3$ to one or another of the following types:

(1) a surface with rational plane sections,
(2) a surface with elliptic plane sections,
(3) a surface with hyperelliptic plane sections.

L. Godeaux([80]) showed that if a surface possesses two pencils of rational curves, then since the curves of one pencil must meet any curve of the other in a rational involution, both pencils are linear and hence the surface is rational.

G. Castelnuovo([28]) showed that in order that a surface shall be rational it is necessary and sufficient that $p_a=0$ and the bigenus $P_2=0$.

Castelnuovo gives the example due to Enriques of a sextic surface with the edges of a tetrahedron for double curve for which $p_a=0$ but $P_2=1$ and which is therefore not rational. He also gives another example of a special F_7 with $p_a=0$, $P_2=1$.

G. Castelnuovo and F. Enriques([21]) showed that a surface for which the series of successive adjoints C', C'', of a system of curves C on the surface terminates is either rational or ruled and that this is the case if the grade n of a system of curves, genus p, on the surface is greater than $2p-2$.

G. Castelnuovo([22]) and ([25]) in proving the rationality of plane involutions shows that a surface is rational on which any linear normal system of curves has its characteristic series complete and on which there exists a linear system of dimension $>$ the genus.

24. *Rational Surfaces with Plane Sections of Genus 3.*—G. Castelnuovo([29]) shows that there are 4 types of surfaces in addition to the 3 types of quartics with no double curve.

(1) F_{16} of S_{14} and its projection, the hyperplane sections being depicted on a plane by quartics.

(2) F_8 in S_6 and its projections, plane depiction by sextics with 7 double points.

(3) F_{16} of S_{14} and its projections, plane depiction by C_6 with a P^4.

(4) F_{16} of S_{14} and its projections, plane depiction by $C_{8-\mu}$ with a $P_{6-\mu}$ and μ consecutive double points ($\mu=0, 1, 2, 3$).

25. *Rational Surfaces Containing an Infinite Number of Conics.*—The case of quartic surfaces was considered by E. E. Kummer([105]). There are 3 cases, namely the quartics with a double line or conic and the Steiner surface. Besides the quintic with a triple line there are the quintics with a double quartic of genus one (A. Clebsch([45])) and with a double quintic of genus 0 with a triple point (E. Caporali([18])). For the sextic surfaces E. Togliatti([136]) finds the surfaces with a 4 fold line; with a double C_7 of genus 5 the plane sections being depicted as quintic curves with a triple point and 10 simple points; with a double curve of order 8 of genus 3 with 2 triple points or a double quintic with a quadric secant triple line or a triple conic and a trisecant double conic. The plane sections are depicted as quartics with a double point and 6 simple points; with a double C_9 with 4 triple points, the plane sections being depicted as cubics through 3 points, see Bordiga([6]).

C. Grimaldi([83]) has classified the rational surfaces of order 7 with an infinite number of conics.

G. Castelnuovo([23]) showed that rational surfaces with hyperelliptic plane sections of genus p are cut by the system of adjoint surfaces of order $n-3$ (of dimension $p-1$) in $p-1$ conics. The plane sections are depicted as hyperelliptic curves either C_{p+3} with a P^{p+1} and a P^2 or $C_{2p+2-\mu}$ with a $P^{2p-\mu}$ and $p-\mu$ double points consecutive to $P^{2p-\mu}$ on different branches where $\mu=0, 1, 2, \ldots, p$.

BIBLIOGRAPHY

1 Armenante, A., Ann. di Mat., (2), 4: 50-72 (1870). [15.

2 Bertini, E., Ann. di Mat., (2), 8: 244-286 (1877). [22.

3 Bonicelli, M., Giorn. di Mat., 40: 184-191 (1902). [12.

4 Bordiga, G., Rom. Acc. L. Mem., (4), 4: 182-203 (1887). [19.

5 ———, Ven. Ist. Atti., (6), 4: 1085-1092 (1885-86). [19.

6 ———, Ven. Ist. Atti., (6), 4: 1461-1501 (1885-86). [19, 25.

7 ———, Ven. Ist. Atti., (6), 5: 1397-1405 (1886-87). [19.

8 ———, Ven. Ist. Atti., (8), 12: 1027-1045 (1909-10). [12.

9 Bottari, A., Ann. di Mat., (3), 2: 277-296 (1879). [22.

10 ———, Giorn. di Mat., 41: 285-320 (1904). [22.

11 Brambilla, A., Giorn. di Mat., 32: 133-140 (1894). [12.

12 ———, Lomb. Ist. Rend., (2), 21: 334-356, 511-520 (1888). [12.

13 ———, Nap. R. A. Atti., (2), 9: no. 14, 28 p. (1889). [7, 12.

14 ———, Nap. R. A. Rend., (3), 5: 21-22 (1899). [12.

15 ———, Torino Atti., 20: 784-790 (1885). [12.

16 Brill, A., Math. Ann., 5: 401-403 (1872). [5.

17 Caporali, E., Coll. Math. Chelini, 144-170 (1881). [18, 8, 12.

18 ———, Ann. di Mat., (2), 7: 149-188 (1875). [10, 25.

19 Cardinaal, J., Amst. K. Ak. Wet. Verh., 1, no. 6, 1-63 (1892-93). [3, 4.

20 ———, Amst. K. Ak. Wet. Versl., (3), 8: 84-146 (1891). [4.

20.1 Castelli, T., Giorn. di Mat., 54: 10-26 (1915). [7.

21 Castelnuovo, G., and Enriques, F., Ann. di Mat., (3), 6: 165-225 (1901). [23.

22 Castelnuovo, G., Math. Ann., 44: 125-155 (1894). [23.

23 ———, Pal. Circ. Mat., 4: 73-88 (1890). [25.

24 Castelnuovo, G., and Enriques, F., Pal. Circ. Mat., 14: 290-302 (1900). [22.

25 Castelnuovo, G., Rom. Acc. L. Rend., (5), 2_2: 205-209 (1893). [23.

26 ———, Rom. Acc. L. Rend., (5), 3: 59-61 (1894). [23.

27 ———, Rom. Acc. L. Rend., (5), 3: 473-481 (1894). [23.

28 ———, Rom. Soc. It. Sc. Mem., (3), 10: 103-123 (1896). [23, 11.

29 ———, Torino Atti., 25: 695-715 (1889-90). [24.

30 Cayley, A., Phil. Mag., 22: 35-88 (1861).

31 ———, London M. S. Proc., 3: 184-202 (1869). [9.

32 ———, Math. Ann., 3: 469-474 (1871). [1, 7.

33 Chasles, M., Corr. l'ec. polyt., 3: 15-16 (1814). [1.

34 ———, Aperçu, 372 (1837). [1.

35 ———, Gergonne's Ann., 18: 305-320 (1827-28). [1.

36 ———, Gergonne's Ann., 19: 157-175 (1828-29). [1.

37 ———, J. de Math., 7: 172 (1842). [1.

38 ———, C. R., 53: 985-996, 1077-1086, 1203-1210 (1861). [1.

39 ———, C. R., 54: 314-324 (1862). [1.

40 Ciamberlini, C., Giorn. di Mat., 32: 69-80 (1894). [1.

41 Clebsch, A., J. f. Math., 65: 359-380 (1866). [2.

42 ———, J. f. Math., 67: 1-22 (1867). [5, 15.

43 ———, J. f. Math., 69: 142-184 (1868). [4.

44 ———, Gött. Nachr., 486-489 (1869). [9.

45 ———, Gött. Ab., 15: 2-64 (1870). [25.

46 ———, Math. Ann., 1: 253-316 (1868). [3, 7, 8, 18.

47 ———, Math. Ann., 1: 277-312 (1869). [6.

48 ———, Math. Ann., 2: 445-446 (1870). [9, 10, 16.

49 ———, Math. Ann., 3: 45-75 (1871). [9, 10, 22.

50 ———, Math. Ann., 5: 1-26 (1872). [16.

51 ———, Math. Ann., 5: 419-421 (1872). [2.

52 ———. Math. Ann., 1: 134-136 (1869). [15.

53 Cremona, L., Coll. Math. Chelini, 413-424 (1881). [6.

54 ———, Ann. di Mat., (2), 1: 248-258 (1867). [15.

55 ———, Bologna Mem., (3), 2: 117-127 (1872). [9.

56 ———, Bologna Rend., (2), 8: 235-250 (1903-04). [16.

57 ———, J. f. Math., 68: 1-133 (1867). [2.

58 ———, Edinb. R. S. Trans., 32: 411-413 (1883-85). [12.

59 ———, J. de Math., (3), 2: 117-127 (1872). [4.

60 ———, Gött. Nachr., 129-148 (1871). [2.

61 ———, Math. Ann., 4: 213-230 (1871). [10, 12.

62 ———, Lomb. Ist. Rend., (1), 4: 15-23 (1867). [5.

63 ———, Lomb. Ist. Rend., (2), 4: 140-149, 152-162 (1871). [4.

64 Dandelin, G. P., Quetel Corr., 3: 9 (1827). [1.

65 Darboux, G., Soc. M. F. Bull., 2: 60-64 (1871). [10.

65.1 ———, Soc. M. F. Bull., 2: 155-158 (1871). [4.

66 Del Re, A., Pal. Circ. Mat., 17: 129-158 (1903). [4.

67 ———, Rom. Acc. L. Rend., (4), 7_2: 11-18, 111-119 (1891). [2.

68 ———, Rom. Acc. L. Rend., (5), 1_2: 170-176 (1892). [20.

69 ———, Rom. Acc. L. Rend., (5), 2_2: 99-107, 138-146, 211-218 (1893). [20.
70 ———, Rom. Acc. L. Rend., (5), 2: 452-460 (1893). [20.
71 ———, Rom. Acc. L. Rend., (5), 3_2: 11-16 (1894). [20.
72 Diekmann, J., Math. Ann., 4: 442-475 (1871). [2.
73 Dimont, F., Nouv. Ann., (3), 15: 318-325 (1896). [2.
74 Enriques, F., Math. Ann., 49: 1-23 (1897). [23.
75 ———, Rom. Acc. L. Rend., (5), 2: 281-287 (1893). [23.
76 ———, Rom. Acc. L. Rend., (5), 4: 311-316 (1895). [23.
77 Frahm, W., Math. Ann., 7: 512-516 (1874). [3.
78 Fresnel, A. J., and Deviel, Corr. l'Éc. Polyt., 1: 76-80 (1805). [1.
79 Geiser, C. F., J. f. Math., 70: 249-257 (1869). [4.
80 Godeaux, L., Enseign. Math., 15: 310-318 (1913). [23.
81 Grassmann, H., Ausdehnungslehre, (1884). [2.
82 ———, J. f. Math., 49: 47-65 (1855). [2.
83 Grimaldi, C., Giorn. di Mat., 54: 341-364 (1916). [25.
84 Guccia, G. B., Pal. Circ. Mat., 1: 165-168 (1884-87). [23.
85 ———, Assoc. Fr. Sc., 191-200 (1880). [20.
86 Hachette, J. N. P., Traité, 225 (1813). [1.
87 Hesse, O., J. f. Math., 49: 279-332 (1855). [2.
88 Hill, J. E., Math. Review, 1: 1-59 (1897). [12.
89 ———, Am. J., 19: 289-311 (1897). [12.
90 Humbert, G., Math. Ann., 45: 428-445 (1894). [18.
91 Jung, G., Ann. di Mat., (2), 15: 277-312 (1887). [6, 21.
92 ———, Pal. Circ. Mat., 4: 253-260 (1890). [21.
93 ———, Rom. Acc. L. Rend., (4), 1: 762 (1885). [20.
94 ———, Rom. Acc. L. Rend., (4), 2: 85-89 (1886). [20.
95 Kalicun, B., Wien. Ber., 120: 1299-1326 (1911). [17.
96 ———, Wien. Ber., 121_{11}: 2351-2360 (1922). [17.
97 Kantor, S., Acta. Math., 19: 115-191 (1895). [22.
98 ———, Wien. Ber., 79_2: 768-786 (1879). [20.
99 Koenigs, G., Soc. M. F. Bull., (2), 12: 28-32 (1888). [5.
100 Korndörfer, G., Math. Ann., 1: 592-626 (1868). [4.
101 ———, Math. Ann., 2: 41-64 (1869). [4.
102 ———, Math. Ann., 3: 496-522 (1870-71). [4.
103 ———, Math. Ann., 4: 117-134 (1871). [4.
104 Kötter, E., Deutsche Math. Ver., 5: 93 (1898). [1.
105 Kummer, E. E., J. F. Math., 64: 66-76 (1865). [25.
106 Laguerre, E., Soc. M. F. Bull., 1: 21-27 (1872-3). [2.
107 Magnus, L. J., Sammlung von Aufgaben, 84 (1873). [2.
108 Majcen, G., Wien. Ber., 117: 631-644 (1908). [2.
109 Milne, W. P., London M. S. Proc., (2), 21: 134-139 (1923). [2.
110 Minea, C., Casopis, 1: 220-230 (1901). [20.
111 Montesano, D., Nap. R. A. Rend., (3), 6: 158-171 (1900). [4, 6.
112 ———, Nap. R. A. Rend., (3), 7: 67-113 (1901). [11.
113 ———, Nap. R. A. Rend., (3), 13: 66-88 (1907). [11.
114 ———, Pal. Circ. Mat., 2: 131-134 (1888). [12.
115 Noether, M., Gött. Nachr., 1-6 (1870). [23.
116 ———, Gött. Nachr., 267-278 (1871). [22.
117 ———, Math. Ann., 2: 293-316 (1870). [18.
118 ———, Math. Ann., 3: 161-227 (1871). [9, 12, 13, 17, 23.

119 ———, Math. Ann., 3: 547-580 (1871). [12.
120 ———, Math. Ann., 33: 546-571 (1880). [6.
121 ———, Erl. Ber., 10: 81-86 (1878). [22.
122 Pensa, E., Ann. di Mat., (3), 6: 249-287 (1901). [11.
123 del Pezzo, P., Pal. Circ. Mat., 1: 241-271 (1884-87). [19.
124 ———, Nap. R. A. Rend., (3), 3: 196-203 (1897). [12.
125 Picard, E., J. f. Math., 100: 71-78 (1885). [23.
126 Plücker, J., J. f. Math., 34: 347-360 (1847). [1.
127 Reye, T., Math. Ann., 48: 113-141 (1896-97). [20.
128 Schröter, H., J. f. Math., 62: 265-280 (1863). [2.
129 Schwarz, H. A., J. f. Math., 67: 23-57 (1867). [15.
130 Segre, C., Pal. Circ. Mat., 1: 217-221 (1884-87). [21.
131 Steiner, J., Systematische Ent., 252-254 (1839). [2.
132 ———, J. f. Math., 1: 38-52 (1826). [1.
133 ———, J. f. Math., 49: 447-465 (1855). [2.
134 ———, J. f. Math., 53: 133-141 (1857). [2.
135 Sturm, R., Geom. Verwandt., 4: 288 (1908). [20.
136 Togliatti, E., Rom. Acc. L. Rend., (5), 24_2: 307-311, 388-392 (1915). [25.
137 Tötössy Béla, Math. Ann., 19: 291-322 (1882). [4.
138 Tummarello, A., Giorn. di Mat., 58: 60-66 (1920). [14.
139 Veronese, G., Ven. Ist. Atti, (6), 2: 1841-1842 (1883-84). [19.
140 ———, Rom. Acc. L. Mem., (3), 19: 344-371 (1884). [19.
141 Visalli, P., Rom. Acc. L. Rend., (4), 22: 80-87 (1886). [20.
142 de Vries, J., Amst. K. Ak. Wet. Versl., 25: 1414-1418 (1916-17). [20.
143 Weierstrass, K., Berlin Monatsber., 38: 324-328, 333 (1863). [5.
144 Wiman, A., Math. Ann., 48: 195-240 (1896-97). [26.
145 Zangli, L., Wien. Ber., 126_{11}: 857-885 (1917). [2.

CHAPTER XIII

THE MAPPING OF A RATIONAL CONGRUENCE ON A PLANE

1. *Introduction.*—A set of ∞^2 lines in S_3 such that through a general point pass m lines and in a general plane lie n lines of the set is a congruence (m, n) of order m and class n. A point through which pass ∞^1 lines lying on a cone of order i is an i-fold singular point, and a plane in which lie ∞^1 lines tangent to a curve is a singular plane. If the lines of a congruence can be put into (1, 1) correspondence with the points of a rational surface, the congruence is rational and can be mapped on a plane. A $(1, n)$ congruence is obviously mappable on a plane by the intersections of its lines with the plane. An (m, n) congruence which has a singular plane such that the lines in the plane are tangent to a curve of class $n-1$ can be mapped on the plane. P. Visalli(28). An (m, n) congruence which has an $m-1$-fold singular point can be mapped by the planes through the singular point. E. Bertini(1). G. Loria(15). The Cremona congruences of T. A. Hirst(8), (9), and (10) formed by the lines joining corresponding points on 2 planes whose points are related by a Cremona transformation and the congruences of E. Caporali(3) where one plane is replaced by a rational surface whose points are in (1, 1) correspondence with the points of the other plane, are both obviously rational. The case of a quadratic transformation is considered by W. Stahl(25), and the lines joining corresponding points of 2 monoids, by A. Del Re(7). I. Conti(4) discusses the lines joining corresponding points on 2 planes related by a (1, 2) correspondence. L. C. Kiefer(11) considers the lines joining corresponding points on a plane and a quadric. The $(2, n)$ congruences with only a finite number of singular points were classified by E. E. Kummer(12). The lines of the congruence are bitangent to a quartic surface, called the focal surface.

The 2 rays through a point of the focal surface coincide. The value of n is between 2 and 7. There are 2 types of (2, 6) congruence, one with a five-fold singular point and the other with four four-fold points.

The $(2, n)$ congruences with no focal surface but with singular curves were classified by R. Sturm(26) and R. Schumacher(22).

2. *Congruences with a Finite Number of Singular Points.*—The first $(2, n)$ congruence to be proved rational was the (2, 2) congruence which is the complete intersection of a linear and a quadratic complex, L. Cremona(5). The linear complex was mapped on S_3 by M. Noether(19) and

S. Lie(13). A second linear complex meets the first in a linear congruence whose image in S_3 is a quadric through a fixed conic. A quadratic complex meets the first linear complex in a (2, 2) congruence whose image is a quartic surface with the fixed conic double, and is therefore rational. For another proof, see E. Caporali(2).

Congruences of type $(8-n, 2)$, where $1 \leqq n \leqq 6$ were studied by T. Reye(21) by aid of the (1, 8) correspondence between 2 spaces S, S' in which the planes of S correspond to ∞^3 quadrics in S' through n fixed points. The vertices of the cones of the system of quadrics lie on a quartic surface K_4, and the images of the cones in S are planes tangent to a corresponding surface F, of class 4 and order $16-2n$.

The lines joining associated points in S' correspond to bitangents of F through the corresponding point. The bitangents form a (28, 12) congruence having the focal surface F. If the quadrics have n basis points then to the n bundles of lines through the basis points correspond n $(8-n, 2)$ congruences which are therefore also mappable on a plane. The $(2, 6)_I$ congruence having 4 four-fold points belongs to a tetrahedral complex with the four points for vertices. When this complex is mapped on S_3 (G. Loria(14)), the congruence is represented by an F_6 with $4P_4$, the lines joining them being double lines. This surface is rational and its plane sections can be mapped on a plane as C_5 with $2P_2$ and $11P_1$. It should be noted that all $(2, n)$ congruences with a quartic focal surface can be mapped on a double plane, the curve of branch points being the intersection of the focal surface with the plane. The double plane is therefore rational and also the congruence. The mapping of these congruences is treated by R. Sturm(27).

3. *Congruences with a Curve of Singular Points.*—D. Montesano(17) considers the (4, 12) congruence consisting of the generators of the cones of a net of quadrics. The vertices of the cones lie on a C_6, $p=3$. If 6 of the 8 points common to the quadrics are chosen so as to coincide in pairs at A_1, A_2, A_3 and a 7th point A_4 is taken on the surface of coincidences of the web of quadrics through the 6 points then the 8th point common to the net of quadrics coincides with A_4 and the sextic is composed of 2 space cubics through $A_1A_2A_3A_4$.

The (4, 12) congruence is composed of two (2, 6) rational congruences each comprising the generators of the cones of the net of quadrics with their vertices on a cubic. The focal surface of each congruence is a quartic surface on which the cubic is double. Each line of the congruence and the cubic determine a quadric and the congruence can be mapped by the quadrics and the system of cones.

If the quadrics of the net are all tangent to a plane π at a point O and pass through P_1, P_2, P_3, P_4 the congruence is composed of the lines through O and a (2, 6) rational congruence. The vertices of the cones of the net lie on a cubic curve of π having O^2 and meeting the 6 lines P_iP_k. The focal surface consists of π and an F_3 having P_i^2 which touches π at O and contains the cubic curve. The cones touch F_3 in a pencil of cubic curves. Through a point on a cubic curve passes one line of the congruence. Hence the congruence can be mapped on a plane.

D. Montesano([16]). The (2, 4) congruence of lines which meet 2 conics which meet in 2 points T, T' is rational because the planes through T or T' and the lines of the congruences are in (1, 1) correspondence.

D. Montesano([18]) shows that the (2, 6) congruence formed by the quartic curve of intersection of two quadrics tangent to one another at a point O can be mapped on a plane because the planes through O and the bisecants are in (1, 1) correspondence.

The secants of a line k and a C_n meeting it in $n-2$ points form a $(2, n)$ congruence which can be mapped on a cone having the genus of C_n. Hence if C_n is rational the congruence is rational. The tangents to a F_n with an $n-2$-fold line k which meet k but are not tangent at a point of k form a $(2, 2(n-1)^2)$ congruence which is rational since the lines of the congruence and the points of F_n which is rational are in (1, 1) correspondence. If between the points of a line k and the planes through k there is a $(2, n)$ correspondence then a $(2, n)$ congruence can be set up by joining any point P to the 2 points Q, R which correspond to the plane kP. The pencils of rays through a point of k can be represented by the points of a C_{n+2} which has L^n and M^2, if the points of k and the planes through k are represented by lines through M and L respectively. The congruence is rational if C_{n+2} has $n-1$ other double points, that is, if the congruence has $n-1$ pencils of double rays.

Given on a quadric cone K vertex O a rational C_n with O^{n-1}. The lines through a point P of C_n lying in the tangent plane to K at P form a $(2, n)$ congruence contained in the quadratic complex of the tangent lines to k. The congruence can be mapped on any tangent plane of the quadric cone by the intersection of its lines with the plane.

4. *Congruences Defined by Quadrics and Projectivities.*—V. Snyder([23]) considers the congruences belonging to the generators of a net of quadrics. A line l of one system of generators meets the generator of the other system through a basis point B of the net in a point P which is in (1, 1) correspondence with l. Hence if BP describes a rational cone, l will describe a rational congruence. V. Snyder([24]) and H. B. Owens([20]) consider congruences of generators of

$$a\lambda^2+b\lambda+c=0, \qquad a\lambda^3+b\lambda^2+c\lambda+d=0,$$

where $a=0$, $b=0$, $c=0$, $d=0$ are quadrics and the two systems of generators are rationally separable.

A. Del Re(6) discusses the lines obtained by the intersection of a plane p through a point S and the plane determined by a point P on a plane s and a line l' on a plane s' when P, p, and l' are projectively related. He also considers the lines joining the points in which 2 lines l' and l'' through 2 points $s's''$ meet a plane p through a point s, when s', s'' and p are projectively related.

BIBLIOGRAPHY

[1] Bertini, E., Rom. Acc. L. Rend., (3), 4: 30-31 (1880). [1.
[2] Caporali, E., Rom. Acc. L. Rend., (3), 2: 749-769 (1878). [2.
[3] ———, Nap. R. A. Rend., 245-249 (1879). [1.
[4] Conti, I., Pal. Circ. Mat., 1: 230-240 (1884-87); 2: 96-101 (1888). [1.
[5] Cremona, L., Rom. Acc. L. Rend., (2), 3: 285-302 (1876). [2.
[6] Del Re, A., Pal. Circ. Mat., 1: 272-283 (1884-87). [4.
[7] ———, Rom. Acc. L. Rend., (5), 2: 452-457 (1893). [1.
[8] Hirst, T. A., London M. S. Proc., 14: 259-301 (1882). [1.
[9] ———, London M. S. Proc., 16: 232-257 (1885). [1.
[10] ———, London M. S. Proc., 17: 287-296 (1886). [1.
[11] Kiefer, L. C., Diss. Strassburg, (1905). [1.
[12] Kummer, E. E., Berlin Ab., 1-20 (1867). [1.
[13] Lie, S., Math. Ann., 5: 145-256 (1872). [2.
[14] Loria, G., Torino Atti., 19: 849-877 (1884). [2.
[15] ———, Torino Atti., 2: 621-632 (1885). [1.
[16] Montesano, D., Torino Atti., 27: 1003-1009 (1891-92). [3.
[17] ———, Rom. Acc. L. Rend., (5), 7: 77-85 (1892). [3.
[18] ———, Pal. Circ. Mat., 7: 159-177 (1893). [3.
[19] Noether, M., Gött. Nachr., 298-306 (1869). [2.
[20] Owens, H. B., Am. J., 35: 323-356 (1913). [4.
[21] Reye, T., J. f. Math., 86: 84-107 (1879). [2.
[22] Schumacher, R., Math. Ann., 38: 298-306 (1891). [1.
[23] Snyder, V., Am. M. S. Trans., 11: 371-387 (1910). [4.
[24] ———, Am. J., 32: 177-185 (1910). [4.
[25] Stahl, W., J. f. Math., 97: 146-164 (1884). [1.
[26] Sturm, R., Math. Ann., 36: 469-472 (1890). [1.
[27] ———, Linien geometric, 2: 98-116. [2.
[28] Visalli, P., Rom. Acc. L. Rend., (5), 4: 33-38; 58-62 (1895). [1.

CHAPTER XIV

INVOLUTIONS ON IRRATIONAL SURFACES

1. *Definitions.*—To the groups of n points of an involution of order n on an algebraic surface F correspond the points of a second algebraic surface F^* which is called the *image* of the given involution I_n and is thus in $(1, n)$ correspondence with F.

Of particular interest are the points P^* of F^* for which two (or more) points P of the n points on F corresponding to P^* coincide. Such a point P^* is called a *branch-point* and P is a *coincidence*. For the purposes of this report, it will be convenient to divide involutions into two classes: (a) those which have infinitely many coincidences forming an algebraic curve, and (b) those which have a finite number (or zero) of coincidences. It is no essential restriction to suppose that no point of either the coincidence locus on F or of the branch locus on F^* is a fundamental point (corresponding to a curve on the other surface). Under these circumstances, the above classification holds also with respect to the branch locus.

We exclude from the following discussion surfaces which have a continuous group of transformations into themselves. These surfaces have been discussed by Castelnuovo-Enriques(**5**) p. 744, p. 759 (elliptic surfaces), Picard-Simart(**63**) p. 517, and by Severi(**69**) Vol. 2, p. 778. Hyperelliptic surfaces are also treated at length in Chap. 17, § 2 of this report.

I. Involutions with a Curve of Coincidences

2. *Relations Between the Invariants.*—F. Enriques has shown (**18**) that, if F and F^* are free from exceptional curves and if the geometric genus of F^* is positive, then, to the canonical curves on F^* correspond on F curves which, added to the curve of coincidences, form canonical curves on F. It is further shown by Enriques, loc. cit., and by F. Severi(**70**) that, if we denote by p_a and $p^{(1)}$ the arithmetic and linear genera of F, and by π_a and $\pi^{(1)}$ those of F^*, then we have

$$p^{(1)} = n(\pi^{(1)} - 1) + \tau + \tfrac{3}{2}\delta$$
$$24(p_a + 1) = 24n(\pi_a + 1) + 6\tau + 3\delta - 2\sigma - 6$$

wherein τ is the genus of the branch-curve K^* on F^*, δ is the number of intersections of K^* with the canonical curves of F^*, and σ is the number of points of F^*, three of whose corresponding points on F coincide.

In this formula, it is supposed that K^* has no multiple component and that there are no fundamental points on F^* which correspond to curves on F. A discussion of the case in which exceptional curves exist, and extensions of the formulas to the case of (m, n) correspondence between two surfaces will be found in Severi, loc. cit. That an involution on a regular surface is regular was shown by G. Castelnuovo(4).

3. *The Double Plane.*—Most of the involutions with a curve of coincidences that have been considered in the literature are rational and cyclic, i. e., are representable as multiple planes, for the most part, indeed, as double planes. The double plane

$$z^2 = F(x, y)$$

is a rather simple, yet fairly general, type of surface. Its properties are easily amenable to treatment and it furnishes excellent examples in various parts of the theory of algebraic surfaces. On the geometric side, it has been quite fully considered by Castelnuovo-Enriques(5). We aim here only to supplement their remarks by some comments on its Picard integrals and the base invariants ρ and σ.

Picard, in his theory of integrals, has turned repeatedly for illustrations to the double, or multiple plane. See, e. g., (59), Chaps. 2 and 3; (63), (57), (61). Most of these cases are also discussed in (63) and to this we shall give more detailed references.

In (63), Vol. 1, Chap. 6, Picard deals with integrals of total differentials of the second and third kinds. As in most of his work, the discussion is based chiefly on algebra and analysis, with relatively little appeal to geometry. In particular, he shows that the integrals in question are reducible, respectively, to the types

$$\int \frac{P dx + Q dy}{f(y) z} \quad \text{and} \quad \int \frac{P dx + Q dy}{f(y) A_1 A_2 \ldots A_n z}$$

where f is a polynomial in y alone and $A_1, A_2, \ldots, A_n$ are polynomials in x, y. The polynomials A are irreducible, prime to F, and cut the surface in two distinct curves whose equations are

$$(1) \qquad A_i = 0, \qquad z = \pm \frac{M_i}{N_i}, \qquad i = 1, 2, \ldots, n$$

where M_i, N_i are polynomials in x, y. Algebraically, this means that A_i is a factor of $M_i^2 - N_i^2 F$. To form all integrals of the third kind it is thus necessary to discover the irreducible factors of polynomials $M^2 - N^2 F$. By reasoning which holds for the general algebraic surface, it follows that there is a maximum number $\rho - 1$ of polynomials A such that there is no corresponding integral of total differentials with *one* of the curves that they cut for logarithmic curves. The number ρ is precisely the Picard-Severi invariant (Chap. 15, No. 20). In the light of Severi's later developments, Picard constructs a base composed of a plane section $y=\text{const.}$, and of $\rho-1$ curves (1). These $\rho-1$ curves correspond to as many distinct polynomials A, for the two curves defined by the two signs in (1) are dependent, their sum being a multiple of the plane sections.

4. *Some Special Double Planes.*—Picard examines some interesting special cases, notably, ([58]), ([60]), also ([63]), Vol. 2, p. 254, the surface

$$z^2=f(x)F(y), \tag{2}$$

where f, F are polynomials of degree 3, $2p+1$ without multiple roots. In general, in this case, the polynomials M^2-n^2fF are irreducible and the polynomials A in the integrals of the third kind can be removed by the subtraction of the logarithm of a rational function. In certain special cases, there are obtained hyperelliptic surfaces of rank two (birational transforms of an elliptic Kummer surface). In this connection see also Chap. 17, No. 22.

Picard also shows, loc. cit., and ([60]), that if, in (2) both f and F are arbitrary polynomials, then $\rho=1$. If we consider the associated curves C, C':

$$u^2=f(v), \qquad u^2=F(v), \tag{3}$$

then $\rho>1$ for (2) demands that the period matrices of their integrals of the first kind (Chap. 15, No. 4) have a common submatrix.

In ([63]), Vol. 2, Chap. 10, Picard determines his invariant ρ_0 (number of double integrals of the second kind) for some of the preceding surfaces.

An interesting connection, compare L. Godeaux([42]), exists between the surfaces (2) and the curves (3). We may establish between them a correspondence by assigning to the pair of points (u, v) of C, and (u', v') of C', the unique corresponding point

$$x=v, \qquad y=v', \qquad z=uu'$$

of the surface and to the point (x, y, z) of the surface the two pairs of points

$$u=\epsilon\sqrt{f(x)}, \quad v=x; \qquad u'=\epsilon'\sqrt{F(y)}, \quad v'=y$$

of the two curves, where ϵ, ϵ' are ± 1, the signs being chosen so that $z=uu'$. The surface (2) is thus the image of an I_2 on the surface ϕ which represents the pairs of points of the curves C, C'. It is quite likely that, by means of the known results on the number ρ for surfaces which represent the pairs of points of two algebraic curves (Chap. 16, No. 6), it would present no great difficulty to determine ρ for all surfaces (2) and also for similar algebraic varieties.

In ([56]), p. 359, Lefschetz has applied his general results on the base to the double plane (1). He has determined the minimum base and the base invariant when one of the components of $F=0$ varies within a sufficiently general linear system of plane curves. This includes most of the double planes considered in the literature.

5. *Cyclic Multiple Plane.*—In addition to the double plane, Picard has repeatedly considered the m-fold multiple plane

$$z^m = x^m + P(y), \tag{4}$$

where P is a polynomial of degree m. In ([63]), Vol. 2, p. 285, he found that the surface has a Severi base composed of $(m-1)^2+1$ straight lines so chosen that only one set of m is coplanar. In Vol. 1, p. 109, he makes use of a special case of equation (4) to prove that non-singular algebraic surfaces possess two-dimensional cycles that are not ~ 0 (Chap. 15, No. 9).

In ([31]), L. Godeaux gives an example of a p-fold plane F (p prime) carrier of an I_p without coincidence points and such that its image F^* has its Severi number $\sigma = p$ (Chap. 15, No. 20). The number σ that he considers is computed on the assumption that the singularities of F have no consecutive two-cycles. Godeaux's example, suitably modified, is likely to lead to an F^* whose σ in the strict sense is p. Incidently, in his transformation T, $z'=z$ must be replaced by $z'=\epsilon^k z$, where $\epsilon^k \neq 1$, for, as it reads, it has coincidence points $(0, 0, z_0)$ where $z_0{}^p$ is the value of ϕ for $x=y=0$.

De Franchis shows, in ([10]), that a double plane whose irregularity $q \geqq 2$ carries an irrational pencil of curves. Comessatti has shown that this is true for every triple plane([7]).

M. Bottasso discusses, in ([2]), the cyclic multiple plane with an irreducible branch-curve and determines certain numerical invariants for a fairly general type of such plane. In ([3]), he treats the case $n=3$ in greater detail.

6. *Other Special Surfaces.*—Given an irreducible curve C, there may be associated with its point pairs two surfaces; the first, F, represents its ordered point pairs, the second, F^*, represents its point pairs with order disregarded. Then F^* is image of an I_2 on F and, from that point of view, has been investigated by M. de Franchis([8]) and by F. Severi([67]), ([68]) p. 35. The curves on F^* are images of the symmetrical correspondences on C (Chap. 16, No. 7) and the degree and genus of F^* curves are to be considered as invariants of the C correspondences. Severi investigated these curves, finds their relation to the correspondences, determines a minimum base on F^*, and shows that ρ is equal to the number of independent symmetrical correspondences on C. Referring to

Rosati's theorems (Chap. 15, No. 7) on these correspondences, this means that $\rho=k+1$, where k in the singularity index of the period matrix of the integrals of the first kind attached to C (Chap. 15, No. 7). Severi gives formulas for the degree and genus of a curve in terms of the numbers which relate it to the minimum base.

Remy, in (64), (65), investigates a surface Φ, image of an I_2 on the surface F^* which represents the non-ordered point pairs of a non-singular plane quartic curve C_4. The groups of the I_2 are such that the four corresponding points on C_4 are collinear. The surface is representable by means of θ-functions in three variables, subject to the vanishing of a certain θ-function. The method is transcendental throughout. Remy examines the curves on Φ ,its various integrals, and determines the related invariant numbers. See also Godeaux(50).

In (1), R. Baldus maps a double ruled surface R^* on a simple one R. If R is not rational, to a generator g^* of R^* corresponds either a generator g of R, taken twice, or two generators g_1 and g_2 of R. In the second case, the locus of invariant points is composed of generators; in the first case, there is a simple or composite curve of coincident points which meets each generator in two points.

O. Chisini(6) discusses an elliptic surface $p_g=0$, $p_a=-1$ which has a unimodular linear pencil $|K|$ of elliptic curves and an elliptic pencil of irrational curves C, such that $[K, C]=n$, determining an I_n on the surface. Particular attention is given to the case where n is composite. The I_n may or may not be cyclical. The surface can be mapped on an n-fold cylinder $\phi(x, y)=0$, with curve of branch-points consisting of one or more plane sections $z=\text{const.}$

F. Severi has considered(66) the I_2 on a quartic surface F_4 which contains a sextic curve C_6 of genus 2. He shows in particular, that the only involutorial birational transformations of such a surface into itself are those generated by the infinite nets of genus two on the surface. He shows further that a quartic surface which is transformed into itself by an involutorial homography depends on, at most, 11 moduli as does also, under the same conditions, a double plane of genus one with a branch sextic.

II. Involutions Having a Finite Number of Coincidences

7. *General Theorems.*—The involutions for which the number of coincidences is finite (or zero) have been studied in particular by L. Godeaux. This author has outlined, in the important paper(51), the chief features of the theory of such involutions.

Let I_n be an involution of order n on an algebraic surface F and let F^* be the image of I_n. We may suppose that F and F^* each lie in an S_3 and that their equations are, respectively,

$$f(x, y, z)=0 \qquad \text{and} \qquad f^*(X, Y, Z)=0.$$

Then there exists a rational transformation

$$X=\psi_1(x,y,z), \qquad Y=\psi_2(x,y,z), \qquad Z=\psi_3(x,y,z),$$

which defines I_n. Moreover, F. Enriques has shown in (12) for surfaces of genera one and later Godeaux in (38), (39) proved without restriction the following fundamental theorem: If an I_n on any algebraic surface has only a finite number (or zero) of coincidences, it is generated by a group of birational transformations of the surface into itself. We shall say that the involution is cyclic, Abelian, etc., according as it is generated by a cyclic, Abelian, etc., group of transformations.

A (simple) point of coincidence P is said to be *perfect* if the group which generates I_n leaves invariant the tangent lines to F at P; otherwise, P in *non-perfect*. The branch-points of F^* are said to be perfect or non-perfect according as the corresponding coincidences on F are, or are not, perfect.

On a normal surface F^*, image of an I_n with n a prime number, we have, from L. Godeaux(38): A perfect branch-point is a conical n-fold point the tangent cone at which is rational; a non-perfect branch-point is a biplanar double point formed by $(n-2)/2$ consecutive double points such that the last is an ordinary biplanar point. On the surface F we have: the perfect coincidences of I_n are $(n-1)$-fold points on the canonical curves of F in composition with I_n.

If we denote the arithmetic and linear genera of F by p_a, $p^{(1)}$, and those of F^* by π_a, $\pi^{(1)}$; if, further, α is the number of perfect, and β the number of non-perfect coincidences of I_n (n a prime), then we have, [L. Godeaux(38)]

$$12p_a=12n\pi_a+(n-1)(n-5)\alpha-(n^2-1)\beta+12(n-1),$$
$$p^{(1)}=n(\pi^{(1)}-1)+(n-2)^2\alpha+1.$$

If, in addition, F^* has an infinite canonical system, not in composition with a pencil, its plurigenera π_i $(i=2, 3, \ldots)$ are connected with the plurigenera P_i of F by the relations:

$$12P_i=12n\pi_i+[(n-1)(n-5)+6i(i-1)(n-2)^2]\alpha-(n^2-1)\beta$$

n=a prime number, $i=2, 3, \ldots.$

8. *Involutions of Given Order.*—In (42) L. Godeaux has determined the necessary and sufficient conditions that a simple, normal surface F^*,

of arithmetic genus $\pi_a \geqq -1$, free from exceptional curves, of order n, with hyperplane sections of genus π, such that $n-\pi > \pi_a+1$, and lying in a space of $n-\pi+\pi_a+1$ dimensions, shall represent an I_2 having a finite number of coincidences. F^* must have a conical double point at each branch-point. The number of these branch points is a multiple of four. Among the hypersurfaces which intersect F^*, in addition to its multiple curve, in the complete linear system of curves expressed as double the hyperplane sections (i. e., such that some curves of the system are constituted of two hyperplane sections) some pass through the branch-points and touch F^* at each point of their common curve. Let the equations of F^*, in non-homogeneous coordinates be:

$$\phi_1(x_1, x_2, \ldots, x_r)=0, \qquad \phi_2(x_1, x_2, \ldots, x_r)=0, \qquad \ldots, \qquad \phi_{r-2}(x_1, x_2, \ldots, x_r)=0$$

and let the hypersurface $f(x_1, x_2, \ldots, x_r)=0$ touch F^* as stated in the theorem. Then the equations of F are:

$$\phi_1=0, \qquad \phi_2=0, \qquad \ldots, \qquad \phi_{r-2}=0, \qquad x_{r+1}=\sqrt{f}.$$

As a particular consequence, we have the result that all the coincidences of an I_2 are perfect.

The I_2 on irregular surfaces F whose image surfaces F^* are regular were studied by L. Godeaux in ([26]) and ([27]). F is the rational transform of a surface S of the same irregularity which belongs to the Picard variety of F and which is transformed into itself by a transformation of second species of that variety. To the points of S invariant for this transformation correspond on F the coincidences of I_2. The number of these coincidences cannot be zero.

The same author has also shown, ([20]), see, also, ([32]), that the algebraic surfaces of genus $p_g>1$ on which the canonical system is not formed of curves of a pencil but such that the bicanonical system is in composition with an I_2 having a finite number of coincidences, have the genera $p^{(1)}=7$, $p_a=0$, $p_g=3$, $P_2=7$, $P_3=19$, etc. As a projective model of these surfaces, one may take a bicanonical double surface of order 12, in S_6, with hyperplane sections of genus 10, having 28 conical double points as branch-points.

The necessary and sufficient conditions that a simple, normal, surface F^* shall be the image of an I_3 having a finite number of coincidences and lying on an algebraic surface were found by L. Godeaux([21]). F^* must have a certain number α of conical triple points at which the tangent cone is rational and a certain number β of ordinary biplanar double points. Among the hypersurfaces determining on F^* the system of curves three-fold to the hyperplane sections, at least one must have three-point contact with F^* along a curve which passes doubly through α_1 of the triple points, simply through the remaining $\alpha-\alpha_1$ triple points, and through the β double points. The number α_1 vanishes simultaneously with β.

It has been pointed out by L. Godeaux([49]) that surfaces exist on which lie involutions having a finite number of coincidences of prime order greater than any preassigned positive number. That involutions of arbitrarily given order, with no coincidences, lie on hyperelliptic surfaces was pointed out by Enriques-Severi([19]).

9. *Involutions of Genera One on Surfaces of Genera One.*—That any involution of genera one ($p_a=P_2=1$) which lies on a surface of genera one, free from exceptional curves, can have only a finite number of coincidences was noticed by F. Enriques([12]). L. Godeaux has classified these involutions ([25]), ([30]), ([34]), ([35]), and ([53]). In making this classification, he first divides these involutions into two species. Those of the first species are cyclic or are generated by cyclic transformations having common coincidences. All the others are of second species. Further the image surface F^* of the involution is simple, normal, of genera one, of order $2\pi-2$ (π arbitrary but sufficiently large). It lies in an S_π, and its hyperplane sections are of genus π. The conditions given are necessary and sufficient, except in (g), that such a surface F^* shall represent an involution of the given type.

Of the first species, he finds the following types:

(a) An I_2 having 8 coincidences. F^* has 8 conical double points. Among the hyperquadrics through these 8 nodes, some touch F^* at each point of their common curve.

(b) Cyclic I_3 having 6 three-point coincidences. F^* has 6 ordinary biplanar points. Among the hypercubics through the double points, some have three-point contact with F^* at each point of their common curve.

(c) Cyclic I_4 with 4 four-point and 4 two-point coincidences. F^* has 2 conical and 4 biplanar, each with 1 consecutive, double points. Among the hyperquartics through the double points, some have four-point contact with F^* at each point of their common curve. This type is discussed in detail in ([25]).

(d) Cyclic I_6 with 2 six-point, 4 three-point, and 6 two-point, coincidences. F^* has 6 double points; 2 conical, 2 ordinary biplanar, and 2 biplanar, each with 2 consecutive. Among the hypersurfaces of order 6 through the double points, some have six-point contact with F^* along their common curve.

(e), (f) I_8 generated by two cyclic birational transformations of order 4 such that $T_1^2=T_2^2=(T_1^2T_2^2)$. In (e) there are 4 eight-point and 4 two-point coincidences and F^* has 1 conical, and 4 ordinary uniplanar, double points. In (f), there are 2 eight-point, and 6 four-point coincidences, and F^* has 2 ordinary uniplanar, and 3 biplanar, each with 1 consecutive, double points. In both cases, 3 families of hyperquadrics touch F^* along curves of genus $\pi-2$ and pass through the uniplanar points in such a way that each of the 3 consecutive nodes at the point lies on the hyperquadrics of a single family. In (f), the hyperquadrics of each family touch, at two biplanar points, the line common to the 2 tangent planes. Further, in each case, $\infty^{2\pi-3}$ sets of three hyperquartics are circumscribed to F^*, have third order contact with it at $2\pi-3$ simple points, and pass through the 5 double points. Each node consecutive to a uniplanar point lies on a single one of them. In (*f*), each hyperquartic touches, at 2 biplanar points, the line common to the 2 tangent planes. These types are discussed in detail in ([30]).

(g) An I_{12} generated by two transformations T_1 and T_2, of periods 4 and 3 respectively, such that $T_1^{-1}T_2T_1=T_2^2$. It has 2 twelve-point, 6 four-point, and 4 three-point, coincidences. The double points of F^* are: 1 ordinary biplanar, 2 biplanar each with 1 consecutive, 2 uniplanar such that 2 of the 3 singular tangents are consecutive.

Of involutions of the second species, he finds the following types:

(h), (i), (j) Involutions of order $2^k(k=2, 3, 4)$ having $8(2^k-1)$ two-point coincidences forming $2^{4-k}(2^k-1)$ groups of the involution. F^* has $2^{4-k}(2^k-1)$ conical double points which may be arranged in k sets of 8 which have, two by two, 4 common points. Among the hyperquadrics which pass through the double points of each set, some touch F^* along their common curve.

(k) Abelian I_9 having 24 three-point coincidences forming 8 groups of I_9. F^* has 8 ordinary biplanar double points which may be arranged in 2 groups of 6 having 4 common points. Among the hypercubics through the points of each group, some have three-point contact with F^* along their common curve.

The same author shows in ([43]) Art. 12 (correcting a statement made by him in ([53])) that there does not exist on these surfaces an I_8 of genera one and species two generated by a transformation of period 4 and one of period 2.

The conditions under which a double plane of genera one is image of an I_3 on a surface of genera one were determined by L. Godeaux([44]) He finds that the branch-curve in the double plane can be reduced to one of the following types:

(*a*) C^6 with 6 cusps which lie on a C^2.

(β) C^8 with two fourfold points, A_1 and A_2, and 6 cusps. Infinitely many C^4 have nodes at A_1, A_2, pass through the cusps, and touch C^8 in 2 variable points.

(γ) C^{10} with a seven-fold point A to which are consecutive 2 triple points, A_1, A_2 and 6 cusps. Infinitely many C^6 have A as a four-fold point, A_1, A_2, as double points, pass through the cusps, and touch C^{10} at four variable points.

(δ) C^{12} having a nine-fold point A, to which are consecutive 3 triple points, A_1, A_2, A_3, and 6 cusps. Infinitely many C^8 have A as a six-fold point, A_1, A_2, A_3, as double points, pass through the cusps, and touch C^{12} in 6 variable points.

A similar list of conditions that a double plane of genera one shall be image of a cyclic I_4 on a surface of genera one were given by the same author in (25).

That any quartic surface, subject only to the conditions of containing 6 ordinary biplanar points, is image of an I_3 lying on a surface of genera one was shown by L. Godeaux(24). He further finds, in (25), conditions under which a quartic surface represents an I_4 lying on a surface of genera one.

L. Godeaux has noticed, in (23) and (52), that, if a surface of genera one is image of an I_2 on a surface of the same genera and if it contains a linear system of curves of odd genus greater than 3, the generic curve not passing through the branch points, then that surface is image of other I_2 on other surfaces of genera one. Further, if a surface of genera one is image of an I_n on a surface of the same genera, and if it contains a linear system of curves of genus $\pi \geq 2$, the generic curve not passing through the branch-points, then that surface contains $n-1$ other linear systems of curves of genus $\pi-2$.

That a quartic surface, subject to the single condition of containing a gauche C^6 of genus 3, contains infinitely many I_2 of genera one was stated by L. Godeaux(45), but these I_2 were shown not to be effective by V. Snyder and F. R. Sharpe in (72), see p. 290. L. Godeaux has determined, in (47), see, also, F. Enriques(17), conditions under which a double plane of genera one shall contain one or more I_2 of genera one.

10. *Involutions of Genera $p_a=P_3=0$, $P_2=1$ on Surface of Genera One.*—F. Enriques has shown in (11) that every surface of genera $p_a=P_3=0$, $P_2=1$, is image of an I_2 on a surface F of genera one and that this I_2 has no coincidences. L. Godeaux has added, (40), and (37), see Chap. 3, that this surface F contains two other I_2. One of these is of genera one; the other is rational and makes it possible to represent F on a double quadric with a branch C^8. The transformations generating

these 3 I_2 are permutable in pairs. He also points out, ([43]), see p. 70, that any surface of genera one which contains an involution of genera zero, bigenus one, also contains one of the same order but of genera one.

In the last named paper, he classifies the involutions of genera $p_a=P_3=0$, $P_2=1$, lying on a surface of genera one, as follows:

(a) I_2 having no coincidences. Every surface of genera $p_a=P_3=0$, $P_2=1$, is image of such an I_2.

(b), (c) I_4 having 8 two-point coincidences. (b) is cyclic; (c) is Abelian and is generated by two transformations of period two.

(d) Abelian I_6 having 6 three-point coincidences.

(e)–(i) I_8 as follows: (e) cyclic, (f) composed of 2 I_4 of which one is of genera one, the other of type (b), (g) composed of an I_4 of genera one and an I_2 of type (a), (h) composed of an I_2 of genera one and an I_4 of type (b), (i) composed of $3I_2$ of which two are of genera one and the third is of type (a). In (e), (f), (g), there are 4 four-point and 4 two-point coincidences. In (h) and (i), there are 24 two-point coincidences.

(j) Abelian I_{12} composed of a cyclic I_6 of genera one and an I_2 of type (a). It has 2 six-point, 4 three-point and 6 two-point coincidences.

(k), (l) I_{16} having 56 two-point coincidences. (k) is Abelian and is composed of $4I_2$ three of which are of genera one and the fourth is of type (a). (l) is composed of $2I_2$ of genera one together with an I_4 of type (b).

The surfaces of genera one which contain a trirectangular group of birational transformations into themselves were studied by L. Godeaux in four notes([48]). He finds five cases and determines the genera of the component involutions in each case.

11. *Involutions of Genera* $p_a=P_3=0$, $P_2=1$, *on Surfaces of the Same Genera.*—The involutions of genera $p_3=P_3=0$, $P_2=1$, which have a finite number of coincidences and lie on a surface of the same genera were also classified by L. Godeaux([37]); see also, ([36]) and ([43]), Art. 11, p. 76. He determines the various types, considers the generating birational transformations, and finds the necessary and sufficient conditions that a simple, normal surface F^*, of genera $p_a=P_3=0$, $P_2=1$, shall represent an involution of the given type, as follows:

(a) Cyclic I_2 having 4 perfect coincidences. F^* is birationally equivalent to a sextic surface having the edges of a tetrahedron as double lines and 4 conical double points. There are ∞^2 quartic surfaces circumscribed to the tetrahedron which pass through the 4 double points and touch F^* along their common curve. It is further shown([42]) that F^* may be represented by a double plane of one of two specified types. These double planes are studied in greater detail in three notes([28]), ([29]). In the second of these notes it is pointed out that F^* is also the image of a second I_2 lying on a surface F' of the same genera as F but birationally distinct from it. The branch-points for the two I_2 coincide.

(b) Cyclic I_3 having 3 non-perfect three-point coincidences. F^* has 3 ordinary biplanar double points. Among the hypersurfaces which intersect F^* in the system three-fold to the hyperplane sections (i. e., in the complete linear system such that some curves of the system are constituted by three hyperplane sections) some pass through the 3 double points and have three-point contact with F^* along their common curve. At the 3 double points, this curve of contact is on one nappe of the surface.

(c) Cyclic I_4 having 2 perfect two-point and 2 non-perfect four-point coincidences. F^* has 1 conical double point and 2 biplanar, each followed by one consecutive, double points. Among the hypersurfaces which intersect F^* in the system of curves four-fold to the hyperplane sections, some pass through the 3 double points and have four-point contact with F^* along their common curve.

(d) Abelian I_4 generated by two transformations of period 2. It has 12 perfect two-point coincidences. F^* has 6 conical double points forming 3 couples. Among the hypersurfaces intersecting F^* in the system of curves double the hyperplane sections, some contain two couples of nodes and touch F^* along their common curve.

(e) Cyclic I_6 having 3 two-point, 2 three-point, and 1 six-point coincidences. The double points of F^* are: 1 ordinary biplanar with 2 consecutive, one ordinary biplanar, 1 conical. Among the hypersurfaces intersecting F^* in the system of curves six-fold to the hyperplane sections, some pass through the 3 double points and have six-point contact with F^* along their common curve.

(f) Abelian I_8 generated by 3 transformations of period 2. It has 28 perfect two-point coincidences. F^* has 7 conical double points. Among the hypersurfaces intersecting F^* in the system of curves double the hyperplane sections, some pass through 4 nodes and touch F^* along their common curve.

The same author shows ([43]), No. 11, that an I_9 of the above genera cannot exist on these surfaces.

12. *Involutions on Other Special Surfaces.*—The involutions on the surfaces which represent the non-ordered pairs of points of a C^4 of genus 3 were studied by L. Godeaux in 4 notes([50]). Such a surface always contains an I_2 with 28 perfect coincidences. The pairs of points of this I_2 correspond to 2 pairs of points of C^4 which are collinear. If the given surface is also transformed into itself by a cyclic transformation of period

greater than two, the C^4 is also transformed into itself by a transformation of the same period. The transformations of a C^4 into itself having been determined by S. Kantor([55]), and A. Wiman([73]), Godeaux is enabled to determine the involutions with a finite number of coincidences on such a surface, as follows:

(α) An I_2 on the surface corresponding to an arbitrary quartic.
(β) An I_3 and an I_6 on the surface corresponding to $x^3y+f_4(y)=0$.
(γ) An I_3 and an I_6 on the surface corresponding to

$$ax^4+bx^2y+cxy^3+dx+ey^2=0.$$

(δ) An I_7 and an I_{14} on the surface corresponding to $xy^3+y+x^3=0$.
(ϵ) An I_9 and an I_{18} on the surface corresponding to $x^4+xy^3+y=0$.

The I_3 of (β) has 6 perfect and 9 non-perfect coincidences; that of (γ) has 3 three-point coincidences. The I_6 of (γ) has 1 six-point, 2 three-point, and 27 two-point coincidences. Image surfaces of some of these involutions are also determined. These involutions are studied further by the same author in ([22]).

An I_3 which has 2 perfect and 5 non-perfect coincidences and lies on a surface of order 5 and genera $p_a=p_g=4$, $p^{(1)}=6$, in S_3, was studied by L. Godeaux([33]). The image surface F^* of this I_3 is of genera $p^{(1)}=2$, $p_a=p_g=2$, $P_i=i(i-1)/2+3$.

A rational involution with a finite number of coincidences which lies on a non-rational surface was found by L. Godeaux([22]). He projects a surface of Veronese from a point not in its S_5 and intersects the resulting $V_3{}^4$ by an hypercubic of the S_6 determined by it. Among the surfaces of section, consider those surfaces F which are invariant under a projective transformation of period 7 having 7 invariant points. Three of these invariant points are shown to lie on F. They constitute the (non-perfect) coincidences of the I_7 on F. The surface F may be transformed into a sextic surface in S_3 having a uniplanar triple point. Its genera are, $p^{(1)}=4$, $p_a=p_g=3$, $P_2=7$. For the genera of I_7 he finds $p_a=P_2=0$; hence the surface is rational. The surface of Humbert (which represents the pairs of points of a C^4 of genus 3 in such a way that to each point of the surface corresponds 2 pairs of points constituting a canonical group) is a special case of the above surface. If C^4 is a Klein quartic, the Humbert surface contains an I_7 with 3 coincidences. Hence, the I_7 having 6 coincidences, on the surface which represents simply the pairs of points of Klein's quartic, is rational.

In ([46]), L. Godeaux has shown that, if a regular surface F^* represents an involution of prime order p, free from points of coincidence and lying on an algebraic surface, then the divisor of F^* is a multiple of p. Illustrations of this theorem are given in the article cited and in ([31]). For further consideration of the number σ for involutions without coincidences, whether on algebraic surfaces or varieties, see S. Lefschetz([56]) and M. de Franchis([9]).

BIBLIOGRAPHY

1 Baldus, R., Math. Ann., 75: 290-318 (1914).
2 Botasso, M., Torino Atti., 44: 12-37 (1908-09).
3 ———, Torino Atti., 44: 169-195 (1908-09).
4 Castelnuovo, G., Rom. Soc. It. Sci. Mem., (3), 10: 82-102 (1896).
5 Castelnuovo, G.-Enriques, F., Encyklopädie III C 6 b.
6 Chisini, O., Rom. Acc. L. Rend., (5), 30: Sem. 2, 172-175, 251-255, 305-308 (1921).
7 Comessatti, A., Pal. Circ. Mat., 31: 369-386 (1911).
8 DeFranchis, M., Pal. Circ. Mat., 17: 104-121 (1903).
9 ———, Pal. Circ. Mat., 48: 384-388, 420-422 (1924).
10 ———, Rom. Acc. L. Rend., (5), 13_1: 688-695 (1904).
11 Enriques, F., Bologna Rend., (2), 12: 40-45 (1908).
12 ———, Bologna Rend., (2), 14: 71-75 (1910).
13 ———, Rom. Acc. L. Rend., (5), 2 Sem., 2: 3-8 (1893).
14 ———, Rom. Acc. L. Rend., (5), 2 Sem., 1: 139-144 (1897).
15 ———, Rom. Acc. L. Rend., (5), 6 Sem., 1: 169-174 (1897).
16 ———, Rom. Acc. L. Rend., (5), 7 Sem., 1: 234-240 (1898).
17 ———, Rom. Soc. It. Sci. Mem., (3), 10: 201-221 (1896).
18 ———, Torino Mem., (2), 44: 171-232 (1893).
19 Enriques, F.-Severi, F., Acta Math., 32: 283-392 (1909); 33: 321-403 (1910).
20 Godeaux, L., Brux. Bull., 855-864 (1919).
21 ———, Brux. Bull., (5), 7: 105-124 (1921).
22 ———, Brux. Bull., (5), 7: 653-665, 694-702 (1921).
23 ———, Brux. Bull., (5), 8: 189-196 (1922).
24 ———, Brux. Bull., (5), 8: 443-456 (1922).
25 ———, Brux. Bull., (5), 9: 75-88, 360-379, 459-483 (1923).
26 ———, Brux. Bull., (5), 10: 434-446 (1924).
27 ———, Brux. Bull., (5), 11: 37-47, 157-166 (1925).
28 ———, Brux. Bull., (5), 12: 527-534 (1926).
29 ———, Brux. Bull., (5), 12: 726-741, 892-904 (1926).
30 ———, Brux. Mém., 7: 3-33 (1924).
31 ———, Bull. Sci. M., (2), 39: 182-185 (1915).
32 ———, Bull. Sci. M., (2), 45: 14-20 (1921).
33 ———, Esp. Soc. M., 6: 7-14 (1917).
34 ———, Ann. Éc. Norm., (3), 31: 357-430 (1914).
35 ———, Ann. Éc. Norm., (3), 36: 51-70 (1919).
36 ———, Soc. M. F. Bull., 41: 178-194 (1913).
37 ———, Soc. M. F. Bull., 43: 89-117 (1915).
38 ———, Soc. M. F. Bull., 47: 1-16 (1919).
39 ———, Rom. Acc. L. Rend., (5), 23_1: 408-413 (1914).
40 ———, Rom. Acc. L. Rend., (5), 23_1: 682-686 (1914).
42 ———, Toulouse, Ann., (3), 5: 289-312 (1913).
43 ———, Ann. Ac. Polyt. Porto, 11: 65-78 (1916).
44 ———, Ann. Ac. Polyt. Porto, 14: 15-23 (1920).

[45] ———, Crac. Bull. Ser. A, 529-547 (1913).
[46] ———, Crac. Bull. Ser. A, 362-368 (1914).
[47] ———, Roum. Ac. Bull., 2: 162-166 (1914).
[48] ———, Roum. Ac. Bull., 2: 300-303 (1914); 3: 49-54 (1914); 4: 165-169 (1915); 4: 281-282 (1916).
[49] ———, Roum. Ac. Bull., 2: 265-267 (1914).
[50] ———, Roum. Ac. Bull., 4: 271-274, 283-286, 373-375, 376-378 (1916).
[51] ———, Rev. Mat. Hisp.-Amer., 4: 97-104 (1923).
[52] ———, R. de Ci. Madr. Rev., 19: 268-272 (1921).
[53] ———, Jassy Ann., 9: 247-254 (1915).
[54] ———, Arxius, 5: 89-107.
[55] Kantor, S., Acta Math., 19: 115 (1895).
[56] Lefschetz, S., Am. M. S. Trans., 22: 327-482 (1921).
[57] Picard, E., Arch. Math. Phys., (3), 1: 209-212 (1901).
[58] ———, Bull. Sci. M., (2), 25: 81-84 (1901).
[59] ———, J. de Math., (4), 5: 135-319 (1889).
[60] ———, C. R., 136: 913-918 (1903).
[61] ———, Ann. Éc. Norm., (3), 18: 397-420 (1901).
[62] ———, Ann. Éc. Norm., (3), 20: 350-377 (1903).
[63] Picard, E.-Simart, G., Théorie des Fonctions Algébriques, vols. 1 and 2.
[64] Remy, L., J. de Math., (6), 4: 1-37 (1908).
[65] ———, Ann. Éc. Norm., (3), 26: 193-258 (1909).
[66] Severi, F., Pal. Circ. Mat., 30: 265-288 (1910).
[67] ———, Torino Atti, 38: 119-134 (1903).
[68] ———, Torino Mem., 54: 1-49 (1903).
[69] ———, P. Rep.
[70] ———, Lomb. Ist. Rend., (2), 36: 495-511 (1903).
[71] Snyder, V.-Sharpe, F. R., Am. M. S. Trans., 16: 62-70 (1915).
[72] ———, Am. M. S. Trans., 19: 275-290 (1918).
[73] Wiman, A., Sven. Vet. Ak. Bih., 21: no. 3 (1895).

CHAPTER XV

TRANSCENDENTAL THEORY

This chapter contains a brief sketch of those parts of algebraic geometry wherein transcendental methods predominate. We pass as rapidly as possible over the classical topics in order to dwell more at length on recent investigations. It will be found that Riemann matrices and topology have been considerably emphasized. Their growing importance in this field and their numerous contacts with others assuredly justify this procedure.

The bibliographical data, admittedly not complete, will naturally be most scanty in the parts already well covered in the literature. At the end will be found a glossary of new notations and terms of most frequent occurrence in this and the next two chapters.

§ 1. Algebraic Curves

1. For the classical theory a long list of standard treatises will be found in Wirtinger(**52**). Two expositions quite in keeping with our point of view are Picard(**23**) vol. 2, Chaps. 13-18, Severi(**34**), Chaps. 7, 8, 9.

2. *Riemann Surfaces.*—We owe to Riemann the clear concept of an algebraic curve C, $f(xy)=0$, as a real continuum ∞^2, with a representation as an m sheeted Riemann surface K_y for $y(x)$ (m degree of f in y). Of greatest importance is the fact that if C' is a birational transform of C with K'_y for its Riemann surface, then K_y and K'_y are in continuous $(1, 1)$ correspondence (homeomorphism). Consequently any entity with homeomorphic invariance as regards K_y also possesses birational invariance as regards C.

Basic Property of K_y.—It is homeomorphic to a *normal* surface K, the two-sided disk with p holes or sphere with p handles. If n is the number of branch points of K_y we have $n=2(p+m-1)$ from which to calculate p. With C in arbitrary position, m, n are its order and class. p is the *genus* and its birational invariance follows from the fact that it is a topological invariant.

By means of $2p$ suitably selected cuts on K, along closed circuits (retrosections) $\gamma, \gamma_2, \ldots, \gamma_{2p}$, it is turned into a two-cell (homeomorph of the interior of a circle) plus its boundary taken twice.

3. A sum of oriented circuits on K is called a *linear cycle* or merely *cycle* of the surface. Two cycles g, g' such that g can be deformed into

g' plus certain pairs of oppositely oriented arcs are *homologous*, this being expressed by a *homology* $g \sim g'$. If 0 stands for a point, then also $g - g' \sim 0$. Such homologies can be combined like ordinary linear equations, the multipliers being always integers. The cycles $g_1, \ldots, g_r$ are dependent or otherwise, accordingly as there exist λ's not all zero such that $\Sigma \lambda_i g_i \sim 0$ or not.

On a two-cell every arc with its end points on the boundary is deformable into an arc on the boundary whose end points are the same. Applying this to the two-cell derived from K by means of the retrosections we have:

I. Every cycle on K is $\sim$ to a sum of multiples of the γ's.

II. On the other hand the γ's are independent. Therefore the maximum number of independent cycles (maximum actually reached) is $2p$.

Given $2p$ independent cycles $g_1, \ldots, g_{2p}$ for every other g there is a relation

$$\lambda g \sim \Sigma \lambda_i g_i, \qquad \lambda \neq 0.$$

If $g'_1, \ldots, g'_{2p}$ is another set of cycles then

$$A_\mu g'_\mu \sim \Sigma A_{\mu\nu} g_\nu, \qquad A_\mu \neq 0$$

and the g''s are independent if and only if $|A| \neq 0$.* There are systems of independent cycles such that $\lambda = 1$ for every g, for example, the system of $2p$ retrosections. A set of such cycles, say $\delta_1, \ldots, \delta_{2p}$ is called a *fundamental set* for linear cycles or simply fundamental set, abridged into F. S.† If δ'_μ are the cycles of another F. S. then

$$\delta'_\mu \sim \Sigma a_{\mu\nu} \delta_\nu, \qquad \delta_\mu \sim \Sigma b_{\mu\nu} \delta'_\nu.$$

Hence

$$\delta_\mu \sim \Sigma b_{\mu\nu} a_{\nu\rho} \delta_\rho$$

and since the δ's are independent these are identical relations. Therefore $\mathbf{ba} = 1$, hence $|b| \cdot |a| = 1$ and as both determinants are integers

$$|b| = |a| = \pm 1.$$

4. *Abelian Integrals.*—By an *abelian integral* attached to C is meant an integral of type

$$v(x) = \int_{x_0 y_0}^{xy} R(xy)\, dx,$$

* We shall greatly simplify matters by making use of matrices. We apply these conventions now quite generally adopted: A matrix of elements, a_{ij} will be called $\mathbf{a}$, its transverse $\mathbf{a}'$, its determinant (when it is square) $|a|$, the matrix of the conjugate elements $\bar{a}_{ij}$, $\bar{\mathbf{a}}$, the unit matrix $\mathbf{I}$. A square matrix $\mathbf{a}$ whose terms are integers with $|a| = \pm 1$ is called *unimodular*.

† Fundamental set = primitive set = minimum base. These terms often found in the literature are all equivalent.

where R is rational and the path of integration is on K. When C is birationally transformed into C', v is changed into an abelian integral v' attached to C' and the singularities of v' are the transforms of those of v and are of the same nature. In particular if v is holomorphic at every point of C, so is v' at every point of C'. v is then called integral of *the first kind,* by far the most important type. We shall denote them by the generic letter u. Any u can be put in the form

$$\int \frac{Q(xy)\,dx}{f'_y},$$

where Q is an adjoint polynomial of order $m-3$ (polynomial with an $(r-1)$-fold point at every r-fold point of f granted that f has only multiple points with distinct tangents; see Chap. 2). There are well known connections with the canonical series on C. A set of integrals $u_1, \ldots, u_k$ is linearly dependent or otherwise accordingly as there do or do not exist constants c_i not all zero, such that $\Sigma c_i u_i$ is constant, or $\Sigma c_i du_i \equiv 0$. There is the basic theorem: The maximum number of linearly independent integrals of the first kind is p. It follows that if $u_1, \ldots, u_p$ are linearly independent every other u is a linear combination of them plus a constant.

Let $g_n{}^r$ be a linear series on C, $G=A_1, \ldots, A_n$ a variable group of it, A a fixed point of C, u an integral of the first kind of the curve. Then

$$\sum \int_A^{A_i} du = \text{const.} \tag{1}$$

This is Abel's classical theorem for integrals of the first kind. It is often written in the form

$$\sum \int_{B_i}^{A_i} du \equiv 0, \quad \text{modd. the periods} \tag{2}$$

where $G'=B_1 \ldots B_n$ is any other group of $g_n{}^r$, and the periods referred to are those of u relatively to the cycles of C. The theorem (2) has been inverted by Clebsch who proved that when it holds for every integral of the first kind then G and G' are equivalent groups, i. e., belong to the same complete $g_n{}^r$.

A related question is Jacobi's *inversion* problem. Given the independent set $u_1, \ldots, u_p$, and the arbitrary constants $v_1, \ldots, v_p$ the problem asks for a group of p points $G=A_1, \ldots, A_p$, such that

$$\sum \int_A^{A_j} du_i = v_i, \quad (i=1, 2, \ldots, p).$$

For the very extensive literature on this topic, see ([52]).

From the geometric point of view the chief result is that for arbitrary v's, G is unique. The exceptional values correspond to $G=$a special group, and then the point $(v_1, \ldots, v_p)$ of S_p belongs to a certain V_{p-2} (non-algebraic). This together with Clebsch's result leads quite rapidly to the Riemann-Roch theorem.

5. There are two other kinds of abelian integrals.—The second is characterized by behavior like a rational function over K, the third by the existence of a finite number of so-called *logarithmic* points in the vicinity of which the integral $v(x, y)$ behaves like a logarithm plus a rational function. More precisely let $A(\alpha, \beta)$ be such a point, t the corresponding local uniformizing variable ($t=(x-\alpha)$ at finite distance, $\frac{1}{x}$ at infinity when axes are arbitrary, $(x-\alpha)^{\frac{1}{q}}$ at a branch point). Then there exists a k such that $t=0$ is at worst a pole for $v-\frac{k}{2\pi i}\log t$. The constant k is the related logarithmic *period* or residue. Let $A_1, \ldots, A_n$ be the logarithmic points, k_i the period for A_i. Then $\Sigma k_i=0$. Conversely if this sum is zero there is a v whose logarithmic points and related periods are the A's and k's.

6. Of very great interest is the theory of periods of integrals of the first kind. For if u is an integral of the first kind and g a cycle. $\int_g du$ the period of u as to g, is a birational invariant of C. Thus the integrals of the first kind appear as the natural tool for deriving quantitative propositions. Let $g_1, \ldots, g_{2p}$ and $u_1, \ldots, u_p$ be independent sets of cycles and integrals and let $\omega_{j\mu}=\int_{g_\nu} du_j$. Then each term of $\boldsymbol{\omega}$ is invariant. Of particular importance is the effect on $\boldsymbol{\omega}$ of changing the sets. If we pass from the g's to the g''s as in No. 2, we have

$$A_\mu \int_{g'_\mu} du_j = \Sigma A_{\mu\nu} \int_{g_\nu} du_j = \Sigma A_{\mu\nu}\omega_{j\nu}.$$

Hence if $A_{\mu\nu}=A_\mu a_{\nu\mu}$ the new period matrix is $\boldsymbol{\omega}\mathbf{a}$. If the two sets are fundamental $|a|=\pm 1$ as before. Let us now pass from the u's to the u''s with

$$u'_j=\Sigma \alpha_{jk}u_k; \qquad |\alpha|\neq 0.$$

Then

$$\int_{g_\mu} du'_j = \Sigma\alpha_{jk}\int_{g_\mu} du_k = \Sigma \alpha_{jk}\omega_{k\mu}.$$

Hence $\boldsymbol{\omega}$ is now changed into $\boldsymbol{\alpha\omega}$. Therefore all possible period matrices are represented by $\boldsymbol{\alpha\omega}\mathbf{a}$; $|a|$, $|\alpha|\neq 0$. Two matrices such as $\boldsymbol{\omega}$ and $\boldsymbol{\alpha\omega}\mathbf{a}$, where $\boldsymbol{\alpha}$ and $\mathbf{a}$ are square, of orders p and $2p$, with determinants $\neq 0$ and a with rational terms, are called *isomorphic*, the passage from one to the other is called an *isomorphic transformation* or an *isomorphism*. When a is unimodular, the matrices are said to be *equivalent*. These notions are due to Scorza([32.1]). Matrices corresponding to two different sets of $2p$ independent cycles are isomorphs; when the two sets are fundamental the matrices are equivalent.

Our chief concern is not so much with the properties of $\boldsymbol{\omega}$ as with the whole class of matrices, equivalent, or more generally isomorphic to it. The properties invariant with respect to isomorphism are then of essential importance. We have first Riemann's fundamental theorems:

I. When the F. S. is composed of the retrosections, the u's can be so chosen as to give ω the form

$$||\delta_{jk};\ \tau_{jk}||, \qquad (j, k=1, 2, \ldots, p)$$

where τ is a symmetrical matrix, and $\delta_{jj}=1$, $\delta_{jk}=0$ $(k\neq j)$.

II. Let $\tau_{jk}=\tau'_{jk}+i\tau''_{jk}$. Then $\Sigma\tau''_{jk}x_jx_k$ is a definite quadratic form.

The equivalent conditions for the general isomorphic ω can be put in the following form: There exists an alternate rational matrix $\mathbf{c}$, such that

III. $\boldsymbol{\omega}\mathbf{c}\boldsymbol{\omega}'=0$.

IV. $i\boldsymbol{\omega}\mathbf{c}\bar{\boldsymbol{\omega}}'$ is a definite positive Hermitian matrix.

In this precise form III, IV, were given explicitly by Baker(2.1) p. 224 (1907), for a period matrix of multiply periodic functions. A substantially equivalent form was given later by Scorza(32), (32.1). Any rational alternate $\mathbf{c}$ satisfying III is called an *alternate matrix* of $\boldsymbol{\omega}$. When it satisfies both III and IV it is called a *principal matrix* of $\boldsymbol{\omega}$.

7. *Riemann Matrices.*—Matrices satisfying III, IV are of considerable importance in various theories and well deserve to be investigated for their own sake as first done by Scorza (1914) who calls them *Riemann* matrices (abridged notation: R. M.). The matrix $\boldsymbol{\omega}$ of No. 6 is the R. M. of the curve C. Scorza's results will be found treated at length in (32.1), (33). See also Krazer-Wirtinger(12) p. 827; Lefschetz(16), and Ch. 17, § 4 (résumés of Scorza's work). Scorza brought out clearly the importance of certain isomorphic invariant integers: (a) The *singularity index* k of $\boldsymbol{\omega}$, or number of linearly independent alternate matrices of $\boldsymbol{\omega}$. (b) The *multiplication index* h, analogous to k when $\mathbf{c}$ is merely a *rational matrix* of $\boldsymbol{\omega}$, i. e., no more constrained to be alternate. (c) The *simultaneity index* λ of two R. M.'s $\boldsymbol{\omega}_1$, $\boldsymbol{\omega}_2$, or number of linearly independent relations $\boldsymbol{\omega}_1\mathbf{c}\boldsymbol{\omega}'_2=0$, $\mathbf{c}$ rational. The index λ is unchanged when both $\boldsymbol{\omega}_1$ and $\boldsymbol{\omega}_2$ are replaced by two isomorphs. As a matter of fact Scorza, wrote $1+h$, $1+k$, where we have h, k. This complicates matters a good deal; h, k as here defined are the actual numbers occurring all the time and are far more convenient in every way.

Scorza has shown that h is also the number of linearly independent *multiplications* of $\boldsymbol{\omega}$, or relations $\boldsymbol{\alpha}\boldsymbol{\omega}=\boldsymbol{\omega}\mathbf{c}$, where $\boldsymbol{\alpha}$ and $\mathbf{c}$ are square of orders p, $2p$, and $\mathbf{c}$ is rational. This is an actual restriction on $\boldsymbol{\omega}$

unless $\mathbf{c}=s\mathbf{I}$, s a rational multiplier and then $\boldsymbol{\alpha}=s\mathbf{I}$ also. In this special case we have *ordinary,* otherwise *complex* multiplication.

Of great importance is the question of reducing $\boldsymbol{\omega}$ to the form $\left\|\begin{matrix}\boldsymbol{\omega}_1 & \mathbf{0}\\ \boldsymbol{\omega}_2 & \boldsymbol{\omega}_3\end{matrix}\right\|$ where $\boldsymbol{\omega}_1$ is an R. M. of genus $<p$ called a *submatrix* of $\boldsymbol{\omega}$. It is then further reducible to the type $\left\|\begin{matrix}\boldsymbol{\omega}_1 & \mathbf{0}\\ \mathbf{0} & \boldsymbol{\omega}_2\end{matrix}\right\|$, Poincaré, Picard, Scorza(32.1), p. 298, Rosati(01.1). The R. M. is then said to be *impure.* In order that two R. M.'s possess a common submatrix (which may coincide with one of them) it is necessary and sufficient that their $\lambda>0$, Scorza(32.1), p. 301. Any R. M. is isomorphic to an essentially unique type

$$\left\|\begin{matrix}\boldsymbol{\omega}_1 & \mathbf{0} & \cdots\\ \mathbf{0} & \boldsymbol{\omega}_2 & \cdots\\ & \cdots & \boldsymbol{\omega}_n\end{matrix}\right\|$$

where the matrices $\boldsymbol{\omega}_j$ are all pure and any two are either identical or non-isomorphic, Scorza(32.1), p. 308.

Scorza has also considered the question of the upper limits of his invariants(32.1), pp. 303, 309. They depend upon the corresponding reduced form. He finds in any case $k\leqq p^2$, $h\leqq 2p^2$, $\lambda\leqq 2p_1p_2$ where p_1, p_2 are the genera of the two matrices whose simultaneity index is λ.

Additional references: Krazer(11), Krazer-Wirtinger(12) for earlier developments. Summaries of Scorza's work are found in (12) p. 822, Lefschetz(16). See also Ch. 17, §§ 4, 5, 6. A treatment more or less parallel to Scorza's is due to Rosati(31), also Ch. 16, No. 13, who applied it mostly to singular correspondences. Applications to hyperelliptic surfaces and abelian varieties will be considered in Ch. 17. In § 4 in particular Scorza's theory is examined more thoroughly and indications as to his method are also given.

§ 2. Linear Systems of Curves on Surfaces and Related Invariants

8. Unless otherwise stated F_m is assumed to be a non-singular irreducible surface in some S_k, or else with only the singularities of the generic projections of such a surface. These singularities may then practically be disregarded. Regarding the reduction of an arbitrary F to this type see Ch. 10.

A linear system $|C|$ of curves of F is by definition a system cut out by a linear family of V_{k-1}'s (hypersurfaces) of its carrying S_k. The *genus* $[C]$ of $|C|$ is the genus of its generic curve, its degree $[C^2]$ the number of intersections of two arbitrary curves of it. Given $|C|$, $|D|$, $[CD]$ represents the number of intersections of arbitrary curves of the two systems. $|C|$ is *complete* if not contained in an ampler linear system. The meaning of $|C+D|$, complete system sum of $|C|$ and $|D|$ is obvious, likewise that of $|C-D|$ whenever there is a C containing some D as a part. When $C-D$ is not *effective*, i. e., does not exist, $C-D$ is said to be a *virtual* curve, Severi([40]). To such curves, to reducible effective curves, or to curves with accidental singularities there may be attached a *virtual* genus and a *virtual* degree or intersection number, Severi([40]) by means of the following two laws: (a) $[CD]$ is a symbol which obeys the distributive law as to C and D, i. e., $[(C+C')D]=[CD]+[C'D]$ and similarly with respect to D. (b) $[C]$ is a symbol such that in all cases

$$[(C+D)]=[C]+[D]+[CD]-1.$$

The two laws hold for effective curves. The proof is immediate for (a), and for (b) has been given by Noether([22]), Picard-Simart([24]), vol. 2, p. 106, Lefschetz([18]), p. 252. See also in this connection ([14]).

An important observation, Lefschetz([13]), p. 19, is that when C, D, F, are properly oriented cycles (see below No. 10), $[CD]$ is actually equal to the algebraic number of suitably signed crossings $(C \cdot D)$, or to the Kronecker-index of the cycles as to F. [Veblen([50]); Weyl([51]); Lefschetz([20])]. When the curves are virtual there are definite cycles corresponding to them for which also $[CD]=(C \cdot D)$, as follows from the fact that the Kronecker-index also obeys (a).

9. The canonical system $|K|$ defined in No. 16, plays an essential part in the theory of F. If p_g-1 is its dimension, p_g is the *geometrical* genus of F. It is also the number of linearly independent adjoint polynomials of order $m-4$ (adjoint polynomial=a polynomial vanishing on the double curve when F_m has no other singularities). The postulation of adjoint polynomials of order $m-4+k$, k sufficiently high (number of linearly independent conditions satisfied by the polynomial) is a polynomial $f(k)$. The number $p_a=f(0)$, *arithmetical* genus of F, is like p_g an important birational invariant of F and their difference $q=p_g-p_a$ is the *irregularity* of F. The surfaces whose $q=0$ are called *regular*.

The canonical system can be defined directly thus: Let $|C|$ be sufficiently general and without base points. There exists a maximal system $|C'|$, such that C' cuts out canonical groups on C, called the adjoint system of $|C|$. Then $|K|=|C'-C|$. The system $|iK|=|iC'-iC|$ may exist whether $|K|$ exists or not and the number P_i of its linearly independent curves is the i-canonical genus. P_i is also an invariant of F and these plurigenera have played an important part especially in the classification of various types of surfaces.

The definition of the canonical and pluricanonical systems indicates plainly that they are invariant systems for F under *all* birational transformations, hence p_g, P_i, are *absolute* invariants, i. e., unchanged by any such transformation. This is also true for q and p_a. There are, however, *relative* invariants or numbers invariant only under certain types of birational transformations. In general when F acquires r' *exceptional* curves (transforms of ordinary points) and loses r'' such curves these invariants increase by a multiple of $(r'-r'')$. The most important and best known is the invariant I of Zeuthen-Segre (references in Castelnuovo-Enriques ([1]), p. 700) defined as follows: Let $|C|$ be an irreducible pencil with $[C^2]$ ordinary distinct base points and let its δ curves of genus $<[C]$ have just one more double point than the generic C, and therefore precisely the genus $[C]-1$. Then $I=\delta-[C^2]-4[C]$. Another important relative invariant is the linear genus $p_1=[K]$. It is a relative invariant because K must always be considered as including the exceptional curves. There are other relative invariants of an algebraic nature but they are expressible in terms of I, p_1 and the absolute invariants.

For further references and details on the various invariants just described, see Castelnuovo-Enriques (4), p. 690 and Baker (2.2).

When $|K-C|$ exists and has the dimension $\nu-1$, $|C|$ is *special* and ν is its *specialization* index. This is true whether C is effective or virtual. In order that C be effective it is sufficient that

$$s=[C^2]-[C]+1+p_a-\nu\geqq 0$$

and then the dimension of the complete system $|C|$ is $\geqq s$. Furthermore $|C|$ belongs to a continuous set ∞^q of such linear systems. With more precision: There exists a V_q, the Picard-variety of F (see Ch. 17, No. 25), such that to each point of it corresponds one and only one $|C|$ varying continuously and birationally with the point.

The preceding theorem is closely related to the proposition on Picard integrals of the first kind recalled in Nos. 16, 17. For complete references on the theory summarized in the present number see Castelnuovo-Enriques([3]), ([4]); Severi([35]).

§ 3. Topology of F

10. The surface F with the simplified singularities assumed in No. 8 and throughout the rest of this chapter, has the important topological property of being representable as a multiply sheeted Riemann four space (complex projective plane) which is a manifold in the sense of Analysis Situs, O. Veblen([49]), p. 88, a property not necessarily shared by arbitrary surfaces.

For the analytical applications the important notion is that of the *k-cycle* Γ_k of F. By that we mean here a closed k dimensional continuum imbedded in F and suitable as integration domain for k-fold integrals. A sphere or an anchor ring more or less deformed and contained in F are simple examples of a Γ_2. There will be local parametric representations for the coordinates of the points of Γ_k as analytical functions of variables $u_1, \ldots, u_k$ whose order determines the orientation of Γ_k. When we transpose the u's an odd number of times Γ_k is considered as changed into $-\Gamma_k$. Given a set of k cycles Γ_k^i $(i=1, 2, \ldots, r)$, we write the homology with integer coefficients:

$$\Sigma\mu_i\Gamma_k{}^i\sim 0, \quad \text{mod. } F,$$

Poincaré([30]), whenever the continuum at the left can be deformed into one of lesser dimension plus certain k dimensional continua each counted a total algebraic number of times equal to zero. The "mod. F" is often omitted. Also $\Gamma_k\sim\Gamma'_k$ stands for $\Gamma_k-\Gamma'_k\sim 0$. The equivalence of the two homologies may equally be justified as in No. 3. When a homology as above exists the cycles $\Gamma_k{}^i$ are *dependent,* otherwise *independent.* These notions are treated more at length by Poincaré([30]), Veblen([49]), Chs. 3, 4; Lefschetz([13]), Ch. 1.

The maximum number of independent cycles is a topological invariant called the k-th *connectivity index* R_k. There are five connectivity indices, corresponding to $k=0, 1, \ldots, 4$. A point is conveniently considered as a zero-cycle and when F is irreducible, if A, B are two points $A-B\sim 0$, for it is the boundary of some arc AB. Hence $R_0=1$. Similarly $R_4=1$ for F is its own unique Γ_4 and it is not ~ 0. There remain R_1, R_2, R_3. But $R_1=R_3$, Poincaré([30]); Veblen([49]), Ch. 4, and their common value is an even integer $2q$. This follows from investigations on Picard integrals (see below), and was also established directly by Lefschetz([15]), ([13]), p. 30. q is in fact the irregularity of F. (See Nos. 17, 18.) There remain then the only distinct indices 1, R_1, R_2. By considering the Euler characteristic, Alexander proved in ([2]):

$$R_2=I+2R_1+2=I+4q+2$$

where I is the Zeuthen-Segre invariant. See also Lefschetz([15]), ([13]), p. 40. The number R_1 is an absolute birational invariant, but R_2 like I, is only a relative invariant (No. 9).

Another interesting set of invariant integers is obtained thus: Let Γ_k be such that $t\Gamma_k\sim 0$ for some $t>1$ but for no lesser integer. Γ_k is called a *zero-divisor* for k-cycles. The number σ_k of non-homologous zero-divisors including one zero-cycle, is finite and a topological invariant. It is the k-th *torsion index,* Poincaré([27]), Lefschetz([13]), Ch. 1. All σ's except σ_1 and σ_2 are unity and these two are equal. We denote their common value by σ. As we shall see this is the same as Severi's base invariant σ([43]).

11. Since any R_k+1 k-cycles are always dependent we may find a set Γ_k^i, $(i=1, 2, \ldots, R_k)$, such that, whatever Γ_k, there is a homology:

$$\lambda\Gamma_k \sim \lambda\Sigma\lambda_i\Gamma_k^i, \qquad \lambda\neq 0.$$

The proof is the same as given by Klein(9), p. 543. The Γ_k^i set may be called an *intermediary* F. S. by analogy with a term used by Severi in his theory of the base. If we agree to neglect zero-divisors, or accept division of the coefficients of a homology by a common factor, they constitute an F. S. It is best to indicate this modified homology by a special sign $\approx$. Then

$$\Gamma_k \approx \Sigma\lambda_i\Gamma_k^i,$$

a relation whose precise meaning is that the difference between the two sides is a zero-divisor. Let G_k^j, $(j=1, 2, \ldots, \sigma_k-1)$ be the actual divisors. Then

$$\Gamma \sim \Sigma\lambda_i\Gamma_k^i + \Sigma\mu_j G_k^j.$$

Hence the cycles Γ_k^i, G_k^j constitute an F. S. for k cycles.

12. There is a series of results due to Picard, Poincaré, Lefschetz (references below), describing the relations between the cycles and linear curve systems. We consider merely a linear system $|H|$ of hyperplane sections, though a more general one could take its place. Let H_{x_1} be an arbitrary section $x_1=\text{const.}$, $((x_1, x_2, \ldots, x_k)$ cartesian spaces for the S_k that carries F), and p the genus of an arbitrary H. With arbitrary axes and infinite H, H will never acquire any other singularity than a double point with distinct tangents. Let this happen for $x_1=a_1, \ldots, a_N$. The genus of H_{a_i} is then $p-1$.

Draw rectilinear cuts aa_i. As x_1 describes aa_i there exists a linear cycle δ_i of H_{x_1} that tends to a point as $x_1\rightarrow a_i$. When x_1 turns around a_i every cycle of H_{x_1} is increased by a multiple of δ_i, or is unchanged (such is in particular δ_i itself).

Concerning linear cycles we have these theorems: Every linear cycle of F is $\sim$ to one on H_{x_1}. The curve H_{x_1} carries $R_1=2q$ invariant cycles which are independent cycles of F, hence also of H_{x_1}. However there may not exist any F. S. composed of them. Finally every cycle of H_{x_1} that is ~ 0 mod. F is $\sim$ to a sum of δ's mod. H_{x_1}.

As regards the two-cycles, let Δ_i be the locus of δ_i when y describes aa_i. Denote generically by (H_a) a two dimensional (open) part of H_a. Then for every Γ_2 there is a homology;

$$\Gamma_2 \sim \Sigma\lambda_i\Delta_i + (H_a), \quad \text{mod. } F, \tag{3}$$

with the related condition

$$\Sigma\lambda_i\delta_i^0 \sim 0, \quad \text{mod. } H_a \tag{4}$$

where δ_i^0 is the initial position of δ_i on H_a. Let x_1 describe any path $a_i x_1$ (x_1 not an a_i) with the δ's occupying final positions called δ_i. Then from (4) and by continuity considerations

$$\Sigma \lambda_i \delta_i \sim 0, \quad \text{mod. } H_{x_1}.$$

It will be observed that the sum in (4) constitutes precisely the boundary of the continuum $-(H_a)$.

The basic references for the last few results are Picard([24]), vol. 1, Chs. 1-4, vol. 2, Ch. 11; Poincaré([26]); Lefschetz([15]), ([13]), Chs. 2, 3.

13. In ([13]), ([19]), ([21]), Lefschetz established a basic connection between the topology of F and the distribution of its systems of curves, or the Severi base theory. The chief feature of his work is the study of the properties of the curves when considered as two-cycles of the surface.

If C', C'' are two curves it will be natural to consider the cycle $C = C' - C''$. Any combination of such cycles is called an *algebraic* cycle. It is always $\sim$ to a difference of two effective curves.

There exists some surface other than F passing through C''—let k be its order. Then if $H_1 \sim kH$ is its complete intersection with F and $H_1 - C'' = E$ the residual of C'' as to it, $C = (C' + E) - (C'' + E) \sim G - kH$ where $G = C' + E$ is effective. Hence of the two effective curves whose difference is C, the second may be chosen as a multiple of the plane sections. When kH varies in its system it remains $\sim$ to itself, hence the particular curve kH may be any one of the linear system $|kH|$.

14. Since any $1 + R_2$ two-cycles are related by a homology, there is a maximum $\rho \leqq R_2$ to the number of independent curves (effective or virtual). ρ is in fact Picard's number([24]), vol. 2, Ch. 9, also called the Picard-Severi base number, as was proved by Lefschetz([21]), ([19]), p. 348; also ([13]) and below, Nos. 16, 21. Since any $\rho + 1$ curves are dependent, the same reasoning as for general cycles will lead to an intermediary F. S. $C_1, C_2, \ldots, C_\rho$, which together with certain zero-divisor curves $D_1, \ldots, D_{\sigma-1}$, will form an F. S., so that for any C we have

$$C \sim \Sigma \lambda_i C_i + \Sigma \mu_j D_j.$$

The number σ is the same as in No. 11, Lefschetz([19]), p. 348.

When $\rho > 1$ there can always be constructed an F. S. whose curves are all effective. For let $\overline{C}_i = e_i C_i + x_i H$, $e_i = \pm 1$. There is an integer ξ such that for every i, when $x_i > \xi$, whatever e_i, $\overline{C}_i$ is effective. Assume then the x's so chosen. We have

$$C \approx \lambda H + \Sigma \lambda_i \overline{C}_i.$$

15. Furthermore there is a relation

$$aH + \Sigma a_i C_i \approx 0, \quad a \neq 0.$$

Since no multiple of H is ~ 0, Lefschetz(13), p. 20, and since $\rho > 1$, we may assume the initial set C_i so chosen that at least *two* integers a_i are $\neq 0$. All that is needed in order to have an effective intermediary F. S. is to be able to express H as $\approx$ to a sum of $\overline{C}$'s. This in turn merely requires that we find e's and positive x's $> \xi$ such that $\Sigma a_i e_i x_i = a - 1$, which can always be done since there are at least two terms at the left.

Since H is then $\approx$ to a sum of $\overline{C}$'s, this is true for every C. Hence $\overline{C}_1, \ldots, \overline{C}_\rho$ constitute an effective intermediary F. S. We have $D_i = \overline{D}_i - s_i H$, where $\overline{D}_i$ is effective and s_i merely higher than a certain limit. Also there is a ν such that

$$\nu H \sim \nu \Sigma \mu_i \overline{C}_i.$$

Take every s a multiple of ν. Then D_i is $\sim$ to a sum of the effective curves $\overline{C}_i$ and $\overline{D}_i$. Therefore every C is $\sim$ to a sum of $\overline{C}$'s and $\overline{D}$'s. Hence $\overline{C}_1, \ldots, \overline{C}_\rho, \overline{D}_1, \ldots, \overline{D}_{\sigma-1}$ is an effective F. S. For $\rho = 1$ a similar reasoning holds with C'_1, H, taking the place of $C_1, \ldots, C_\rho$, and the intermediary F. S. will be composed of two effective curves, the F. S. of $\sigma + 1$ effective curves.

The treatment here is related to Severi's construction of intermediary and minimum bases, (38), (43), (35), for his relation expressing algebraic dependence.

§ 4. Integrals of F

16. Abelian integrals have been generalized in two different directions by Noether and Picard; see Castelnuovo-Enriques(4), p. 714. The simple integrals, or integrals of total differentials, or also Picard integrals, are due to Picard. F being assumed projected on an S_3 into $f(xyz) = 0$, these integrals are of the form

$$(5) \qquad \int R dx + S dy, \quad \frac{\partial R}{\partial y} = \frac{\partial S}{\partial x},$$

where R, S are rational in x, y, z. (5) is of the first kind if it is holomorphic on F. The number of linearly independent integrals of the first kind is q [Castelnuovo, Enriques, Severi]. References and complete discussion: Castelnuovo-Enriques(4), p. 723, to which may be added Lefschetz(13), Ch. 4; Severi(35), p. 769: also (45.1), where he indicates in outline that an integral of the first kind (5), called for the present u, may be completely characterized by the set of double points of the loci $u = \text{const}$.

Noether introduced double integrals of F of the type

$$(6) \qquad \iint R(xyz)\, dx dy, \quad R \text{ rational}.$$

It is of the first kind if finite over any analytical continuum of integration on F. Noether showed that in that case (6) is of the form

$$\iint \frac{Q(xyz)\,dxdy}{f'_z},$$

Q being an adjoint polynomial of order $m-4$. The number of linearly independent double integrals of the first kind is p_g. The curves which the Q's cut out on F outside of its double curve constitute the canonical system $|K|$. References Castelnuovo-Enriques(4), pp. 696, 716.

Picard has also introduced simple integrals of the second and of the third kinds. (5) is of the second kind when it behaves everywhere like a rational function. The number of linearly independent integrals of the second kind is $R_1=2q$, Picard(24), vol. 1, p. 150, additional references Castelnuovo-Enriques(4), p. 721. Finally (5) is of the third kind if it possesses logarithmic curves. There may be found on F sets of curves that cannot be logarithmic curves of any integral (5). The maximum ρ of their number is the Picard-Severi base number, No. 14. Further references Castelnuovo-Enriques(4), p. 727, to which may be added Lefschetz(15), (13), Note I; Severi(35), p. 773.

Picard also considered and extensively investigated double integrals of the second kind(24), vol. 2. The number ρ_0 of linearly independent integrals of that kind (in a sense that need not be described in detail here) is an absolute invariant given by

$$\rho_0=I+4q-\rho+2,$$

where I is as before the Zeuthen-Segre invariant number. Lefschetz(15), (13), Note I, greatly simplified the Picard theory. In particular he puts his formula in the form $\rho_0=R_2-\rho$, and in (17) shows that it is only a special case of a very general similar formula for algebraic varieties. Further references Castelnuovo-Enriques(4), p. 731.

§ 5. Distribution of the Curves on F and the Bases

17. The best attack on the whole transcendental theory is due to Poincaré(25), (28), (29). His work was considerably simplified and then extended by Lefschetz(13), (19), (18), (21), who brought into play Analysis Situs and linked it with Severi's work on the theory of the base(35), (38), (42), (43), (45), (46). This outline is substantially as given in Lefschetz(18).

To H_y there belongs a set of p linearly independent integrals of the first kind

$$u_j=\int\frac{Q_j(xyz)\,dx}{f'_z}, \quad (j=1, 2, \ldots, p)$$

where the Q's are adjoint polynomials and of degree $m-3$ in x, z. Among them, however, there are actually $p-q$ say the first $p-q$, of degree $m-3$ in x, y, z, while the remaining q are of degree $m-2$. Let $A_1, \ldots, A_m$ be the points at infinity of H_y (section $y=$const.)—they are its fixed points. Let C be an algebraic curve on F and $B_1, \ldots, B_r$ be a group of points on H_y that includes all points on C and also some of the A's each counted with arbitrary multiplicity. We consider with Poincaré the sums

$$v_j(y) = \Sigma \int_{A_1}^{B_h} du_j.$$

Upon investigating their behavior we find that they have this form:

$$v_j(y) = \frac{1}{2\pi i} \Sigma\lambda_k \int_A^{A_k} \frac{\Omega_{jk}(Y)\,dY}{Y-y}, \qquad j \leqq p-q;$$

$$v_{p-q+j} = a_j.$$

Here the a's are the critical values of y, the α's constants and the λ's certain integers.

As a matter of fact when the Q's are not further restricted than as stated, the v's do not have quite the simple aspect here indicated. The situation is at first this: The R. M. of H_y is impure for $q>0$ and isomorphic to

$$\left\| \begin{matrix} \boldsymbol{\omega}_1 & 0 \\ 0 & \boldsymbol{\omega}_2 \end{matrix} \right\|$$

where $\boldsymbol{\omega}_2$ is of genus q, the cycles corresponding to its columns being precisely the invariant cycles and its terms all constant. As to $\boldsymbol{\omega}_1$ its columns correspond to cycles which are dependent upon the δ's. Its terms are then functions of y, and it has been shown by Severi that for y arbitrary it is pure([39]), p. 530. Now the special choice of u's leading to the form given for the v's corresponds precisely to the case when $\boldsymbol{\omega}$ is in the above reduced form, Lefschetz([13]), p. 74.

At this point we connect up with Picard's integrals of the first kind. For to every u_{p-q+h} corresponds an integral of the first kind (5) that reduces to it on H_y, i. e., such that

$$R = \frac{Q_{p-q+h}}{f'_z}$$

and conversely. Therefore F possesses precisely q linearly independent Picard integrals of the first kind. This is essentially Poincaré's proof of that theorem already mentioned in No. 16. See Poincaré([29]), p. 91; Severi([45]); Lefschetz([13]), pp. 74, 76.

18. To each C there is attached a definite set of v's corresponding to the case where the B's include all the intersections of C with H_y and no

other points. The set is of course only defined modd. the periods. At the very outset there is this proposition: If C' has the same abelian sums as C then it belongs to the complete linear system $|C|$; Poincaré([29]), p. 98. He imposes the condition that C' is of same order as C, which is superfluous, Lefschetz([13]), p. 77. Hence in order that C and C' belong to the same complete linear system it is necessary and sufficient that they cut H_y in two groups of points that belong to the same linear series; Severi([36]), p. 55; Lefschetz([13]), p. 77. See also Severi's closely related extensions of Abel's theorem to Picard integrals([36]), ([41]), ([45]) and his inversion theorem for them ([47.1]).

We shall consider in a moment the necessary conditions on the v's in order that they be the abelian sums of some curve. For the moment let us just state that they do not involve the α's. In the v's corresponding to a given C let us vary the α's. There results for each choice of them a linear system of curves $|D|$ whose limit as the α's tend to their values for C may be not C but $C+kH$, $k>0$. We conclude: Given any C there exists a k such that $D=C+kH$ determines a complete continuous system of curves which constitutes an irreducible algebraic set of ∞^q linear systems. See Lefschetz([13]), p. 80. This brings us quite close to the very precise result embodied in the theorem at the end of No. 9. For the present we return to the functions v_i and shall show how they lead very rapidly to the Severi bases.

We shall now be particularly concerned with the part played by the λ's. Poincaré in ([29]), ([25]), gave a set of equations linear in the Ω's and λ's. Working in a different direction Lefschetz obtained in ([19]), p. 343; ([13]), p. 66, the results to be described presently. If the v's are to be functions as obtained in No. 14 a first condition is that the λ's must be the coefficients of the Δ's in some Γ_2 of the reduced form of No. 12. Furthermore, ([19]), p. 346; ([13]), p. 71, the period of every double integral of the first kind as to that Γ_2 must be equal to zero. Hence these theorems:

I. The necessary and sufficient condition in order that the λ's shall correspond to a C or to a sum of multiples of the points A_i is that they define a reduced Γ_2 and that all the corresponding periods of double integrals of the first kind are zero.

II. The algebraic cycles or virtual curves, are completely characterized by the property that the corresponding periods of double integrals of the first kind are all zero.

This last proposition gives rise to the only systematic method for obtaining ρ. We can so choose the u's that the integrands of the first

p_g be the same as for certain double integrals of the first kind. The Γ_2 just considered will be determined up to a multiple of H_a by the relations

$$\Sigma\lambda_k\Omega_{hk}\equiv 0, \qquad (h=1, 2, \ldots, p-q).$$

We can determine a set of independent relations such as these, calculate the period of the double integral attached to u_h, $h\leqq p_g$,

$$\Sigma\lambda_k\int_a^{a_k}\Omega_{hk}(y)\,dy$$

corresponding to the Γ_2, form the matrix of these periods. Its rank is precisely ρ. This is manifestly a method of little practical value—a defect shared by many explicit solutions of mathematical problems.

19. *Equivalence of Curves and the Severi Bases.*—Two curves C, D, on F are *algebraically equivalent* or simply *equivalent,* and we write $C=D$, whenever a curve E can be found such that $C+E$ and $D+E$ are effective and that the complete linear systems $|C+E|$, $|D+E|$ are members of one and the same irreducible set of linear systems of curves on F. When F is regular this amounts to demanding coincidence of the two linear systems. It will be observed that this does not demand that C and D shall be effective. This all important relationship was introduced by Severi([38]), who made a complete study of it in this paper and also in ([43]), ([42]), ([35]). He was led to it by his investigations of the surface image of point pairs of two algebraic curves([46]). See also Albanese([1]), ([2.2]) for an investigation of various types of continuous systems of curves.

If $C=D$, then $C+A=D+A$, whatever A. For in the definition E may be replaced by any other curve obtained by adding to it an arbitrary effective curve. Hence we can so choose E that $C+A+E$ and $D+A+E$ will also behave in accordance with the definition. From this the desired equivalence follows.

We conclude that $C=D$ and $C-D=0$ express one and the same thing. There follows also a definite meaning for the relation

$$\mu_1C_1+\ldots+\mu_rC_r=0$$

where the μ's are integers, expressing the *dependence* of the curves C_i.

20. The bridge between Severi's investigations and the theory discussed so far is the following theorem, Lefschetz([19]), p. 347, ([13]), p. 81: The two relations

$$\Sigma\mu_iC_i=0, \qquad \Sigma\mu_iC_i\sim 0,$$

are completely equivalent. This gives at once: (a) A maximal set of independent algebraic cycles is also a maximal set of independent curves in the sense of Severi and in particular is composed of exactly ρ curves.

This is Severi's *ordinary* base, ([38]), p. 209. It implies also that the Picard-Severi ρ is the same as our ρ. See also Albanese([1,2]). (b) There exists an F. S. for the Severi relation composed of $\rho+\sigma-1$ curves. They may all be chosen effective for $\rho>1$. For $\rho=1$ it may be necessary to take $\sigma+1$ effective curves in the set.

By II, No. 18, if Γ_2 is a zero divisor it is algebraic. Hence σ is the number of distinct curves C such that $C\neq 0$, $\mu C=0$, for some $\mu>1$. This actually identifies σ with the σ of Severi([43]), which he defines thus: If C_1 is effective and sufficiently general there exist $\sigma-1$ effective curves $C_2, \ldots, C_\sigma-1$ such that $C_i\neq C_j$, $\mu C_i=\mu C_j$ for some $\mu>1$. Examples of F with $\sigma>1$ have been given by Severi, Godeaux, Lefschetz, references in ([18]), p. 248.

Conclusion.—The curves on F are distributed into discrete continuous sets of linear systems. To each set corresponds a definite set of characteristic integers λ in the v's and to each component linear system of each set a definite set of q characteristic constants a_j. The sets are derivable from a certain minimum base $C_1, \ldots, C_\nu$ ($\nu=\rho+\sigma-1$ or $\sigma+1$) in the following manner essentially due to Albanese([1]), p. 203: For any C there is a set of positive μ's such that $|C+\mu H+\Sigma\mu'_iC_i|$, $|\mu H+\Sigma\mu''_iC_i|$ are complete and belong to the same continuous set of ∞^q linear systems.

21. Severi has given this equivalence criterium for two curves A, B, ([38]): In order that $\lambda A=\lambda B$ it is necessary and sufficient that A and B be of the same order and that

$$[A^2]=[AB]=[B^2].$$

He demands also that their common value be positive but this is not essential, Albanese([1]), p. 201. From the criterium follows that in order that $C_1, \ldots, C_r$ be dependent it is necessary and sufficient that

$$\left\|\begin{matrix}[HC_1] & [HC_2] & \cdots & [HC_r]\\ [C_iC_1] & [C_iC_2] & \cdots & [C_iC_r]\end{matrix}\right\|$$

be of rank $<r$, Severi([38]).

From this condition Severi derives the following result: In order that there exist a Picard integral of the third kind whose logarithmic curves are $C_1, \ldots, C_r$, it is necessary and sufficient that the curves be algebraically dependent. Hence ρ as defined by Picard (maximum number of curves not logarithmic for any integral (5), see No. 16) is also the maximum number of algebraically distinct curves on F in the sense of Severi, or of non-homologous curves in the sense of Lefschetz. This leads rapidly to the various Severi bases (intermediary, minimum), ([43]), ([42]). For further references and details see Castelnuovo-Enriques([4]), p. 728, Severi([35]), p. 764.

The value of ρ has actually been determined for certain important classes of surfaces, Castelnuovo-Enriques([4]), p. 730; Lefschetz([19]), p. 358, ([13]), Ch. 5.

§ 6. Extension to V_r

22. The chief differences are already present in the passage from $r=2$ to $r=3$, hence it will be sufficient to consider the latter case; for the general r see Lefschetz([19]).

Severi has made a searching investigation([41.1]) of the linear systems of surfaces on a V_3 and the related genera. The results are much more inconclusive than for F_m, and also much more difficult to derive. Castelnuovo-Enriques have shown([4.1]) that the linear cycles and Picard integrals of total differentials of the first kind of V_3 and of its hyperplane sections are the same. Further references along this line in ([4]), p. 761.

Let $f(x, y, z, t)=0$ be the equation of V_3, assumed with no other singularities than those of the projection of a non-singular variety from a higher space. Its topology is in many respects similar to that of F and has been treated at length by Lefschetz([13]), Ch. 5.

The surfaces of V_3 can be related to one another by homologies or by Severi equivalences, Severi([43]), Lefschetz([19]), ([13]). The chief results pertain to the relation between the surfaces and their sections by a given hyperplane section (or a surface variable within a sufficiently general linear system), and are practically all due to Lefschetz, loc. cit. Let H_z be a generic section $z=$const. and call AB the intersection of the surfaces A, B of V_3. The relations

$$A=B; \qquad AH_z=BH_z \text{ on } H_z;$$
$$A\sim B, \text{ mod. } V_3; \qquad AH_z=BH_z, \text{ mod. } H_z,$$

are all equivalent, Lefschetz([19]), p. 352; ([13]), p. 104; also for the first two relations Severi([47]), p. 635. This leads at once to the bases of various types, which are the same whether $\sim$ or $=$ are considered. In particular there are related numbers ρ, σ.

To the critical values of z (those for which $z=$const. is tangent to V_3) correspond vanishing two-cycles of H_z. If one of these is algebraic so are all the rest. When none are algebraic every F. S. for the curves of H_z is the intersection with H_z of an F. S. for the surfaces of V_3; Lefschetz([19]), p. 354, ([13]), p. 106. In many cases this makes possible the determination of an F. S. for surfaces in an S_3. The vanishing cycles will certainly not be algebraic when V_3 possesses a triple integral of the first kind whose periods are not all zero, Lefschetz([19]), p. 357.

The determination of ρ, σ and the bases can be carried out in some fairly simple but important cases, in particular for complete intersections.

See Klein([10]); Fano([5]), ([6]); Severi([44]); Lefschetz([19]), p. 358. In ([48]) Sisam extends Klein's theorem of ([10]) on the rectilinear congruences of S_3 to any S_r. This is equivalent to finding the F. S. for the hypersurfaces of a certain V_k whose points correspond to the lines of S_r, hence to determining its ρ and σ. As in Klein's case the F. S. is composed of a unique V_{k-1} image of a linear congruence of S_r. This has been extended to the congruences of linear spaces in any S_r by Severi([37]). In ([19]), p. 362, Lefschetz determines the numbers ρ, σ for the variety image of the point pairs of two varieties. In the same paper, also in ([13]), Ch. 6, he has determined ρ, σ and the F. S. for abelian varieties, for which see Ch. 17.

Severi([41.1]) has defined the arithmetic genus of any V_r. Lefschetz([14]) solves this question: Given $C_1, \ldots, C_n, D_1, \ldots, D_k$, hypersurfaces of V_r, with $\lambda_i D_i = \Sigma \lambda_{ij} C_j$, to determine the arithmetical genera of the intersections of the D's in terms of those of the C's and of the λ's. Albanese([1.1]) treats special cases of that problem and also determines the Zeuthen-Segre invariant of a sum of two surfaces on V_3 in terms of the other invariant characters of the two surfaces. Severi([41.1]) indicates a number of unsolved problems concerning V_r.

GLOSSARY OF NEW TERMS, AND NOTATIONS OF MOST FREQUENT OCCURRENCE

$$\mathbf{a} = \|a_{ij}\|; \qquad \mathbf{a}' = \|a_{ji}\|; \qquad \bar{\mathbf{a}} = \|\bar{a}_{ij}\|.$$

When $\mathbf{a}$ is square its determinant is $|a|$. When furthermore the terms are all integers and $|a| = \pm 1$, a is called unimodular. $\mathbf{I}$ = the matrix unity.

F. S. = fundamental set = primitive set = minimum base.
R. M. = Riemann matrix.
h = multiplication, k = singularity: index of an R. M.
λ = simultaneity index of two R. M.'s.

It is not necessary to repeat here the definitions of the preceding terms, nor of what is meant by principal matrix, rational matrix, rational alternate matrix of an R. M., nor by isomorphic or equivalent R. M.'s. They will be found in No. 7. The R. M. of C^p is the period matrix of p independent integrals of the first kind relatively to an F. S. of cycles. In reading the literature it is important to bear in mind that our h, k, are Scorza's $1+h$, $1+k$.

Special notations for F_m: $|C|$ = linear system of curves. $[C]$ genus of the generic C; $[CD]$ number of intersections of curves C and D;

$[C^2]=[CC]$. p_g=geometric genus; p_a=arithmetic genus; $q=p_g-p_a$= irregularity; P_i=i-canonical genus; I=Zeuthen-Segre invariant; p_1=linear genus; ρ=Picard-Severi number, σ=Severi number=number of distinct virtual curves such that $C\neq 0$, $\mu C=0$, $\mu>1$. ρ and σ are also called *base* invariants. Γ_i=i-cycle. $\sim$ is the homology symbol with zero divisors included; $\approx$ the same with zero-divisors omitted. R_i=i-th connectivity index with $R_1=2q$. These terms are all defined in Nos. 9-14. Most of them are also applied to a V_r.

BIBLIOGRAPHY

1 G. Albanese, Ann. di Mat., (3), 24: 159-234 (1915). [5.

1.1 ———, Ann. di Mat., (4), 4: 153-184 (1927). [6.

1.2 ———, Rom. Acc. L. Rend., (6), 5_1: 481-488 (1927). [5.

2 J. W. Alexander, Rom. Acc. L. Rend., (5), 23: 55-62 (1914). [3.

2.1 H. F. Baker, An introduction to the theory of multiply periodic functions, Cambridge, 1 vol. (1907). [1.

2.2 ———, London M. S. Proc., (2), 12: 1-40 (1913). [2.

3 Castelnuovo-Enriques, note at end of 24. [2.

4 ———, Encyklopädie III C 6 b, [2, 4, 5, 6.

4.1 ———, Ann. Éc. Norm., (3), 23: 339-366 (1906). [6.

5 G. Fano, Torino Atti, 39: 596-613 (1904). [6.

6 ———, Torino Atti, 44: 633-648 (1909). [6.

7 G. Humbert, J. de Math., (4), 10: 169-201 (1894). [2.

8 ———, C. R., 117: 361-363 (1893). [2.

9 F. Klein, Theorie der elliptischen Modulfunctionen, Leipzig, vol. 2 (1892). [3.

10 ———, Math. Ann., 22: 234-242 (1883). [6.

11 A. Krazer, Lehrbuch der Thetafunktionen, Leipzig (1903). [1.

12 Krazer-Wirtinger, Encyklopädie II B 7, [1, 2.

13 S. Lefschetz, L'Analysis Situs et la géométrie algébrique, Paris, (1924). [3, 4, 5, 6.

14 ———, Ann. of Math., (2), 17: 197-212 (1916). [2, 5.

15 ———, Ann. of Math., (2), 21: 225-258 (1920). [3, 4.

16 ———, Bull. Sci. M., (2), 47: 120-128 (1923). [1.

17 ———, J. de Math., (9), 3: 319-343 (1924). [4.

18 ———, Am. M. S. Bull., 29: 242-256 (1923). [2, 5.

19 ———, Am. M. S. Trans., 22: 327-482 (1921). [3, 5, 6.

20 ———, Am. M. S. Trans., 28: 1-49 (1926). [1.

21 ———, Rom. Acc. L. Rend., (5), 26: 228-234 (1917). [3.

22 M. Noether, Acta. Math., 8: 161-192 (1886). [2.

23 E. Picard, Traité d'analyse, vol. 2, Paris, 3d edition (1925). [1.

24 Picard-Simart, Théorie des fonctions algébriques de deux variables indépendantes, Paris, vol. 1 (1897); vol. 2 (1906). [2, 3, 4.

25 H. Poincaré, Berlin Sitzungsber, 10: 28-55 (1911). [5.

26 ———, J. de Math., (6), 2: 135-190 (1906). [3.

27 ———, London M. S. Proc., 32: 277-308 (1900). [3, 5.

28 ———, C. R., 149: 1026-1027 (1909). [5.

29 ———, Ann. Éc. Norm., (2), 27: 56-108 (1910). [5.

30 H. Poincaré, J. Éc. Polyt., (2), 1: 1-121 (1895). [3.
31 C. Rosati, Ann. di Mat., (3), 25: 1-32 (1916). [1.
31.1 ———, Torino Atti, 50: 1-12 (1914-15). [1.
32 G. Scorza, Pal. Circ. Mat., 36: 386-395 (1913). [1, 5.
32.1 ———, Pal. Circ. Mat., 41: 263-379 (1916). [1.
33 ———, Pal. Circ. Mat., 45: 1-204 (1921). [1.
34 F. Severi, Vorlesungen über algebraische Geometrie, Trans., by Löffler, Leipzig (1921). [1.
35 ———, P. Rep. II, 2: ch. 33 (1922). [2, 3, 4, 5.
36 ———, Ann. di Mat., (3), 12: 55-79 (1906). [5.
37 ———, Ann. di Mat., (3), 24: 89-120 (1915). [6.
38 ———, Math. Ann., 62: 194-225 (1906). [3, 5.
39 ———, Math. Ann., 74: 515-544 (1913). [5.
40 ———, Lomb. Ist. Rend., (2), 38: 859-865 (1905). [2.
41 ———, Pal. Circ. Mat., 21: 257-282 (1906). [5.
41.1 ———, Pal. Circ. Mat., 28: 33-87 (1909). [6.
42 ———, Pal. Circ. Mat., 30: 265-288 (1910). [5.
43 ———, Ann. Éc. Norm., (3), 25: 449-468 (1908). [3, 5, 6.
44 ———, Rom. Acc. L. Rend., (5), 15_2: 691-696 (1906). [6.
45 ———, Rom. Acc. L. Rend., (5), 30: 163-167, 204-208, 231-235, 273-280, 296-301, 328-332, 365-367 (1921). [5.
45.1 ———, Rom. Acc. L. Rend., (6), 2: 521-526 (1925). [4.
46 ———, Torino Mem., 54: 1-49 (1904). [5.
47 ———, Ven. Ist. Atti, 65: 625-643 (1905). [6.
47.1 ———, Ven. Ist. Atti, 72: 765-772 (1913). [5.
48 C. H. Sisam, Torino Atti, 46: 481-487 (1911). [6.
49 O. Veblen, Colloquium Lectures on Analysis Situs, New York (1920). [3.
50 ———, Am. M. S. Trans., 25: 540-550 (1923). [2.
51 H. Weyl, Math. Soc. Esp. Rev., 43 p. (1923). [2.
52 W. Wirtinger, Encyklopädie II B 2. [1.

For a list of treatises on the topics covered in this chapter see the Encyklopädie articles.

The following books, not mentioned in our list have appeared since these articles were written:

A. Brill. Ebene algebraische Kurven und algebraische Functionen. Braunschweig (1925), 1 volume.

Enriques-Chrisini. Introduzione alla teoria geometrica degli equazioni. Bologna (1915-1924), 3 volumes.

H. W. E. Jung. Algebraische Flächen. Hannover (1925), 1 volume.

H. W. E. Jung. Einführung in der Theorie der algebraischen Functionen einer Veränderlichen. Berlin (1923), 1 volume.

H. W. E. Jung. P. Rep. I_2: 17 (1927). [1.

F. Severi. Trattato di geometria algebrica. Vol. I. Bologna (1927).

H. Weyl. Die Idee der Riemannschen Flächen. Leipzig (1921), 2d edition.

CHAPTER XVI

CORRESPONDENCES BETWEEN ALGEBRAIC CURVES

I. Generalities

1. An algebraic correspondence $T(m, n)$ between the points of two distinct curves C, C', or between those of the same curve C, is the same thing as an algebraic transformation of C into C' or of C into itself, whereby to each point of C correspond n points of C', to each point of C', m points of C. T may or may not impose restricting conditions upon the R. M.'s of the curves. In the first case the correspondence is *singular*, in the second, *non-singular*. A non-singular correspondence between the points of C is also known as correspondence with *valence*. If A is any point of C, G the group of points TA, there exists an integer γ, the *valence* of T, such that $G+\gamma A$ is a group of a fixed linear series on C. This type of correspondence, the earliest known, has already been treated in Chapter 7.

The groups of points of a $(1, n)$ correspondence between C' and C define an *involution* of order n, I_n, on C. It is rational or irrational accordingly as C' is rational or otherwise. Rational involutions, present on every C, are treated in Chapter 7.

An involution of order *one* defines a birational transformation of C into C', or of C into itself when $C'=C$. The various types of correspondences that we have just rapidly described will now be considered separately in detail. The notations and terminology of this chapter are those of the glossary at the end of Chapter 15.

§ 2. Singular Correspondences on C^p

2. Historically speaking they have been introduced by Abel in connection with complex multiplication of elliptic functions but without much regard to the geometric aspects of the question. A. Hurwitz as we shall see must be regarded as the true creator of the theory. Before going into it we shall find it instructive to consider an elliptic carrier C, for example the cubic whose representation by Weierstrass functions is

$$x=\wp(u;\,\omega_1,\,\omega_2)\,; \qquad y=\wp'(u;\,\omega_1,\,\omega_2),$$

and examine the relation of the *multiplication* or *transformation* of elliptic functions with algebraic correspondences on C.

If the correspondence $(m, n): T$ assigns to the parametric point u the points $u_1, \ldots, u_n$, then by a familiar argument Σu_i is an integral of a rational function of $\wp, \wp'$, holomorphic over the period parallelogram hence of the form au, and $\wp(au)$, $\wp'(au)$ are rational functions of $\wp(u)$, $\wp'(u)$. Conversely if a is a constant such that $\wp(au)$, $\wp'(au)$ are rational functions of $\wp(u)$, $\wp'(u)$, there exists a correspondence T attached to it, namely the correspondence assigning to u the points au. Now the discovery of constants a with the property in question is the precise object of the theory of *transformation* or *multiplication* of elliptic functions, upon which we are thus thrown back. Referring to that theory; on every elliptic carrier there exist correspondences for which a is an integer: they are the *ordinary* correspondences, or correspondences with *valence*, the valence being in fact $-a$. In the exceptional case where the period ratio $\frac{\omega_2}{\omega_1}$ is a number of a quadratic imaginary domain there are present correspondences of a different type, the *singular* correspondences. There appears to be no reason for considering them here, as they have received, under one name or another, ample treatment elsewhere. Severi ([74.1]), p. 273, ([79]), has discussed them geometrically. For further details and references see Fricke([25]), Krazer([44]), p. 183, Fueter([25.1]).

3. Leaving aside the case $p=1$, singular correspondences on an algebraic curve C have been introduced and thoroughly investigated by A. Hurwitz, first in connection with a special C^3 that arises in the study of modular equations([30]), then in the general case (1886), ([32]), ([34]). There is a very complete treatment in Klein-Fricke([40]), vol. 2, p. 518, and a more brief one in Baker([4]), p. 637. See also Zeuthen([96]), p. 285, Berzolari([7]), p. 347.

4. On the algebraic curve of genus p, C^p, let $\delta_1, \ldots, \delta_{2p}$ be an F. S. of cycles, $u_1, \ldots, u_p$, a set of independent integrals of the first kind, $\boldsymbol{\omega}$ the corresponding R. M. If x and y are corresponding points in an (m, n) correspondence T on C^p, then as (x) describes δ_μ the corresponding points $y_1, \ldots, y_n$ describe paths whose sum is a cycle $\sim\Sigma a_{\mu\nu}\delta_\nu$. The matrix $\mathbf{a}$ defines the transformation on the cycles corresponding to T. There is also between the integrals of the first kind a relation

$$(1) \qquad \sum_j u_k(y_j) = \sum_s a_{ks} u_s(x) + g_k$$

which leads to the matrix equation

$$(2) \qquad \boldsymbol{\alpha\omega} = \boldsymbol{\omega}\mathbf{a}.$$

Thus T defines a multiplication of $\boldsymbol{\omega}$ (Ch. 15, No. 7). Conversely for every multiplication (2), we know from the theory of Jacobi's inversion problem that there exists an attached correspondence T whose index

$n=p$. The relation (2) is an actual restriction for $\boldsymbol{\omega}$ if and only if the multiplication is complex. It is then and only then that T is a singular correspondence. In the non-singular case the integer matrix a is a multiple of the identity, or $\mathbf{a}=-\gamma\mathbf{I}$. Then from (2) follows $\boldsymbol{\alpha}=-\gamma\mathbf{I}$ and (1) becomes

$$\gamma u_k(x)+\Sigma u_k(y_j)=\text{const.}$$

These relations show that for γ positive, $\gamma x+\Sigma y_i$ is a group of a fixed linear series $g_{n+\gamma}$, while for γ negative γx plus some fixed points is a coresidual of the group Σy_j. This identifies T as a correspondence with the valence γ. Conversely the preceding relation subsists for any T with valence γ. Hence non-singular correspondences and valence correspondences are the same.

Valence correspondences, and with valence γ an arbitrary integer, exist on every C as follows from the remark made above and as first shown by Hurwitz([34]). Severi has shown([76]) that it is the only type present on a C^p with general moduli. From what we have said follows that it is the only type present whenever there is no complex multiplication, Scorza([70]), Rosati([60]). Severi's theorem is then an indirect proof that the R. M. of a general C is without complex multiplication. See, however, a recent discussion by Zariski([95.1]), also Lefschetz([50.1]), and Severi([77.3]).

5. Hurwitz([34]) has defined correspondences as independent or not, accordingly as their multiplications are or are not linearly independent and showed that the maximum number that are independent is $\leqq 2p^2$. It is, in fact, the invariant h of $\boldsymbol{\omega}$. Klein expressly introduces an F. S. for correspondences([40]), vol. 2, p. 543. Hurwitz([34]) also gives for the number of coincidences of T the expressions

$$U=m+n-\Sigma a_{\mu\mu} \tag{3}$$

$$U=m+n+\Sigma c_i\lambda_i \tag{4}$$

where the λ's are the coefficients in the expression of T in function of the correspondences of the F. S. and the c's are integers that depend upon those correspondences but not upon T itself. When there is valence $a_{\mu\mu}=-\gamma$ and (3) reduces to the Cayley-Brill formula (Ch. 7, No. 6). A formula that includes (4) has also been derived by Severi([79]), ([74.1]), p. 240. From (3) Hurwitz also deduces an expression for the number of groups common to two correspondences. This method is based upon theta functions. Klein([40]), vol. 2, Baker([4]) replace these by the so-called prime functions.

Recently purely topological methods have been applied to coincidence formulas which have been shown to hold for correspondences that are merely continuous. Lefschetz has investigated the most general type of continuous correspondences on an n dimensional manifold and obtained formulas which reduce to those of Hurwitz in the case of algebraic curves, ([46]), ([47]), ([50]), ([48]). See also Chisini([16]); Enriques-Chisini([23]), vol. 3, p. 427; Alexander([1]).

6. Severi has systematically developed (1903), ([79]), ([74.1]), Ch. 6, the geometric aspects of the Hurwitz theory. Let $T(x)$ denote the group of points assigned by T to x. He introduces the sum $T+T'$ and product TT' of two correspondences. The latter has its natural meaning, the former is defined as assigning to x, the group $T(x)+T'(x)$. If T, T' have valences γ, γ', TT' and $T+T'$ also have valences, respectively equal to $-\gamma\gamma'$ and $\gamma+\gamma'$. Operations of a somewhat different type have been considered by A. A. Bennett([5]), ([6]). Severi also defines T^{-1} as assigning to x every y such that x belongs to $T(y)$. A relation $\Sigma\lambda_i T_i=0$ implies that the abelian sums of $\Sigma\lambda_i T_i(x)$ are all zero or congruent to periods, which is written $\Sigma\lambda_i T_i(x)\equiv 0$ (linear equivalence), with a well-known interpretation in terms of linear series. Any correspondence with valence γ satisfies the relation $T+\gamma I=0$. The existence of a minimum base follows then from the Hurwitz results. Severi shows that if U_i is the group of coincidence points of T_i then from the above equation between the T's follows the linear equivalence $\Sigma\lambda_i(T_i(x)+T_i^{-1}(x)-U_i)\equiv 0$, and therefore $\Sigma\lambda_i(m_i+n_i-U_i)=0$ where m_i, n_i are the indices of T_i and U_i its number of coincidences. From this to the expression (4) is but a step.

In the second part of ([79]) Severi applies the very suggestive notion of the surface F image of the non-ordered point pairs of two algebraic curves C^p, C^q to the study of their correspondences. T is then represented by an algebraic curve Γ on the F corresponding to the case when C^q is merely a copy of C^p. Known birational invariants of algebraic curves on a surface give rise to two new invariants of T: the *genus* and *degree* (or similar elements of Γ for F). See also De Franchis([22]), Severi([77.4]), Lefschetz([48]). Further references to papers on the surface F are given by Severi([79]), p. 5. To dependent correspondences and a minimum base in the sense of Hurwitz correspond dependent curves and a minimum base for the curves of F in the sense of Severi. In fact it is here that Severi found the starting point for his investigations on the bases for surfaces (Ch. 15, No. 19).

In the same paper Severi considers special correspondences, in particular the so-called symmetric type (T and T^{-1} coincide) and also the associated surface, image of an involution of order two on F. In ([74.1]), p. 281, ([78]), he shows that if μ is the degree of T, then

$$2p(2mn-\mu)^2 \geqq (U-m-n)^2$$

where the equality holds only when T has the valence $(U-m-n)/2p$.

This follows also from an inequality due to Hurwitz([34]), p. 580. See also Castelnuovo([12]); Rosati([66]); Roth([67]). Severi has studied in ([76]), ([77]) correspondences on an algebraic curve which varies within a sufficiently general linear system on an algebraic surface F. When F is regular every such T has valence. This is to be understood in the sense that exceptional curves of the system may well have singular correspondences, but not so its generic curve. Recently discussing Severi's theorem, Zariski([95.1]) has shown that a sufficient condition in order that it hold is that the linear system on the regular F be *simple* (i. e., its curves through an arbitrary point of F must not pass ipso facto through other variable points). Severi shows([76]), p. 521, that on C^p all correspondences of given indices m, n, are distributed into a finite number of continuous families.

In his latest book ([74.1]), Ch. 6 (1927), which has very complete historical and bibliographical data, Severi makes a searching investigation of algebraic correspondences and goes as far as it is possible to go without transcendental methods. His treatment is very complete on the algebraical side. He makes use of the same surface F as above and of the degree and genus of T. He associates with T its *complementary* correspondence T' defined as follows: Let g^n_{n+p} be the complete non-special series determined by a definite group $T(x_0)$ plus p arbitrary points of C^p. It will assign a unique residual group of p points to every group $T(x)$, and this group is by definition $T'(x)$. Manifestly $T+T'=0$. Severi uses T' in the proof of the Cayley-Brill formula, and investigates the properties of T' in relation to those of T. See also Severi([78]), ([77.2]).

7. A systematic study of singular correspondences and related numerical invariants has been made by Rosati(1913), ([66]), ([66.1]), ([60]), ([63]). Rosati in fact developed the theory of Riemann matrices about the same time as Scorza (1915) and by a projective method dual to his, but with the treatment of singular correspondences as his chief object. Let two correspondences T, T' be called *equivalent* if the abelian sums for $T(x)$ and $T'(x)$ differ only by constants, i. e., if $T=T'$ in Severi's notation. Geometrically speaking T and T' are equivalent if $T-T'$ is a correspondence with zero valence. Then to a class of equivalent correspondences corresponds a unique multiplication (2) and conversely. To a sum or product of correspondences correspond the sum or product of the multiplications of their two classes. Hence classes of correspondences and multiplications may be designated by a common symbol to be combined according to the ordinary laws of addition and multiplication. If $T_1, \ldots, T_n$ are any number of such symbols, any polynomial $F(T_1, \ldots, T_n)$ whose coefficients are integers is the common symbol for a certain correspondence and its associated multiplication. It follows that the multipli-

cation algebra (Ch. 17, § 6) goes over into an algebra for correspondences, and it is this algebra which Rosati systematically investigates in his work.

Rosati ([66]), like Severi notably in ([74.1]), considers *virtual* correspondences, corresponding to virtual curves on the surface F defined in No. 6, and their characters are considered by both authors. Rosati defines and investigates at length ([66]) the *simultaneous* invariant of T, (m, n) and T', (m', n') as $mn'+nm'-\nu$ where ν is the number of their common couples. The simultaneous character K of T and T^{-1} is $m^2+n^2-(U+2d)$ where d is the number of common couples of T and T^{-1}. It satisfies the inequality $-\Omega \leqq K \leqq \Omega$, $\Omega = 2mn-\mu$. The equalities correspond respectively to $T=-T^{-1}$ or $T=T^{-1}$, ([66]) p. 389. Here T^{-1} is the inverse in the more general sense considered below.

Rosati ([66.2]), ([66.3]), introduces the notion of characteristic and minimum equations for a correspondence (see Ch. 17, § 6), and investigates their properties. He greatly generalizes the notion of *multiple valences* introduced by Severi ([76]), p. 533. The valences of T as defined by Rosati are the roots of the equation $|\alpha+xI|=0$, or the negative multipliers of the multiplication T. A Rosati valence may also be defined ([66.3]) as any number γ such that there exists an integral of the first kind u having the property that $\gamma u(x)+\Sigma u(y_j)$ is constant. When the valences are integers, but not all equal, they become the Severi *plurivalences*.

The inverse T^{-1} of T already mentioned above plays an important part in Rosati's work. As he understands it, it is the class individualized by the Severi inverse (see No. 6), *not* the class belonging to the inverse of the multiplication of T. Let the F. S. on C^p be the retrosections, so that the fundamental principal form of $\boldsymbol{\omega}$ is $\mathbf{c}=\begin{Vmatrix} 0 & \mathbf{I} \\ -\mathbf{I} & 0 \end{Vmatrix}$. Then if (2) is the multiplication T, the multiplication T^{-1} (associated with the correspondence T^{-1}) will have in place of $\mathbf{a}$ the matrix $-\mathbf{c}\mathbf{a}'\mathbf{c}$. This was stated by Hurwitz ([34]) and proved analytically by Rosati ([61]), p. 38, and topologically by Lefschetz ([48]). If T, T' are such that TT' has valence, i. e., if $TT'+\gamma I=0$, Rosati ([60]), p. 15, calls them *complementary*. The matrices $\boldsymbol{\alpha}$, $\mathbf{a}$ of T' are then $-\gamma\boldsymbol{\alpha}^{-1}$, $-\gamma\mathbf{a}^{-1}$. T is said to be *symmetric* if $T=T^{-1}$, *skew-symmetric* if $T=-T^{-1}$. To each rational alternate matrix of $\boldsymbol{\omega}$ corresponds a class of symmetric correspondences, to each symmetric rational matrix a class of skew symmetric correspondences. Hence the numbers of linearly independent symmetric and skew symmetric correspondences are $\mu_1=k$, $\mu_2=h-k$. These numbers are both $\leqq p^2$, Rosati ([60]), p. 14.

The valences of T and T^{-1} are conjugate of one another and roots of equal multiplicity of their respective equations. The valences of a symmetric T are all real, those of a skew symmetric T are all pure complex or zero. Their minimum equations have no multiple roots.

On various occasions, notably in ([62]), Rosati considered correspondences S, T, such that $lS=f(T)$, where f is a polynomial, its coefficients and l being integers. For a given T let S vary within the total set of similar correspondences. Then its j-th valence varies within a definite Dedekind order of the algebraic domain $K(\gamma_j)$, where γ_j is the j-th valence of T. The units of the order correspond to birational transformations of the Jacobi V_p of C^p, which under certain circumstances may have an infinite discontinuous abelian group G. When T^{-1} is not a rational function of T its order is in general different from that of T and $G'\neq G$. When the two orders, hence also the two groups coincide, certain cyclic transformations of V_p determined by the orders become Hermitian and therefore by a theorem due to Comessatti ([21]), they give rise to a birational transformation of C^p into itself. Rosati examines at length the application of this theory to C^2. See in this connection Ch. 17, No. 63.

When $\boldsymbol{\omega}$ is impure the matrix $\boldsymbol{\alpha}$ may be of rank $q<p$. Then there are two special sets of integrals associated with T. The first consists of q integrals having the property that all abelian sums $\Sigma u(y_j)$ are linear combinations of these q integrals. The second consists of $p-q$ integrals for which the sums are constant. A more general circumstance would be that these sums are equal to the corresponding integral multiplied by an integer γ, which is a Severi plurivalence. The number of independent correspondences corresponding to assigned pairs of sets of integrals is $\leqq 2q^2$. The sets are intimately bound up with the reductions of $\boldsymbol{\omega}$ to its normal form in the sense of Ch. 15, No. 7; Ch. 17, No. 39. The presence of the first set indicates that there is an isomorph $\begin{Vmatrix} \boldsymbol{\omega}_1 & 0 \\ 0 & \boldsymbol{\omega}_2 \end{Vmatrix}$ such that when $\boldsymbol{\omega}$ is reduced to it, (2) becomes a multiplication $\boldsymbol{\alpha}_1\boldsymbol{\omega}_1=\boldsymbol{\omega}_1\mathbf{a}_1$ on $\boldsymbol{\omega}_1$ alone. In a slightly more general case we would have $\boldsymbol{\alpha}_1\boldsymbol{\omega}_1=\boldsymbol{\omega}_1\mathbf{a}_1$, $\gamma\boldsymbol{\omega}_2=\boldsymbol{\omega}_2\gamma$, where γ is a plurivalence. As this last relation is identically verified in $\boldsymbol{\omega}_2$, γ is an arbitrary integer. See in this connection Severi ([76]), Rosati ([60]), ([66.2]).

8. Castelnuovo has shown ([10]) that if C^p possesses a continuous system ∞^1 of correspondences with one index m then $p\leqq\binom{m+1}{2}$; when equality holds, C^p is the birational transform of a non-singular plane curve of order $m+2$. This holds with the restriction that the ∞^2 groups of points which by means of the system correspond to all points of C^p must not be compounded with the groups of an involution.

Correspondences on hyperelliptic curves have been studied by Wiman ([94]), first paper; Kantor ([38]); Holmquist ([28]); Torelli ([83]).

Correspondences on a C^2 have been studied by Rosati(60), p. 18, (64).

The most interesting special investigations on singular correspondences have originated in modular equations, Klein(43); Hurwitz(30), (33). Extensive elaboration with further developments by Klein(40), vol. 2, pp. 597-704. If in place of the general modular group, congruence subgroups modulo m are considered, the fundamental polygon of the complex variable, say Z, is depictable on a Riemann surface whose genus p depends upon m. Let C be an algebraic curve parametrically representable, in uniform manner by automorphic functions of Z attached to the polygon. In the Z plane the relation $Z'=nZ$ (n integer prime to m), defines a transformation giving rise on C^p to an algebraic correspondence whose indices are both $n\Pi\left(1+\frac{1}{q_i}\right)$, where the q's are the distinct prime factors of n(40), vol. 2, p. 599. In many cases the correspondence turns out to be singular, for example when $m=11$ and n is not a quadratic residue of 11(40), vol. 2, p. 618. This is closely related to the theory of modular equations. There is worth while work to be done by way of investigating these singular correspondences, whose properties would help to enlighten the general theory. Further types of singular correspondences arise also in connection with curves invariant under a cyclic group, § 6.

§ 3. CORRESPONDENCES BETWEEN C^p AND C^q

9. Their theory is in close parallel with the case already considered. If $\boldsymbol{\omega}$, $\boldsymbol{\Omega}$ are the R. M.'s of the curves, to the correspondence T is attached a relation

$$\boldsymbol{\alpha}\boldsymbol{\omega}=\boldsymbol{\Omega}\mathbf{a} \tag{5}$$

analogous to (2). This is not a restriction on the matrices when and only when $\boldsymbol{\alpha}=0$, in which case also $\mathbf{a}=0$. Then if x is a generic point of C^p the group $T(x)$ on C^q belongs to a fixed linear series and similarly on C^p. Such a correspondence is said to be *ordinary* and may be set up between any pair of curves. When (5) is an actual condition it implies that $\boldsymbol{\omega}$, $\boldsymbol{\Omega}$ have a common submatrix (Ch. 15, No. 7). The correspondence is then *singular*. The number of independent singular correspondences is equal to the simultaneity index λ of $\boldsymbol{\omega}$ and $\boldsymbol{\Omega}$, Hurwitz(34); Severi(78).

Rosati(61), (63), has more especially investigated what happens when the R. M.'s are impure. In that case (5) may actually be reduced to a multiplication of a common submatrix $\boldsymbol{\omega}_1$ of genus p_1 of $\boldsymbol{\omega}$ and $\boldsymbol{\Omega}$. There are $p-p_1$ integrals of the first kind invariant on the groups of T on C^p and similarly $q-p_1$ on C^q. The curve Γ image of $T(m, n)$

(image of the pairs of associated points of the two curves) carries two involutions I_n, I'_m whose images are C^p and C^q. Conversely a curve Γ with two distinct involutions can be considered as the image of a correspondence between two other curves in the manner just described. Rosati studies these involutions and their products especially in relation to the R. M. of Γ. In particular he finds that if they have the same genus they cannot leave the same set of integrals of the first kind invariant. Each is then individualized by a definite mode of decomposition of the R. M. To this we return in No. 10. Rosati also considers([61]), p. 48, ([62]), p. 8, the correspondences $T^{-1}T$ on C^p and TT^{-1} on C^q. Their valences are real and $\leqq 0$, those $\neq 0$ being equal and of same multiplicity. On Γ the products of I, I' in one or the other order give rise to correspondences whose valences are also real, those $\neq 0$ being again the same as for TT^{-1} and with same multiplicity.

The algebraic aspect of correspondences between two curves has been searchingly investigated by Severi([74.1]), Ch. 6, especially in regard to their representation by one or two equations. See also Hurwitz([34]), Snyder-Sharpe([82]) who give examples. Severi, loc. cit., gives a very complete discussion of various properties of correspondences, also very complete historical and bibliographical indications.

Further references: Severi([78]), ([77.1]), where he shows that the degree of T, of indices m, n, is $\leqq 2mn$, the maximum being reached only when T is ordinary.

10. Regarding special types, Chisini in ([14]) has completely classified all (2, 2) correspondences. See also § 4 for special $(1, n)$ types, involutions and Klein's regular Riemann surfaces.

R. Torelli([87]), ([88]) has made a special study of correspondences between curves of same genus p with the chief object of determining when they are birationally equivalent. His work has recently been summarized by L. Godeaux([26]). His chief result is that if two curves of the same genus have the same R. M. relatively to their retrosections they are birationally equivalent. That the period matrices must correspond to retrosections is essential when the moduli of the curves are not general [Rosati([64])]. Severi showed in a supplement to ([21]), p. 452, that when the singularity index k of the period matrix ω of a C^p is unity (non-singular matrix) the Jacobi variety V^p entirely individualizes C^p. The central point in his proof is the fact that the hypersurface cut out on V^p by a θ of order one is an F. S. for all hypersurfaces of V_p. This is a special case of a general theorem due to Lefschetz([49]), p. 389. See also Ch. 17, § 3. Comessatti has proved this general theorem([21]): When there exists between the Jacobi varieties V_p, V'_p of C^p, C'^p a hermitian birational transformation

T, or a transformation into one another of the θ's of order one corresponding to their normal period matrices, there is associated with T a definite birational transformation of the curves. This theorem was proved for $p=2$ by Enriques-Severi([24]), p. 375. It does not imply any restrictions on the R. M.'s. A θ of order one cuts out on V_p a hypersurface V_{p-1} image of all sets of p points of C^p that belong to an assigned g^{p-1}_{2p-1}. Geometrically then the theorem is that if TV_{p-1} is a similar hypersurface for V'_p, T determines a birational transformation of C^p into C'^p. Comessatti gives also a new proof of Torelli's theorem and shows that any singular birational hermitian T of V_p into itself induces a birational transformation of C^p into itself. See also in this connection Lefschetz([49]), p. 459.

§ 4. Irrational Involutions and Related Topics

11. Let C^q be the image of the involution I_n, $q>0$. It defines an $(n, 1)$ correspondence between C^p and C^q or a rational transformation from the first curve into the second. We are particularly concerned, however, with all such correspondences for an assigned C^p, and variable C^q. There are also the related questions of regular Riemann surfaces and reducible abelian integrals to be examined in their turn.

Weber has shown([92]) that when two curves of same genus $p>1$ are in simply rational correspondence the correspondence is actually birational. See also Picard([57]), p. 542 (also previous two editions of same work); Enriques-Chisini([23]), p. 456. This means that on C^p an I_n of genus p is of order *one*. This has been proved topologically by J. W. Alexander([1]), p. 175. His stated restriction that I_n has for image C^p itself is not used in his proof. Rational transformations of an elliptic curve into itself have been completely classified by Torelli([85]) and Scorza([69]). In ([54]) Painlevé has made a thorough investigation of a rational transformation T of C^p into C^q, $q>0$. Following a method applied by Picard to birational transformations (§ 6), he considers the effect of T on differentials of the first kind, then takes their ratios, obtains Weber's theorem and proves that the number of transformations with $q>1$ is at best finite. For $q=1$ there may be a discreet set and only a continuous family when $q=0$. See also Humbert([29]); Painlevé([54.1]). ([55]); DeFranchis([22]); Berzolari([7]), p. 350; Severi([74.1]), p. 288; Enriques-Chisini([23]), p. 477. DeFranchis and Enriques-Chisini gave this very simple proof of the Painlevé theorem: (a) From Zeuthen's expression $U=2p-2-n(2q-2)$ for the coincidences (Ch. 7) they derive $n\leqq \dfrac{p-1}{q-1}$ when $q>1$. (b) For a given n the algebraic nature of the problem insures that the number of distinct continuous families is finite.

(c) When $q>0$ each family consists of a single I_n for the number of common pairs of transform points for two involutions in the same family is found on computation to be negative. From this to the theorem is but a step. Further references on involutions are Rosati([61]), ([62]); Severi ([74.1]), Ch. 6 and this chapter No. 9.

12. *Regular Riemann Surfaces.*—The inverse and closely related problem consists in the determination of all curves C^p which carry an I_n whose image is C^q. A Riemann surface K^p for C^p is derived from a surface K^q attached to C^q about like an ordinary Riemann surface from the complex plane. We draw on K^q non-severing cuts issued from a fixed point, take n copies of the configuration and unite them along the borders of the cuts. The cuts and mode of uniting are governed by the sole condition that the final configuration be itself a connected Riemann surface. K^p so obtained has been called *regular Riemann surface* by Klein, who appears to have been first to construct that type. The groups of its I_n are the sets of n points of K^p that correspond to one and the same point of K^q. For the main references along this line see Krazer-Wirtinger([45]), p. 843, to which may be added Enriques-Chisini([23]), p. 427. In ([56]) Poincaré has considered the question from the point of view of automorphic functions and the related fundamental polygon. For polygons corresponding to modular functions see Klein([40]), vol. 1, p. 333. Of interest is the observation that I_n may well be without coincidences, Poincaré([56]), p. 318; Enriques-Chisini([23]), p. 430. Then the cuts on K^q are all closed circuits. Chisini([17]), ([18]), and Enriques-Chisini([23]), pp. 430-456, have investigated at length the case where the permutations of the n copies of K^q along the cut circuits are powers of one and the same cyclic permutations. Then if $f(x, y)=0$ represents C^q, C^p can be represented as $f(x, y)=0$, $z^n=F(x, y)$, where F is a polynomial.

13. *Reduction of Abelian Integrals.*—As applied to integrals of the first kind it consists in the discovery of a set of $q<p$ integrals of the first kind of C^p whose period matrix is of the form $\|\boldsymbol{\Omega}; 0\|$ where $\boldsymbol{\Omega}$ is a submatrix of genus $q<p$ of the R. M. $\boldsymbol{\omega}$ of C^p. In other words the question is to find if $\boldsymbol{\omega}$ is pure or impure and in the latter case to find its submatrices. As a more special problem we may wish to find if the q integrals are rationally transformable into a set of integrals of a certain C^q. More precisely given $C^p: f(x, y)=0$ we wish to know if it possesses a set of $q<p$ integrals of the first kind $v_i(x, y)$ $(i=1, 2, \ldots, q)$ such that a rational transformation

$$(6) \qquad x=R(x', y'), \qquad y=S(x', y')$$

reduces them to q distinct integrals attached to the transformed curve C^q; $F(x', y')=0$. C^q represents an involution of genus q on C^p, so that there is an intimate connection between this problem and the determination of irrational involutions. The periods of the v's are linear combinations of $2q<2p$ of their number and those relatively to a suitably chosen set of $2q$ cycles give rise to a submatrix $\mathbf{\Omega}$ of $\boldsymbol{\omega}$ isomorph of the R. M. of C^q.

It must be kept in mind that the reduction in the general sense does not imply reduction by means of a transformation (6). In other words when $\boldsymbol{\omega}$ is impure it does not follow that its submatrix $\mathbf{\Omega}$ is the actual isomorph of the R. M. of a C^q, $q<p$ or, if we please, it does not follow that C^p carries an irrational involution of genus $<p$. This was first pointed out by Wirtinger, also mentioned by Poincaré and Severi. See ([45]), p. 843.

It is in this connection that Poincaré and Picard proved for the first time the fundamental theorem on impure matrices, according to which, if $\boldsymbol{\omega}$ is reducible to

$$\left\|\begin{matrix}\boldsymbol{\omega}_1 & 0\\ \boldsymbol{\omega}_3 & \boldsymbol{\omega}_2\end{matrix}\right\|$$

where $\boldsymbol{\omega}_1$ is an R. M., then it is also reducible to the form

$$\left\|\begin{matrix}\boldsymbol{\omega}_1 & 0\\ 0 & \boldsymbol{\omega}_2\end{matrix}\right\|,$$

where $\boldsymbol{\omega}_2$ is another R. M., the first complete proof of which was given by Scorza([70]), p. 298, where further references are found.

For additional references on reducible abelian integrals see Krazer-Wirtinger([45]), p. 838; on combinations of systems of reducible integrals also Severi([74]), Chap. 10.

§ 5. Irrational Series

14. Their study is closely related to that of correspondences between two curves. The series, assumed on C^p, is usually denoted by γ_n^1, where n is the number of points in each group (all variable unless otherwise stated). The *index* ν is the number of groups of γ_n^1 which contain a given point, the *genus* q that of the algebraic curve C^q image of γ_n^1. The curves C^p, C^q are in (n, ν) correspondence. Here, however, C^q is a variable curve and the object is to investigate properties common to all such correspondences for C^p fixed. The necessary and sufficient transcendental condition for the existence of a γ_n^1 with C^q for image is that the R. M.'s of C^p and C^q have a common submatrix.

R. Torelli has investigated these series([86]), ([84]). He defines a class of series: it consists of a maximal set whose groups can be paired off so as to give constant abelian sums or differences. Any two such groups of associated series of the same class either have a sum which is a group of a fixed linear series, or else are coresidual relatively to such a series. Torelli considers the effect produced on the Jacobi variety V_p when the curve possesses series of genus $<p$, in particular as to the congruence of subvarieties that V_p may thereby acquire. The existence of such a series when not contained in a g_n^r implies the existence on V_p of abelian varieties of genus $<p$, hence the R. M. is impure. Hence if it is pure, in particular if it has general moduli, there can be no γ_n^1 of genus $<p$ on C^p. This is also contained in Severi's theorem ([76]) according to which on a C^p with general moduli every correspondence has valence. Torelli also investigates the behavior of γ_n^1 regarding linear equivalence of its groups, examines classes containing an involution or a series composed with an involution and conditions under which an involution is cyclic (generated by a cyclic birational transformation). See also Godeaux ([26]). There are a number of investigations referring to the comparison of γ_n^1 with a g_m^r. Schubert has proved the relation

$$Z = m\nu\binom{n-1}{k} - \binom{n-2}{k-1}d - \mu\binom{n}{k+1},$$

where $2d$ is the number of double points of γ_n^1, $k+1(\leqq r)$ the number of conditions imposed upon a group of g_m^r in order that it contain a generic group of γ_n^1, and finally Z is the number of groups of the latter which impose k or less conditions on the linear series. Schubert's proof is reproduced by Segre in ([72]), ([71]), § 13. For further related developments see also Castelnuovo ([11]). In the same reference will also be found the proof of the relation

$$z = \nu(n+p-1) - d,$$

where z is the *defect of equivalence* of γ_n^1 or number of its groups partially contained in a non-special g^{n-1}_{n-1+p}. Severi ([75]) has shown that if the ν groups of γ_n^1 which contains a given point belong to a fixed g_n, then so does γ_n^1 itself. Castelnuovo showed ([13]) that in order that γ_n^1 belong to a g_n it is necessary and sufficient that $z=0$. For these properties see also Berzolari ([7]), pp. 350-352, Severi ([74]), pp. 191-196. Extensions of the Castelnuovo criterium to multiple series have been given by Torelli ([89]), ([90]). The transcendental interpretation for that criterium is that the p abelian sums for γ_n^1 must be constant. This Comessatti generalizes in ([20]), by introducing the character Z_r, $r\leqq p-1$, $Z_0=z$, number of sets of r groups of γ_n^1 contained in a non-special $g^{\mu}_{\mu+p}$, $\mu=(r+1)(n-1)$. He shows that in order that precisely $p-r$ integrals of the first kind give rise to constant abelian sums it is necessary and sufficient that $Z_r=0$, $Z_s\neq 0$ for $s<r$. He derives various relations between these integers, but

obtains their explicit expression only for an involution of genus q when $Z_r = n^{r+1}\binom{q}{r+1}$.

Let T be the symmetrical correspondence derived from γ_n^1 by assigning to every point of C^p all those which belong to a group that contains it. By investigating sets of common zeros of intermediary functions attached to the R. M., Castelnuovo([13]) showed that the Z's are the coefficients of the characteristic equation of $-T$. In ([65]) Rosati derives in a new manner Comessatti's result, relations between the Z's and invariants of certain correspondences associated with γ_n^1, also these results: If the representative curve Γ of the series and C itself have general moduli they are birationally equivalent. The Comessatti invariants of the two series determined by a correspondence between two curves on the curves are the same.

Characters analogous to the above but for multiple irrational series have been considered by Allen([2]), ([3]). He also applies them to the extension of a formula given by Jonquières for linear series.

§ 6. Birational Transformations

15. We shall consider exclusively those of a C into itself. When $p>0$ they give rise to excellent examples of curves with singular correspondences.

Schwarz has shown, ([68]) (1879), that when the set of transformations is continuous C^p is parametrically representable by meromorphic functions of an integral of the first kind if any exist, hence $p=0$ or 1. Klein showed by means of automorphic functions that the set cannot even be discrete when $p>1$, ([39]), p. 64; ([41]). Of these results Picard has given very simple proofs, readily extended to surfaces, by means of ratios of differentials of the first kind, ([59]), ([58]), ([57]), p. 524 (also previous editions). Proofs of a more geometric character: Hettner([27]), completed by M. Noether([52]), ([53]); Segre([71]); DeFranchis([22]); Chisini([15]); Weierstrass([93]), p. 285; Severi([74]), p. 143; ([74.1]), p. 173; Enriques-Chisini([23]), p. 309. Important generalizations by Painlevé([54]), including proofs of the same theorems also. In ([35]), ([31]) (1887), Hurwitz has treated the question of determining all curves with birational transformations. With a given transformation T there is associated a linear transformation T' of the integrals of the first kind. The groups of T and T' are isomorphic for $p>1$. T' determines in fact a multiplication of the R. M. $\boldsymbol{\omega}$ of C^p, which is a complex multiplication unless the curve is hyperelliptic and T the birational transformation that permutes the points of each pair of its g_2^1. It follows that C^1, $p>2$, does not in general possess a birational transformation into itself. The multiplication is shown to be a so-called *principal transformation,* leaving invariant the fundamental alternate form of $\boldsymbol{\omega}$, or, if we please, a set of associated theta functions (Ch. 17, No. 50). It follows that T' is cyclic and so is T. A very similar result holds for simply rational transformations of C^p into itself, (35), p. 301. That the total number of birational transformations is finite Hurwitz (36) proves by considering their effect on the Weierstrass points whose number he shows to be $\geqq 2p+2$ and comparing with the number of fixed points of T whose number is at most $2p+2$ with special considerations for the hyperelliptic case. If n is the order of T, C^p is the birational transform of a curve $f(x^n, y)=0$. In any case for $p>1$, $n \leqq 10(p-1)$ and the total order of the birational group is $\leqq 84(p-1)$, in the hyperelliptic case in fact $\leqq 4p+2$. See also Klein (40), vol. 2, p. 554-556; Wiman, who lowers n down to $4p+2$ for all cases, (94), first paper; Berzolari (7), p. 352; Severi (74), p. 145, (74.1), p. 175; Wirtinger (95). For the related investigations of Torelli, Severi, Comessatti, Rosati, see No. 11.

For birational transformations of C^1 see Fricke (25), Segre (73), Chisini (16.1), Severi (74.1), p. 185.

Birational transformations for $p=2$ have been investigated by Bolza (8), (9), summary of his results in Enriques-Severi (24), p. 378. Reducing C^2 to the form $y^2=f(x)$ where f is a polynomial of degree 6. the problem is to determine all such polynomials invariant under a projective group and their groups. He finds six types of essentially distinct curves and determines their groups. See also Kantor (38); Wiman (94), first paper.

Klein has studied the group of the quartic of genus three that bears his name (42); (40), vol. 1, pp. 692-731. The curve can be reduced to the form $x^3y+y^3z+z^3x=0$. It is a normal curve for $p=3$, not hyperelliptic. Its total group is the well-known projective G_{168} and a thorough discussion of it is given especially in relation to the subgroups of G_{168} and their effect upon the configuration of the inflections and double tangents both real and complex. Ciani in (19), has investigated the projective groups of all plane quartics. Enriques-Chisini (23), pp. 312-339, have done likewise for the quartic normal curve for the genus three. By considerations of fixed points they show that the order of any transformation of the group (which is then the complete birational group), can only be 2, 3, 7. See also Kantor (37); Wiman (94), first paper.

Birational transformations for $p=4$, 5, 6, have been considered by Wiman (94), second paper. For $p=4$ also by Miss VanBenschoten (91); for $p=5$ by McKelvey (51); for $p=6$ by Snyder (80). Their method is to investigate the projective transformations of the normal curve. Snyder has also investigated all plane quintics invariant under a projective group and determined their groups([81]). In ([49]) part II, Ch. IV, Lefschetz examines the birational group of

$$y^q = \Pi(x-a_i)^{\alpha_i}; \quad q \text{ odd prime},$$

when the a's are arbitrary, the type including many found in the literature. The genus $p=\frac{1}{2}r(q-1)$ where $r+2$ is the number of critical points of $y(x)$. He determines a set of integrals of the first kind and the corresponding R. M. By considering the effect of a birational transformation upon the critical points and also upon the R. M., he determines the complete group of the curve. The case $r=1$ demands a special treatment. The groups G are cyclic of order q, unless the curve is hyperelliptic when G is of order $2q$, or reducible to the curve

$$x^m y^n + y^m z^n + z^m x^n = 0, \qquad m^2 - mn + n^2 = q,$$

when G is of order $3q$, except when $q=7$ where the curve is Klein's quartic already considered. Several curves of these last types have been discussed previously, notably for $q=13$, $m=4$, $n=7$, $p=6$ by Snyder([81]). He also considers another curve of similar equation, with $m=2$, $n=3$, $q=7$, but its genus $p=5$ and as $2p=10$ is not divisible by $q-1=6$ it does not belong to the type considered by Lefschetz, so that although its group is of order $3q$ as shown by Snyder, there is no contradiction. The curves just considered are noteworthy also in that they give simple examples, for as large p as we please, of curves with pure R. M. possessing singular correspondences. Lefschetz determines the invariants h, k for all cases, hence the base numbers for the various correspondences, symmetrical or otherwise.

BIBLIOGRAPHY

1 J. W. Alexander, Am. M. S. Trans., 25: 173-184 (1923). [2, 4.

2 E. S. Allen, Pal. Circ. Mat., 37: 345-370 (1914). [2, 5.

3 ———, Rom. Acc. L. Rend., (5), 22: 424-428 (1913). [5.

4 H. F. Baker, Abel's Theorem and the Allied Theory, Cambridge (1897). [2.

5 A. A. Bennett, Ann. of Math., (2), 18: 200-208 (1916-17). [2.

6 ———, Ann. of Math., (2), 21: 299-305 (1919-20). [2.

7 L. Berzolari, P. Rep., 2: (1910). [2, 4, 5.

8 O. Bolza, Am. J., 10: 47-70 (1888). [6.

9 ———, Math. Ann., 30: 546-552 (1887). [6.

10 G. Castelnuovo, Scritti matematici offerti a d'Ovidio, Torino, 164-174 (1918). [2.

11 ———, Rom. Acc. L. Rend., (4), 7_2: 294-299 (1891). [5.

12 ———, Rom. Acc. L. Rend., (5), 15: 337-344 (1906). [2.

13 ———, Rom. Acc. L. Rend., (5), 30: 355-359 (1921). [5.

14 O. Chisini, Ann. di Mat., (3) 24: 121-156 (1915). [3.

15 ———, Lomb. Ist. Rend., (2), 47: 346-349 (1914). [6.

16 ———, Lomb. Ist. Rend., (2), 57: 480-496 (1924). [2.
16.1 ———, Lomb. Ist. Rend., (2), 59: 529-548 (1926). [6.
17 ———, Rom. Acc. L. Rend., (5), 24_1: 153-158 (1915). [4.
18 ———, Rom. Acc. L. Rend., (5), 29_2: 172-175, 251-255, 305-309 (1921). [4.
19 E. Ciani, Pal. Circ. Mat., 28: 217-232 (1909). [6.
20 A. Comessatti, Pal. Circ. Mat., 36: 35-58 (1913). [5.
21 ———, Torino Atti, 50: 439-455 (1915). [2, 3.
22 M. De Franchis, Rom. Acc. L. Rend., (5), 12: 303-310 (1903). [2, 4, 6.
23 Enriques Chisini, Introduzione alla teoria geometrica delle equazioni, Bologna, 3: (1924). [2, 4, 6.
24 Enriques-Severi, Acta Math., 32: 284-391 (1909). [3, 6.
25 R. Fricke, Encyklopädie II B 3. [2, 6.
25.1 R. Fueter, Vorlesungen über die singulären Moduln und die komplexe Multiplikation der elliptischen Funktionen, 1 vol., Berlin (1924). [2.
26 L. Godeaux, Bull. Sci. M., (2), 50: 258-264, 272-284 (1926). [3, 5.
27 G. Hettner, Gött. Nach., 386-398 (1880). [6.
28 Holmquist, Lund thesis, (1900). [2.
29 G. Humbert, Ann. Éc. Norm., (4), 10: 174-201 (1894). [4.
30 A. Hurwitz, Gött. Nach., 350-363 (1883). [2.
31 ———, Gött. Nach., 85-107 (1887). [6.
32 ———, Leipz. Ab., 10-38 (1886). [2.
33 ———, Math. Ann., 25: 157-196 (1884). [2.
34 ———, Math. Ann., 28: 561-585 (1887). [2, 3.
35 ———, Math. Ann., 32: 290-308 (1888) (reprint of (31)). [6.
36 ———, Math. Ann., 41: 403-442 (1893). [6.
37 S. Kantor, Acta Math., 19: 115-193 (1895). [6.
38 ———, Pal. Circ. Mat., 9: 565-578 (1896). [2, 6.
39 F. Klein, Über Riemann's Theorie der algebraischen Functionen. [6.
40 ———, Elliptische Modulfunctionen, 2 vols. (1893). [2, 4, 6.
41 ———, Acta. Math., 7: p. 16 (1886). [2, 6.
42 ———, Math. Ann., 14: 428-471 (1879). [6.
43 ———, Math. Ann., 17: 62-70 (1880). [2.
44 A. Krazer, Lehrbuch der Thetafunktionen, 1 vol., Leipzig (1903). [2.
45 Krazer-Wirtinger, Encyklopädie II B 7. [4.
46 S. Lefschetz, Wash. Nat. Ac. Sc. Proc., 9: 90-93 (1923). [2.
47 ———, Wash. Nat. Ac. Sc. Proc., 12: 737-739 (1926). [2.
48 ———, Ann. of Math., (2), 28: 342-354 (1926). [2.
49 ———, Am. M. S. Trans., 22: 327-482 (1921). [3, 6.
50 ———, Am. M. S. Trans., 28: 1-49 (1926). [2.
50.1 ———, Am. J., 50: (1928). [2.
51 J. V. McKelvey, Am. J., 34: 115-136 (1912). [6.
52 M. Noether, Math. Ann., 20: 59-62 (1882). [6.
53 ———, Math. Ann., 21: 138-140 (1883). [6.
54 P. Painlevé, Leçons sur la théorie analytique des équations différentielles professées à Stockholm, Paris, 1 vol. (1897). [6.
54.1 ———, Ann. Éc. Norm., (3), 8: 103-140 (1891). [4.
55 ———, Ann. Éc. Norm., (4), 10: 203-206 (1894). [4.
56 H. Poincaré, Pal. Circ. Mat., 47: 281-336 (1909). [4.
57 E. Picard, Traité d'analyse, vol. 2, 3d edition, Paris (1925). [4, 6.
58 ———, Bull. Sci. M., 21: 1-3 (1893). [6.

59 ———, C. R., 103: 517-520 (1886). [6.

60 C. Rosati, Ann. di Mat., (3), 25: 1-32 (1915). [2.

61 ———, Ann. di Mat., (3), 28: 35-61 (1918). [2, 3, 4.

62 ———, Ann. di Mat., (3), 31: 1-50 (1921). [2, 3, 4.

63 ———, Ann. di Mat., (4), 3: 109-132 (1926). [2, 3.

64 ———, Pal. Circ. Mat., 44: 1-29 (1920). [2, 3.

65 ———, Pal. Circ. Mat., 49: 1-36 (1925). [5.

66 ———, Rom. Acc. L. Rend., (5), 22_2: 385-390, 431-436 (1913). [2.

66.1 ———, Rom. Acc. L. Rend., (5), 24_2: 182-184 (1915). [2.

66.2 ———, Torino Atti, 51: 991-1014 (1916). [2.

66.3 ———, Torino Atti, 53: 5-22 (1917). [2.

67 G. Roth, Wien. Ber., 129: 491-521 (1920). [6.

68 H. A. Schwarz, J. f. Math., 37: 139-145 (1879). [6.

69 G. Scorza, Rom. Acc. L. Rend., (5), 27: 171-175 (1918). [4.

70 ———, Pal. Circ. Mat., 41: 263-380 (1916). [2, 4.

71 C. Segre, Ann. di Mat., (2), 22: 41-143 (1894). [5.

72 ———, Rom. Acc. L. Rend., (4), 3_2: 3-6, 149-153 (1887). [5.

73 ———, Torino Atti, 24: 734-756 (1889). [6.

74 F. Severi, Vorlesungen über algebraische Geometrie, Leipzig (1922). [5, 6.

74.1 ———, Trattato di geometria algebrica, vol 1, Part 1, Bologna (1927). [2, 3, 4, 6.

75 ———, Ann. di Mat., (3), 12: 55-81 (1906). [5.

76 ———, Math. Ann., 74: 515-544 (1913). [2.

77 ———, C. R., 156: 287-290 (1913). [2.

77.1 ———, Rom. Acc. L. Rend., (6), 1: 562-568 (1925). [3.

77.2 ———, Rom. Acc. L. Rend., (6), 2: 521-526 (1925). [2.

77.3 ———, Rom. Acc. L. Rend., (6), 6: 435-441 (1927). [2.

77.4 ———, Torino Atti, 38: 185-200 (1903). [2.

78 ———, Torino Atti, 48: 660-674 (1912-13). [2, 3.

79 ———, Torino Mem., 54: 1-49 (1904). [2.

80 V. Snyder, Am. J., 30: 325-336 (1908). [6.

81 ———, Am. J., 31: 1-9 (1909). [6.

82 Snyder-Sharpe, Am. M. S. Trans., 22: 31-40 (1921). [3.

83 R. Torelli, Pal. Circ. Mat., 19: 297-304 (1905). [2.

84 ———, Pal. Circ. Mat., 37: 25-36 (1914). [3, 5.

85 ———, Rom. Acc. L. Rend., (5), 21_1: 453-457 (1912). [4.

86 ———, Rom. Acc. L. Rend., (5), 22_1: 772-775 (1913). [5.

87 ———, Rom. Acc. L. Rend., (5), 22_2: 98-103, 437-441 (1914). [3.

88 ———, Rom. Acc. L. Rend., (5), 24_1: 1079-86, 1101-1108 (1915). [3.

89 ———, Torino Atti, 42: 86-99 (1907). [5.

90 ———, Ven. Ist. Atti, 67: 1323-1335 (1908). [5.

91 A. L. Van Benschoten, Am. J., 31: 213-252 (1909). [6.

92 H. Weber, J. f. Math., 76: 345-348 (1873). [3, 4.

93 K. Weierstrass, Werke, vol. II. [6.

94 A. Wiman, Sven. Vet. Ak. Bih., 21: 2 papers, 23 p., 43 p. (1895). [2, 6.

95 W. Wirtinger, Encyklopädie II. B 2. [6.

95.1 O. Zariski, Am. J., 50: (1928). [2.

96 H. G. Zeuthen, Encyklopädie III. C 3. [2.

CHAPTER XVII

HYPERELLIPTIC SURFACES AND ABELIAN VARIETIES

By abelian varieties are understood varieties that admit of a parametric representation by means of abelian functions. They are highly interesting and important both intrinsically and because they are found in the background of many questions of algebraic geometry. It is our purpose here to give an account of their theory. The chief topics to be treated are the geometry on abelian varieties and the questions that center around complex multiplication.

Abelian functions proper have been fully covered in the Encyklopädie article of Krazer-Wirtinger([48]), and in practically all non-geometric questions we shall merely refer to them. This will notably be the case in § 1 where we give a rapid résumé of the foundations. As the contributions of the earlier mathematicians, Hermite, Weierstrass, Riemann and others, have largely been along that line it will cause little surprise to find scarcely any mention of them in the chapter and none in the bibliography.

The notations and terminology are those of the Glossary at the end of Chapter 15.

§ 1. Multiply Periodic and Related Functions

1. *Theta-Functions.*—Consider the matrix

$$\mathbf{N}\ \|\delta_{jk};\ \tau_{jk}\|$$

$$(\tau_{jk}=\tau_{kj};\ \delta_{jj}=1,\ \delta_{jk}=0 \text{ when } k\neq j;\ j,\ k=1,\ 2,\ \ldots,\ p).$$

Let $\tau_{jk}=\tau'_{jk}+i\tau''_{jk}$ and assume that for arbitrary real values of the x's:

$$(1)\qquad \Sigma\tau''_{jk}x_jx_k>0.$$

This characterizes $\mathbf{N}$ as an R. M. in the normal form with the principal matrix $\left\|\begin{matrix}0 & 1\\ -1 & 0\end{matrix}\right\|$, (Ch. 15, Nos. 6, 7).

We shall repeatedly consider functions of p variables $u_1,\ \ldots,\ u_p$ and shall designate them by the abridged notation $f(u)$, $\phi(u)$, etc.

Let us introduce the pair of forms

$$L(m;\ u)=\Sigma m_ju_j;\ Q(\tau;\ m)=\Sigma\tau_{jk}m_jm_k.$$

By means of them we define the θ function of order one as:

$$\theta(u)=\Sigma e^{\pi i(Q+2L)}$$

where the summation is over all integral values of the m's. From (1) it follows that the series is convergent for all u's and therefore $\theta(u)$ is an entire function. Let us write $\theta(u+\delta_k)$, $\theta(u+\tau_k)$ for the values of the function at $u_1+\delta_{1k}, \ldots, u_p+\delta_{pk}$ or $u_1+\tau_{1k}, \ldots, u_p+\tau_{pk}$. Then θ satisfies the periodicity relations

$$\theta(u+\delta_k)=\theta(u),$$
$$\theta(u+\tau_k)=e^{-\pi i(\tau_{kk}+2u_k)}\theta(u).$$

For this reason the elements of $\mathbf{N}$ are called the periods of θ, and $\mathbf{N}$ its *period matrix*, or R. M.

2. In conjunction with a period matrix $\boldsymbol{\omega}$, whether of the form $\mathbf{N}$ or not it is convenient to consider the fundamental parallelepiped U_{2p} or set of points u given by

$$u_j=\Sigma t_\mu\omega_{j\mu}, \qquad 0\leqq t_\mu<1.$$

In all cases to be considered $\boldsymbol{\omega}$ will be an R. M., hence $\left|\begin{matrix}\omega\\ \bar{\omega}\end{matrix}\right|\neq 0$. Thus for $\mathbf{N}$, the determinant is $|\tau''|\neq 0$ owing to (1). There is then a unique set of real values of the t's corresponding to any set of u's. To increase the u's by the periods $\omega_{j\mu}$ is equivalent to replacing t_μ by $t_\mu+1$. Hence modd. the periods there is a unique point of U_{2p} corresponding to every point u.

Let now $\boldsymbol{\omega}$ be $\mathbf{N}$ itself and let $(c_1, \ldots, c_p)$ be an arbitrary fixed point of U_{2p} with $h_1, \ldots, h_p, g_1, \ldots, g_p$, as the corresponding t's. We introduce a more general type of function defined by

$$\theta\begin{bmatrix}g\\h\end{bmatrix}(u)=\Sigma e^{\pi i[Q(\tau;\, m+g)+2L(m+g;\, u+h)]}$$

and called function θ of characteristic $\begin{bmatrix}g\\h\end{bmatrix}$. We verify immediately that

$$\theta\begin{bmatrix}g\\h\end{bmatrix}(u)=e^{\pi i[Q(\tau;\, g)+2L(g;\, u+h)]}\cdot\theta(u+c).$$

Hence the periodicity relations are

$$\theta\begin{bmatrix}g\\h\end{bmatrix}(u+\delta_k)=e^{2\pi i g_k}\cdot\theta\begin{bmatrix}g\\h\end{bmatrix}(u),$$
$$\theta\begin{bmatrix}g\\h\end{bmatrix}(u+\tau_k)=e^{-\pi i(\tau_{kk}+2u_k+h_k)}\cdot\theta\begin{bmatrix}g\\h\end{bmatrix}(u).$$

Of particular interest is the case where $\mathbf{N}$ is the R. M. of a set of normal integrals of the first kind of a C^p with respect to its retrosections. When the u's are replaced by the integrals with a common upper limit of integration (x, y), θ becomes a function of (x, y) on C^p, and its investigation opens up various interesting questions and in particular leads to the solution of the Jacobi inversion problem. See Krazer-Wirtinger([48]), Krazer([47]), Ch. 9.

3. By means of the θ function of No. 2 we can describe the most general function of order n whose periods are

$$(\mathbf{N}') \quad \left\| \frac{\delta_{jk}}{e_k}\ ;\ \tau_{jk} \right\|$$

where the e's are positive integers with $e_1 = 1$, e_k a factor of e_{k+1} and the δ's and τ's as before. We observe at first that the functions built with the τ's all multiplied by ν:

$$f_\rho(u) = \theta \begin{bmatrix} g+\rho \\ q \\ h \end{bmatrix} (\nu u)$$

($\nu = ne_p$ and n an integer; $q_j = \dfrac{\nu}{e_j}$; $0 \leqq \rho_j < q_j$ and ρ_j an integer), have a periodicity behavior as regards $\mathbf{N}'$, independent of the integers ρ; (the subscript ρ in f_ρ stands for the complex $\rho_1\rho_2 \ldots \rho_p$). Hence

$$\Theta_n \begin{bmatrix} g \\ h \end{bmatrix} (u) = \Sigma C_\rho f_\rho(u)$$

has a behavior independent of the constants c and described by the equations

$$\Theta\left(u + \frac{\delta_k}{e_k}\right) = e^{-2\pi i g_k} \Theta(u)$$

$$\Theta(u + \tau_k) = e^{\pi i (\nu \tau_{kk} + 2\nu u_k + h_k)} \Theta(u).$$

For this reason the function is called: Θ *of order* n and characteristic $\begin{bmatrix} g \\ h \end{bmatrix}$ attached to $\mathbf{N}'$. By direct comparison of coefficients it is found that for arbitrary c's the function has no other periods than sums of multiples of the elements of $\mathbf{N}'$, Wirtinger([123]), p. 97. Owing to this $\mathbf{N}'$ is said to represent a *fundamental set* or F. S. of periods.

The necessary and sufficient condition for the existence of our functions is again that (1) hold and $\mathbf{N}'$ is also an R. M. with an attached principal matrix $\mathbf{c}$, such that $c_{k,\,p+k} = -c_{p+k,\,k} = e_k$, the terms of different type being all zero.

The functions f corresponding to the different systems ρ are linearly independent and their linear combinations are the most general entire functions behaving as they do. They constitute a linear system whose dimension is

$$q_1 \ldots q_p - 1 = \frac{(ne_p)^p}{e_1 \ldots e_p}.$$

When all the e's are unity $\mathbf{N}'$ reduces to $\mathbf{N}$ and its θ's of order n become θ's of order n attached to $\mathbf{N}$.

Remarks.—I. We have denoted the theta functions belonging to $\mathbf{N}$ or $\mathbf{N}'$ by the respective symbols θ, Θ. In the sequel they shall all be denoted by the same letter θ and the particular type specified when necessary by a statement concerning the R. M. When nothing is said in regard to the latter it is assumed to be of the more general type $\mathbf{N}'$.

II. While $\mathbf{N}$ has the same aspect as the R. M. of a set of normal integrals of the first kind of a C^p relatively to the retrosections, the latter R. M. is very special. Indeed when $p>1$ it depends upon $3p-3$ independent parameters, while the first contains $\frac{p(p+1)}{2}$ arbitrary constants, and $\frac{p(p+1)}{2}-(3p-3)>0$ for $p>3$. Hence when $p\geqq 4$ the C^p matrix is not the most general.

For $p=4$ there is then a single relation between the normal periods τ_{jk} corresponding to a C^4. This relation has been explicitly determined by Schottky and Poincaré, references and further details Krazer([47]), p. 418, Krazer-Wirtinger([48]), p. 786. We are as yet completely in the dark as to the similar relations for $p>4$.

The most complete treatment of theta functions is by Krazer([47]).

4. *Intermediary Functions.*—An intermediary function $\phi(u)$ (Jacobi function according to some authors) with an F. S. of periods

$$\boldsymbol{\omega}=||\omega_{j\mu}||, \qquad (j=1, 2, \ldots, p;\ \mu=1, 2, \ldots, 2p)$$

is an entire function such that

$$\phi(u+\omega_\mu)=e^{-2\pi i(\Sigma\epsilon_{j\mu}u_j+\eta_\mu)}\cdot\phi(u),$$

where $\epsilon_{j\mu}$, η_μ, are constants. We impose the condition that the points

$$u_j=\Sigma t_\mu\omega_{j\mu}, \qquad 0\leqq t_\mu<1$$

fill up an actual domain U_{2p} of S_{2p}, which is finite. This amounts to exacting that $\begin{vmatrix}\omega\\ \bar{\omega}\end{vmatrix}\neq 0$. All θ's behave thus and are therefore intermediary functions.

Since adding ω_μ then ω_ν or vice versa must give the same result we have a set of relations

$$\Sigma(\epsilon_{j\mu}\omega_{j\nu}-\epsilon_{j\nu}\omega_{j\mu})=m_{\mu\nu},$$

the second member being an integer, with $\mathbf{m}$ an alternate matrix. The H. C. F. n of these integers is the *order* of ϕ. For a θ it reduces indeed to the order as defined in No. 3.

Frobenius has shown([28]) that $|m|>0$. Hence $\mathbf{m}^{-1}=\mathbf{M}$ is uniquely defined. He shows then by direct analysis that $\boldsymbol{\omega}$ is an R. M. with $\mathbf{M}$ as an attached principal matrix.

If we apply to the u's a linear transformation

$$(2) \qquad u'_j=\Sigma\lambda_{jk}u_k, \qquad |\lambda|\neq 0,$$

and at the same time adopt a new F. S. of periods, $\boldsymbol{\omega}$ is replaced by an equivalent matrix $\boldsymbol{\lambda\omega}\mathbf{a}$, $|a|=\pm 1$, $\boldsymbol{\varepsilon}$ by $\boldsymbol{\lambda}'^{-1}\boldsymbol{\varepsilon}\mathbf{a}$, $\mathbf{m}$ by $\mathbf{a}'\mathbf{m}\mathbf{a}$ and n is unchanged. Let us choose $\mathbf{a}$ such as to reduce $\mathbf{m}$ to its canonical form $\mathbf{N}'$. Then it is found that $||\epsilon_{j\mu}||$, $(j, \mu=1, 2, \ldots, p)$ is symmetrical. Hence upon multiplying ϕ by a suitable function $e^{Q(u)}$, where Q is a quadratic polynomial, the similar matrix for the new functions is zero, and there is a transformation of variables (2) reducing $\boldsymbol{\omega}$ to $\mathbf{N}'$ with $\phi e^{Q}=\psi$ behaving like a Θ of order n.

Conversely any R. M. $\boldsymbol{\omega}$ is equivalent to a certain matrix $\boldsymbol{\lambda\omega}\mathbf{a}$, $|a|\equiv\pm 1$, of type $\mathbf{N}'$. An arbitrary $\Theta(u')$ attached to the latter when expressed in terms of the u variables gives rise to an intermediary function $\phi(u)$ with $\boldsymbol{\omega}$ corresponding to an F. S. of periods.

Thus the necessary and sufficient condition for the existence of intermediary functions attached to $\boldsymbol{\omega}$ is that it be an R. M. If then $\mathbf{M}$ is any principal matrix of $\boldsymbol{\omega}$ there is a whole class of intermediary functions attached to $\boldsymbol{\omega}$ for which the matrix $\mathbf{m}=q\mathbf{M}^{-1}$, q a rational number. The whole class is reduced to a class of Θ's by a linear transformation of the u's that reduces any particular one of them.

By reference to the dimension of the associated Θ's we find that the intermediary function ϕ belongs to a linear system of functions with the same behavior as to $\boldsymbol{\omega}$ whose dimension is $\sqrt{|m|}-1$. There are important enumeration formulas regarding the number of common solutions of sets of intermediary functions, due to Poincaré, Wirtinger, Lefschetz. See Krazer-Wirtinger([48]), p. 837, also No. 34 of this chapter.

5. *Multiply Periodic Functions.*—By a $2p$-ply periodic function $f(u)$ with $\boldsymbol{\omega}$ for matrix of an F. S. of periods is meant a meromorphic function such that $f(u+\omega_\mu)=f(u)$. The existence of a single function $f(u)$ not a function of a smaller number of linear combinations of the u's, implies the existence of $p-1$ more of type $f(u+c)$, functionally independent of f and any others are algebraic functions of these p, rational functions of $p+1$ suitably chosen, Krazer-Wirtinger([48]), p. 824. In fact we may choose as the p additional functions the first partials f'_{u_j}; Wirtinger([123]), p. 90. On this basis it can be proved that $\boldsymbol{\omega}$ is again an R. M., a theorem due to Riemann and Weierstrass.

6. Conversely to any R. M. $\boldsymbol{\omega}$ correspond $2p$-ply periodic functions for which $\boldsymbol{\omega}$ represents an F. S. of periods. For let $\phi(u)$ be an arbitrary

intermediary function attached to $\boldsymbol{\omega}$, and W its locus of zeros in the u space. In the vicinity of any point u^0 of W, $\phi \equiv \phi_1 \cdot \phi_2$ the two factors being holomorphic at u^0, with $\phi_2(u^0) \neq 0$, so that $\phi_1 = 0$ represents W about u^0. Then ϕ_1 cannot be a function of $q < p$ linear combinations of the u's for every point of W. For a linear transformation (2) will not alter the situation, hence we may assume that ϕ_1 is a function of the variables $u_1, \ldots, u_q$ alone. Then the functions

$$\frac{\partial^2 \log \phi}{\partial u_h \partial u_{q+k}},$$

being periodic and finite, are necessarily constant. As a consequence, by expanding $\log \phi$ in power series about a point not on R we find

$$\log \phi = Q(u_1, \ldots, u_p) + f(u_1, \ldots, u_q),$$

where Q is a quadratic polynomial. Hence

$$\phi = e^Q \cdot \Phi(u_1, \ldots, u_q).$$

Hence ϕ has infinitesimal periods and $\boldsymbol{\omega}$ does not correspond to an F. S. of periods for the function. Therefore ϕ_1 cannot behave as assumed.

It follows that for arbitrary constants c_{jk}

$$f(u) = \Sigma c_{jk} \frac{\partial^2 \log \phi}{\partial u_j \partial u_k}$$

is a periodic function of $u_1, \ldots, u_p$ and of no smaller number of variables. In fact, in general $\boldsymbol{\omega}$ defines an F. S. of periods for $f(u)$, Wirtinger([123]), p. 96. This shows then the existence of periodic functions, of the desired type attached to $\boldsymbol{\omega}$. Furthermore, every other such periodic function is a rational function of $f(u)$, f'_{u_j} and among these $p+1$ there is a set of p linearly independent. These functions are all obtained by a linear transformation of variables from ratios of $p+2$ suitably selected θ's of the same order and characteristic. It follows that every $2p$-ply periodic meromorphic function of p variables is obtained by a linear transformation of the variables from a ratio of θ functions of the same order and characteristic.

7. In our presentation we have followed the same lines as most authors. Appell in ([1]) proceeded in a distinctly different manner and while the details are only given for $p=1, 2$, it is very likely that his method can be extended to any p. Starting directly with a multiply periodic function $f(u)$ attached to $\boldsymbol{\omega}$ he shows that it may be written as the quotient of two intermediary functions with the same periods. In connection with the work of Frobenius([28]), Appell's method contains all the elements of a very elegant exposition of the subject.

Additional references of a general character on the topics of this section are Krazer-Wirtinger([48]); Poincaré([70]); Wirtinger([120]) (the last two are summaries of the work of the authors on abelian functions); Osgood([58]); Baker([1.1]), Chs. 7, 8; Scorza([93.1]), ([102]); Castelnuovo([10]); Lefschetz([51]), Part II, Ch. 1, ([55]), ([50]), Ch. 6; Jung([45.1]).

8. *Definition of Abelian Varieties.*—Let $\phi_0, \phi_1, \ldots, \phi_n$ be linearly independent intermediary functions with the R. M. $\boldsymbol{\omega}$ and behaving alike relatively to it, i. e., with the same constants $\epsilon_{j\mu}, \eta_\mu,$ of No. 4. Then the matrix $\mathbf{m}$ is the same for all and they belong to the same linear system of intermediary functions with like behavior, whose dimension is $\sqrt{|m|}-1$. The total system can be simultaneously reduced by a linear change of variables to a set of θ's of the same order and characteristic.

Let $n>p+1$. Since the ratios $\dfrac{\phi_i}{\phi_0}$ are periodic functions, from the theorems recalled in Nos. 5, 6, it follows that any $p+2$ of the ϕ's are related by a homogeneous algebraic equation. Hence the locus of the system of points of S_n given by

$$\rho x_i = \phi_i(u) \tag{3}$$

is an algebraic V_q of S_n, called *abelian variety* attached to $\boldsymbol{\omega}$. There are two possibilities: (a) $q<p$. Then the functional matrix of the ϕ's is of rank $q<p$, and V_q is actually an abelian variety attached to a submatrix of genus q of $\boldsymbol{\omega}$, which is then impure. We shall say that V_q is a *degenerate* abelian variety attached to $\boldsymbol{\omega}$. (b) $q=p$. The functional matrix is now of rank p. To each u point corresponds a single point of V_p and to an arbitrary point of V_p correspond a finite number r of points of U_{2p}, or points u distinct modd. the periods. r is called the *rank* of the variety.

It has been proved explicitly by Lefschetz([51]), p. 367, that (3) may be so chosen that V_p is without singularities, and in (1, 1) *non-exceptional* correspondence with U_{2p}. In particular its rank $r=1$. In place of (3) he takes θ functions and bases the whole argument on the fact that sufficiently general θ's have $\mathbf{N}'$ as F. S. of periods. The V_p thus obtained is not unique, but any two are birational transforms of one another. The birational transformation between them is here in fact a homeomorphism. In particular there are no fundamental subvarieties on the two V_p's which are transformed into one another. Whenever we shall speak of the abelian variety attached to $\boldsymbol{\omega}$ we shall mean unless otherwise indicated, any variety of the class just defined. The varieties of various ranks attached to $\boldsymbol{\omega}$ are images of involutions on the V_p of rank one.

For $p=1$ the variety is merely an elliptic curve. To investigate it from the present point of view amounts to studying its properties in relation to its parametric representation by elliptic functions. There is nothing to be added along that line to Fricke([26]), Fueter([29]). The case $p=2$ leads to hyperelliptic surfaces, the name being due to their close connection with curves of genus two. These surfaces have been of paramount importance in algebraic geometry and we shall therefore consider them at length in the following section.

§ 2. Hyperelliptic Surfaces

9. *Generalities.*—A hyperelliptic surface F on an S_3, is representable as

$$(4) \qquad x_i=\theta^i(u, v), \qquad (i=0, 1, 2, 3)$$

where the θ's are of the same order and characteristic and belong to a matrix $\mathbf{N}'$ of genus $p=2$ and might also be replaced by intermediary functions.

The Kummer surface is the first hyperelliptic surface to appear in the literature and it was Klein who first pointed out that it could be parametrically represented by means of θ functions in two variables. From that point of view it has been studied by numerous writers to whose papers references will be found in Krazer-Wirtinger([48]), p. 747. Hudson([31.1]) has given a very thorough and complete exposition of the Kummer surface. Another closely related special hyperelliptic surface is the Weddle surface, Krazer-Wirtinger([48]), p. 751. As these two special surfaces are adequately covered in ([48]) we shall consider them only incidentally.

The Kummer and Weddle surfaces are both of rank *two,* for they admit parametric representations (4) by thetas that are *even* functions of u, v for the first, odd functions for the second. Thus the surface points (u, v), $(-u, -v)$ coincide. They are distinctly special hyperelliptic surfaces.

A surface of rank r represents an involution I_r on a surface of rank one belonging to the same period matrix. More generally surfaces of ranks r, s, belonging to the same matrix are in (r, s) correspondence. Hence the study of surfaces of the various ranks and of the involutions on the surface of rank one are equivalent problems and we do not separate them in the sequel.

10. The study of hyperelliptic surfaces of rank one was initiated by Picard in connection with his investigations on integrals of total differentials (Picard integrals). In ([62]), p. 312, ([69]), Ch. 14, ([64]), ([66]), ([68]), he gives necessary and sufficient conditions in order that $F(xyz)=0$ shall represent a hyperelliptic surface of rank one. Every Picard integral of the first kind is reducible to the form $au+bv$, hence the irregularity

$q=2$. The surface must then possess two functionally independent Picard integrals of the first kind. Every double integral of the first kind is reducible to $a\int\int du\,dv$, hence the geometric genus $p_g=1$. F must then have a single adjoint polynomial of order $m-4$ (m order of F) and Picard proves that, outside the multiple curve, the surface defined by that polynomial cuts F in a set of rational curves. These are sufficient conditions in order that F be hyperelliptic. Enriques has shown ([20]) that these conditions are equivalent to the following conditions on the genera: $p_a=-1$, $p_g=P_4=1$. See also Enriques-Severi([22]), p. 303. Observe in passing that the rational curves arise from the fact that the θ's of the representation (4) vanish simultaneously at some isolated points whose neighborhoods the rational curves represent. The curves are exceptional curves of the surface. The representation may be so chosen that none will be present, Enriques-Severi([22]), p. 321.

In ([63]), p. 223, Picard proves that when a surface is invariant under a continuous birational group in $s>1$ parameters, necessarily $s=2$, the group is abelian and F hyperelliptic of rank one. See also Painlevé([59]), p. 255; Castelnuovo-Enriques([11]), p. 746; Severi([104]), p. 778, ([108.1]).

11. The systematic study of hyperelliptic surfaces with a matrix **N** was undertaken by Humbert([32]) to ([44]), to whom we owe much of what is known about the surfaces, and also a secure foundation for later progress in abelian functions.

In conformity with Humbert and most subsequent writers we take **N** as

$$\begin{Vmatrix} 1 & 0 & g & h \\ 0 & 1 & h & g' \end{Vmatrix}, \qquad g_1g'_1-h_1^2>0,$$

where g_1, g'_1, h_1, are the complex parts of g, g', h. It is also a period matrix for a set of normal integrals of the first kind of a C^2, relatively to four retrosections. The related surface of rank one, F, is a birational transform of the surface image of the point pairs of C^2. To every pair of points of C^2 corresponds a unique point of F, and conversely with the exception that to a certain definite point of F correspond all the groups of the canonical g_2^1 on C^2. F is called the *Jacobi surface of* C^2. The quantities g, h, g' are the *moduli* of C^2 or F. There is a more general type of surface corresponding to a matrix **N**$'$ which we shall take here as

$$\begin{Vmatrix} 1 & 0 & g & h \\ 0 & \frac{1}{e} & h & g' \end{Vmatrix}, \quad e>1; \qquad g_1g'_1-h_1^2>0.$$

It is called a *Picard surface.* Humbert in all his investigations considers exclusively Jacobi surfaces. Actually, however, the two types may merge:

His *singular* surfaces, or surfaces whose **N** has a singularity index $k>1$ may well also correspond to matrices **N**′ for a different F. S. of periods.

12. Humbert's first paper, ([39]) (1893), is mainly taken up with the properties of general Jacobi and Kummer surfaces. A résumé of his results is given by Hudson([31.1]). At the outset Humbert proves, however, by means of a theorem due to Appell([1]), this basic proposition: On any hyperelliptic F the parametric equation of every algebraic curve is of the form $\phi(u, v)=0$, where ϕ is an intermediary function. Conversely every such equation represents an algebraic curve on F. In his proof he restricts himself to general (i. e., non-singular) R. M.'s, but this restriction is superfluous and he was well aware of it, as his later work shows.

In ([39]), Part II, Humbert makes an extensive investigation of the Kummer surface based on the parametric representation, due to Weber([116]), which is (4) when the functions at the right are four linearly independent even θ's (the maximum number) of order two and characteristic zero. Humbert classifies the curves on the surface. They are all of even order, and with 32 families for each value of the order. He examines the mode of obtaining them by means of intersecting surfaces, the properties of these surfaces, the singular points of the curves, their genera. He shows that any two curves of the same family constitute the complete intersection of the Kummer surface with a surface of S_3. These investigations are the most important on the distribution of families of curves on a surface before Severi's work on the base (Ch. 15, No. 20).

In Part III of ([39]) Humbert considers the Jacobi surface F, proves the Picard necessary and sufficient conditions (see No. 10) in order that a surface of S_3 be a Jacobi surface and extends them to the Kummer surface. He then studies the curves of F. The θ's of order m and given characteristic cut out a linear system whose dimension is m^2-1, degree $2m^2$, genus m^2+1, and which varies within a continuous system ∞^2 when the characteristic varies. This anticipates the theorem of Castelnuovo, Enriques, Severi, Picard, Ch. 15, No. 9. Humbert also considers curves with singular points and their intersections, sections of F by various surfaces, notably its adjoints, examines the special F of the minimum degree 8 and its curves. This surface, however, is not in (1, 1) unexceptional correspondence with the points (u, v), (modd. the periods) or points of U_4 in the sense of No. 2, for the surface carries exceptional curves. Enriques-Severi([22]), p. 321, have shown that the Jacobi surface of least degree without exceptional curves is an F_{18} in S_9. See also Krazer-Wirtinger([48]), p. 871.

In ([39]) Part IV Humbert proves that every surface of rank two belonging to **N** is a birational transform of the Kummer surface. He examines the possible orders, observes that quartic surfaces other than Kummer's surface are among them, for example Weddle's surface, whose relation to Kummer's surface he also investigates.

Humbert's results have been extended to Picard surfaces by Traynard ([112]).

13. In the short notes ([32]) to ([38]) and in an extended paper ([41]), ([42]), ([43]), Humbert investigates from all sides the singular matrix **N**, its abelian functions and Jacobi surfaces. **N** will possess, besides its principal matrix $\begin{Vmatrix} 0 & 1 \\ -1 & 0 \end{Vmatrix}$, another rational alternate matrix. The relation between the moduli takes the explicit form

$$Ag+Bh+Cg'+D(h^2-gg')+E=0, \tag{5}$$

where the coefficients are integers. Humbert says that the functions or surfaces are l-ply *singular* when there are l linearly independent relations (5). If k is the singularity index, $l=k-1$, hence $l\leqq 3$. Humbert considers particularly the case where $k=2$ and (5) is the only relation between the moduli, so that the multiplication index $h=2$ also.

Among the transformations of **N** into an isomorphic matrix of the same form, some do not, some do, imply relations between the moduli. The first are called *ordinary*, the second *singular* transformations. The latter are to be found only when $k>1$, i. e., when (5) is not trivial. Under ordinary equivalence transformations (5) is changed into a similar relation between the new periods but the number $\Delta=B^2-4AC-4DE$, called the *invariant* of (5), is unchanged. Δ is always positive and $\equiv 0$ or 1 mod. 4. If we think of (5) as a relation $\mathbf{NcN}'=0$, $\frac{\Delta}{4}$ or $\frac{\Delta-1}{4}$ is the pfaffian of **c**, Bagnera-deFranchis ([2]), which explains its invariance. See Scorza ([99]), first paper; also No. 35 of this chapter. **N** is pure or impure accordingly as Δ is not or is a perfect square. When **N** is impure F carries elliptic curves and conversely. Hence the necessary and sufficient condition for the presence of elliptic curves on F is that Δ be a perfect square.

By applying an ordinary equivalence to **N**, it may be reduced to a similar matrix with the corresponding relation (5) in the simplified form,

$$g+\beta h+\gamma g'=0, \qquad \Delta=\beta^2-4\gamma. \tag{6}$$

Furthermore we may assume g_1 and $g'_1>0$. Then given any two integers λ, κ such that

$$\delta=\lambda^2+\beta\kappa\lambda+\gamma\kappa^2>0, \qquad \lambda>0, \tag{7}$$

there are associated intermediary functions $\phi(u, v)$ not reducible to θ's by an ordinary transformation and behaving in accordance with the relations, Humbert([41]), p. 262:

$$\phi(u+1, v)=\phi(u, v+1)=\phi(u, v),$$
$$\phi(u+g, v+h)=e^{2\pi i(\lambda u-\kappa\gamma v+\nu)}\cdot\phi(u),$$
$$\phi(u+h, v+g')=e^{2\pi i[\kappa u+(\lambda+\kappa\beta)v+\nu']}\cdot\phi(u).$$

The determinant of the matrix $\mathbf{m}$ of No. 4 attached to ϕ is here equal to δ^2. Therefore if $C_{\kappa\lambda}$ denotes the algebraic curve $\phi=0$, the dimension of $|C_{\kappa\lambda}|$ is $\delta-1$, its degree 2δ and its genus $\delta+1$. More generally for the intersections of curves of two distinct systems, Humbert([41]), p. 270,

$$[C_{\kappa\lambda}C_{\kappa'\lambda'}]=2\lambda\lambda'+\beta(\kappa\lambda'+\lambda\kappa')+2\gamma\kappa\kappa'.$$

In ([41]) Humbert also investigates *singular* Kummer surfaces, i. e., attached to a singular $\mathbf{N}$, in particular he determines the properties of their curves. See also Hudson([31.1]), Ch. 18.

14. In ([41]), Part 4, Humbert deals with modular equations related to singular hyperelliptic surfaces. If we set

$$z^2=f(x)=x(1-x)(1-\xi x)(1-\eta x)(1-\zeta x),$$

the coordinates of a point on the Kummer surface can be expressed rationally in terms of $x, y, \sqrt{f(x)}, \sqrt{f(y)}$. If the surface carries algebraic curves that do not exist when ξ, η, ζ, are arbitrary there exist algebraic relations between these parameters, which are the algebraic moduli of the Kummer surface. When $k=2$ and there is only one relation between g, h, g' there is only one relation $F(\xi, \eta, \zeta)=0$, that is the point (ξ, η, ζ) is on a certain algebraic surface. This surface is invariant for all ordinary transformations of g, h, g', or change in moduli corresponding to a change of retrosections on the curve $z^2=f(x)$, or if we take the moduli such as to correspond to a singular relation (6) for a narrower set of transformations, constituting a group G. Thus certain algebraic functions of ξ, η, ζ, will be automorphic functions of, say g, h, invariant under the transformations of a certain modular group. This aspect of the question is not considered by Humbert in the paper in question, nor is it as yet of importance for our present topic. What is of interest, however, is that he actually forms the equation $F=0$ for some special values of Δ, and gives a theoretical (very geometric) method for obtaining it in the general case, in terms of certain families of curves related to the hexagon $y+2ax+a^2=0$, $f(a)=0$.

For related investigations but with little geometrical content see Krazer-Wirtinger([48]), p. 860, Hecke([30]).

15. In ([42]) Humbert considers all questions that center around transformations of multiply periodic functions. Let F be the same Jacobi

surface as before, F' another with parameters U, V. In dealing with these questions and more in conformity with our habitual procedure, we do not restrict the R. M.'s to the type $\mathbf{N}$ and denote them by $\boldsymbol{\omega}$, $\boldsymbol{\Omega}$.

Geometrically the question is to determine whether F' is the image of an involution on F, or a rational transform of F. The method used by Hurwitz for correspondences on curves (Ch. 16, No. 3) gives for the equations of the transformation

$$(8) \qquad U = a_{11}u + a_{12}v + \beta_1; \qquad V = a_{21}u + a_{22}v + \beta_2$$

and the matrix equation $\boldsymbol{\alpha\omega} = \boldsymbol{\Omega}\mathbf{a}$ where $\mathbf{a}$ is a matrix of order four whose terms are integers. Humbert obtains this relation by the analytical condition that the periodic functions attached to $\boldsymbol{\Omega}$ be rational in those attached to $\boldsymbol{\omega}$. If the matrix relation imposes no restrictions on $\boldsymbol{\omega}$, nor on the moduli g, h, g', when it is reduced to the form $\mathbf{N}$, the transformation is *ordinary*, otherwise it is *singular*.

The problem might be slightly generalized by considering merely algebraic correspondences instead of involutions or algebraic transformations instead of rational transformations. Then the resulting matrix equation would be the same. See in this connection § 4 of this chapter.

Humbert shows that singular transformations exist when and only when $\boldsymbol{\omega}$ is singular. The order δ of the involution I_δ on F is then equal to $|a|$, ([42]), p. 291, a number always $\geqq 0$. To each solution λ, κ, of (7) corresponds a definite transformation with associated I_δ singular when $\kappa \neq 0$. The transformation is only defined up to the product by an ordinary transformation. Humbert gives the explicit coefficients a in (8), ([42]), p. 305. The effect on the θ's of order m is to change them into intermediary functions whose indices are $m\lambda$, $m\kappa$.

Of particular interest are the transformations whose $\delta = 1$, or birational transformations of the Jacobi surface, Humbert([42]), p. 313. Their equations may be given the simple form

$$(9) \qquad U = \lambda u - \gamma\kappa v; \qquad V = \kappa u + (\lambda + \beta\kappa)v.$$

The ordinary transformations correspond to $\kappa = 0$. For (7) we can also substitute the Pellian equation

$$(10) \qquad \xi^2 - \Delta\kappa^2 = 4; \qquad \xi = 2\lambda + \beta\kappa;$$

where due to the nature of Δ, if ξ is an integer so is λ and conversely. Thus to every solution of (10) corresponds a definite transformation (9) and conversely. By a well-known property of (10) it follows that all transformations (9) are powers (positive or negative) of the transformation T, corresponding to the least positive solution ξ_1, κ_1 of (10). Moreover if $T=T_1^n$ its characteristics ξ, κ are given by

$$\frac{\xi+\kappa\sqrt{\Delta}}{2}=\left(\frac{\xi_1+\kappa_1\sqrt{\Delta}}{2}\right)^n.$$

Of course Δ must not be a perfect square. This we may well assume for in that case there are no birational transformations. The most general birational transformation is obtained by replacing U, V, in (9), by $\epsilon U+\beta_1$, $\epsilon V+\beta_2$; $\epsilon=\pm 1$; β_1, β_2 arbitrary constants.

Humbert makes also the very interesting observation, ([42]), p. 317, that two Jacobi surfaces, F, F_0, may be birationally equivalent without its being the case for the associated curves of genus two. When that happens $\mathbf{N}$ is a period matrix with respect to retrosections for one of the curves but not for the other, Rosati([86]), Comessatti([17]). Humbert gives this precise theorem([42]), p. 324: When x^2+xy+y^2 can represent -1 there is a unique C^2 associated with F, and all surfaces T_1^nF have the same moduli. When the form cannot represent -1 there are two birationally distinct curves with the respective Jacobi surfaces $T_1^{2n}F$ and $T_1^{2n+1}F$.

16. Humbert was the first to study complex multiplication for the genus two in all its generality, ([42]), p. 327. The situation is as in No. 15 and (8) are the equations of the problem, with the matrix condition $\boldsymbol{\alpha\omega}=\boldsymbol{\omega}\mathbf{a}$. Geometrically we have to find an I_δ on F having for image F itself. The more general problem is to find algebraic correspondences between the points of F. But we shall take the situation as considered by Humbert and assume that the correspondence is an involution.

In conformity with most writers we call a multiplication or transformation *ordinary* when $\boldsymbol{\alpha\omega}=\boldsymbol{\omega}\mathbf{a}$ is satisfied whatever $\boldsymbol{\omega}$. The equations (8) are in this case:

$$U=au+\beta_1; \qquad V=av+\beta_2$$

where a is an integer. All other multiplications will be described as *complex,* or the transformations as *singular.* Humbert, however, calls "ordinary transformation" the somewhat more general type, characterized by the property of leaving invariant the set of all θ's. To be precise it has then the property that if the change of variables (8) be applied to $\theta(U, V)$, the function becomes a $\theta(u, v)$. That type of transformation was called *principal* by Frobenius([27]), *hermitian* by Enriques-Severi([22]), p. 369. In terms of matrices the transformation is *principal* if there exists a principal matrix $\mathbf{c}$ of $\boldsymbol{\omega}$ such that $\mathbf{aca}'=l\mathbf{c}$, where l is a rational multiplier. Then $\mathbf{c}$ is also a principal matrix of $\boldsymbol{\omega}\mathbf{a}$.

Humbert shows that complex multiplication is present, when and only when, $\boldsymbol{\omega}$ is singular. He considers the three cases of 1, 2, 3, singular relations (5) and finds five principal categories according to the number of distinct singular relations and additional relations between the moduli, ([42]), p. 357. He thus determines all possible values of the singularity index k and related matrices and indirectly lays down the basis for the determination of h, but this last index actually introduced by Scorza, was not considered by Humbert.

The indices h, k were determined for Jacobi surfaces by Rosati([82]), Part II, under another name and in connection with his investigations on correspondences. He also showed the existence of the various types. As every R. M. has an isomorph $\mathbf{N}$, Rosati's determination holds for Picard surfaces as well. See also Scorza([94]), p. 77, Lefschetz([51]), p. 432. Scorza in particular enumerates 9 precise types. It must be remembered that his h, k are less by one unit than ours. Lefschetz gives also explicit forms for the three types of pure R. M.'s. The impure types can be constructed automatically. See also in this connection Cotty([19]), Hecke([31]), Krazer-Wirtinger([48]), p. 857, and this chapter No. 55.

17. An important type of multiplication corresponds to birational transformations of the Jacobi surface. Humbert([42]), p. 369, first seeks principal transformations and reduces the problem to finding curves $y^2 = f_6(x)$, such that the roots of $f_6(x) = 0$ are invariant under a linear group. This question has been solved by Bolza([7]), who showed that f_6 belongs to any one of six types. Humbert then considers non-principal birational transformations and determines them for every value of the singularity index k. In particular when $k=2$, and in the notations of No. 13, any T is represented by equations (9) with arbitrary constants added on the right-hand side. Let $\boldsymbol{\alpha}$ be the matrix of the u, v coefficients, and $\boldsymbol{\alpha\omega} = \boldsymbol{\omega}\mathbf{a}$ the associated multiplication. Since the terms of $\boldsymbol{\alpha}$ are real we have $\boldsymbol{\alpha}\bar{\boldsymbol{\omega}} = \bar{\boldsymbol{\omega}}\mathbf{a}$ with $\mathbf{a}$ unimodular. Hence

$$\left\| \begin{matrix} \boldsymbol{\alpha} & 0 \\ 0 & \boldsymbol{\alpha} \end{matrix} \right\| \cdot \left\| \begin{matrix} \boldsymbol{\omega} \\ \bar{\boldsymbol{\omega}} \end{matrix} \right\| = \left\| \begin{matrix} \boldsymbol{\omega} \\ \bar{\boldsymbol{\omega}} \end{matrix} \right\| \cdot \mathbf{a}; \qquad |a|^2 = |a| = 1;$$

$$|a| = \lambda^2 + \beta\kappa + \gamma\kappa^2 = \pm 1.$$

The related birational transformations of the Kummer surface are again represented by (9) with arbitrary half periods added at the right. They are the products of the powers of a single singular birational transformation multiplied by the 16 ordinary birational transformations of the surface (its six projectivities when in an S_3). Therefore the total group of the surface is discontinuous. This constitutes in fact the first example of a surface with an infinite discontinuous group; it is due to Humbert([42]), p. 372; ([32]). See also Painlevé([60]), Castelnuovo-Enriques([11]), p. 752, Enriques([21]).

For further information on the groups of Jacobi surfaces see Comessatti ([17]), Rosenblatt ([91]) and the papers of Enriques-Severi and Bagnera-de Franchis to be considered in the sequel. Scorza ([94]), Part II, determines the complete group for every one of the types of hyperelliptic surfaces in his list p. 340. For his type V ($k=2$, $h=4$), Lefschetz in ([51]), p. 440, has derived his result by a direct arithmetical method.

18. We now come to the important investigations of Enriques-Severi (1907), which are contained in two preliminary notes ([24]) and in a comprehensive paper in two parts ([22]), ([23]). The authors first observe that every Picard surface F is representable by θ's with an R. M. $\mathbf{N}'$, whose e is the *divisor* of F. The Jacobi surface corresponds to the divisor unity. Every hyperelliptic surface of divisor e and rank r is the image of an I_{er} on a Jacobi surface, or of an I_r on the Picard surface corresponding to the same matrix. Among the birational transformations of F of rank one there are two types present whatever the matrix; they are the ordinary transformations: (a) of the first kind or

$$u+u'=\text{const}; \qquad v+v'=\text{const}.$$

(b) of the second kind or

$$u-u'=\text{const}; \qquad v-v'=\text{const}.$$

The latter form a continuous two-parameter, abelian group. Those of the first kind form a cyclic binary group when the constants are zero. The transform of (u, v) is then $(-u, -v)$ and the two points together are a group of an I_2 whose image is a generalized Kummer surface with 16 nodes corresponding to u, v both equal to half periods. Transformations of the second kind give rise to cyclic groups of any assigned order, when the constants are $\dfrac{1}{n}$-th periods and the associated I_n are enumerated for the Jacobi surface ([22]), p. 293. The image of such an I_e on a Jacobi surface is a Picard surface whose divisor is e.

The curves of lowest genus carried by a surface of rank one are particularly important ([22]), p. 304. Rational curves occur only in finite number and can be removed by a birational transformation ([22]), p. 322. If there exists a single elliptic curve there are two irrational pencils of them. Each is characterized by the constancy of a linear combination $au+bv$ on its curves. Therefore in general there are no elliptic curves. An interesting Jacobi surface with elliptic curves is the image of the point pairs of two curves of genus one. The surfaces that carry elliptic curves are completely characterized by the presence of an irrational pencil. Both pencil and curves are then necessarily elliptic.

A Picard surface F always carries curves of genus two. They are obtained from an arbitrary linear system $|C|$, of genus p, by imposing $p-2$ double points on C. The number of such curves in $|C|$ is finite. By applying the ordinary transformations of the second kind to them there is obtained a continuous system ∞^2 whose image is F itself. When F is a Jacobi surface the curves have no double points and are birationally equivalent to the basic C^2 associated with the surface. The continuous system is then of degree two and index two (i. e., there are two curves of it through an arbitrary point of F). This property completely characterizes Jacobi surfaces. Every Picard F of divisor e carries an ∞^2 system of curves of genus two with $e-1$ double points, the degree and index being both $2e$. Both for the Picard and Jacobi surfaces these systems are cut out by the θ's of lowest order. On the Jacobi surface the system is complete, on the Picard surface it belongs to a continuous system ∞^2 of linear system $|C|$, whose dimension is $e-1$, genus $e+1$, degree $2e$. On the most general surfaces (i. e., with general period matrix) every curve belongs to a linear system $|mC|$. This is merely another way of stating the fact that it is cut out by a θ. For $e>3$ the consideration of these linear systems shows that F is birationally transformable into a surface of degree $2e$ on an S_{e-1}, carrying no exceptional curves. The authors henceforth assume that F has no exceptional curves. Strictly speaking they are considering a suitable birational transform in some space, (1, 1) image of the points of the fundamental domain U_4 (No. 8).

19. The authors now proceed to classify involutions on the Jacobi F, or, which is the same, surfaces of rank >1 belonging to a matrix **N**. The only types are rational or elliptic surfaces, or F of divisor >1 and rank one attached to some R. M., or else regular surfaces such that either $p_g=P_3=0$, $P_2=1$, or $p_g=P_4=1$ ([22]), p. 324. Invariance under ordinary transformations leads to a classification according to the values of the irregularity q.

Of more importance is the number of coincidences of an I_n on any F of rank one. When it is infinite, I_n is rational or has for image an elliptic ruled surface. When there are no coincidences, I_n has for image a surface of rank one or an elliptic surface (surface with an elliptic group). Hence the most interesting case corresponds to an I_n with a finite, non-zero, number of coincidences with $P_2\neq 0$, ([22]), p. 329.

Enriques-Severi give this basic theorem on involutions ([22]), p. 331: Every hyperelliptic I_r on F of rank one is generated by a birational group G_r of order r of the surface F. An equivalent statement is that every surface of rank r is the image of an I_r on the related surface F of rank one, generated by a birational G_r of F. The group of I_r to which any assigned point P of F belongs consists of the transforms of P by the operations of G_r. The operations of G_r are of the form

$$(11)\quad u'=a_i u+b_i v+c_i;\qquad v'=a'_i u+b'_i v+c'_i;\qquad (i=1, 2, \ldots, r).$$

With the transformations of G_r there are associated definite multiplications of the R. M. $\boldsymbol{\omega}$. If $\boldsymbol{\omega}$ is arbitrary they can only be ordinary transformations. Hence the generic F of rank one carries only involutions of order two representable by a generalized Kummer surface of the same divisor as F, or else involutions representable by surfaces of different divisor but of rank one. The generalized Kummer surfaces are of degree $4e$ on an S_{2e} with 16 double points and singular hyperplanes giving rise to a configuration similar to Kummer's and these properties completely characterize them ([22]), p. 351.

The next important and simple type of I_r has the irregularity $q=1$ and its image is therefore an elliptic surface, and is completely characterized by $p_a=-1$, $p_g=0$, $P_i=0$, 1; ([22]), p. 361. The authors also determine the divisors for these surfaces and their period matrices ([22]), p. 361. See also Bagnera-de Franchis ([4]).

20. In the classification of involutions associated with a G_r Enriques-Severi limit themselves to Jacobi surfaces. Like Humbert ([42]) they base themselves on Bolza's work on $f_6(x)$, ([7]), and thus obtain eleven different families ([22]), p. 392, the first being composed of Kummer surfaces. In ([23]) they construct models of the remaining ten types which are of order $2r$ in S_{r+1} characterized by definite configurations of singular points and hyperplanes tangent along certain rational curves. The singular points correspond to coincidences of the related I_r on the Jacobi surface and are fully described in each case, the authors giving also schemes analogous to Humbert's ([39]), to represent the associated configurations of singular points and tangent hyperplanes.

The surfaces of the eleven types can all be projected into quartic surfaces of S_3, or double planes $z^2=f_6(x, y)$, with characteristic configurations of singular points and tangent planes in the first case, multiply tangent curves to $f=0$ in the second ([23]), p. 397. See also Piazzola-Beloch ([61.1]), ([61.2]).

21. Bagnera-de Franchis in ([2]) to ([6]), most of which appeared at the same time as the work of Enriques-Severi (1907-09), investigated all questions concerning hyperelliptic surfaces from the most general point of view. They made relatively little use of special geometric properties and considered directly the R. M. Owing to this their methods have lent themselves exceptionally well to extensions to abelian varieties. Their chief contribution is the complete classification of all hyperelliptic involutions and the determination of the Picard-Severi number ρ for all hyper-

elliptic surfaces. Their starting point[6] is the theorem of Enriques-Severi on the group G_r associated with any hyperelliptic I_r (No. 17). Let $\boldsymbol{\omega}$ be the R. M., F the corresponding surface of rank one. Let T be a cyclic transformation of order n, such as (11), but in the more convenient form (8). We have (No. 17), $\boldsymbol{\alpha\omega}=\boldsymbol{\omega}\mathbf{a}$, where $\mathbf{a}$ is unimodular. Since $\boldsymbol{\alpha}^s$, $\mathbf{a}^s$ correspond to T^s like $\boldsymbol{\alpha}$, $\mathbf{a}$, to T (Ch. 15, No. 7), $\boldsymbol{\alpha}^n=\mathbf{a}^n=\mathbf{I}$. Therefore the roots of the characteristic equations $|\alpha-xI|=0$, $|a-xI|=0$, are n-th roots of unity. From this it follows that with a suitable linear change of variables the equations of T are $u'=\lambda u+v$, $v'=\mu v$, where λ, μ, called multipliers of T, are n-th roots of unity and $n\nu$ is a period, Bagnera-de Franchis[6], p. 265. When $\lambda=\mu=\pm 1$, T is an ordinary transformation—a type which need not be considered. When neither is $+1$ the equations can be given the simpler form $u'=\lambda u$, $v'=\mu v$. Since $|a-xI|=0$ is a biquadratic whose roots are n-th roots of unity we must have $\phi(n)\leqq 4$, where ϕ is the Euler function of number theory. It follows that n can only have the values 2, 3, 4, 5, 6, 8, 10, 12.

Another important result is obtained by considering the effect of T on the integrals of the first kind[6], p. 263. When λ and μ are both $\neq 1$, T leaves no Picard integral of the first kind invariant and the associated I_n is regular. Returning to the group G_r of operations (11), its I_r is regular if there exists no linear change of variables reducing simultaneously all its operations to the form

$$u'=u+\nu_i; \qquad v'=\lambda_i v.$$

Should such a transformation exist and one of the multipliers λ_i be $\neq 1$, I_r is irregular with $q=1$. When all the λ's are 1, $q=2$ and I_r has for image a hyperelliptic surface of rank one.

The effect of T on $\iint du\,dv$ is to change it into $\lambda\mu\iint du\,dv$. Hence when $\lambda\mu=1$, I_n has its $p_g=1$, when $\lambda\mu\neq 1$, $p_g=0$.

Regarding I_r when $\lambda\mu=1$ for every operation of G_r, $p_g=1$, otherwise $p_g=0$. Thus by very simple considerations Bagnera-de Franchis go far towards classifying the involutions. By examining the relation $\boldsymbol{\alpha\omega}=\boldsymbol{\omega}\mathbf{a}$ for the possible operations T they obtain all $\boldsymbol{\omega}$'s corresponding to a cyclic T and the required classification of surfaces of rank $r>1$. They obtain explicit equations of a birational model for each type, in particular a double plane when the surface is regular. At the end of [6] there is a complete list of matrices and groups determined in the paper. For the construction of the various types see also Verzi[114].

22. In [2] Bagnera-de Franchis determine the Picard-Severi number ρ and the number ρ_0 of linearly independent double integrals of the second kind for all hyperelliptic surfaces. They apply throughout Humbert's

theorem([39]) on the representation of the curves by intermediary functions. When the rank $r=1$ they find by direct examination that $\rho=k$, the singularity index of $\boldsymbol{\omega}$, and they completely characterize the classes of intermediary functions. Hence ρ can have the values 1, 2, 3, 4. For surfaces whose $r>1$ they consider the number $[\rho]$ analogous to ρ when the isolated singular points are not counted as infinitesimal curves, and the number which they call ρ, when they are so considered. Both numbers are determined for all cases. For the generalized Kummer surfaces $[\rho]=k$ also, but $\rho=16+k$. See also Picard-Simart([69]), pp. 256, 447, Severi([108]).

The various invariants can be determined by the topological methods due to Lefschetz([50]), ([51]). His theorems described in Chapter 15, lead at once to the relation $\rho=k$ for surfaces of rank 1, 2.

In ([56]) Lefschetz examines real hyperelliptic surfaces and their invariants h, k. He defines a new invariant ρ', the number of algebraically distinct real curves for a real surface of rank one, or number of real curves between whose associated two cycles there is no homology. He finds six distinct types corresponding to as many sets of values h, k. See also for these real surfaces Comessatti([16]).

Comessatti([17]) in line with a certain theorem due to Severi([107]) has connected the theory of the base and the systems of curves on a surface of singularity $k=2$ with the group of automorphisms of the quadratic form whose coefficients are the intersection numbers of the two curves of the minimum base. The form is closely related to, but not identical with that considered by Humbert in ([42]). Comessatti's exposition is especially interesting from the geometric point of view. In particular he considers certain projective models of the surfaces and examines their properties.

23. *Additional References.*—Brief treatments are given by Castelnuovo-Enriques([11]), p. 748; Severi([104]), p. 779; more extensive discussions by Hudson([31.1]), Chs. 16, 18; Krazer-Wirtinger([48]), pp. 855, 870. Bonomi([7.1]) considers surfaces with an elliptic pencil. Scorza([95]), examines the intersections of the elliptic curves when any are present. Humbert([34]), Hudson([31.1]), Spampinato([109]), consider the elliptic Kummer surface. In a number of the following papers, interesting special cases will be found treated, notably birational transforms of the Kummer surface (in the strict sense or generalized) in the form of quartics with 13, 14 or 15 nodes: Traynard([111]); Rémy([75]), ([76]), ([78]), ([79]), ([81]); Chillemi([13]), ([13.1]); Piazzola-Beloch([61]). In connection with his work on Cremona groups Coble([14.2]), p. 354, has extensively investigated a projective model of the Jacobi surface, an F_{18} in S_8 invariant under a collineation group of order 162 independent of the moduli.

§ 3. Abelian Varieties. General Properties

24. Their definition has already been given in No. 8. That they exist follows from the Riemann-Weierstrass theorem on the algebraic relation between $p+1$ periodic functions with the same periods. The first step beyond that was taken by Klein who in 1886-87 discussed the analogue of the Kummer surface for any p, represented in a certain S_n by setting the homogeneous coordinates proportional to even θ's, or even intermediary functions. See No. 62.

25. In ([67]), ([65]), Picard extended to any p the theorem described in No. 10, and showed that a V_p with a continuous abelian group in p parameters is an abelian variety of genus p.

The essential step of connecting abelian varieties with algebraic surfaces was taken by Castelnuovo([8]), ([9]). See also Castelnuovo-Enriques ([11]), p. 709; Severi([104]), p. 762. On a surface F of irregularity q let $|C|$, $|C_1|$, $|C_2|$ be three elements of a sufficiently general maximal (complete) set of linear systems, W. Then $|C+C_1-C_2|$ is also an element of W, from which Castelnuovo concludes that W is reproduced by the addition of C_1-C_2 to its curves. That means that W is invariant under a continuous abelian group in q parameters, hence it is an abelian V_q. The variety is independent of C provided it is sufficiently general; it has been called the *Picard variety* of F. Referring to Ch. 15, No. 17, V_q corresponds point for point to the sets of constants a_j, modd. the periods of the Picard integrals u_{p-q+j}. Construct for the linear cycles an intermediary F. S. (Ch. 15, No. 11) of which the first $2q$ cycles, $\Gamma_1^1, \ldots, \Gamma_1^{2q}$ are independent and the rest zero-divisors, and let $\omega_{j\mu}$ be the period of u_{p-q+j} as to Γ_1^{μ}. Then the a's are determined up to sums of multiples of the terms of $\boldsymbol{\omega}$ and V_q is the most general abelian variety attached to $\boldsymbol{\omega}$. A similar V_q may be defined for any algebraic variety of linear connectivity index $R_1=2q$. The Picard variety of an algebraic curve of genus p, C^p, $(q=p)$ is called the Jacobi variety of the curve. The counting of moduli shows that a Jacobi V_p is a very special abelian variety. Not so, however, the Picard V_q of an algebraic surface or algebraic variety of irregularity q. Indeed from theorems on the reductions of linear cycles of a variety to a surface or subvariety that it carries, Lefschetz([50]), Ch. 5, it follows that any abelian variety is the Picard variety of some algebraic surface or variety that it carries, for example, sections by suitable linear spaces. Furthermore any V_n, $n\geqq 3$, has the same Picard variety as its hyperplane sections. Hence the Picard variety is a perfectly arbitrary abelian variety. In this connection there is also an interesting theorem due to Severi([106]) according to which a surface without an irrational pencil is the rational transform of a surface of the same irregularity, carried by its Picard variety, a theorem which he has also extended to varieties.

Another important series of investigations are those on irrational series on a C^p, Ch. 16, § 5, notably those of Torelli and Comessatti. Many of the questions taken up by them are of interest not only for the Jacobi V_p of C^p but for the general abelian V_p as well. A similar remark applies to much of the work of Rosati, discussed at various places in Ch. 16. It is indeed plain that the properties of the Jacobi variety of C^p are of basic importance for the *singular* correspondences of the curve.

26. The comprehensive study of the general abelian variety V_p, $p>2$, was actually initiated by Scorza in 1914, and is largely the work of Scorza, Rosati, Lefschetz. Scorza's investigations bear mostly on Riemann matrices and complex multiplication, and Rosati's are along parallel lines with singular correspondences on a C^p chiefly in view. The work of Lefschetz leans more to geometry and function theory; in many places the three authors overlap. The most profitable thing to do here will be to give a comprehensive picture of the theory as it now stands.

27. *Hypersurfaces of V_p. Topology.*—By a *hypersurface* of V_p is meant any V_{p-1} on the variety. We take the V_p attached to $\boldsymbol{\omega}$ and propose to describe its continuous systems of hypersurfaces in relation to the matrix. In this we follow Lefschetz([51]), Part II, Ch. 1; ([50]), Ch. 6, whose method is based on Analysis Situs considerations of a fairly simple nature. For the notions of Analysis Situs involved, see Ch. 15, § 3.

Topologically V_p is homeomorphic to the generalized ring derived from U_{2p} of No. 2 by bringing into coincidence the boundary points whose u's differ by a set of periods—a familiar procedure for elliptic functions. Let P_0 be the vertex of U_{2p} whose u's or real coordinates t_μ of No. 2 are all zero, P_μ the vertex whose $t_\mu=1$, the others being again zero. There is a definite k dimensional face of U_{2p} carrying the simplex (generalized tetrahedron) $P_0P_{i_1}\ldots P_{i_k}$. This face sensed by means of the simplex as indicatrix, ([50]), p. 3, is denoted by $(i_1\ldots i_k)$. The face is the image of a definite k-cycle of V_p also designated by that symbol.

A set of $\binom{2p}{k}$ k-cycles so obtained, no two composed of the same set of i's, constitutes an F. S. for k-cycles and as the cycles of the F. S. are independent, the connectivity and torsion indices of V_p are

$$R_k=\binom{2p}{k}, \qquad \sigma_k=1.$$

In particular Severi's invariant $\sigma=\sigma_1=1$. Therefore any intermediary F. S. for the hypersurfaces of V_p is also an actual F. S.

Observe that $(12\ldots 2p)$ is a $2p$-cycle which coincides with V_p or $-V_p$ depending upon the orientation of the variety.

28. Given two cycles Γ_h, Γ_k, there exists a definite cycle of $h+k-2p$ dimensions, their intersection, written $\Gamma_h \cdot \Gamma_k$. When $h+k=2p$ we consider the Kronecker-index $(\Gamma_h \cdot \Gamma_{2p-h})$ or algebraic number of intersections. The latter is computed on the basis that each crossing is affected with $+1$ or -1 and the sum of these units is the index. All doubtful cases are settled by the following theorems: (a) Intersection cycle or index obey the distributive laws with respect to both intersecting cycles. (b) The intersection cycle is ~ 0 or the index zero when one of the intersecting cycles is a zero divisor. This can all be extended to several intersecting cycles. See Lefschetz([50]), Ch. 1, ([52]); Veblen([113]); Weyl([118]).

When V_p is sensed like $(12\ldots 2p)$ the intersections and indices of the cycles of No. 27 are, ([50]), pp. 10, 114:

$$(i_1 i_2 \ldots i_k i_{k+1} \ldots i_h) \cdot (i_1 \ldots i_k i_{h+1} \ldots i_{2p}) \sim (-1)^n (i_1 i_2 \ldots i_k), \tag{12}$$

$$((i_1 i_2 \ldots i_k) \cdot (i_{k+1} \ldots i_{2p})) = (-1)^n, \tag{13}$$

where the i's are the sequence $1, 2, \ldots, 2p$ with n transpositions. If at the left in (13) the i's are not all distinct, the index is zero. When V_p is not sensed as above the right sides must be changed in sign.

29. Let $A_1, \ldots, A_p$ be hypersurfaces of V_p intersecting at the point Q. We assume that they have no singularity there and that their tangent S_{p-1}'s intersect at Q only. They are then cut out in the space that carries V_p by certain polynomials and in the u space by certain intermediary functions $\phi_h(u)$ which have a common zero Q', (u^0), corresponding to Q, with a non-vanishing functional determinant at Q'. Set $\phi_h = \phi'_h + i\phi''_h$; ϕ'_h, ϕ''_h real. The locus of the u space where all functions ϕ', ϕ'' are zero except ϕ'_h is an arc on which we take a point Q'_{2h-1} in the direction of increasing ϕ'_h and very near Q'. There is also a similar point Q'_{2h} for ϕ''_h. We now assign to the u space the indicatrix $Q'Q'_1 \ldots Q'_{2p}$ and to the image A'_h of A_h the indicatrix obtained by removing Q'_{2h-1} and Q'_{2h} from the preceding one. Naturally the variety and A_h will receive for indicatrices the images of the preceding. Under the circumstances by reference to the situation at Q' in the u space, we find that the contribution of Q to the Kronecker index $(A_1 \cdot A_2 \cdot \ldots A_p)$ is always $+1$.

30. Given two sets of analytical functions behaving respectively at R, S, like the ϕ's, together with the sets of associated manifolds of the u space, treat each exactly as above. There is a continuous deformation of the space bringing R into S and the neighborhood of the first set of manifolds about R into that of the second at S with all orientations

preserved. When the functions of the sets are all linear the verification is immediate, and as the situation merely depends upon the first degree terms in the Taylor expansions, it holds in all cases. Hence the indicatrices so defined for all points of our manifolds are consistent, and moreover independent, for any one of them, of all the others. The only condition to be fulfilled is that the functional determinant of the functions in any set be $\neq 0$. The same conclusion is valid as regards the orientation of the hypersurfaces of V_p: The orientations of all the indicatrices of No. 29 for a given A are in agreement and the cycle A is therefore a perfectly definite oriented spread.

31. When one of the sets of functions of No. 30 is the set $u_1, \ldots u_p$ there is obtained an indicatrix deformable into $PP_1 \ldots P_{2p}$ of No. 27 if and only if

$$\Delta = \frac{(-1)^{\frac{p(p+1)}{2}}}{(2i)p} \left| \begin{matrix} \omega \\ \bar{\omega} \end{matrix} \right| > 0. \tag{14}$$

When $\boldsymbol{\omega}$ is replaced by $\boldsymbol{\omega}\mathbf{a}$, Δ is replaced by $\Delta|a|$. If $\Delta < 0$ we can take a new F. S. of periods corresponding to a matrix $\mathbf{a}$, $|a| = -1$ and Δ will have become >0. Hence we can and shall assume that $\boldsymbol{\omega}$ is so chosen that (14) holds. As a consequence $V_p \sim (1\ 2 \ldots 2p)$ and formulas (12), (13), are valid as they stand.

32. Consider a definite analytical function $f(u)$ whose derivatives do not all vanish at R. Through R draw in the u space a small oriented planar element E not tangent to $f=0$. On E draw a small positive circuit ζ surrounding R. Then

$$\frac{1}{2\pi i} \int_{\zeta} d \log f = (E \cdot f), \tag{15}$$

where the index is that of E and $f=0$. For both sides are ± 1 and it is only a question of showing that their signs are the same. The true sign to be put at the right is unchanged under a deformation such as described in No. 30 combined with a deformation of E during which it never becomes tangent to f. This enables us to take $u_1 = f$ and for E an element of $u_2 = \ldots = u_p = 0$, so oriented that both sides in (15) are $+1$. Therefore (15) holds in all cases.

33. We shall now apply the preceding results to the hypersurfaces of V_p. A first remark of considerable importance: the index $(A_1 \ldots A_p)$ is greater than 0 and equals the virtual number of intersections $[A_1 \ldots A_p]$ in the sense of Ch. 15, No. 8. It is in fact the actual number of intersections whenever they are distinct and each counts for one in the total number. For by applying transformations $u = u' + \text{const.}$ to the A's we may replace them if necessary by hypersurfaces in the same continuous systems, hence homologous to them, and therefore with the same

$(A_1 \ldots A_p)$ and $[A_1 \ldots A_p]$, whose intersections will be isolated and as in No. 29. Since each intersection contributes then $+1$ to each of these numbers, the latter are equal.

34. Let again ϕ be an intermediary function attached to $\boldsymbol{\omega}$. The second derivatives of $d \log \phi$ are periodic, hence rational point functions on V_p. Therefore $\phi = 0$ is the parametric equation of a hypersurface A of V_p. Since it is also a $2p-2$ cycle we have

$$A \sim \Sigma(-1)^n m_{\mu\nu}(\sigma \ldots \kappa), \tag{16}$$

where $\mu, \nu, \sigma, \ldots, \kappa$, is the sequence $1, 2, \ldots, 2p$ with n transpositions and the sets $\sigma, \ldots, \kappa$ are all different. By (13), (15), and as in ([50]), p. 120, we find in the notations of No. 4

$$m_{\mu\nu} = (A \cdot (\mu\nu)) = \Sigma(\epsilon_{j\mu}\omega_{j\nu} - \epsilon_{j\nu}\omega_{j\mu}).$$

Hence the integers $m_{\mu\nu}$ are the same as in No. 4. In any case by (13), whether A is cut out by an intermediary function or not, the first equality sign is valid, and the interpretation of $m_{\mu\nu}$ as a Kronecker-index is to the point. But this is not very important for as we shall see A is always cut out by some ϕ.

Having definite homologies for the hypersurfaces we can apply (12), (13), to obtain homologies or indices for their intersections. Let

$$A_j \sim \Sigma(-1)^n m^j_{\mu\nu}(\sigma \ldots \kappa),$$

where in case A_j is cut out by the intermediary function $\phi_j = 0$, $m^j_{\mu\nu}$ is the $m_{\mu\nu}$ of the function. Then

$$A_1 A_2 \ldots A_s \sim \frac{1}{2^s \cdot (2p-2s)!} \Sigma(-1)^n m^1_{\mu\nu} \ldots m^s_{\lambda\rho}(\sigma \ldots \kappa), \tag{17}$$

while for $s = p$ the number of intersections is

$$[A_1 \cdot A_2 \ldots A_p] = \frac{1}{2^p} \Sigma(-1)^n m^1_{\mu\nu} \ldots m^p_{\sigma\kappa}. \tag{18}$$

The sequence $\mu, \ldots, \kappa$ is the same as $1, 2, \ldots, 2p$ with n transpositions.

Every A determines a continuous system such that there are always sets of two, three, $\ldots, p$, intersecting in cycles of $2p-4, \ldots, 2$ dimensions or in isolated points. For the related cycles and intersections, conveniently denoted by $A^2, \ldots, A^{p-1}, A^p$, (17), (18) lead to especially simple expressions.

Let π be the pfaffian of the skew-symmetric matrix m, Kowalewski([46]), p. 140. In its expression let $M_{\mu \ldots \rho}$ be the coefficient of the term $m_{\mu\nu} \ldots m_{\lambda\rho}$. It may be considered as a certain minor of π. Then

$$A^s \sim s! \Sigma M_{\mu \ldots \rho}(\mu \ldots \rho). \tag{19}$$

$$(A^p) = [A^p] = p! \pi. \tag{20}$$

35. If as in No. 31 we replace $\boldsymbol{\omega}$ by $\boldsymbol{\omega}\mathbf{a}$, $|a|=\pm 1$, π is replaced by $\pi|a|$ and therefore $\pi\Delta$ is unchanged. We have so chosen $\boldsymbol{\omega}$ that $(A^p)>0$, therefore $\pi\geqq 0$. Thus as a consequence of the choice of $\boldsymbol{\omega}$ the pfaffian of the $\mathbf{m}$ of every intermediary function is positive. Whatever the choice of $\boldsymbol{\omega}$ all the pfaffians have the same sign.

Maintaining our convention we have $\pi=+\sqrt{|m|}$. Therefore the dimension of the complete linear system to which ϕ belongs is $\pi-1$.

In order to verify whether $\boldsymbol{\omega}$ has been correctly chosen it is then merely necessary to compute a single pfaffian. When $\boldsymbol{\omega}$ is reduced to the form $\mathbf{N}'$ of § 1 some intermediary function becomes a θ_n and its pfaffian is $\dfrac{(ne_p)^p}{e_1 e_2 \ldots e_p}>0$ hence $\boldsymbol{\omega}$ is then such that (14) holds. Consequently any proper $\boldsymbol{\omega}$ is of the form $\mathbf{N}'\mathbf{a}$, $|a|=+1$. This is absolutely in line with Humbert's conventions, ([41]), ([42]), also those of Bagnera-de Franchis([2]); see also Scorza([99]), first paper.

Special cases of (18), (20), are due to Poincaré([70]) and Wirtinger([120]), ([124]), their treatment, especially Poincaré's, being related to ours. We have gone into the matter at some length as our procedure reduces most questions of enumeration on a V_p to an arithmetical algorithm. Further references are Baker([1.1]), Ch. 10, ([1.2]); Lefschetz([51]), p. 381, where the whole set of formulas was first given in its completeness, ([50]), Ch. 6; Castelnuovo([10]), where interesting applications to correspondences on a C^p are also given. For $p=2$ see also Humbert([41]), Comessatti([17]). In ([51]), p. 389, Lefschetz gives various enumerative formulas for the arithmetical genera of the hypersurfaces of V_p and of their complete intersections.

36. Starting with (16) we can derive the theorem of Frobenius given in No. 4, according to which if ϕ is a non-degenerate intermediary function attached to $\boldsymbol{\omega}$, $\mathbf{m}^{-1}=\mathbf{M}$ is a principal matrix, hence $\boldsymbol{\omega}$ is an R. M., Poincaré([70]), Lefschetz([51]), Ch. 1, ([50]), Ch. 6. This can be inverted: Given $\boldsymbol{\omega}$ with the principal matrix $\mathbf{M}$, $\mathbf{m}$ is defined (up to an integer factor) and there is a whole class of related intermediary functions of every order $n>0$. They cut out hypersurfaces all algebraically equivalent (homologous) to a multiple of a fixed A (cut out by functions of order one of the class). From this Lefschetz derives, loc. cit.:

I. The intermediary functions of like behavior, hence with same $\mathbf{m}$ and π, cut out a complete linear system $|A|$ whose dimension is $\pi-1$ and which belongs to a continuous system ∞^p.

II. The base number ρ of V_p is equal to the singularity index k of $\boldsymbol{\omega}$.

III. Every hypersurface of V_p is cut out by an intermediary function.

IV. There exists in all cases an F. S. composed of ρ effective hypersurfaces.

The precise construction of the F. S. is as follows: We can find k linearly independent principal matrices $M_1, \ldots, M_k$ of $\boldsymbol{\omega}$ and construct the class of intermediary functions for each. In each class we take an intermediary function of order one. It corresponds to an $\mathbf{m}=l\mathbf{M}^{-1}$ with the integer factor l so chosen that the terms of $\mathbf{m}$ are relatively prime integers. The k functions $\phi_1, \ldots, \phi_k$ so chosen cut out the hypersurfaces of the F. S. Incidentally in every complete linear set of intermediary functions there exists a definite one of the form $\Pi\phi_i^{n_i}(u+c)$ where the n's are positive or negative integers.

These theorems were known for $p=2$. See Humbert([39]), Bagnera-de Franchis([2]), this chapter, Nos. 12, 22; also Castelnuovo([10]) for the general case.

37. Just as for hyperelliptic surfaces the integrals of the first kind of various multiplicities (integrals of total differentials and their generalizations) can be expressed in terms of the u's. The s-fold integrals are all linear combinations of the integrals $\int\ldots\int du_{i_1}\ldots du_{i_s}$ and their periods are minors of order s of $\boldsymbol{\omega}$. Of chief interest are the simple integrals which are merely the u's. For the enumeration questions connected with these integrals see Lefschetz([51]), p. 373.

§ 4. Impure Matrices and Their Varieties

38. The most important work along that line is due to Scorza. To give a good account of it we will find it advantageous to consider in some detail his very suggestive projective method, since some of its best applications have been in the theory of impure matrices. The chief references for Scorza's work are his notes([99]) and major paper([94]), Part I; Krazer-Wirtinger([48]), p. 827; Lefschetz([54]) (brief summary). For impure matrices see ([48]), p. 838, to whose bibliographical list should be added Krazer([47]), Ch. 11; Severi([105]), Ch. 10. There is an important parallel development initiated by Rosati([82]), ([88]), almost simultaneously with Scorza (1914). In his work, based on a dual of Scorza's method, he considers throughout, period matrices of algebraic curves, and his chief aim is to investigate singular correspondences. There are many points of contact between the two authors but naturally Rosati's is far more special. An account of his investigations on correspondences has already been given in various parts of Chapter 16.

Given an R. M. $\boldsymbol{\omega}$, consider with Scorza the S_{p-1}, σ, of S_{2p-1}, carrier of the p points whose projective coordinates are $x_\mu=\omega_{j\mu}$, together with its conjugate $\bar{\sigma}$ carrier of the points whose coordinates are $x_\mu=\bar{\omega}_{j\mu}$. An alternate matrix $\mathbf{c}$ of $\boldsymbol{\omega}$ gives rise to a rational null system

$$\Sigma c_{\mu\nu}x_\mu y_\nu=0$$

with respect to which all point pairs of σ (or of $\bar{\sigma}$) are conjugate. It may also be considered as defining a line complex of S_{2p-1} carrying all lines on σ (hence also those on σ). When $\mathbf{c}$ is principal there are no real lines of the complex intersecting both σ and $\bar{\sigma}$. As ordinary rational $\mathbf{c}$ of $\boldsymbol{\omega}$ defines a rational reciprocity ρ with respect to which all point pairs of σ (or of $\bar{\sigma}$) are conjugate. The indices h, k are the numbers of linearly independent rational reciprocities or null systems of the preceding type.

Let us consider a definite principal matrix $\mathbf{c}_0$ with its reciprocity ρ_0 which is not degenerate since $|c_0|\neq 0$. Then $\eta=\rho\rho_0^{-1}$ is a rational projectivity leaving σ and $\bar{\sigma}$ invariant. There is a corresponding matrix relation $\boldsymbol{\alpha\omega}=\boldsymbol{\omega}\mathbf{a}$; $\mathbf{a}=\mathbf{c}\mathbf{c}_0^{-1}$, or multiplication of $\boldsymbol{\omega}$ (Ch. 15, No. 7). Conversely such a multiplication defines an η, hence a rational reciprocity $\rho=\eta\rho_0$ of the above type. This proves Scorza's theorem([94]), p. 280: The numbers of linearly independent rational reciprocities and of rational projectivities leaving σ invariant are equal. Their common value is the index of multiplication h. It is also the number of linearly independent multiplications (see below No. 46).

39. Let now $\boldsymbol{\omega}$ be isomorphic to $\left\|\begin{matrix}\boldsymbol{\omega}_1 & 0\\ \boldsymbol{\omega}_3 & \boldsymbol{\Omega}_2\end{matrix}\right\|$, where $\boldsymbol{\omega}_1$ is an R. M. of genus $p_1<p$, i. e., let $\boldsymbol{\omega}$ be impure. Since isomorphs have the same σ there exists a rational S_{2p_1-1} whose intersection with σ is an S_{p_2-1}, σ_1, called an *axis* of $\boldsymbol{\omega}$. Of course S_{2p_1-1} intersects $\bar{\sigma}$ in $\bar{\sigma}_1$. The theorem of Poincaré, Picard, and Scorza([94]), p. 298, Krazer-Wirtinger([48]), p. 838, on the isomorphism of $\boldsymbol{\omega}$ to $\boldsymbol{\omega}_0=\left\|\begin{matrix}\boldsymbol{\omega}_1 & 0\\ 0 & \boldsymbol{\omega}_2\end{matrix}\right\|$ signifies that to each axis σ_1 of $\boldsymbol{\omega}$ corresponds a *complementary* axis σ_2 (belonging to $\boldsymbol{\omega}_2$), and the two do not intersect.

It is important to observe, Scorza([95]), p. 222, that $\boldsymbol{\Omega}_2$ like $\boldsymbol{\omega}_1$, $\boldsymbol{\omega}_2$, is also an R. M., but not necessarily complementary to $\boldsymbol{\omega}_1$. Furthermore up to a transformation into an isomorph, $\boldsymbol{\omega}_2$ is uniquely determined by $\boldsymbol{\omega}_1$, Scorza([94]), p. 304. This does not mean that σ_1 necessarily determines its complementary axis σ_2, but that all submatrices $\boldsymbol{\omega}_2$ corresponding to these σ_2's are isomorphic. From this follows the reduction of $\boldsymbol{\omega}$ to an essentially unique normal form

$$\left\|\begin{matrix}\boldsymbol{\Omega}_1 & & \\ & \ddots & \\ & & \boldsymbol{\Omega}_k\end{matrix}\right\|, \qquad \boldsymbol{\Omega}_j=\left\|\begin{matrix}\boldsymbol{\omega}_j & & \\ & \ddots & \\ & & \boldsymbol{\omega}_j\end{matrix}\right\|$$

where $\boldsymbol{\omega}_j$ is pure and $\boldsymbol{\omega}_k$, $\boldsymbol{\omega}_j$ are non-isomorphic for $k\neq j$. The number of essentially distinct modes of reductions to the normal form (number of distinct sets of pure complementary axes) is finite when the $\boldsymbol{\Omega}$'s are all pure, infinite otherwise, Scorza([94]), p. 307.

When $\boldsymbol{\omega}$ is impure and possesses an axis σ_1 its representative space is invariant under a degenerate rational reciprocity ρ which assigns to every point of σ an S_{2p-2} passing through σ_1 and conversely when such a ρ exists σ_1 is an axis of $\boldsymbol{\omega}$ and $\boldsymbol{\omega}$ is impure. Hence in order that $\boldsymbol{\omega}$ be impure it is necessary and sufficient that it possess a degenerate rational reciprocity, or a rational matrix $\mathbf{c}$ whose determinant $|c|=0$. This remains true when reciprocities are replaced by null systems, or projectivities. In particular $\boldsymbol{\omega}$ is impure if and only if there is a multiplication $\boldsymbol{\alpha\omega}=\boldsymbol{\omega}\mathbf{a}$ with $|a|=0$. See Scorza([94]), p. 278.

40. Scorza([94]), p. 301, determines the invariants of $\boldsymbol{\omega}$ in terms of those of the submatrices of its normal form. Let first $\boldsymbol{\omega}$ be isomorphic to $\boldsymbol{\omega}_0$ above, with $h_j, k_j;\ h, k;\ \lambda$, as the invariants of $\boldsymbol{\omega}_j$ and $\boldsymbol{\omega}$, and simultaneity index of $\boldsymbol{\omega}_1$, $\boldsymbol{\omega}_2$ (Ch. 15, No. 7). Then

$$h=h_1+h_2+2\lambda, \qquad k=k_1+k_2+\lambda.$$

Since the λ of two isomorphs is their common h the repeated application of these formulas gives for $\boldsymbol{\omega}$ when in the normal form the indices:

$$h=\Sigma n_j^2 h_j; \qquad k=\Sigma\left(\left(n_{j_2}\right)h_j+n_j k_j\right),$$

where n_j is the number of components $\boldsymbol{\omega}_j$ in $\boldsymbol{\Omega}_j$ and h_j, k_j the invariants of $\boldsymbol{\omega}_j$.

Owing to our definition of h, k, these formulas are decidedly simpler than Scorza's. See also Lefschetz([51]), p. 394.

In ([94]), Part I, §§ 6-9, Scorza examines at length the various possibilities for h, k. In any case $h\leqq 2p^2$, $k\leqq p^2$ and for a pure $\boldsymbol{\omega}$: $h\leqq 2p$, $k\leqq 2p-1$. The indices possess lacunary values but for every p there are R. M.'s for which they simultaneously assume their extremes $2p^2$, p^2, and for all others $h<2p^2$, $k<p^2$. The R. M.'s with maximal indices are reducible to the form $\mathbf{N}$ of No. 1, with $\tau_{jk}=\delta_{jk}\cdot i\sqrt{d}$, where d is a positive integer, Scorza([94]), p. 312, de Franchis([25]), Lefschetz([51]), p. 400. The special R. M. with $d=1$ has an isomorph whose terms differ by as little as we please from those of any assigned $\boldsymbol{\omega}$, Poincaré, further details and references in Krazer-Wirtinger([48]), p. 831. The R. M.'s of type $\mathbf{N}$ with $\tau_{jk}=\delta_{jk}\tau$ (τ an imaginary number) possess a number of interesting properties. In particular they have an infinite number of distinct elliptic submatrices $\|1, \tau\|$. The abelian V_p is the image of the groups of p points, each on one of p elliptic curves with the common modulus τ.

41. *Varieties with Impure R. M.*—Let again $\boldsymbol{\omega}$ be isomorphic to $\boldsymbol{\omega}_0$ of No. 39, with $\boldsymbol{\omega}_i$ of genus p_i. The initial variables u_j can be linearly transformed into new variables v_j such that the periods of $v_1, \ldots, v_{p_1}$ as to the last $2p_2$ columns of $\boldsymbol{\omega}_0$ and of $v_{p_1+1}, \ldots, v_p$ as to the first $2p_1$ columns are all zero. Expressed in terms of $\boldsymbol{\omega}$ and the u's, there are

p_i linear combinations of these variables with only $2p_i$ distinct periods. In view of No. 37, this explains the common description of such a set as "regular system of reducible integrals."

Another property of an impure $\boldsymbol{\omega}$ is that there are associated intermediary functions factorable into a product of functions of a smaller number of variables, all linear combinations of the primitive variables, Wirtinger([119]), Krazer([47]), p. 500, Krazer-Wirtinger([48]), p. 839.

Let V_{p_i} be the abelian variety of $\boldsymbol{\omega}_i$. V_p carries a congruence of varieties V'_{p_i} birationally equivalent to V_{p_i}, their equations being $v_{p_1+j}=$const. for $i=1$, $v_j=$const., $j\leq p_1$ for $i=2$. These congruences have been investigated by Castelnuovo([8]), ([9]); Scorza([93]), ([95]); and on a Jacobi variety by Torelli([110]), Comessatti([16]) in connection with their work on singular correspondences, Ch. 16, No. 13. Scorza in his papers determines the number n of intersections of V'_{p_1} with V'_{p_2}. It can also be obtained by writing down homologies analogous to (16) for V'_{p_1} and V'_{p_2} and applying (13). V_p itself is in $(1, n)$ correspondence with the variety V'_p that represents the point pairs of V_{p_1} and V_{p_2} and whose R. M. is $\boldsymbol{\omega}_0$. It is birationally equivalent to V'_p when and only when $\boldsymbol{\omega}$ is equivalent, not merely isomorphic, to $\boldsymbol{\omega}_0$. To every reduction of $\boldsymbol{\omega}$ to its normal form corresponds a definite involution on V_p.

§ 5. Transformations of V_p

42. There are two phases to the question: Geometric or transformation theory proper, and arithmetic or complex multiplication. We propose to consider the geometric phase first and to show its connection with complex multiplication. In § 6 we shall then treat the latter without concerning ourselves with its geometric implications.

Let T be an algebraic correspondence on V_p. By the same reasoning as that of Hurwitz for correspondences on a C^p (Ch. 16, No. 4), Scorza shows, ([93]), ([94]), Part II, that T leads to a multiplication

$$\boldsymbol{\alpha\omega}=\boldsymbol{\omega}\mathbf{a}, \tag{21}$$

the terms of $\mathbf{a}$ being here also integers. He considers various properties of these correspondences, in particular the valence when (21) reduces to an ordinary multiplication. He also shows that every multiplication determines a whole class of transformations, among which there are always rational transformations. Multiply if necessary both sides of the multiplication by an integer so chosen that $\mathbf{a}$ is as above a matrix with integer terms and let (21) be the form then assumed by the multiplication. The rational T will then have for equations:

$$u'_j=\Sigma a_{jk}u_k+\beta_j \tag{22}$$

similar to Humbert's system (8) of No. 14. The β's are of course arbitrary constants.

The ordinary transformations are here also those for which (21) is identically verified. They are classified as for hyperelliptic surface (No. 18) in transformations of the first kind

$$u_j + u'_j = \text{const.}$$

and transformations of the second kind

$$u_j - u'_j = \text{const.}$$

The latter alone form a continuous group of birational transformations, which is a p-parameter abelian group.

Let x_μ ($\mu = 1, 2, \ldots, 2p$) be the differences of the real coordinates (defined as in No. 2) of two points u corresponding by (22) to given u'. Then

$$\Sigma a_{jk}\omega_{k\mu}x_\mu = \Sigma \omega_{j\mu}a_{\mu\nu}x_\mu$$

must be a set of periods. The number n of distinct points of V_p corresponding to the point u' is then the number of distinct sets of rational x's between 0 and 1 (0 included) such that the expressions $\Sigma a_{\mu\nu}x_\mu$ are all integers. Hence $n = |a|$, Krazer(47), p. 58. Therefore T is $(1, n)$, Humbert(42), p. 289, Scorza(94), (93). T has $|a - I|$ coincidences (when their number is finite), Scorza(93).

The n points of V_p derived by T from a given point u of V_p constitute an I_n whose image is V_p itself. The u's of any two of them differ by constants. Therefore the n points are obtained from a given one of them by the operations of a birational group of order n of V_p whose transformations are all ordinary of the first kind. T may be considered as a birational transformation of V_p into another copy V'_p of itself taken as the image of I_n.

43. When we have a multiplication (21) we can deduce a whole class from it by multiplying the terms of $\boldsymbol{\alpha}$ and $\mathbf{a}$ by the same rational number m, the matrices being then $m\boldsymbol{\alpha}$, $m\mathbf{a}$. Hence we can always assume that the terms of $\mathbf{a}$ in (21) are relatively prime integers. Such a multiplication is called *primitive*. To every rational T associated with the class of multiplications will correspond a multiplication whose matrices are $m\boldsymbol{\alpha}$, $m\mathbf{a}$, where m is now an integer, and the index n of T will be $m^{2p}|a|$. For the matrices of all rational transformations there exists an F. S., $\mathbf{a}_1, \ldots, \mathbf{a}_h$, where h is the singularity index of $\boldsymbol{\omega}$. Every $\mathbf{a}$ whose terms are integers is then of the form $\Sigma x_i\mathbf{a}_i$ (x_i integer), Scorza(94), p. 283. All possible indices are $|\Sigma x_i a_i| = f(x_1, \ldots, x_h)$ where f is a homogeneous form of degree $2p$, with integer coefficients, $\geqq 0$ for all x's not all zero,

and in fact >0 when $\boldsymbol{\omega}$ is pure, Scorza([93]), p. 6. A more accurate result is given by Scorza([96]), p. 198. (See below No. 59). In order that T be a birational transformation it is necessary and sufficient that $|a|=1$.

Additional reference on transformations of V_p when $\boldsymbol{\omega}$ is impure: Cherubino([12.1]). For $p=2$, see No. 15.

The preceding discussion can be extended to a rational transformation T of V_p^2 attached to $\boldsymbol{\omega}_2$, into V_p^1 attached to $\boldsymbol{\omega}_1$. In place of (21) we have now a relation $\boldsymbol{\alpha}\boldsymbol{\omega}_1=\boldsymbol{\omega}_2\mathbf{a}$, with $\pm|a|$ as the order of T. Here T is birational if, and only if, $|a|=\pm 1$ that is if $\boldsymbol{\omega}_1$ and $\boldsymbol{\omega}_2$ are equivalent. Assume $|a|\neq 0$. Then $\boldsymbol{\omega}_1$ and $\boldsymbol{\omega}_2$ are isomorphic and to the multiplication $\boldsymbol{\beta}\boldsymbol{\omega}_1=\boldsymbol{\omega}_1\mathbf{b}$ of $\boldsymbol{\omega}_1$ corresponds $(\boldsymbol{\alpha}\boldsymbol{\beta}\boldsymbol{\alpha}^{-1})\boldsymbol{\omega}_2=\boldsymbol{\omega}_2(\mathbf{a}\mathbf{b}\mathbf{a}^{-1})$ of $\boldsymbol{\omega}_2$. Thus the multiplications of $\boldsymbol{\omega}_1$ and $\boldsymbol{\omega}_2$ correspond to one another but the primitive multiplications, i. e., the fundamental sets, do not in general unless $|a|=\pm 1$, when V_p^1 and V_p^2 are birationally equivalent.

To determine all rational transformations of V_p or all multiplications (21) with the terms of $\mathbf{a}$ integers, we may first obtain a set of h independent multiplications. From this set, an F. S. for the more special type just mentioned is readily derived by well-known arithmetical processes, Scorza([94]), p. 282.

§ 6. Complex Multiplication

44. We have already examined the case $p=1$ in Ch. 16, No. 2, and may refer for more extensive information to Fricke([26]), Fueter([29]), Weber([117]). The case $p=2$ has also been considered at length in § 2 and we have shown there how Humbert in ([42]) laid down the foundations for many of the later developments. Humbert-Lévy([45]); Lévy([57]); have developed the properties of a certain geometric configuration attached to an R. M. of genus three in relation to its multiplications. This representation is somewhat related to that used by Scorza([94]) or Rosati([82]).

Previous to Humbert, only complex multiplications corresponding to *principal* transformations had been considered. For references see Krazer-Wirtinger([48]), p. 852; also Krazer([47]), p. 220, where these transformations are treated in detail. Here as in No. 16, and with Scorza([94]), p. 287, a transformation is *principal* whenever there is a principal $\mathbf{c}$ of $\boldsymbol{\omega}$ such that $\mathbf{a}\mathbf{c}\mathbf{a}'=l\mathbf{c}$. The study of the most general multiplication was initiated by Scorza in ([99]) and the chief contributions are due to him, to Lefschetz and to Rosati. The latter has always in view his main problem: singular correspondences on a C^p. His work has been considered in various parts of Ch. 16.

45. From now on we associate with the multiplication (21) the uniquely defined rational transformation (22) with constants β all zero,

and apply to one or the other the same symbol T. Let T_1, T_2, be two multiplications with $\boldsymbol{\alpha}_1$, $\mathbf{a}_1$ and $\boldsymbol{\alpha}_2$, $\mathbf{a}_2$ as $\boldsymbol{\alpha}$, $\mathbf{a}$ matrices. Then $m_1\boldsymbol{\alpha}_1+m_2\boldsymbol{\alpha}_2$, $m_1\mathbf{a}_1+m_2\mathbf{a}_2$ (m_1, m_2 rational) belong to a multiplication to be called $m_1T_1+m_2T_2$ and $\boldsymbol{\alpha}_1\boldsymbol{\alpha}_2$, $\mathbf{a}_1\mathbf{a}_2$ to a multiplication called T_1T_2. When Σm_iT_i corresponds to matrices $\boldsymbol{\alpha}$, $\mathbf{a}$, both zero (they always vanish simultaneously) we say that the T's are *dependent* and write $\Sigma m_iT_i=0$. The maximum number of linearly independent T's is the multiplication index h, Nos. 38, 46. Hence the T's are combined like the numbers of a linear associative algebra, in fact a matrix algebra, in h units over the field of rational numbers. These properties are all due to Scorza([94]), ([96]) and he has made a systematic study of the algebra in ([96]). We return to it in No. 59.

46. We have already considered in No. 38, Scorza's important theorem according to which the numbers of linearly independent rational projectivities and rational reciprocities are both h. It implies that the number of linearly independent multiplications is also h. Scorza's proof actually shows, however, that the reciprocities or the rational matrices of $\boldsymbol{\omega}$ determine the multiplications, and vice versa. It is worth while giving a proof of this important proposition, more in line with the order of ideas of this Report where we have made the matrices themselves the central elements.

Let then $\mathbf{c}$ be a principal matrix of $\boldsymbol{\omega}$ so that $\boldsymbol{\omega}\mathbf{c}\boldsymbol{\omega}'=0$, $|c|\neq 0$. Then from (21) follows $\boldsymbol{\alpha}\boldsymbol{\omega}\mathbf{c}\boldsymbol{\omega}'=\boldsymbol{\omega}\mathbf{a}\mathbf{c}\boldsymbol{\omega}'=0$, therefore $\mathbf{d}=\mathbf{a}\mathbf{c}$ is an ordinary matrix of $\boldsymbol{\omega}$. Conversely with $\boldsymbol{\Omega}=\left\| \begin{matrix} \boldsymbol{\omega} \\ \overline{\boldsymbol{\omega}} \end{matrix} \right\|$, we have, given $\mathbf{d}$ and since $\mathbf{c}$ is principal,

$$\boldsymbol{\Omega}\mathbf{c}\boldsymbol{\Omega}'=\left\| \begin{matrix} 0 & \boldsymbol{\gamma} \\ \overline{\boldsymbol{\gamma}} & 0 \end{matrix} \right\|; \qquad \boldsymbol{\Omega}\mathbf{d}\boldsymbol{\Omega}'=\left\| \begin{matrix} 0 & \boldsymbol{\delta} \\ \overline{\boldsymbol{\delta}} & 0 \end{matrix} \right\|,$$

where $\boldsymbol{\gamma}$, $\boldsymbol{\delta}$ are square, of order p. Since $|\gamma|$, $|c|$ are $\neq 0$, $\boldsymbol{\alpha}=\boldsymbol{\delta}\boldsymbol{\gamma}^{-1}$, $\mathbf{a}=\mathbf{d}\mathbf{c}^{-1}$, are uniquely defined and we find

$$\mathbf{A}\boldsymbol{\Omega}=\boldsymbol{\Omega}\mathbf{a}; \qquad \mathbf{A}=\left\| \begin{matrix} \boldsymbol{\alpha} & 0 \\ 0 & \overline{\boldsymbol{\alpha}} \end{matrix} \right\|,$$

from which (21) and the mutual dependence of $\mathbf{d}$ and $\mathbf{c}$ upon one another follow.

47. The problem that naturally comes next is the classification of R. M.'s of a given genus according to their multiplications. The two chief contributions on that topic are due to Scorza([97]), ([100]), ([96]) and Lefschetz([53]), ([51]). They are found in the major papers ([96]) and ([51]), both published in (1921), the others being merely preliminary short

notes. Scorza's main object is to develop the general theory of the multiplication algebra and particularly the properties of a new invariant, the rank ρ of an R. M. His typical theorems bear on the possible relative values of h, k, ρ, p. Lefschetz on the contrary attacked directly the classification and in particular made a thoroughgoing investigation of two special but very important types of R. M.'s that include all those found in the literature. We shall therefore report on his work and at the same time on related special contributions. In No. 59 we shall then give an account of Scorza's work on the multiplication algebra.

48. Consider first the multiplication T represented by (21). The characteristic equations of $\boldsymbol{\alpha}$ and $\mathbf{a}$ are respectively

$$\phi(x) = |\alpha - xI| = 0; \qquad F(x) = |a - xI| = 0.$$

The roots of ϕ are the *multipliers* of T. When (21) arises from a correspondence on a C^p they are the negatives of the Rosati valences ([89]), ([90]). In the notation of No. 46, since $\mathbf{a} = \boldsymbol{\Omega}^{-1}\mathbf{A}\boldsymbol{\Omega}$, necessarily

$$F(x) = |a - xI| = |\alpha - xI| \cdot |\bar{\alpha} - xI| = \phi(x) \cdot \bar{\phi}(x).$$

Therefore if x_1 is a root of ϕ, $\bar{x}_1$ is a root of $\bar{\phi}$ and with the same multiplicity. Hence the real roots of F are all of even multiplicity and its complex roots are associated in pairs of conjugate roots of same multiplicity. This was proved for principal transformations by Frobenius([27]), and for the general case by Scorza([94]), p. 284. See also Krazer([47]), p. 224, Krazer-Wirtinger([48]), p. 853.

Incidentally $|a| = |\alpha| \cdot |\bar{\alpha}| \geqq 0$ and can only be zero when one of the multipliers is zero, and then $\boldsymbol{\omega}$ is impure. Conversely when $\boldsymbol{\omega}$ is impure it always possesses a multiplication whose $|a| = 0$ (No. 39).

T is an ordinary multiplication when all the multipliers are equal and then they are rational.

T satisfies the *characteristic* equation $F(T) = 0$, but its *minimum* equation $f(T) = 0$, or equation of least degree q that it satisfies, may be different, although the roots of $F(x) = 0$, $f(x) = 0$ are the same. Here q, called the *degree* of T, is the maximum number of linearly independent powers of T. See in this connection Rosati([89]), ([90]).

49. Let $\boldsymbol{\omega}$ be impure and in the normal form of No. 39. All its $\mathbf{d}$ matrices (No. 46) are linear combinations of those of degenerate type which involve one or two pure components of the same $\boldsymbol{\Omega}_i$, both of which types are known when the $\mathbf{d}$ matrices for the pure components are known. Hence the determination of all multiplications of $\boldsymbol{\omega}$ can be reduced to that of all multiplications of the pure components $\boldsymbol{\omega}_i$, and to find all possible multiplications and all R. M.'s that possess complex multiplications we may limit ourselves to pure R. M.'s.

Now for a pure R. M., f is always irreducible and $F \equiv f^r$. Hence when $\boldsymbol{\omega}$ is pure the degree q of f is always a factor of $2p$. It is furthermore a factor of h; and these circumstances narrow the field down a good deal. Observe that F may well have the preceding form for an impure $\boldsymbol{\omega}$. See Lefschetz([51]), Part II, Ch. 2, § 1; Scorza([100]), ([96]), Part II, §§ 2, 3, 4.

50. When there is a T of degree $q>1$ such that $F\equiv f^r$, with the roots of f all simple, $\boldsymbol{\omega}$ can be reduced to a form exhibiting the presence of its multiplication. If the multipliers are $\mu_1, \ldots, \mu_p$, the u's can be so chosen that the equations of the transformation become

$$u'_j = \mu_j u_j \tag{23}$$

and (21) gives then

$$\mu_j \omega_{j\sigma} = \Sigma \omega_{j\nu} a_{\nu\sigma}. \tag{24}$$

Hence if $\omega_{j\sigma}$ is any set of periods so is $\mu_j{}^s\omega_{j\sigma}$. This enables us to reduce $\boldsymbol{\omega}$ to the form

$$\mathbf{H} = \|\xi_{j\sigma}\mu_j{}^s\|, \quad (s=0,1,\ldots,q-1; \quad \sigma=1,2,\ldots,r; \quad j=1,2,\ldots,p),$$

which is especially convenient for our purpose because it exhibits plainly the elements independent of T: the terms of the matrix ξ. Given T and its characteristic equation when $r>1$, ξ is indeed in general variable, an observation first made by Frobenius([27]) for the special case of a principal transformation. For $r=1$, $\mathbf{H}$ was first given by Scorza([94]), p. 24, for $r>1$ by Lefschetz([51]), p. 397. Earlier examples for $p=2$ and cyclic T occur in Bagnera-de Franchis([2]).

Among the notable R. M.'s of type $\mathbf{H}$ are those which correspond to principal transformations. They have been repeatedly considered by various authors. See Koenigsberger([45.2]); Kronecker([49]); Weber([115]); Frobenius([27]); Scorza([94]), p. 287; ([96]), p. 153; Lefschetz([51]), p. 399; Krazer-Wirtinger([48]), p. 853. Frobenius showed that the multipliers of such a T have all the same absolute value $\sqrt{\rho}$, where ρ is a rational number. Scorza showed that if $\rho=1$, T is a cyclic birational transformation, and that conversely when T has that property it is principal with its $\rho=1$, ([94]), p. 289. In ([96]), p. 153, he observes that when $\boldsymbol{\omega}$ is non-singular ($k=1$) all its transformations are principal.

Rosati showed ([89]), p. 1007 (see Ch. 16, No. 7) that a symmetric correspondence on a C^p is associated with a T of the R. M. of C^p whose minimum equation has only simple roots all real. When the correspondence is skew symmetric the multipliers are pure imaginary, Rosati([90]), p. 14. When the R. M. is pure it falls in the two categories investigated by Lefschetz in ([51]) and considered below Nos. 55, 56.

51. Lefschetz has extensively investigated, ([51]), Part II, Ch. 2, the case $r=1$. Then $F\equiv f$, $q=2p$ and the matrix is

$$\mathbf{K}\equiv||\mu_j{}^s||,\quad (s=0,1,\ldots,2p-1;\quad j=1,2,\ldots,p).$$

The roots of $F=0$ are all distinct and imaginary and the multipliers consist of a set of p of these roots no two of which are conjugate. The properties of $\mathbf{K}$ depend partly upon the equation $F=0$, partly upon the particular choice of roots for multipliers.

Given, a priori, $F(x)=0$ with the same properties as above, select p roots $\mu_1, \mu_2, \ldots, \mu_p$, that do not include a conjugate pair and write down $\mathbf{K}$. In order that it be an R. M. it is necessary and sufficient that the group G of F permute no pair of roots μ_i, μ_j with a conjugate pair. There are $p+t$ pairs that cannot be derived from the pairs of p roots selected by the operations of G. The invariants of $\mathbf{K}$ are $h=2k-2(p+t)$. In order that $\mathbf{K}$ be pure it is then necessary that $t=0$ (No. 40) a condition which is also found to be sufficient. Hence a pure K has the indices $h=2k=2p$.

In G there is a unique substitution (binary) S permuting every root with its conjugate. The necessary and sufficient conditions in order that $\mathbf{K}$ be pure are: (a) S must be permutable with G. (b) It must be possible to derive by means of operations of G every non-conjugate pair of roots from some pair of multipliers. Of these two conditions (b) alone depends upon the choice of multipliers. When (a) is fulfilled every choice of p roots, no two of which are conjugate, gives rise to an R. M. $\mathbf{K}$ (pure or impure).

When $\mathbf{K}$ is pure it is isomorphic to an R. M. of the same type with corresponding

$$F(x)\equiv\prod_{j=1}^{p}(x^2-2\xi_j x+R(\zeta_j)),$$

where R is a polynomial of degree $p-1$ with rational coefficients and the ζ's all real, satisfy an irreducible equation with rational coefficients and are such that $\zeta_j{}^2<R(\zeta_j)$ for every j. The special case where R is a rational number has been considered by Scorza([103]).

52. More precise results can be given when G is an abelian group. Then $\mathbf{K}$ is an R. M. provided the ζ's include no pair of conjugate roots. The necessary and sufficient condition in order that the R. M. be pure is that the group G contain no subgroup (other than the identity) leaving the set $\mu_1, \ldots, \mu_p$ invariant. Let there exist such a subgroup and let n be its order. Then $\mathbf{K}$ is isomorphic to an R. M. in the normal form (No. 39) consisting of a single $\mathbf{\Omega}_1$ with n component submatrices each of type $\mathbf{K}$ and genus $\frac{p}{n}$, with the associated equation analogous to $F=0$

also abelian. The invariants are $h=2k=2np>2p$, another indication that the matrix is impure.

In order that two different matrices **K** derived from $F(x)=0$ be isomorphic it is necessary and sufficient that G contain some operation (not the identity) permuting their sets of multipliers. The number of essentially distinct R. M.'s belonging to $F(x)=0$ is therefore the same as its number of classes of transitively permutable sets of p roots containing no conjugate pair.

An especially interesting case is that in which T is cyclic and the μ's are roots of unity.

53. In ([51]), Part II, Ch. 4, Lefschetz investigated the R. M.'s of the curves

$$y^q=(x-a_1)^{a_1}(x-a_2)^{a_2}(x-a_3)^{a_3},$$

where q is an odd prime not a factor of the exponents but a factor of their sum. The genus $p=\frac{1}{2}(q-1)$. The R. M. is pure of type **K** unless the curve is a birational transform of

$$y^q=(x-1)(x-\eta)^{\lambda}(x-\eta^2)^{\lambda^2}$$

with $\dfrac{q-1}{3}=$ an integer; $\eta=e^{\frac{2\pi i}{3}}$; $\lambda^3\equiv 1$, mod q. In this case $h=2k=6p$. See the end of Chapter 15. A number of special cases of matrices **K** occur in the literature. For early references see Krazer-Wirtinger([48]), p. 830. To them must be added Scorza([92]), Raciti([71]), ([73]). Some special matrices **K** of genus two are also given by Enriques-Severi([22]), Bagnera-de Franchis([2]), Scorza([94]), Part II.

In ([96]), p. 164, Scorza considers R. M.'s whose rank (see No. 59) $\rho=2p$ is maximum. He shows that when they are pure they are isomorphic to a matrix **K**.

54. In ([51]), p. 413, ([55.1]), Lefschetz considers the general properties of the R. M. **H** with irreducible minimum equation $f(x)=0$. The multipliers can be assumed in such an order that those that are equal are all consecutive. Let the root μ_j of $f(x)$ occur r_j times among them and call ξ_j the matrix in r_j rows and r columns made up of the corresponding terms of ξ. To a rational matrix **c** of **H** corresponds a square matrix whose terms are polynomials in μ_j, μ_k with rational coefficients, $\gamma(\mu_j, \mu_k)$, such that $\xi_j \cdot \gamma(\mu_j, \mu_k) \cdot \xi'_k=0$. When **c** is alternate

$$\gamma(\mu_j, \mu_k)=-\gamma(\mu_k, \mu_j).$$

In order that **H** be an R. M. it is necessary and sufficient that it possess an alternate **c** such that the hermitian form $\frac{i}{2}\Sigma\gamma_{\rho\sigma}(\mu_j, \bar{\mu}_j)x_\rho\bar{x}_\sigma$ be positive for all sets of x's linear combinations of the rows of ξ_j. If $|\gamma(\mu_j,\mu_k)|=0$ for any one pair of roots, **H** is impure.

When the multiplications of a given R. M. are all permutable, they depend upon the powers of a unique one T, and if q is its degree, $h=q$. The Scorza algebra is then permutable and its elements correspond term for term to the numbers of an algebraic domain of degree q. When the multiplications are not permutable, whatever T, $h>q$. In order that the first case occur for a pure matrix it is necessary that for any T the multipliers be either all real, or else all complex, the group G of $f(x)$ being permutable with the operation S that interchanges conjugate roots.

55. When the multipliers are all real, Lefschetz([51]), p. 420, the integers r_j are all equal to $\frac{1}{2}r$. The necessary and sufficient conditions in order that **H** be an R. M. are: There exists a square matrix of order r, $\boldsymbol{\gamma}(\mu_j)$, whose terms are polynomials in μ_j with rational coefficients, such that $\xi_j \cdot \boldsymbol{\gamma}(\mu_j) \cdot \xi'_j=0$ while $i\xi_j \cdot \boldsymbol{\gamma}(\mu_j) \cdot \xi'_j$ is a definite positive hermitian matrix. In short the ξ's behave like R. M.'s whose principal matrices have for terms numbers of the domain $R(\mu_j)$ obtained by adjoining μ_j to the ordinary domain of rationality. The general theorems on R. M.'s can be paraphrased here, giving rise to new theorems obtained by replacing rational matrices by matrices whose terms are in $R(\mu_j)$. In particular there are indices h_1, k_1, analogous to h, k and when $r>2$, $h=qh_1$, $k=qk_1$, while for $r=2$, $h=k=q$. The matrix **H** is pure, if and only if, the ξ's are pure. In order that the multiplication T be of maximum degree it is necessary and sufficient that $h_1=1$ and since $0<k_1\leqq h_1$, $k_1=1$, hence $h=k=q$. Given $f(x)$, **H** depends upon $\frac{1}{4}p(r+2)$ essential parameters.

The case $r=2$ gives rise to a theory analogous to elliptic functions, with modular functions, thetas, and applications to number theory. There are important investigations along that line due to Hecke([30]). See also Krazer-Wirtinger([48]), p. 857.

56. Lefschetz examines in detail ([51]), p. 426, the case of a matrix **H** whose f has a group G permutable with the substitution S of No. 54. In this case the integers r_j may all be different. Let r_j, $\bar{r}_j$ correspond to μ_j, $\bar{\mu}_j$. Then for any choice of these integers such that $r_j+\bar{r}_j=r$ (for every j) there exists a corresponding R. M. and it depends upon $\Sigma r_j\bar{r}_j$ continuous parameters, the sum in this expression including only one term for each conjugate pair of roots. Lefschetz gives explicitly the values of h, k. When G is abelian and $r>2$, $h=2k=q$ and all multiplications of **H** depend upon the powers of T. One type discussed, for which $r_j=r'_j=\frac{1}{2}r$, and in general $h=2q$, $k=\frac{1}{2}q$, includes a noteworthy class discovered by Scorza([97]), ([96]), p. 179. It is characterized by $q=2$, $k=1$,

$h=4$. Humbert([42]), had shown that for $p=2$, complex multiplication could only occur when the R. M. is singular, i. e., when $k>1$. Scorza showed by his example that this is true only for $p=2$, the special R. M.'s that he discovered being, however, the only exception.

57. The R. M.'s of the curves

$$y^q=\Pi(x-a_i)^{\alpha_i}, \qquad q \text{ an odd prime}$$

fall under the type **H**. Their properties are investigated at length by Lefschetz([51]), Part II, Ch. 4, under the assumption that the a's are arbitrary numbers. Let $r+2$ be the number of branch points of $y(x)$. The curve may be birationally transformed into one of the same type with $r+1$ factors at the right and neither the α's nor their sum divisible by q. The Jacobi variety V_p possesses a cyclic birational transformation T:

$$u'_j=\epsilon^{n_j}u_j\,; \qquad \epsilon=e^{\frac{2\pi}{q}},$$

the exponents n_j being readily determined when the α's are known. The degree of T is $q-1$ and the R. M. is of type **H** with

$$\mu_j=\epsilon^{n_j}, \qquad f(x)\equiv 1+x+\ldots+x^{q-1},$$

and the present r playing the same part as before. We have already considered $r=1$ in No. 53. For $r>1$ the R. M. is pure with $h=2k=q-1$, unless the curve is birationally transformable into

$$y^q=\frac{x-a_1}{x-a_2}\left(\frac{x-a_3}{x-a_4}\right)^a,$$

when it is impure, with $h=2k=6p$. Lefschetz determines the complete birational group of the curve and of V_p. With the one exception just mentioned, the multiplications all depend upon the powers of T, just as for the general R. M. of the class. This is a remarkable fact in view of the restrictions here present. For we may always take $a_1=0$, $a_2=1$, hence there are only $r-1$ essential parameters which is always less than $\Sigma r_j\bar{r}_j$, the number of parameters of the general matrix of the class.

58. In ([51]), p. 432, Lefschetz applies his general results to the classification of R. M.'s with $p=2$, 3 into classes of non-isomorphic matrices. They all fall under the types **H**, **K**, just described, or else have no complex multiplications, when $h=k=1$. These general R. M.'s may also be thought of as of type **H** with $q=1$. The matrices are described in each case, the number and range of the essential parameters given and the values of h, k determined. The one R. M. of genus two of type **K** ([51]), pp. 405 and 432, may be reduced to a simpler form. For if $\phi(x)$ is any polynomial with rational coefficients there is a T whose multipliers are

$\phi(\mu_j)$, with a corresponding new isomorphic form for the matrix. By choosing ϕ properly, the multipliers can be taken as

$$\mu_1 = i\sqrt{n+m\sqrt{d}}, \qquad \mu_2 = i\sqrt{n-m\sqrt{d}},$$

where m, n, d are positive integers such that m, n are prime to one another, d has no square factor, and n^2-dm^2 is not a perfect square. See also for $p=3$, Scorza(98), (96), p. 167.

59. In (100), (96), Scorza introduces various linear algebras, related to the multiplications of an R. M. $\boldsymbol{\omega}$, and investigates their chief properties. These algebras are isomorphically invariant. The most important one is the algebra over the rational domain whose elements are the multiplications T. Scorza defines the *rank* ρ of $\boldsymbol{\omega}$ (not to be confused with the rank of an attached V_p). ρ is the maximum of the degree of T, or degree of the generic T, and is an isomorphic invariant integer. In any case $\rho \leqq 2p$. Scorza examines the nature of the algebra in relation to the normal form of No. 39, and finds that its properties are known whenever those of the individual components $\boldsymbol{\Omega}_i$, and ultimately those of the pure components $\boldsymbol{\omega}_i$, are known. He gives a number of theorems on ρ. For a pure matrix it is a factor of $2p$ and also of h. Furthermore $\frac{h}{\rho}$ is a factor of ρ, hence $h \leqq \rho^2$. He also proves in this connection that when $\boldsymbol{\omega}$ is pure h is a factor of $4p^2$. If ρ is prime $h=\rho$ or ρ^2. Scorza also examines the special cases where ρ is maximum ($\rho=2p$), $\rho=2$, or ρ is prime, also the case where p is prime and gives the corresponding values of h, k. He investigates at length (96), p. 179, the non-singular matrices mentioned in No. 56, proves their existence for every $p>2$, shows that their $\rho=2$, while their invariants h, k can assume the sets of values $k=1, 2$; $h=2, 4$ in all four possible combinations, with $k=1$ possible only when $p\neq 2$. In the same paper, p. 198, Scorza also proves this theorem: If $\boldsymbol{\omega}$ is pure, its V_p can only possess rational $(1, n)$ transformations, whose n can be represented by integer values of the variables s, in the form $f(x_1, \ldots, x_h)^{\frac{2p}{\rho}}$, where f is a certain homogeneous polynomial of degree ρ. The form f itself can represent either 1 or -1 and when $\frac{2p}{\rho}$ is odd and $\boldsymbol{\omega}$ pure the form is >0 when the variables are not all zero.

As closely related to the properties of ρ see the theorems on the degree q of T given by Lefschetz(51), p. 395.

§ 7. Additional Topics

60. *Jacobi Varieties.*—We have already observed (No. 3) that they are very special abelian varieties, owing to the fact that they depend upon $3p-3$ moduli only instead of $\frac{p(p+1)}{2}$ like the general abelian V_p. The actual effect of this reduction in the number of moduli is as yet wholly obscure and there is no inkling of anything like a theory of the Jacobi V_p, but only scattered results. They bear mostly upon the interrelations of the properties of the period matrix and the singular correspondences on the related C^p, and in this connection have been treated in Chapter 16. In regard to the connections between the hermitian birational transformations of V_p and those of C^p see Comessatti([18.1]); Rosenblatt([91]); also Ch. 16, No. 15.

61. *Varieties of Rank* >1.—Here also we have scattered results rather than a general theory. This is partly due to our ignorance of the properties of algebraic varieties of more than two dimensions. However, there would appear to be no difficulty in extending a good part of the results due to Enriques-Severi, and Bagnera-deFranchis (§2 of this chapter). For example we believe that by rather simple considerations of Analysis Situs it is possible to prove the following extension of the important theorem of Enriques-Severi([22]), p. 331, this chapter No. 19: An I_n on a V_p of rank one whose coincidence locus has at most $p-2$ dimensions is generated by a finite birational group of V_p. However, although there is no complete theory for the varieties whose rank >1, there are a number of results which we shall now describe.

62. The generalized Kummer variety W_p, image of the I_2 generated on V_p of rank one by the binary transformation $u'_j=-u_j$, has been investigated by Wirtinger([121]), ([122]). He found that it has most properties of the Kummer surface. See also Coble([14.1]), especially for the case $p=3$, in connection with his investigations on modular functions. In ([19]) Wirtinger, by proving the functional independence of the thetas of order two, has indirectly shown that the general Kummer variety can be parametrically represented by making the homogeneous coordinates proportional to the elements of a maximal linearly independent set of even θ's of order two. The representation is (1, 1) without exception. The carrying space of W_p is on an S_n $n=2^p-1$ and its order is $2^p \cdot p!$. For further references see Krazer([47]), p. 368; Krazer-Wirtinger([48]), pp. 669, 753; Raciti([72]).

63. In ([51]), Part II, Ch. 3, Lefschetz examines cyclic birational transformations of a V_p of rank one, and related varieties of higher rank. If

the multiplications of V_p all depend upon the powers of a single one, the discovery of the birational transformations is reducible to that of the units of a certain Dedekind order (ring) of a definite algebraic domain. This is applied to the case $p=2$. See in this connection Spampinato([109.1]), ([109.2]) and for a related discussion for correspondences on curves, Rosati([84]), p. 37.

Any cyclic transformation T of V_p has parametric equations

$$u'_j = \epsilon^{n_j} u_j + \eta_j, \qquad \epsilon = e^{\frac{2\pi i}{n}},$$

where n is the order of T. The resulting variety W_p of rank n, image of the I_n whose groups are made up of any point of V_p and its transforms, has $F(1)$ singular points, where $F(T)=0$ is the characteristic equation of T. Lefschetz also determines the connectivity and torsion indices of W_p. When there are no isolated singular points some of the latter are >1 and the surfaces which they carry yield simple examples of surfaces whose $\sigma>1$. He also examines the number of linearly independent integrals of the first kind of various multiplicities and proves this theorem: When $\boldsymbol{\omega}$ is pure and the characteristic equation reducible W_p cannot be a locus of linear spaces. This he also extends to the case where the characteristic equation is irreducible and the order n is a prime number >7.

64. *Real Abelian Varieties.*—In ([56]), Lefschetz examines their properties having especially in view the number and properties of the real folds. By considering the effect upon the linear cycles of the (1, 1) transformation S of V_p which permutes its pairs of conjugate points he finds that the R. M. $\boldsymbol{\omega}$ is isomorphic to the form

$$\mathbf{L} \equiv ||\Omega_{j1}, \ldots, \Omega_{jp}; i\Omega_{j,\,p+1}, \ldots, i\Omega_{j,\,2p}||$$

where all quantities $\Omega_{j\mu}$ are real. The double of every period system of $\boldsymbol{\omega}$ is expressible as a sum of multiples of those just written, so that $2\boldsymbol{\omega}$ is equivalent to $\mathbf{L}$. If there are $p-s$ distinct periods whose imaginary part is $\frac{i}{2}\Sigma r_\mu \Omega_{h,\,p+\mu}$ the variety possesses 2^s real folds. An example is given to show that there are actual varieties for every s from 0 to p, extremes included. The folds are all continuously and birationally transformable into one another on V_p and constitute therefore homologous p-cycles. Topologically they are p dimensional rings. Lefschetz applies this to real Jacobi varieties and obtains a new proof of Harnack's theorem according to which a real curve of genus p can have at most $p+1$ real branches. He also determines the number ρ' of algebraically distinct real hypersurfaces of V_p. It may be considered as the analogue of the Picard number ρ when only real curve systems are considered. These systems are defined by the condition that the transformation S defined above, leaves them invariant. The value of ρ' is equal to the number of linearly independent rational matrices attached to **L** that are of the form $\begin{Vmatrix} 0 & \mathbf{m} \\ -\mathbf{m}' & 0 \end{Vmatrix}$. When **L** is pure $\rho' \leqq p$. This is to be compared with the limit resulting for ρ itself from the fact that $\rho = k \leqq 2p$. Lefschetz examines also the possible real types of hyperelliptic surfaces and the corresponding values of h, k, ρ'.

In (12) Cherubino shows that in order that there exist a real V_p attached to $\boldsymbol{\omega}$ it is necessary and sufficient that $\boldsymbol{\omega}$ and $\bar{\boldsymbol{\omega}}$ be equivalent. They satisfy then a relation $\boldsymbol{\alpha}\boldsymbol{\omega} = \bar{\boldsymbol{\omega}}\mathbf{a}$, $|a| = (-1)^p$. He investigates the isomorphisms between $\boldsymbol{\omega}$ and $\bar{\boldsymbol{\omega}}$ (whether there exists a real V_p or not) and develops for them a theory quite similar to Scorza's for ordinary multiplication. He also obtains the number of distinct classes of real varieties belonging to a non-singular $\boldsymbol{\omega}$ $(k=1)$, distinct in the sense that they are not reducible to one another by real birational transformations.

Comessatti has considered the same questions, (15), (15.1), (18), and showed that the number in question is 2^{s+1}, where s is the same integer as introduced previously by Lefschetz (56). Among the classes in question only two have real points. In (15.1) Comessatti also considers various types of hyperelliptic surfaces and generalized Kummer varieties, particularly from the point of view of their distribution into distinct real classes in the above sense. See also Cherubino (12.2).

65. *Additional References*.—A number of questions more or less closely related to those treated in this chapter are taken up in Krazer-Wirtinger (48), pp. 843, 853. There are also a number of papers due to Rosati not described in this chapter, but listed under his name in the bibliography at the end, likewise for some of the papers, mostly short notes, listed under Scorza. The former author's work is directly or indirectly applicable to our topic and much of it has been described in Chapter 16. The material in Scorza's short notes is always contained in his major papers to which we have largely confined the quotations of this chapter.

Humbert (40); Rémy (74), (77), (78); Godeaux (29.1), (29.2), (29.3) have investigated the surface F obtained as follows: to every point pair of a curve C^3 corresponds a point of F, to every point of F two point pairs constituting a canonical group of C^3. F is on a generalized Kummer V_3 whose other surfaces are studied by Rémy.

Krazer has given a very complete treatment of ordinary transformations (47), Ch. 5, also Krazer-Wirtinger (48), with very complete references. To the references given by Krazer-Wirtinger (48), p. 648, on half-integer characteristics should be added Coble([14]), ([14.2]), where there is found a very comprehensive representation by certain points on a finite projective space.

BIBLIOGRAPHY

1 P. Appell, J. de Math., (4), 7: 157-219 (1891). [2.

1.1 H. F. Baker, An introduction to the theory of multiplicity periodic functions, Cambridge, 1 vol. (1907). [1, 3.

1.2 ———, London M. S. Proc., (2), 10: 353-395 (1911). [3.

2 Bagnera-deFranchis, Pal. Circ. Mat., 30: 185-238 (1910). [2, 3, 6.

3 ———, C. R., 145: 747-749 (1907). [2.

4 ———, Rom. Acc. L. Rend., (5), 16: 492-498, 596-603 (1907). [2.

5 ———, Rom. Congr. di Mat., 2: 242-248, 249-256 (1909). [2.

6 ———, Rom. Soc. It. Sc. Mem., (3), 15: 251-344 (1908). [2.

7 O. Bolza, Am. J., 10: 47-70 (1888). [2.

7.1 C. Bonomi, Pal. Circ. Mat., 43: 47-70 (1919). [2.

8 G. Castelnuovo, C. R., 140: 220-222 (1905). [3, 4.

9 ———, Rom. Acc. L. Rend., (5), 14_1: 545-556, 593-598, 655-663 (1905). [3, 4.

10 ———, Rom. Acc. L. Rend., (5), 30_1: 50-55, 99-103, 195-200, 355-359 (1921). [1, 3.

11 Castelnuovo-Enriques, Encykl. III C 6b, (1914). [2, 3.

12 S. Cherubino, Giorn. di Mat., (3), 60: 65-94 (1922); 61: 47-68 (1923). [7.

12.1 ———, Rom. Acc. L. Rend., (5), 26_1: 538-542 (1917). [5.

12.2 ———, Rom. Acc. L. Rend., (6), 6: 197-201, 274-278 (1927). [7.

13 G. Chillemi, Pal. Circ. Mat., 29: 164-168 (1910). [2.

13.1 ———, C. R., 148: 1091-93 (1909). [2.

14 A. B. Coble, Ann. of Math., (2), 17: 101-112 (1916). [7.

14.1 ———, Am. M. S. Bull., 28: 329-364 (1922); 30: 301-313 (1924). [7.

14.2 ———, Am. M. S. Trans., 14: 241-276 (1913); 18: 331-372 (1917). [2, 7.

15 A. Comessatti, Ann. di Mat., (4), 2: 67-106 (1924). [7.

15.1 ———, Ann. di Mat., (4), 3: 27-71 (1925). [7.

16 ———, Pal. Circ. Mat., 36: 35-58 (1913). [2, 4.

17 ———, Rom. Soc. It. Sc. Mem., (3), 21: 45-71 (1919). [2, 3.

18 ———, Ven. Ist. Atti, 85: 471-494 (1925). [7.

18.1 ———, Torino Atti, 50: 439-455 (1915). [7.

19 G. Cotty, Toulouse Ann. (Paris thesis), (3), 3: 209-376 (1911). [2.

20 F. Enriques, Pal. Circ. Mat., 20: 61-72 (1905). [2.

21 ———, Rom. Acc. L. Rend., (5), 15_2: 665-669 (1906). [2.

22 Enriques-Severi, Acta Math., 32: 283-292 (1909). [2, 6, 7.

23 ———, Acta Math., 33: 321-403 (1910). [2.

24 ———, Rom. Acc. L. Rend., (5), 16_1: 443-453 (1907); (5), 17_1: 4-9 (1908). [2.

25 M. deFranchis, Pal. Circ. Mat., 38: 192 (1914); 40: 187-193 (1915). [4.

26 R. Fricke, Encykl. II B 3. [1, 6.

27 G. Frobenius, J. f. Math., 95: 264-296 (1883). [2, 6.

28 ———, J. f. Math., 97: 16-49, 188-224 (1884). [1.

29 R. Fueter, Vorlesungen über die singulären Moduln und die komplexe Multiplikazion der elliptischen Funktionen, Berlin, 1 vol. (1924). [1, 6.

29.1 L. Godeaux, Brux. Bull., 653-665, 694-702 (1922). [7.
29.2 ———, Bull. Sci. M., (2), 45: 14-20 (1921). [7.
29.3 ———, Roum. Ac. Bull., 4 p. (1925). [7.
30 E. Hecke, Math. Ann., 71: 1-37 (1912); 74: 465-510 (1913). [2, 6.
31 ———, Gött. Nachr., 85-112 (1915). [2.
31.1 R. W. H. T. Hudson, Kummer's Quartic Surface, Cambridge (1905). [2.
32 G. Humbert, C. R., 126: 394-397, 508-510, 814-817, 882-884 (1898). [2.
33 ———, C. R., 128: 857-862 (1899). [2.
34 ———, C. R., 129: 640-642, 955-956 (1899). [2.
35 ———, C. R., 130: 483-486 (1900). [2.
36 ———, C. R., 132: 72-74 (1901). [2.
37 ———, C. R., 133: 425-429 (1901). [2.
38 ———, C. R., 134: 876-882, 1261-1266 (1902). [2.
39 ———, J. de Math., (4), 9: 27-171, 361-475 (1893). [2, 3.
40 ———, J. de Math., (5), 2: 263-293 (1896). [2.
41 ———, J. de Math., (5), 5: 233-350 (1899). [2, 3.
42 ———, J. de Math., (5), 6: 279-386 (1900). [2, 3, 5, 6.
43 ———, J. de Math., 7: 97-123, 395-417 (1901). [2.
44 ———, Am. J., 16: 221-253 (1894). [2.
45 Humbert-Lévy, C. R., 158: 1609-1616 (1914). [6.
45.1 H. W. E. Jung, P. Rep. I_2, Ch. 18 (1927). [1.
45.2 Koenigsberger, J. f. Math., 65: 335-359 (1866). [6.
46 G. Kowalewsky, Einführung in die Determinantentheorie, Leipzig, 1 vol. (1909). [3.
47 A. Krazer, Lehrbuch der Thetafunktionen, Leipzig (1903). [1, 4, 5, 6, 7.
48 Krazer-Wirtinger, Encykl. II B 7. [1, 2, 4, 6, 7.
49 L. Kronecker, J. f. Math., 68: 273-285 (1868); Reproduced in Werke, I, 143-162. [6.
50 S. Lefschetz, L'Analysis Situs et la géométrie algébrique, Paris, 1 vol. (1924). [1, 2, 3.
51 ———, Am. M. S. Trans., 22: 327-482 (1921). [1, 2, 3, 4, 6, 7.
52 ———, Am. M. S. Trans., 28: 1-49 (1926). [3.
53 ———, C. R., 168: 758-762 (1919). [6.
54 ———, Bull. Sci. M., (2), 47: 120-128 (1923). [4.
55 ———, Rom. Acc. L. Rend., (5), 30_1: 48-50 (1921). [1.
55.1 ———, Comptes Rendus du Congrès de Strasbourg, 300-308 (1921). [6.
56 ———, Wash. Nat. Ac. Sc. Proc., 5: 103-106; 296-298 (1919). [2, 7.
57 P. Lévy, J. de Math., (8), 4: 357-382 (1921); (9), 1: 255-334 (1922). [6.
58 W. F. Osgood, The Madison Colloquium, New York, Lecture 1 (1913). [1.
59 P. Painlevé, Leçons sur la théorie analytique des équations différentielles professées à Stockholm en 1895, Paris, 1 vol. (1897). [2.
60 ———, C. R., 126: 512-515 (1898). [2.
61 M. Piazzola-Beloch, Pal. Circ. Mat., 47: 182-192 (1923); 48: 41-96 (1924); 50: 285-297, 345-367 (1926); 51: 49-69 (1927). [2.
61.1 ———, Pal. Circ. Mat., 51: 70-76 (1927). [2.
61.2 ———, C. R., 184: 61-63 (1927). [2.
62 E. Picard, J. de Math., (4), 1: 281-346 (1885). [2.
63 ———, J. de Math., (4), 5: 135-321 (1889). [2.
64 ———, Math. Ann., 19: 569-577 (1882). [2.
65 ———, Pal. Circ. Mat., 9: 244-255 (1895). [3.

66 E. Picard, C. R., 103: 730-732 (1886). [2.

67 ———, C. R., 120: 658-660 (1895). [3.

68 ———, Soc. M. F. Bull., 28: 17-25 (1900). [2.

69 Picard-Simart, Théorie des fonctions algébriques de deux variables indépendantes, Paris, vol. 2 (1906). [2.

70 H. Poincaré, Acta Math., 26: 48-98 (1902). [1, 3.

71 C. Raciti, Rom. Acc. L. Rend., (5), 28_1: 347-350 (1919). [6.

72 ———, Rom. Acc. L. Rend., (5), 28_1: 377-380 (1919). [7.

73 ———, Torino Atti, 51: 443-449 (1918). [6.

74 L. Rémy, J. de Math., (6), 4: 1-37 (1908). [2, 7.

75 ———, C. R., 142: 386-388 (1906). [2.

76 ———, C. R., 143: 767-769 (1906). [2.

77 ———, C. R., 144: 412-414, 623-625 (1907). [7.

78 ———, C. R., 149: 1057-1060 (1909). [2.

79 ———, C. R., 150: 677-680 (1910). [2.

80 ———, Ann. Éc. Norm., (3), 26: 193-258 (1909). [7.

81 ———, Soc. M. F. Bull., 35: 53-69 (1907). [2.

82 C. Rosati, Ann. di Mat., (3), 25: 1-32 (1915). [2, 4, 6.

83 ———, Ann. di Mat., (3), 28: 35-61 (1918). [7.

84 ———, Ann. di Mat., (3), 31: 1-50 (1921). [7.

85 ———, Ann. di Mat., (4), 3: 109-132 (1925). [7.

86 ———, Pal. Circ. Mat., 44: 307-335 (1920). [2.

87 ———, Pal. Circ. Mat., 49: 165-200 (1925). [7.

87.1 ———, Rom. Acc. L. Rend., (5), 24_2: 182-184 (1915). [7.

88 ———, Torino Atti, 50: 1-12 (1915). [4.

89 ———, Torino Atti, 51: 991-1014 (1916). [6.

90 ———, Torino Atti, 53: 5-22 (1917). [2, 6.

91 A. Rosenblatt, Crac. Bull., 193-206 (1918). [2.

92 G. Scorza, Acc. Gio. Atti, (5), 10: 7 p. (1917). [6.

93 ———, Acc. Gio. Atti, 11: 19 p. (1917). [4, 5.

93.1 ———, Pal. Circ. Mat., 36: 386-395 (1913). [1.

94 ———, Pal. Circ. Mat., 41: 263-380 (1916). [2, 4, 5, 6.

95 ———, Pal. Circ. Mat., 43: 213-239 (1918). [4.

96 ———, Pal. Circ. Mat., 45: 1-204 (1921). [5, 6.

96.1 ———, C. R., 160: 392-394 (1915). [7.

97 ———, C. R., 165: 497-498 (1917). [6.

98 ———, C. R., 167: 454-455 (1918). [6.

99 ———, Rom. Acc. L. Rend., (5), 23_2: 566-572 (1914); 24_1: 412-418, 645-654 (1915); 24_2: 279-284, 333-338, 393-400, 446-453, 603-610 (1915); 25_1: 289-296 (1916). [4, 6.

100 ———, Rom. Acc. L. Rend., (5), 26_2: 176-182 (1917). [6.

101 ———, Rom. Acc. L. Rend., (5), 27_1: 171-175 (1918). [7.

102 ———, Rom. Soc. It. Sc. Mem., (3), 19: 139-183 (1916). [1.

103 ———, Torino Atti, 53: 1008-1017 (1918). [6.

104 F. Severi, P. Rep. II_2, Ch. 33. [2, 3.

105 ———, Vorlesungen über algebraische Geometrie, Translated by Löffler, Leipzig, 1 vol. (1922). [4.

106 ———, Ann. di Mat., (3), 20: 201-215 (1913). [3.

107 ———, Pal. Circ. Mat., 30: 265-288 (1910). [2.

108 ———, Ann. Éc. Norm., (3), 25: 449-468 (1908). [2.

108.1 ———, Ven. Ist. Atti, 67: 409-419 (1908). [2.

109 N. Spampinato, Ann. di Mat. (3), 30: 257-274 (1921); 31: 127-148 (1922). [2.

109.1 ———, Rom. Acc. L. Rend., (6), 5: 105-107, 251-254, 427-431 (1927). [7.

109.2 ———, Pal. Circ. Mat., 51: 199-208 (1927). [7.

110 R. Torelli, Pal. Circ. Mat., 37: 25-36 (1914). [4.

111 E. Traynard, C. R., 138: 339-342 (1904); 139: 718-719 (1904); 140: 218-219, 931-932 (1905). [2.

112 E. Traynard, Ann. Éc. Norm., (3), 24: 77-177 (1907). [2.

113 O. Veblen, Am. M. S. Trans., 25: 540-550 (1923). [3.

114 M. Verzi, Pal. Circ. Mat., 46: 265-307 (1922); 48: 209-267 (1924). [2.

115 H. Weber, Ann. di Mat., (2), 9: 126-166 (1878). [6.

116 ———, J. f. Math., 84: 332-354 (1878). [2.

117 ———, Encykl. I C 6. [6.

118 H. Weyl, Rev. Mat. Hisp. Amer., 42 p. (1923). [3.

119 W. Wirtinger, Untersuchungen über Thetafunctionen (Gekroente Preisschrift), Leipzig, 1 vol. (1895). [4.

120 ———, Acta Math., 26: 133-156 (1902). [1, 3, 7.

121 ———, Gött. Nach., 474-480 (1889). [7.

122 ———, Monatsh. f. Math., 1: 113-128 (1890). [7.

123 ———, Monatsh. f. Math., 6: 69-98 (1895). [1.

124 ———, Monatsh. f. Math., 7: 1-25 (1896). [3.

SELECTED TOPICS IN ALGEBRAIC GEOMETRY—II

Supplemental Report of the Committee on Rational Transformations[1]

VIRGIL SNYDER, AMOS H. BLACK AND
LEAMAN A. DYE

---

[1] The members of this committee of the Division of Physical Sciences are: Virgil Snyder, *Chairman*, Professor of Mathematics, Cornell University, Ithaca, New York; Arthur B. Coble, University of Illinois, Urbana, Illinois; Arnold Emch, University of Illinois, Urbana, Illinois; Solomon Lefschetz, Princeton University, Princeton, New Jersey; F. R. Sharpe, Cornell University, Ithaca, New York; C. H. Sisam, Colorado College, Colorado Springs, Colorado.

CHAPTER I

CURVES AND RULED SURFACES

VIRGIL SNYDER

1. Rational curves. (*a*) *Cubics.* Condition for contact of the plane rational cubic can be expressed parametrically. For ordinary intersection with a C_n each symmetric function S_i appears to the first power. For contact, powers higher than the first appear (Griffiths [61]). Further properties of involutions on the plane rational cubic can be expressed by means of a pencil of conics. One group (on a composite conic) consists of collinear points (de Vries [155]).

A problem in closure, defined by the first tangential point of a given point on a rational cubic is solved by Allen.[3] If $x=3t^2$, $y=3t$, $z=t^3+1$ is the parametric representation, then the nth tangential will coincide with the given point $t=-t^{(-2)n}$ and $2^n-4\equiv 0 \pmod{12}$. If n is odd, $2^n+1\equiv 0 \pmod 3$. Further applications of involutions are given by Pomey.[111] Let a rational plane cubic curve be defined by a double point and six simple points, to find its point of inflexion.

The nodal cubic is also studied by Cattaneo.[19] The same author [21] discusses the circular conchoid. *See also* Campedelli [18] and Biggiogero.[13]

Given a rational space cubic by means of its plane projection, to find its osculating planes through a given point. The same procedure applies to plane quartic curves.

(*b*) *Quartics.* Properties of the tricuspidal quartic are obtained by quadratic transformation. The invariants are derived and covarient curves are obtained by Longhi.[80] Properties obtained from the method of generation without the use of transformations are given by Villa.[149] Contact conics, including those having two points of four point contact, were found by Joliffe.[70] That eight point contact of a conic is impossible except at a singular point was shown by Hilton and Green.[63] The polar properties expressed by means of an appropriate coordinate system are given by Rutter.[124] A generation of trinodal C_4 by means of two projective pencils of conics with three common base points is given by Deaux.[26] A quartic with an oscnode is rational. The condition that a C_4 has an oscnode is expressed as a binary relation, when the triangle of reference is suitably chosen (Neelley [102]). Moore and Neelley [96] derive the condition that a tacnodal or an oscnodal C_4 can be a plane section of a given quartic surface. Further properties are given by Neelley[103] with interpretations in hyperspace. James[69] corrects a statement in an earlier paper concerning involutions on a normal C_4. Further involutions of points on a trinodal C_4 are derived by de Vries.[156]

(c) *Curves of higher order.* Winger[161] discusses self-projective rational C_7. A C_4 in S_4 has seven trisecant planes in general position. When the planes are given, ten C_4 can be found meeting each in 3 points (Babbage[7]). Morton[100] discusses inscribed tetrahedra and other forms which fulfill one extra condition. Another representation of C_4 of S_4 is given by Semple[133]. A quadric primal of S_5 is mapped on S_4 and the points of V are interpreted as lines of S_3. Associated quartics are determined by the reciprocal transformation $x'_i = 1/x_i$.

Libois[78] derives conditions for rational C_6 in S_3. The C_6 may have $\infty^1 s_5$ (lines meeting it in five points) or one or none; in the latter case it must have six s_4 and lies on a cubic surface.

A rational C_5 of S_3 can be generated by various projective forms (Deaux[27]). A C_5 with two double points lies on a quadric cone, but has only one quadri-secant, thus correcting a statement made by Loria. Arnold[5] derives various binary relations between the rational C_5 of S_3 and the plane rational C_4, when the former lies on a quadric surface. James[68] provides algebraic machinery for correspondences on rational curves, with simple geometric applications.

Conditions for closure of polygons circumscribing one conic and inscribed in another were derived by Burnside.[15] It is a recurring equation of the form

$$(p_n - p_1)^2 p_{n-1} p_{n+1} = (p_n p - 1)^2.$$

If the point x describes a curve C, x' a curve C', then $\xi = x + ax'$ divides the line segment joining x, x' in a constant ratio. Consequences of this for various relations between x and x' are found by Papillon,[107] largely by differential methods.

Given a rational plane quintic with three cusps, it is possible to effect a classification of possible forms by means of the sequence of double points. Ten schemes are possible (Peterson[108]).

An elaborate study of involutions on rational curves was made by Pomey,[111, 112] by means of linear systems of transversal curves, with base points on the given curve. A large number of theorems is included. Among those already known, the proofs are different from those by which they were first established. Gambier[43, 44] discusses a generalization in which one of the conics is replaced by a curve of class m.

The condition that a binary f_4 shall be a line section of the Jonquières C_4 is given by Arnold.[4] The form must be apolar to the fundamental involution of the curve.

The rational C_5 was also studied by Siefert.[132] Further properties of the lemniscate were found by Marck.[87] Cattaneo [19] derives the line equation of a rational plane curve when the parametric form of the point equation is given.

A generalization of the theorem of Bertini, that the three principal chords of a rational C_4 of S_3 are concurrent, can be extended to rational curves in S_{2n+1} (Berzolari [12]). By means of stereographic projection of a quadric surface, polarity as to a quadric, various properties of circles, pencils and nets of circles is found by Biggiogero.[13]

The points of contact of an algebraic curve and its envelope are discussed at the points of higher multiplicity, by means of the methods of differential geometry (Gambier [44]). Colucci [24] extends parametric representation to the complex plane and applies it to the geometry of the circle. Many results already known are confirmed by this method.

Rindi [120] obtains relations among n points of a line, squares of first members of their equations. Vaidyanathaswami [145] expresses a rational normal curve C_n in terms of its quadratic manifolds without any use of transformations.

The generation of plane algebraic curves by means of points in the complex plane defined by $z=x+iy=f\left(\frac{\alpha+\lambda\beta}{\gamma+\lambda\delta}\right)$ is given by Abramesco.[1]

2. Correspondences. (*a*) *General.* A few extensions to correspondences between two one-dimensional carriers have been made but mostly to particular problems. If a geometric problem admits, in general, of no solution, then the existence of one solution in a special case implies the existence of an infinite number. If the existence of one solution implies the existence of an infinite number, then, in general, there is no solution. This is applied to space cubics touching four lines by Grace.[57] Thus Vaidyanathaswami [146] applies the general (m, n) correspondence to C_r in S_r. Elementary algebraic formulation. In a later paper the same author [147] applies the (2, 1) case to circles. Apolarity is used. A third paper considers (3, 2) correspondences on circles.[148] Emch [34] considers the symmetric (n, n) correspondences. Given $f(x, y)=0$, let $x=\lambda+\mu$, $y=\lambda\mu$, where λ, μ is the parameter pair of a point on a rational curve. Various problems in closure result. The principle is applied to the correspondence between the tangents of two plane curves, with a number of illustrative examples by Barce.[10]

Terracini [140] discusses the problem of the reducibility of the correspondence defined by a point P on a curve of S_4 and the residual points in which the osculating hyperplane meets the curve. In S_3 there are two cases; one is furnished by W-curves, in which one component may define a collineation, and those belonging to a linear complex. The asymptotic lines on ruled surfaces having two rectilinear directrices are always of the latter class. In S_2, if one component defines a collineation, every one must.

Geometrical applications of binary (n, n) correspondences expressed in the form $\Sigma a_i\lambda^{n-i}=0$, $\Sigma b_i\mu^{n-i}=0$, in which a_i, b_i are linear in point coordinates are discussed by Emch.[27] If $\phi(\lambda, \mu)=\phi(\mu, \lambda)=0$ is given, a $(1, n^2)$ transformation is defined in which $C_m(x)\sim\Gamma_{n^2m}(x)=0$. A Cremona I_6 in S_2 is also obtained in this way. Gambier [44] applies this and other principles to the intersection of two plane algebraic curves at a point k-fold on each. The derivation of k^2+m such points is partly synthetic and partly transcendental.

The coincidences of a multiple correspondence between the elements of two varieties is brought to a reduced form by Ursell.[144] The general formula was first derived by Lefschetz.[77]

A (2, 2) correspondence between the points of two planes may be periodic. The reduction of such cases to a canonical form has been made by Edwards [33]; *see also* Meyer.[92]

Applications of the theory of forms have been made by Moore,[98] Turnbull,[142] Piedvache [110] and McCoy.[86] Jonzon [74] gives a systematic analysis of the Cremona transformations under which particular elliptic and hyperelliptic curves remain invariant, thus supplying the details of the earlier work by Wiman; *see also* Baker [9] and Gherardelli.[49]

Applications of Zeuthen's formula for coincidences have been made by Marletta and by Brigaglia.[15]

Extensions to correspondence in hyperspace were done by Roth [122] and by James.[68, 69]

Applications to lines on quadrics were made by Sapienza and to congruences of lines by Hulin,[66] while Segre [129, 130] applies the principle to pairs of points of a given entity.

(*b*) *Cubics.* Ciani [23] proposes the canonical form $x_1^2x_2+x_2^2x_3+x_3^2x_1+2kx_1x_2x_3=0$ for the plane cubic curve.

Properties of the synthetic pencils of cubics obtained from the Hesse form and from its dual, without use of transformations, are given by Joliffe.[71, 72]

The base points of a pencil of plane C_3 were discussed by Emch.[35] A C_3 given by three asymptotes is discussed by Rao [116] and earlier by Durairajan.[30] Polarity as to a given plane C_3 is applied to furnish proofs of a number of theorems previously given without proof by Genoino.[48]

Triangles in-and-circumscribed to two cubic curves are unique unless some particular relation exists between the two given curves (Bell [11]).

The circular C_3 can appear as the envelope of a particular system of circles (Green and Prior [59]).

The general (2, 1) correspondence between two real variables was discussed by Purushottam.[114] Pencils of $|C_3|$ having 4 given points for common points of inflexion are studied by Turner.[143] They can be arranged into groups according to the number of imaginary basis points. The 24 triangles in-and-circumscribed to a general C_3 are discussed by Ogg [105] by methods partly algebraic and partly by means of elliptic functions. An equianharmonic C_3 is invariant under a certain G_{108} of Cremona transformations. The Hesse configuration on a plane C_3 is associated with desmic tetrahedra $(12_4, 16_3)$ (Mathews [90]). This is a generalization of Coble's checkerboard arrangement, further extended in.[91]

Pencils of equianharmonic curves were first constructed by Chisini [22] and further developed by Yerushalmy.[164, 165] Each curve of the pencil is invariant under the Cremona transformation $u'=u+a$, $[\omega^3=1]$. The pencil contained in a net through six points, vertices of two triangles in-and-circumscribed to a cubic ϕ are three fold perspective from the vertices of the third triangle of base points. Sets of points on C_3 with rational coordinates are treated by Logsdon.[79]

(*c*) *Quartic curves.* The bicircular C_4 can be expressed in terms of power coordinates. This is done in a new way by Wilson.[159] The entire system is invariant under inversion. A particular such curve is treated by D'Ocagne.[28] The tritangent circles of a circular C_4 are obtained by Richmond [117] by projecting a $C_6=F_3$, F_2 from a double point on F_3, simple on F_2. The C_4 obtained by the Geiser projection is studied from the stand point of its systems of bitangents by Blakey.[14] The C_4 with all real bitangents is further discussed by Du Val.[31] Systems of in-and-circumscribed triangles are constructed by Edge [32] and by Milne,[95] using the same method. The system of bitangents of the uni-nodal C_4 is further developed by Richmond,[118] without further use of transformations.

The $|C_6|:9P_2$ of Halphen are constructed by Godeaux by means of mapping an F_3 on S_2. Construct a conic touching a C_3 in three points and pass a pencil $|F_2|$ through it. Various particular cases are given by Godeaux.[51, 54, 55] Ogg [104] describes a Γ_{256} under which a C_4, $p=1$, of S_3 remains invariant, by means of elliptic parameters. Telling [138] discusses reality conditions for the rational quartic curves of S_3 by means of projections of the normal C_4 of S_4 upon S_3, and the dual operation.

(*d*) *Various curves.* Given a γ_k^m on $C_n^{(p)}$ of S_r, to determine the number of triads of points, such that each will belong to the same group of γ_k^m with the same multiplicity. The problem was solved by Longhi,[81] who made applications to ruled surfaces, circles and spheres.

If a curve C contains a complete g'_k, then C has a g_{n+k} containing a complete g_n^r, without a fixed residue. The necessary and sufficient condition that g_n^r be contained partially in g_{n+k} is that g'_k exists, but not a g'_{k-1} (Maroni [89]).

3. Linear series. Segre [128, 129] showed that continuous systems of plane curves exist belonging to every compatible set of solutions of Plücker's numbers; *compare* Crone [Acta Math. 2: 81-96 (1883)]. The postulate of Lefschetz is not always true. The virtual dimension δ' may not be the same as the effective dimension δ^*. The equality of Plücker's numbers for two curves does not insure that both curves belong to the same continuous system. The two C_6 with six cusps, one set on a conic, the other not, furnish an example.

The theorem of residuation of Brill and Noether is established in a new way by Gambier [45] and applied to mapping quartic surfaces with one or two double lines upon a plane. Some theorems concerning excess are established; *see also*.[46]

Fano [39] considers algebraic curves containing series that are self residual as to the canonical series, and illustrates with C of $p=5$ containing a self residual g_4^1; partial generalization to g_{p-1}^r. Some generalizations of conics by methods of differential geometry applied to trajectories are given by Jonescu.[73]

4. Curves invariant under Cremona transformations. A purely arithmetic derivation of consistent Plücker numbers is given by Meyer.[92] D'Ocagne [29] proposes a generalization of polar curves by replacing a conic by a pair of curves largely metric. A particular case is studied still further by Harmegnies.[62] Kapferer [75] submits a proof that only linear (1, 1) correspondences can exist between two non-singular curves of order >3. An alternate proof of the same theorem is given by Wirtinger [160] by means of Zeuthen's theorem. This theorem was proved by Küpper in 1892 and a much broader one which includes it by Snyder [133.1] A recent alternate proof for non-singular plane curves was given by Segre,[131] with an extension to surfaces and varieties by Severi. The number of trisecants and quadrisecants of a curve in S_3 is found by Welchman [157] by considering the curve replaced by a system of lines.

5. Trisecant surfaces of space curves. The order of the surface of trisecants is found to be

$$2\binom{n-1}{3}-(n-2)p-1/3\binom{r}{2}(8n-r-4),$$

and the number of quadrisecants is

$$\frac{1}{2}(n-3)\binom{n-2}{3}-\binom{n-3}{2}p+\binom{p}{2}$$
$$-\binom{r}{2}\left[\frac{1}{6}(n-3)(3n-2r-8)-p\right].$$

If the curve has actual multiple points these formulas must be modified. If C has h apparent double points and v_i points of multiplicity i, there are

$$\frac{1}{24}\left[12h(h-4n+11)-n(n-2)(n-3)(n-13)\right]+\sum_{i=3}^{i=m}\frac{1}{24}v_i i(i-1)(i-2)(4n-3i-15)$$

four fold secants not through any multiple point (Schaake [127]).

Wong [163] obtained enumerative properties of C_n in S_4 by replacing C_n by n lines with $n+p-1$ intersections and employing known formulas for transversals of lines of this system. Of the results obtained, those previously known are confirmed.

The number of tritangent planes of C_n in S_r is

$$2^r\sum_{i=0}^{q}\binom{n-q-i}{q-i}\binom{p}{i}.$$

The number of $(r+1)$ secant S_{r-1} of C_n in S_{2r} is

$$\sum_{i=1}^{q}(-1)^i\binom{n-r-2i}{r+1-2i}\binom{p}{i}.$$

Baker [9] shows that if C_n, genus p, with k cusps, lies on a quadric of S_r, and its tangents, osculating planes, . . . , osculating S_{n-1} all lie on the quadric, then $2[n+2h(p-1)]-2k$ is the number of osculating S_n to C_n on the quadric. The rational C_6 of S_3 having six points of inflexion are of two distinct types, according as the curve does or does not belong to a linear complex (Gherardelli [49, 50]). Both of the curves are self-dual.

6. Plane sextics with six cusps. Similarly, the plane curves of order 6 with six cusps belong to distinct families, according to whether the cusps lie on a conic or not. This was first mentioned by Del Pezzo without proof; cf. Segre,[129] Godeaux [54] and Villa.[153] Cusps may be at the vertices of a complete quadrilateral.

Another C_6, $p=3$ is obtained by Emch [36] by means of the symmetric (n, n) correspondence. It is the locus of the vertex of the cones in a net of quadrics and is inscribed in a complete quadrilateral.

Schaake [125, 126] considers an extension of the theorem of Poncelet.

7. Morley configurations. The configuration of Morley-Peterson was obtained by elementary methods by Mussel [101]; various forms of the configuration are discussed by Labrousse.[76] A variation of the theorem of Miquel is given by Pacquement.[106]

A further contribution of the inversion geometry of the plane bicircular C_4 is made by Ramamurti,[115] by means of four mutually orthogonal circles consisting of the axes, the unit circles centers at the origin and the circle with same center orthogonal to it. The C_4 invariant under a non cyclic G_4 is then

$$(x^2+y^2)^2+ax^2+by^2+1=0.$$

Richmond [117] gives an alternate proof, by means of symmetric functions, of the Morley theorem concerning circles and points. Lob [78.1] proved the same theorem by means of projection.

8. Special correspondences. Longhi [81] determines the number of triads of elements conjugate two by two in a symmetric non singular correspondence of grade α, valence γ and a ∞^1 algebraic entity of genus p. The number is $2\binom{\alpha}{3}+2\binom{\gamma}{3}p-\gamma(\alpha-\gamma)p$.

It was shown by Gherardelli [51] that every symmetric $(n-1,\ n-1)$ correspondence of valence 1, having a cycle of n distinct points, is cyclic and can be generated by a $g^1{}_n$.

Morley and Patterson [99] determined the invariants of biquadratic curves, and their possible components, by means of inversion. The C_4 is written as a (2, 2) correspondence between a complex variable and its conjugate.

For the application to the theory of binary forms, compare Moore.[98] The condition for composition was determined by Segre.[130] *See also* McCoy.[81]

9. Hessian of a plane curve. In a series of papers Villa [150, 151] shows that the Hessian may have multiple points, even if the given curve does not. The necessary and sufficient conditions that this may happen are given. Properties analogous to those of the Hessian are generalized to apply to the rth polars by Takami.[137]

10. Incidence theorems. The theorem of Segre states that the lines of S_4 which meet four independent planes also meet a fifth. This was extended by James.[68] An $|S_2|$ meeting a C_4 of S_4 in three points and meeting a given line also meets a second C'_4. Also if $[C_4,\ C'_4]=6$, $|S_2|$ meets a third. The configuration is symmetric in the three curves.

Another extension is: given in S_4 two conics and two lines meeting both, the $|S_2|$ meeting all will also meet another C_2. This system is symmetric: three such congruences $|S_2|$ meet $3c_2$ and $2c_1$.[138] Various generalizations of the theorem: if $[S_2, C_4]=3$, $[S_2, C'_4]=1$, then a third C''_4 exists such that $[S_2, C''_4]=1$. Proof similar to that given by Segre for lines. That the three quartics have symmetric positions as to transversal planes was proved by another method by Telling.[137]

This incidence theorem can be generalized in various ways. The theorem is: in S_4 given two C_4, C'_4 meeting in six points; the $|S_2|$ meeting C_4 in three points, C'_4 in one, also meet a third C''_4 through the six points. The generalization made by Babbage [7] is: In S_{2n}, given C_{2n}, C'_{2n} $[C_{2n}, C'_{2n}]=2n+2$; the $|S_n|$ meeting C_{2n} in $n+1$ points and C'_{2n} in one point meet a third C''_{2n} through the same common points. A different extension of Segre's original incidence theorem is made by Room.[121] No use is made of transformations.

It was shown by Hodge [64] that there is only a finite number of birationally independent systems of plane curves of genus π with ordinary base points. On a ruled surface for which $-p_a=p>0$, any system of curves of genus π belongs to one of a finite number of simply infinite families of linear systems of curves of genus π.

Telling [139] generalizes the incidence theorem of three related quartic curves of S_4. Given a C_n of S_n; its points, tangents, osculating planes, etc., form a system; the number of points in an S_{n-1} (order), the number of tangents meeting an S_{n-2} (first rank). . . . For $n=4$, these numbers are (4, 6, 6, 4). Project from any point of S_4 upon S_3. The result is (4, 6, 6). Dually, cut system by general S_3. The result is (6, 6, 4). The first is C_4 of S_3, the second (dual) is its developed surface. C_4 lies on a quadric surface; the planes of the developable all touch a second quadric. Generalized to C_n in S_n.

11. Various theorems. Schubert's formula, $M_{n,\,r}=(r+1)(n+rp-r)$ and Castelnuovo's criterion are extended to series having variable multiple points by Maroni.[89]

Piazolla-Beloch [109] proved that every curve of order n in S_3, having the maximum index $(n-2)$ and the maximum number of circuits, has the circulation equal to the index.

Cubic and biquadratic constructions are carried out by means of two conics, first in terms of their common self-polar triangle, then in particular for circles and equilateral hyperbolas, from which the general case can be obtained by projection (Sobotka [135]). The same author [136] discusses an interesting configuration determined by a space cubic curve and four of its bisecants.

An I_3 on C_5 of S_3, $p=2$, determined by the bisecants which meet a given bisecant p generate a ruled surface R_6 and cut on C_5 an I_3 of the first class, with 8 coincidences. R_6 has C_5 and p each triply. Each p

determines an I_3 (Mikan[93]). The curve can be constructed by knowing 11 points, which must lie on a quadric surface.

A particular metric C_4 is discussed by Goormaghtigh.[56]

Longo[83] derives properties of multisecant conics of space curves by means of composite curves.

Application of Noether's fundamental theorem to the plane C_5 with three double points A, B, C and ten simple points shows that they will have three remaining points in common, provided they do not all lie on a conic through A, B, C.

Similarly given $C_6: 8P_i^2$; there are $\infty^4 C_9$ having $8P_i^3$ and through B, the ninth base point of $|C_3|: 8P_i$; these curves also pass through an arbitrary point on C_6 (Thalberg[141]).

BIBLIOGRAPHY

1. Abramesco, N. Boll. U. M. It. 11: 10-13 (1931).
2. Albanese, Maria. Lomb. Ist. Rend. (2) 67: 121-128 (1934).
3. Allen, F. E. Am. J. 49: 456-461 (1927).
4. Arnold, H. E. Am. M. S. Bull. 36: 212 (1930).
5. ———. Am. J. 52: 601-606 (1930).
6. Babbage, D. W. Camb. Ph. Soc. Proc. 28: 421-426 (1931).
7. ———. London M. S. J. 8: 9-11 (1932).
8. Baker, H. F. Principles of Geometry, vol. IV (1925).
9. ———. Edinb. M. S. Proc. (2) 1: 19-30 (1927-9).
10. Barce. Nouv. Ann. (5) 2: 201-212 (1924).
11. Bell, C. Tôhoku M. J. 36: 34-40 (1931).
12. Berzolari, L. Boll. U. M. It. 12: 121-123 (1933).
13. Biggiogero, G. Per di Mat. (4) 12: 17-39 (1931).
14. Blakey, J. London M. S. J. 5: 50-54 (1930).
15. Brigaglia. Pal. Circ. Mat. 53: 310-312 (1928).
16. Brill, A. Kurven (1925).
17. Burnside, W. Camb. Ph. Soc. Proc. 22: 163-166 (1924).
18. Campedelli, L. Per. di Mat. (4) 13: 221-223 (1933).
19. Cattaneo, P. Lomb. Ist. Rend. (2) 66: 978-982 (1933).
20. ———. Boll. U. M. It. 12: 139-142 (1933).
21. ———. Per. di Mat. (4) 13: 296-300 (1933).
22. Chisini, O. Pal. Circ. Mat. 41: 69-93 (1916).
23. Ciani, E. Ann. S. N. Pisa (2) 1: 215-228 (1932).
24. Colucci, A. Giorn. di Mat. 62: 62-90 (1924).
25. Coolidge, J. L. Algebraic Curves (1931).
26. Deaux, R. Mathesis 39 :166-169 (1925).
27. ———. Enseign. Math. 29: 262-278 (1930).
28. D'Ocagne, M. Nouv. Ann. (5) 3: 153-155 (1925).
29. ———. Enseign. Math. 31: 31-49 (1932).
30. Durairajan, N. Ind. M. S. 18: 60-64 (1929).
31. Du Val, P. Camb. Ph. Soc. Proc. 25: 433-435 (1928-9).
32. Edge, W. L. London M. S. Proc. (2) 34: 492-525 (1932).
33. Edwards, B. A. London M. S. J. 2: 72-79 (1927).

34. Emch, A. Comm. Math. Helv. 1: 64-68 (1929).
35. ———. Am. J. 54: 285-292 (1931).
36. ———. Comm. Math. Helv. 4: 65-73 (1931).
37. ———. Monatsh. f. Math. 40: 193-200 (1933).
38. Enriques, F. and Chisini, O. Funzioni algebriche, vol. 3 (1924).
39. Fano, G. Lomb. Ist. Proc. (2) 63: 949-967 (1930).
40. Field, P. Math. Zschr. 37: 402-404 (1933).
41. Gambier, B. C. R. 178: 837 (1924).
42. ———. C. R. 179: 1241-1243 (1924).
43. ———. C. R. 180: 897-899 (1925).
44. ———. Ann. Éc. Norm. (3) 42: 217-291 (1925).
45. ———. C. R. 194: 578-580 (1931).
46. ———. C. R. 195: 1057-1059 (1932).
47. Ganguli, S. Curves, 2d ed., vol. 1 (1925).
48. Genoino, G. Giorn. di Mat. 63: 180-202 (1925).
49. Gherardelli, G. Rom. Acc. L. Rend. (5) 33: 335-339 (1924).
50. ———. Rom. Acc. L. Rend. (6) 11: 173-179 (1930).
51. ———. Boll. U. Mat. It. 9: 176-177 (1930).
52. Godeaux, L. Mathesis 39: 154-157 (1925).
53. ———. Trans. birat. (1927).
54. ———. Brux. Bull. (5) 16: 576-580 (1930).
55. ———. Mathesis 46: 333-340 (1932).
56. Goormaghtigh, R. Gaz. Mat. Buk. 36: 93 (1930).
57. Grace, J. H. Camb. Ph. Soc. Proc. 25: 421-432 (1928-9).
58. Graustein, W. C. Higher Geom. (1930).
59. Green, H. G. and Prior, L. E. J. Éc. Polyt. (2) 28: 167-171 (1930).
60. ———. Am. Math. Month. 39: 16-18 (1932).
61. Griffiths, L. W. Am. M. S. Bull. 31: 312-317 (1925).
62. Harmegnies, R. Enseign. Math. 31: 50-57 (1932-3).
63. Hilton, H. and Green, Margaret. London M. S. Proc. (2) 24: 305-318 (1925).
64. Hodge, W. V. D. London M. S. Proc. (2) 29: 145-167 (1928).
65. Hudson, Hilda P. Cremona Trans. (1927).
66. Hulin, G. Liège S. R. Mém. (3) 14: No. 19 (1928). 14 p.
67. James, C. G. F. Camb. Ph. Soc. Proc. 21: 664-684 (1923).
68. ———. Mess. of Math. 54: 129-138 (1924).
69. ———. Camb. Ph. Soc. Proc. 22: 24-25 (1924).
70. Joliffe, A. E. Mess. of Math. 54: 88-96 (1925).
71. ———. London M. S. Proc. (2) 23: xxxiii-xxxvi (1925).
72. ———. Mess. of Math. 57: 17-37 (1927).
73. Jonescu, D. V. C. R. 194: 2006-2009 (1931).
74. Jonzon, B. Upsala, Diss., 1930. 65 p.
75. Kapferer, H. Münch Ber. 155-175 (1931).
76. Labrousse. A. Rev. de Math. Spec. 42: 313-316, 345-347 (1931).
77. Lefschetz, S. Am. M. S. Trans. 28: 1-49 (1926).
78. Libois, P. Liège, S. R. Mém. (3), 15 (1930). 17 p.
78.1 Lob, H. Camb. Ph. Soc. Proc. 29: 45-51 (1933).
79. Logsdon, M. I. Am. M. S. Trans. 27: 474-490 (1925).
80. Longhi, A. Giorn. di Mat. 63: 209-247 (1925).
81. ———. Boll. di U. M. It. 11: 269-273 (1932).
82. ———. Pad. Sem. Rend. 4: 1-26 (1933).

83. Longo, Sebastiana. Cat. Acc. Gio. Atti (5) 16, No. 13 *bis* (1928-9). 19 p.
84. Loria, G. Curve sghembe, I (1925); II (1925).
85. ———. Curve piane I, II (1930).
86. McCoy, N. H. Am. M. S. Bull. 37: 42 (1931).
87. Marek, R. Časopis 62: 4-9 (1932).
88. Maroni, A. Rom. Acc. L. Rend. (6) 17: 265-267 (1933).
89. ———. Mem. Acc. It. 3 (1931). 13 p.
90. Mathews, R. M. Am. M. S. Trans. 28: 502-522 (1926).
91. ———. Am. M. S. Trans. 30: 19-23 (1928).
92. Meyer, W. F. Math. Zschr. 35: 746-759 (1931).
93. Mikan, M. Časopis 57: 114-118 (1928).
94. ———. Časopis 59: 33-39 (1930).
95. Milne, W. P. London M. S. J. 8: 211-216 (1933).
96. Moore, T. W. and Neelley, J. H. Am. J. 50: 467-472 (1928).
97. Morley, F. and Morley, F. V. Inversive Geometry (1933).
98. Moore, T. W. Ann. of Math. (2) 30: 92-100 (1928).
99. Morley, F. and Patterson, B. C. Am. J. 52: 413-424 (1930).
100. Morton, V. C. London M. S. Proc. (2) 30: 379-400 (1930).
101. Mussel, M. Rev. de Math. Spec. 42: 369-371 (1931).
102. Neelley, J. H. Am. M. S. Bull. 36: 269-273 (1930).
103. ———. Tôhoku M. J. 35: 51-59 (1931).
104. Ogg, F. C. Ann. of Math. (2) 29: 355-368 (1928).
105. ———. Tôhoku M. J. 30: 340-345 (1929).
106. Pacquement, A. Rev. de Math. Spec. 43: 450-451 (1932).
107. Papillon, P. Enseign. Math. 31: 219-228 (1933).
108. Peterson, O. J. Am. M. S. Bull. 36: 221 (1930).
109. Piazzolla-Beloch, Margarete. Boll. d. N. M. It. 12: 145-147 (1933).
110. Piedvache. Ann. of Math. (2) 30: 281-284 (1928).
111. Pomey, L. Jour. Éc. Polyt. (2) 27: 1-84 (1929).
112. ———. Jour. Éc. Polyt. (2) 28: 108-165 (1931).
113. ———. C. R. 197: 881-883 (1933).
114. Purushottam, S. Ind. M. S. 17: 129-133 (1928).
115. Ramamurti. Ind. M. S. 19: 177-181 (N. and Q.) (1932).
116. Rao, A. N. Ind. M. S. 18: 65-67 (1929).
117. Richmond, H. W. Edinb. M. S. Proc. (2) 3: 46-52 (1931).
118. ———. Edinb. M. S. Proc. (2) 1: 31-38 (1927-29).
119. ———. Camb. Ph. Soc. Proc. 29: 165-172 (1933).
120. Rindi, S. Lomb. Ist. Proc. (2) 65: 472-478 (1931).
121. Room, T. G. London M. S. Proc. (2) 31: 445-486 (1933).
122. Roth, L. Camb. Ph. Soc. Proc. 29: 178-183 (1933).
123. Roy, G. Mathesis 46: 462 (1932).
124. Rutter, D. Edinb. M. S. Proc. (2) 3: 30-36 (1931).
125. Schaake, G. Nieuw Arch. (2) 14: 325-329 (1924).
126. ———. Nieuw Arch. (2) 15: 68-70 (1925).
127. ———. Amst. Ak. Proc. 32: 1398-1402 (1929).
128. Segre, B. Rom. Acc. L. Rend (6) 10: 31 (1929).
129. ———. Rom. Acc. L. Rend. (6) 10: 557-560 (1929).
130. ———. Boll. d. U. M. It. 9: 1-4 (1930).
131. ———. Math. Ann. 109: 1-3 (1933).
132. Seifert, L. Časopis 61: 146-152 (1932).

133. Semple, J. G. London M. S. J. 7: 266-271 (1931).
133.1 Snyder, V. Am. J. 30: 337-346 (1908).
134. Snyder, V. and others. National Research Council, Bull. 63 (1928).
135. Sobotka, J. Bull. Intern. 26: 74-86 (1925).
136. ———. Bull. Intern. 27: 500-503 (1926).
137. Takami, M. Tôhoku M. J. 35: 136-141 (1931).
138. Telling, H. G. Camb. Ph. Soc. Proc. 28: 403-415 (1931-2).
139. ———. Camb. Ph. Soc. Proc. 29: 195-206 (1933).
140. Terracini, A. Pal. Circ. Mat. 56: 112-143 (1931).
141. Thalberg, O. M. Norske Vid. Ak. Avh. (1932), 7 p.
142. Turnbull. London M. S. Proc. (2) 30: 469-480 (1928).
143. Turner, B. M. Am. J. 47: 149-162 (1925).
144. Ursell, H. D. Camb. Ph. Soc. Proc. 25: 39-41 (1928-9).
145. Vaidyanathaswami, R. Camb. Ph. Soc. Proc. 23: 109-119 (1923-5).
146. ———. Camb. Ph. Soc. Proc. 23: 233-261 (1925-7).
147. ———. Camb. Ph. Soc. Proc. 23: 631-652 (1925-7).
148. ———. London M. S. J. 5: 249-253 (1930).
149. Villa, M. Giorn. di Mat. 68: 81-91 (1930).
150. ———. Lomb. Ist. Rend. (2) 65: 449-460 (1932).
151. ———. Lomb. Ist. Rend. (2) 65: 625-641 (1932).
152. ———. Lomb. Ist. Rend. (2) 65: 129-142 (1932).
153. ———. Lomb. Ist. Rend. (2) 65: 1013-1024 (1932).
154. ———. Boll. d. U. M. It. 12: 15-18 (1933).
155. de Vries, J. Amst. K. Ak. Wet. Proc. 35: 1119-1121 (1932).
156. ———. Amst. K. Ak. Wet. Proc. 35: 1121-1123 (1932).
157. Welchman, W. G. Camb. Ph. Soc. Proc. 28: 206-208 (1932).
158. White, H. S. Curves third order (1925).
159. Wilson, R. Edinb. M. S. Proc. 43: 25-34 (1925).
160. Winger, R. M. Am. M. S. Bull. 30: 16 (1924).
161. ———. Am. J. 47: 207-223 (1925).
162. Wirtinger, W. Münch. Ber. 145-147 (1932).
163. Wong, B. C. Am. M. S. Bull. 40: 431-432 (1934).
164. Yerushalmy, J. Am. J. 53: 319-332 (1931).
165. ———. Am. J. 54: 279-284 (1932).

CHAPTER II

THE MAPPING OF SYSTEMS OF CURVES

LEAMAN A. DYE

1. Systems of lines. The mapping of systems of curves was begun by Klein [8] who mapped the lines of S_3 on the points of a quadric primal $V_4^2 \equiv p_{23}p_{14}+p_{31}p_{24}+p_{12}p_{34}=0$ in the S_5 of the six p_{ik}. A purely geometrical method of establishing this correspondence was given by Turnbull [45] and supplemented by Wren.[78] The lines of S_3 may be mapped on S_4 by projecting the V_4^2 of S_5 from a point on it into S_4. Semple [38] represents the V_4^2 on S_4 by taking for the p_{ik}, quadric primals V_3^2 through a rational normal quartic curve C_4 in S_4. The points of one of the V_3^2 correspond to the points of a prime section of the V_4^2. By considering the special cases where C_4 breaks up into a cubic and a line, or three lines and a common transversal, Semple obtains a simple representation on S_4 of congruences of lines in S_3 of the second and third orders.

Schaake [34] uses the representation of lines of S_3 upon a V_4^2 in S_5 to determine the number of bilinear congruences of lines. He [32] also uses this representation to discuss and map line complexes and congruences. Todd [42] maps the lines of S_n upon a manifold in S_N, where $N=n(n+1)/2-1$. The mapping is accomplished by the use of Plücker line coordinates; by means of the mapping the properties of doubly infinite line systems are studied.

De Vries [74] represents any line of S_3 by the focus F and the vertex V of a parabola in a fixed plane π. A line l meets π in V, and a line parallel to l through a fixed point O meets π in F; thus a (1, 1) correspondence between the lines of S_3 and the parabolas of a fixed plane may be established. A variant of this mapping is also given in the same paper. Let d, F be the directrix and focus of a parabola in a fixed plane π. Let H be a quadric surface and O a point not in π. The polar plane of F with respect to H meets the plane Od in a line l, image of the parabola determined by F and d. Conversely the polar of a line l meets π in F, and d is determined by the intersection of π and the plane Ol. The lines of a linear complex are mapped by Fischer [5] upon a conic of a fixed plane in the following manner. Let a line l of the complex meet two fixed planes π_1, π_2 in the points P_1, P_2, which are projected into the points P'_1, P'_2 in a third fixed plane π', thus mapping the line l on π'. Fix P'_1 and hence P_1; the lines of the complex through P_1 meet π_2 in points P_2 on a line, therefore P'_2 moves on a line l'_2. As P'_1 moves on a line l'_1, the line l'_2 passes through a point Q'; if Q' lies on l'_1, then l'_1 envelopes a conic which maps the complex.

Staehlin [39] establishes a correspondence between the tangent lines of a quadric cone and the points of S_3 by a suitable choice of the parameters of the tangents.

The bisecants of a rational twisted curve C_n are mapped by de Vries [50] on the points of a plane by setting up a (1, 1) correspondence between the points of the C_n and the points of a conic. The image of a bisecant of C_n is the point of intersection of the tangents to the conic at the points corresponding to the intersections of the C_n and the bisecant. de Vries [56] represents the congruence of common secants of two rational space curves on the points of a plane by the use of a projectivity between the points of the curves and the lines of two pencils of lines in a fixed plane. The map of a common secant of the two curves is the point of intersection of the two lines, one from each pencil, which correspond to the points in which the secant line meets the two curves.

A complex of lines in S_5 which meet five linear three-spaces is discussed by van Kol [24] and mapped on one of the three-spaces S_3 by relating the line to the point in which it pierces S_3. Given a point of S_3, there is just one line through it which meets the other four three-spaces. Aprile [1] discusses the systems of lines of dimension ∞^{r-1} in S_r, such that only one line of the system passes through a general point of S_r. The lines may be mapped on the points of S_r.

2. Systems of conics. The conics of S_3 are mapped on the points of S_{19} by Todd.[44] A line p_{ij} meets a plane $\Sigma a_i x_i = 0$ in the point $x_i = \Sigma a_j p_{ij}$. If the point lies on the conic $\Sigma a_i x_i = 0$, $\Sigma b_{ij} x_i x_j = 0$, then the line p_{ij} belongs to the quadratic complex $\Sigma d_{ij,\,kl} p_{ij} p_{kl} = 0$, where $d_{ij,\,kl} \equiv b_{ik} a_j a_l + b_{jl} a_i a_k - b_{il} a_j a_k - b_{jk} a_i a_l$. The 21 coefficients $d_{ij,\,kl}$ are connected by the relation $d_{12,34} + d_{31,24} + d_{23,14} = 0$, and, subject to this condition, may be taken as the coordinates of the conic in S_3 or as the coordinates of a point in S_{19}. The locus of points representing complexes of lines meeting a conic is a V_8^{92}. The equations of this manifold may be taken to be the 21 linearly independent quadratic relations among the $d_{ij,\,kl}$ which define 21 quadric primals through the V_8^{92}, together with some cubic identities of the same sort. A linear complex of conics is defined as a set of ∞^7 conics containing one member in a general pencil of conics and is given analytically by a linear relation $\Sigma a_{ij,\,kl} d_{ij,\,kl} = 0$.

A correspondence between the ∞^4 conics through a fixed point P in a plane π and the primes of S_4 is established by Schaake.[36] To the conics through P and a point Q correspond the primes through a point Q' in S_4. As Q covers the plane π, Q' describes a ruled cubic surface F_3. The tangent planes at two points Q' of F_3 meet in a point R which corresponds to the two points Q in π, which are images of the two points Q'. Conversely through a point R of S_4 there are two planes which are tangent to F_3 at two points Q' which correspond to two points Q of π. Thus the unordered pairs of points in a plane are mapped upon the points of S_4.

Schaake [29] maps the complex of conics meeting five lines l_i on the points of S_3. To a conic corresponds the point which is the pole of the plane of the conic with respect to a fixed quadric H. To a point of S_3 corresponds the conic in the polar plane of the point with respect to the quadric H which passes through the points of intersection of the lines l_i with the plane. de Vries [52] studies a similar complex of conics in which two of the five lines are replaced by a fixed conic. Sweerts [41] discusses the complex of conics touching 5 fixed planes.

The congruence of conics through a fixed point and meeting a rational space quartic curve in four points can be represented on the points of a fixed plane (de Vries [66]). In the plane of each conic of the congruence, the chord which meets two fixed lines maps the conic by its intersection with a fixed plane. The congruence may also be mapped by establishing a (1, 1) correspondence between the planes of the conics and the points of a fixed plane. The congruence of conics through a fixed point, with a fixed secant, and meeting a twisted cubic in three points is mapped on the triads of points in a fixed plane by the intersections with the plane of the common bisecants of the cubic and the conics (de Vries [70]).

The conics through a fixed point which meet two fixed conics each in two points form a congruence which de Vries [72] maps by the quadruples of points in a fixed plane. Each set of four points in the plane determines four common secants of the two fixed conics and hence a conic of the system. If the two fixed conics are replaced by four lines, de Vries [71] maps the conics by the intersections with a fixed plane of the chords of the conics which are also transversals of a selected pair of the fixed lines. The same method of mapping is used by de Vries [77] for a congruence of conics determined by a fixed point, two fixed lines as secants, and a fixed conic meeting each conic of the system in two points. In another paper de Vries [47] deals with the case where the conics are restricted to orthogonal hyperbolas meeting four lines.

de Vries [46] sets up a trilinear correspondence between the ranges of points (A), (B), (C) lying on three lines a, b, c. Through each triad of points A, B, C passes a conic meeting a fixed conic twice. The conics form a congruence which may be mapped on a plane in any one of several ways. If a (1, 1) correspondence is established between the conics of a bilinear congruence and the lines of a congruence of lines meeting two fixed lines, then the congruence of conics can be mapped on a fixed plane by the piercing points of the lines of the line congruence (de Vries [55]).

van Kol [23] uses a representation of a congruence of conics in a linear S_4, on the points and lines of a plane in studying systems of such congruences. Complexes of conics in S_4 are mapped on the points of S_3 and formulas of characteristics are derived by the methods of enumerative geometry by James.[7] The scheme is generalized by Miglio.[27] All the V^r_{r-1} of S_r passing through r given S_{r-2}, represent the primal sections of $V^{r!}_r$ of S_n $(n=2^r-1)$ having r pencils of $V^{(r-1)!}_{r-1}$. The rV^r_{r-1}, passing through rS_{r-2} have in common, apart from these, a $V^{r!}_0$, that is $r!$ points. They are the images of the primal sections of $V^{r!}_r$, of S_n.

3. Systems of cubics. The cubics of a plane are mapped upon the points of manifolds in S_9 by Yerushalmy,[79] the order and dimensionality of a manifold depending upon the nature of the system of cubics; for example, the cuspidal cubics are mapped upon the points of a V_7^{24} in S_9.

There are ∞^4 space cubics C_3 which pass through two fixed points and have two fixed lines l_i as bisecants. This system of cubics is mapped upon the points of S_4 by van Kol.[10] Take two quadric primals H_i in S_4 each having a double line and establish a projective correspondence between the planes of H_i and the points of the lines l_i. The map of a C_3 is the point of intersection of the four tangent primes to the primals H_i at the planes which correspond to the points of intersection of C_3 with l_i. van Kol [9] establishes a (1, 4) correspondence between the curves of a complex of space cubics and the points of S_3. The complex of cubics is made up of those through two fixed points D_1, D_2 and having a line a_1 as a secant and two lines a_2, a_3 as bisecants. A (1, 1) correspondence is set up between the points of the three lines a_i and the planes through three other fixed lines a'_i. Let a cubic of the complex meet a_1 in A_1, a_2 in A_2, A'_2 and a_3, in A_3, A'_3 and call the planes corresponding to these points α_1, α_2, α_2', α_3 and α_3'. To this cubic is associated the quadruple determined by the intersections of $\alpha_1\alpha_2\alpha_3$, $\alpha_1\alpha_2\alpha_3'$ $\alpha_1\alpha_2'\alpha_3$, and $\alpha_1\alpha_2'\alpha_3'$. Conversely, take a point P of S_3, then the planes Pa'_i determine the points A_i on the lines a_i. Let H be the quadric through the points

D_1, D_2, A_1 and the lines a_2, a_3. The image of P in the (1, 4) correspondence is that cubic which lies on H and has a_2, a_3 as bisecants and passes through D_1, D_2, A_1, A_2 and A_3.

The complex of twisted cubics through two fixed points, with three fixed lines as secants and having another fixed line as a bisecant, can be represented on the points of S_3 (van Kol [18]). Set up a (1, 1) correspondence between the points of the three secants and the planes of three pencils of planes. The representation on S_3 of a curve of the complex is the point of intersection of the three planes, one from each pencil, which correspond to the three points in which the cubic meets the three fixed secant lines.

The ∞^2 space cubics through five points is called the congruence of Reye. Each line through a basis point of the congruence is the bisecant of one and only one curve of the system. This property may be used to map the congruence on the points of a fixed plane (de Vries [54, 69]). The congruence may also be mapped by the intersections with a fixed plane of the tangents to the cubics at one of the basis points (de Vries [62]). In the same paper de Vries maps the congruence by the residual intersections of the curves with a fixed plane through two of the basis points. This congruence is also discussed by de Vries.[49, 59]

The congruence of cubics through four points and having a fixed line l as a bisecant can be represented on the points of a fixed plane π through l. Schaake [30] establishes a (1, 1) correspondence between the points of l and the points of a fixed conic in π. The map of a cubic of the congruence is the point of intersection of the tangents to the conic at the points which correspond to the points of intersection of the cubic and l. de Vries [75] maps this congruence by means of the intersections of the tangents to the cubics at one of the fixed points with a fixed plane. de Vries [63, 48] maps the cubics of a doubly infinite system having a common bisecant l on the points of a plane through l. In another paper he [64] represents the cubics through four points and meeting a fixed conic twice by the poles of the common chords of the cubics and the conic.

Schaake [31] maps the ∞^2 cubics through four fixed points and with two fixed lines l_i as secants on the points of a plane π. Take two points P_i in π and establish a projectivity between the points of l_i and the lines on P_i. The congruence is mapped by the intersections of pairs of lines from the two pencils which correspond to pairs of points on the secant lines and also on the cubics. This same method of mapping is used by de Vries [51] in representing the system of cubics through three fixed points, with two fixed lines as secants and with a third line as a bisecant. The congruence discussed by Schaake [31] is also mapped by de Vries [76] on the points of a plane by the lines joining the intersections of the cubics with the fixed secant lines.

If two pencils of quadric surfaces have a basis line in common, then the residual intersections of pairs of quadrics, one from each pencil, make up a congruence of twisted cubics. This congruence is sometimes called the congruence of Veneroni and is bilinear. de Vries [60] maps it on a fixed plane by the intersections of the bisecants of the cubics passing through a fixed point. Hoestra [6] also considers this congruence along with those of Reye and of Stuyvaert (the cubics through three points and having three fixed lines as bisecants).

The congruence of cubics through a fixed point and with four fixed lines as bisecants is mapped on the points of a plane by van Kol.[20] He establishes a projectivity between the points of one of the fixed chords and the tangent lines to some fixed conic. The intersections of pairs of tangents to the conic is used to map the congruence. de Vries [61] and van Kol [21] use a similar method to map the system of cubics having five fixed lines as bisecants. The system of ∞^2 cubics determined by one point and five lines, two of which are secants and three being bisecants, is discussed and mapped by de Vries [57] on the points of a plane, by establishing a projectivity between the points of the two secants and the lines of two pencils in the mapping plane.

4. Systems of quartics. There are ∞^6 rational space quartics through five points P_i of S_3. Take a point Q in S_4, and on the lines QP_i take five points O_i as the vertices of the simplex of reference. By projection, a quartic through the five points P_i is associated with a quartic through the five points O_i, which is transformed into a line in S_4 by the transformation $x_i x'_i = 1$. Conversely a line of S_4 goes into a quartic through the five O_i which corresponds to a quartic through the five P_i. In this way Schaake [35] maps the system on the lines of S_4. van Kol [11] maps the ∞^6 rational quartics through five points of S_4 by the same device and then goes on to consider simpler systems of quartics. The congruence through six points and meeting a fixed plane is mapped on a plane through three of the points. The congruence through five points and having a fixed line as a bisecant is mapped by the use of a projectivity between the points of the bisecant and the points of a fixed conic. The intersections of pairs of tangents map the congruence on the points of the plane of the conic. The congruence of quartics through five points and having a pair of fixed lines as secants is mapped on the points of a plane by the aid of a projectivity between the points of the secants and the lines of two pencils in the mapping plane.

The congruence of rational quartics through seven points P_i is mapped on the points of a plane by Schaake.[33] Any line l through P_7 determines, with the remaining $6P_i$, a cubic cone on which l is a double line and the six lines P_7P_i are simple. The line l and the six points P_i also determine a quadric which meets the cubic cone in a residual quartic of the congruence obtained by considering all lines l through P_7. The mapping is accomplished by the use of the intersections of l with a fixed plane.

van Kol [19] discusses a congruence of rational space quartics through six points P_i and having a fixed line through P_1 as a trisecant. He also considers the congruence of rational quartics through four points and having one fixed line as a secant, another as a bisecant and a third as a trisecant. These congruences may be mapped on the points of a fixed plane through the trisecant line. The same mapping may be used for the congruence of rational quartics through two points and having two fixed lines as bisecants and two others as trisecants (de Vries [67]).

The ∞^4 elliptic quartics through six points are determined by the intersections of pairs of quadrics, one from each of two nets of quadrics having six points in common. The quartics are mapped upon the points of S_4 by van Kol,[13] who sets up a (1, 1) correspondence between the quadrics of the two nets in S_3 and the planes of two nets through two lines in S_4. A quartic of the system is represented by the point of intersection of the two planes which correspond to the two quadrics which determine the quartic curve.

A pencil of quadrics $H_1: 5P_i+l$ and a web of quadrics $H_2: 5P_i+Q$ determine a system of ∞^4 elliptic quartics through the five points P_i and having the line l as a bisecant. van Kol [25] establishes a (1, 1) correspondence between the quadrics of the pencil H_1 and the $|S_3|$ through a plane π of S_4 and also sets up a correspondence between the web of quadrics H_2 and the ∞^3 lines through some point O of S_4. The points of intersection of the lines through O and the $|S_3|$ through π are the maps of the quartics determined by the corresponding quadrics.

van Kol [17] studies the system of ∞^4 elliptic quartics through four points P_i and having two lines l_i as secants and a third line l_3 as a bisecant. He establishes a (1, 1) correspondence between the points of l_1, l_2 and the primes of two pencils in S_4. The points of l_3 are made projective with the tangent primes to a quadric primal $V_3{}^2$ having a double line. To a quartic of the system corresponds the point of intersection of the four primes which correspond to the points common to the quartic and the three lines l_i. Another system of ∞^4 elliptic quartics is considered by van Kol.[15]

The complex of elliptic quartics through five points and having three fixed lines as secants is mapped on the points of S_3 by means of a (1, 1) correspondence between the points of the secants and the planes of three pencils of planes (van Kol [22]). The complex of elliptic quartics through five points, with a line l_1 as a secant, and a line l_2 as a bisecant, is mapped upon the points of S_3 by van Kol.[12] He uses a (1, 1) correspondence between the points of l_1 and the tangent planes to a fixed quadric cone, and another such correspondence between the points of l_2 and the planes of a pencil through a fixed line. In another paper van Kol [16] discusses the complex of quartic curves in S_4 which have five fixed points as basis points and meet a fixed plane in a fixed line.

Two pencils of quadric surfaces have the elliptic quartic curves C_4 and C'_4 as basis curves, respectively. The intersections of two quadrics, one from each pencil, is an elliptic quartic of a congruence which is mapped by de Vries [58] on the points of a plane. Let P be a point of C_4, and P' a point of C'_4. The tangent planes to the quadrics at P, P' constitute two pencils of planes (p), (p'), which are made projective with two pencils of lines in a fixed plane. The map of a quartic of the congruence is the point of intersection of the two lines of the two pencils which correspond to the pair of planes, which are tangent to the quadrics determining the quartic curve. de Vries [73] maps the bilinear congruence of elliptic quartics determined by two pencils of quadrics, by the pairs of intersections with a fixed plane of the pairs of bisecants of each quartic which pass through a fixed point.

Another congruence of elliptic quartics through five fixed points and with two fixed lines as bisecants is given by the intersections of quadrics chosen from two pencils each having the five fixed points and one of the fixed lines as basis elements. This congruence is discussed by de Vries [65]; it may be mapped by the preceding method if the fixed point is taken on one of the fixed bisecants. van Kol [26] studies a congruence of elliptic quartics through six points and meeting a fixed conic twice. This congruence can be mapped on the plane of the conic.

5. Miscellaneous. de Vries [53] discusses a congruence of plane monoids made up of those plane curves of order n having an $(n-1)$-fold point at a fixed point P, and having $2n$ lines of S_3 as secants. The congruence can be mapped by its planes.

Two points O, O' and $(2n-1)$ points P_i determine two pencils of cones K, K' of order n through the lines OP_i and $O'P_i$, respectively, and each having the line OO' as an $(n-1)$-fold line. The two pencils determine a congruence of C_{2n-1} of genus zero, with O, O' as $(n-1)$-fold points and with the $(2n-1)$ P_i as simple basis points. de Vries[68] maps the congruence by the residual intersections with a fixed plane through the line OO'.

van Kol[14] maps a complex of planes of S_4 upon the points of S_3. Rao[28] obtains a representation of the planes of S_5 upon a manifold V_9 in S_{19}. In Rao's paper three points A, B, C in S_5 determine the plane whose coordinates p_{ijk} are the 20 third order determinants of the matrix $(a_ib_ic_i)$. The p_{ijk} are interpreted as point coordinates in S_{19}, and the 30 linearly independent quadratic relations among them may be taken as the equations of the V_9 on which the planes are mapped.

The S_k's of S_n may be considered as points of a Grassmanian variety $V_{(n-k)(k+1)}$ in S_N, where $N=\binom{n+1}{k+1}-1$. Brown[2] shows that those S_k's of S_n which are chordal $(k+1$ secant) to a rational normal C_n in S_n are represented by the points of a Veronese variety V_{k+1} lying in a Grassmanian variety, where the V_{k+1} is the map on S_N of a S_{k+1} by means of its primals of order $n-1$.

Edge[3] maps the quadrics of a net on the points of a plane, thus establishing a (1, 1) correspondence between a $C_6(p=3)$ and a quartic curve in the plane. He[4] generalizes these results by mapping a net of quadrics in S_n on the points of a plane π. He then maps the base curve θ of the net on the points of a curve in the plane π, and determines its order and genus.

Sorrentino-Giordano[40] represents the points of S_4 by ordered pairs of points in a plane π by means of two lines l_i. If P is a point of S_4, then the planes Pl_i meet π in two points P_i; conversely, given two points P_i in π, the planes P_1l_1 and P_2l_2 meet in a point P.

Todd[43] discusses the configuration defined by six lines in S_3. There are ∞^4 quartic surfaces passing through six arbitrary lines in S_3. Since the points of S_3 may be mapped on a $V_3{}^4$ of S_4, the points of any quartic surface of the system corresponds to a prime section of the $V_3{}^4$. The $V_3{}^4$ has 36 nodes, is determinantal and is capable of projective generation in 15 ways. To the interchange of conjugate systems of curves and surface on $V_3{}^4$ corresponds a Cremona transformation in S_3 of order 19.

Semple[37] studies a linear system of cubic surfaces $\Phi\equiv\Sigma_0{}^r\lambda_iF_i=0$, of freedom $r\geqq 4$. Let the coordinates of a point P in S_r be proportional to F_i; then to a point Q of S_3 corresponds a point P of S_r, and as Q describes the S_3, P describes a $V_3{}^n$, where n is the grade of Φ. To the points of a surface of Φ corresponds a prime section of $V_3{}^n$. Cremona transformations are obtained by projection of $V_3{}^n$ upon another S_3.

BIBLIOGRAPHY

1. Aprile, G. Giorn. di Mat. 70: 196-216 (1932).
2. Brown, L. M. London M. S. J. 5: 168-176 (1930).
3. Edge, W. L. London M. S. Proc. (2) 34: 492-525 (1931).
4. ———. Edinb. M. S. Proc. (2) 3: 259-268 (1933).
5. Fischer, A. Deutsche Math.-Ver. 37: 263-272 (1928).
6. Hoestra, H. P. Utrecht, Diss., 1928. 80 p.
7. James, C. G. F. Camb. Ph. Soc. Proc. 22: 621-629 (1923-25).
8. Klein, F. Math. Ann. 5: 257-258 (1872).
9. van Kol, J. W. A. Amst. K. Ak. Wet. Versl. 36: 1066-1072 (1927).
10. ———. Amst. K. Ak. Wet. Versl. 37: 156-160 (1928).
11. ———. Amst. K. Ak. Wet. Versl. 37: 334-339 (1928).
12. ———. Amst. K. Ak. Wet. Versl. 37: 772-777 (1928).
13. ———. Amst. K. Ak. Wet. Versl. 37: 907-910 (1928).
14. ———. Wis-en Nat. Verh. 4: 83-89 (1928-29).
15. ———. Wis-en Nat. Verh. 4: 90-94 (1928-29).
16. ———. Wis-en Nat. Verh. 4: 147-153 (1928-29).
17. ———. Amst. K. Ak. Wet. Proc. 32: 495-500 (1929).
18. ———. Amst. K. Ak. Wet. Proc. 32: 625-632 (1929).
19. ———. Amst. K. Ak. Wet. Proc. 32: 978-984 (1929).
20. ———. Nieuw Arch. (2) I, 16: 37-39 (1929).
21. ———. Nieuw Arch. (2) I, 16: 40-41 (1929).
22. ———. Nieuw Arch. (2) I, 16: 84-98 (1929).
23. ———. Nieuw Arch. (2) II, 16: 31-38 (1929).
24. ———. Amst. K. Ak. Wet. Proc. 33: 488-490 (1930).
25. ———. Amst. K. Ak. Wet. Proc. 33: 707-710 (1930).
26. ———. Nieuw Arch. (2) III, 16: 80-84 (1930).
27. Miglio, M. Cat. Acc. Gio. Atti. (5) 16, No. 2 (1927-28).
28. Rao, C. V. H. Camb. Ph. Soc. Proc. 26: 72-81 (1929-30).
29. Schaake, G. Amst. K. Ak. Wet. Versl. 32: 626-634 (1923).
30. ———. Amst. K. Ak. Wet. Versl. 35: 142-152 (1926).
31. ———. Amst. K. Ak. Wet. Versl. 35: 374-382 (1926).
32. ———. Amst. K. Ak. Wet. Versl. 37: 311-324 (1928).
33. ———. Amst. K. Ak. Wet. Proc. 32: 109-114 (1929).
34. ———. Nieuw Arch. (2) I, 16: 42-54 (1929).
35. ———. Amst. K. Ak. Wet. Proc. 33: 332-336 (1930).
36. ———. Amst. K. Ak. Wet. Proc. 34: 1116-1123 (1931).
37. Semple, J. G. Camb. Ph. Soc. Proc. 25: 145-167 (1929).
38. ———. London M. S. Proc. (2) 35: 294-324 (1933).
39. Staehelin, H. Enseign. Math. 25: 288-290 (1926).
40. Sorrentino-Giordano, R. Giorn. di Mat. 69: 149-161 (1931).
41. Sweerts, L. Nieuw Arch. (2) II, 17: 137-143 (1931).
42. Todd, J. A. London M. S. Proc. (2) 33: 328-352 (1931).
43. ———. Camb. Ph. Soc. Proc. 29: 52-68 (1932).
44. ———. London M. S. Proc. (2) 36: 172-206 (1933).
45. Turnbull, H. W. Camb. Ph. Soc. Proc. 22: 694-699 (1925).
46. de Vries, J. Amst. K. Ak. Wet. Proc. 22: 641-644 (1920).
47. ———. Amst. K. Ak. Wet. Proc. 22: 943-947 (1920).
48. ———. Amst. K. Ak. Wet. Proc. 26: 126-128 (1923).

49. ———. Amst. K. Ak. Wet. Proc. 28: 459-460 (1925).
50. ———. Amst. K. Ak. Wet. Proc. 29: 664-666 (1926).
51. ———. Amst. K. Ak. Wet. Proc. 29: 1045-1047 (1926).
52. ———. Amst. K. Ak. Wet. Proc. 30: 268-271 (1927).
53. ———. Amst. K. Ak. Wet. Proc. 30: 566-568 (1927).
54. ———. Amst. K. Ak. Wet. Proc. 30: 850-852 (1927).
55. ———. Amst. K. Ak. Wet. Proc. 30: 988-990 (1927).
56. ———. Nieuw Arch. (2) III, 15: 217-218 (1927).
57. ———. Nieuw Arch. (2) III, 15: 219-224 (1927).
58. ———. Nieuw Arch. (2) III, 15: 225-229 (1927).
59. ———. Amst. K. Ak. Wet. Proc. 31: 241-243 (1928).
60. ———. Amst. K. Ak. Wet. Proc. 31: 244-247 (1928).
61. ———. Amst. K. Ak. Wet. Proc. 31: 454-458 (1928).
62. ———. Amst. K. Ak. Wet. Proc. 31: 626-628 (1928).
63. ———. Amst. K. Ak. Wet. Proc. 31: 629-636 (1928).
64. ———. Amst. K. Ak. Wet. Proc. 32: 100-104 (1929).
65. ———. Amst. K. Ak. Wet Proc. 32: 251-253 (1929).
66. ———. Amst. K. Ak. Wet. Proc. 32:562-564 (1929).
67. ———. Amst. K. Ak. Wet. Proc. 32: 1146-1148 (1929).
68. ———. Amst. K. Ak. Wet. Proc. 32: 1149-1151 (1929).
69. ———. Arch. Néerl. (3) 12: 140-142 (1929).
70. ———. Nieuw Arch. (2) II, 16: 39-42 (1929).
71. ———. Nieuw Arch. (2) II, 16: 43-46 (1929).
72. ———. Amst. K. Ak. Wet. Proc. 33: 223-226 (1930).
73. ———. Nieuw Arch. (2) IV, 16: 1-4 (1930).
74. ———. Amst. K. Ak. Wet. Proc. 35: 156-157 (1932).
75. ———. Amst. K. Ak. Proc. 37: 129-132 (1934).
76. ———. Amst. K. Ak. Proc. 37: 201-203 (1934).
77. ———. Amst. K. Ak. Proc. 37: 204-206 (1934).
78. Wren, T. L. Camb. Ph. Soc. Proc. 23: 386-390 (1926-27).
79. Yerushalmy, J. Am. J. 54: 129-144 (1932).

CHAPTER III

SYSTEMS OF LINES IN S_n; IRREGULAR SURFACES

VIRGIL SNYDER

1. Complexes. The linear line complex in S_3 can be generated by the method of Sylvester and by that of Chasles. The former has been generalized to S_n. Weiss [82] generalizes that of Chasles, the transversals of the pair of lines forming an involution on a regulus. This is extended to S_5 and then to S_{2p+1}.

Segre [69] has continued his study of the general quadratic complex in S_4. Todd [78] finds a locus in hyperspace which represents the lines of S_4, similar to the quadric primal in S_5 for the lines of S_3. He applies it to the linear complexes.

Lines of hyperspace are determined by two points or by systems of primals passing through them. The word complex is applied to a system of ∞^3 or more of lines, the exact meaning not being always evident from the context. A linear complex is said to be of the first, second, third, etc., species, according to the dimensionality of the points through which pass ∞^2 (or more) lines.

Thus, if a complex Γ of lines of S_4 is of the third species the locus of its singular points is a surface ξ. If ξ is a plane, the lines of Γ in any S_3 through ξ is a congruence γ without singular lines. Aprile [2] considers a Cremona transformation between ξ and another plane ξ' in the same S_3. The lines joining corresponding points describe a primal of order m, on which ξ is of multiplicity $m-1$. If the Cremona transformation is linear, the locus of points which determine the projectivity is a curve h, which has $h-1$ points on ξ. Relations among various complexes, linear or of orders $k_1, k_2, \ldots$ are obtained, having in ξ a curve of singular points of orders k_i-1. The Cremona transformation has the same characteristics in each S_3 through ξ; *see* the earlier paper by Aprile.[1]

Segre [68] defines a complex Ω as an aggregate of ∞^{2n-3} lines of S_n. The order m is the number of lines of Ω in an arbitrary pencil. A pencil is called tangent if two of the m lines coincide. An $S_{n'}$ in S_n contains $\infty^{2n'-3}$ lines, which constitute the section of Ω by $S_{n'}$. A projective correspondence exists between the tangent cones along a line and the points of the line. The envelope of singular S_{n-1} and the locus of singular points are the same primal.

A 4-complex Γ is the system of ∞^4 lines in (1, 1) correspondence with the points of S_4. Sgroi [70] classifies them into 13 types. A singular point is of the i-species if through it passes ∞^i lines of Γ. A plane is an i-parasite if every line in it is i fold for Γ. Every focal plane touches the focal primal in three points. A 4-complex of type (1, 0, 1) is generated by the lines incident to a given line and a given S_3. (0, 1, 2) is generated by lines incident to two given S_3 and a given S_2, (0, 0, 4) by lines incident to four given S_3.

Todd [77] generalizes the method of generating the tetrahedral complex in S_3 to apply in hyperspace. Consider lines meeting k given S_{r-1} of S_r in points projective with a given system of k points. Proved directly for $r=2$, 3, 4, then by induction. The ∞^2 lines of S_3 meeting five fixed planes in a given range form the axes of a cubic developable containing the five planes. The lines meeting six planes of S_3 projectively generate a quadric touching them.

A complete classification of complexes of S_4 of the third species, including those of order 2 was made by Miglio.[45] Seven types are obtained, making use of the methods and results of Marletta. The same author applies a somewhat similar procedure to the study of ∞^3 complexes of plane curves of order $n>1$ in S_4. The order of the complex is the number of its curves through a point; its class is the number in a primal S_3. The easiest case is that in which the curves are arranged in ∞^2 planes. If Γ consists of curves in ∞^3 planes, the discussion is limited to complexes of order 1 or 2 and to class 1. Various methods of generating such complexes are given.

Marletta [41] proposes a method of studying $(r-1)$-complexes of lines in S_r. Given two superposed S_r, projectively related with distinct fixed points. A complex Γ is defined as the totality of lines joining corresponding points. For $r=3$, the tetrahedral complex results; for $r=4$, that studied by Aprile [Mem. Acc. Zelanti (3), 6 (1908)]. The complex Γ is rational and invariant under Ω; it is of order $r-1$, and can be mapped on the points of S_{r-1} by incidence. Lines of a cone of Γ are transformed into the points of a rational curve. These curves form a complex Λ having the fixed points of Ω for its only base points. It is always possible by means of a Cremona transformation to transform the $\infty^{2(r-1)}$ rational curves through $r+1$ fixed points of Ω into lines of S_r (Bennati [10]). Parisi [49] describes some cubic line complexes of S_4. These have a surface δ of singular points. The congruence in any S_3 has a singular curve. The surfaces of S_4 which generate complexes of order 3 are: rational ruled R_4 with two directrix conics not in same S_3, R_4 with C_3 and C_1 not in same S_3, projection of the Veronese surface. All these are generated by chords. Three such complexes are generated by lines in planes of conics obtained by projecting from a generator of a normal cubic scroll a curve C upon it. The curve is section having as trisecants a C_5 with three generators as chords, and C_4 meeting each once. Every sextic ruled surface R_6, $p=1$ of S_5 has either just two (distinct or coincident) plane cubic curve sections or an infinite number. Every elliptic R_6 with ∞^1 plane cubic curves is the locus of the line centers of three distinct nets of linear complexes (Fano [21]). The same author [22] showed that the ∞^3 variety of the base lines of an ∞^4 linear system of linear complexes of S_5 is birationally equivalent to a general cubic variety of S_4. The Grassmannian M_8^{14} of S_{14} of the lines of S_5 is met by any general S_{14-i}, $i<6$ in M_{8-i}^{14} on which the prime sections constitute a minimum base.

Complexes of space cubic curves which cover S_4 simply have been determined by Privitera.[51.1]

Gherardelli [25.1] proved that if a system of plane curves of dimension $r>1$ is linear, both as a point locus and as a line envelope, the system is composed of conics. There are three types:

$\infty^2|C_2|$ with a common self-polar triangle.

$\infty^3|C_2|$ with a given point and given line for pole and polar.

$\infty^5|C_2|$ of S_2.

Marletta [41.1] discusses complexes of curves in S_r and proves the following theorem: In S_r every $(r-1)$ irreducible complex of order 1 of rational normal curves $|C_r|$ such that $\infty^r S_k$ meet the curves in $k+1$ points for every value of k is generated by those rational normal curves which pass through $r+1$ given independent points and meet a given S_{r-2} in $r-1$ points.

Sapienza [66] maps an $(r-1)$-complex on the points of S_{r-1}. It is shown that the focal surface of order $2(r-1)$ coincides with the envelope of the S_{r-2} of the system. Earlier papers on complexes of lines in S_4 by Marletta are Atti Acc. Gioenia Cat. (3), 3, No. 5 (1894); *ibid.* (5), 3 (1909); 6, No. 1 (1913). Rosenblatt [59] considers varieties determined by matrices (Grassmannian). To lines on a Grassmann variety correspond pencils of linear spaces. On a Grassmann variety are two families of linear spaces. One is composed of S_{k+1} and S_k of S_{k+1} in S_r. The other is composed of S_{r-k} bundles of axis S_{k-1} of S_r. The first has $\infty^{(r-k-1)(k+2)}S_{k+1}$ and the second contains $\infty^{k(r-k+1)}S_{r-k}$.

2. Ruled surfaces and other systems. By means of the degeneration method the formula is obtained for the number of planes which satisfy a certain set of conditions in S_n. The number of planes in S_{2n} meeting $6S_{n-1}$ is $n(n+1)(n^2+n+4)/12$. This agrees with the particular interpretation of the more general formula obtained by Giambelli (Todd [75]). Weitzenböck [83] considers the transversals of four planes in S_5. Three lines meet all of them. The equations of the transversal lines are determined and some invariants concerning them derived. Welchman [86] generalizes the generation of a quadric or more generally a ruled surface in S_3 to study figures generated by $\infty^1 S_2$ in S_4. Details are given for orders 3, 4 and 5. These forms are all rational. This author [84,85] also generalizes the incidence theorem of Segre (1888), which states that the planes of S_4 which meet four lines also meet a fifth. The planes of S_4 meeting a rational normal C_4 in three points and also meeting a line meet a second C'_4 (James [32.1]). Other cases were given by Pieri. Consider systems of $\infty^2 S_2$ in S_4, two through each point. If more pass through a point it is singular. Given two C_4 meeting in six points. The congruence of planes trisecant to one and meeting the other once is examined and all possible degenerate cases given.

Aprile [3] considers line complexes Γ of order 1 in S_r defined by the transversals meeting $x_i S_i$, where i may be any number $\leqq r-2$, and $\Sigma(r-i-1)x_i = r-1$. These are replaced in part by certain rational varieties. A Cremona transformation is defined in S_{r-1} by means of this complex. Given S_{r-1}, S'_{r-1} in S_r. A point P in S_{r-1} has a single line g of Γ passing through it, and this meets S'_{r-1} in the image point P'. It is a direct generalization of the Steiner construction for a quadratic inversion between two planes π, π' in S_3 by means of two skew lines u, v.

Panello [47] applies Segre's representation of lines to S_3, in particular ∞^2 lines of a pencil of quadrics; they are mapped on the ∞^1 points of an elliptic curve C_7 in S_5 in which the pairs of incident lines generate a g^1_2; the four double points are the images of the bundles whose vertices are the vertices of the cones of the pencil.

The same author [48] also considers a pencil of quadrics expressed in terms of the common self polar tetrahedron. The line coordinates p_{ik} of each generator are obtained and the linear complex to which each belongs is determined, then these are interpreted as points of a quadric primal in S_5. The locus is expressed by the vanishing of the determinants of a matrix. The modifications necessary to apply to other pencils are outlined. A representation of S_k contained in an S_m on the points of $[(m-k)(k+1)]$ by means of Grassmann coordinates is expressed by the vanishing of certain determinants of a matrix by Room.[58]

van Slooten [79] derives further properties of inflexional tangents, conflecnodal curves on a general surface of order n in S_3, generalizations of the 27 lines on a cubic surface.

Weiss [81] gives an alternative proof of the invariant defined by four planes in [5]. For the geometry of S_k contained in S_n, *see* Morin.[46]

Algebraic surfaces containing an infinite number of conics, particularly quintic surfaces, are given by Marletta.[41.3]

There are three kinds of non-ruled surfaces of order n in S_{n-1}: (1) non-normal, obtained by projecting surfaces of order n in S_n from an external point; (2) surfaces of Veronese and of Segre projected from two points not on the surface—including the Steiner surface of S_3; (3) normal surfaces with spatial sections of genus 2. The only non-ruled surfaces of order n in S_{n+1} are those of Veronese and of Segre, or those of order n in S_n discussed by del Pezzo. An $(n-1, n)$ can be projected into a quartic surface with a double line in S_3 (del Re [52]). The double points of algebraic primals in S_h are discussed by the same author.[53] Forms of multiple points are discussed by means of polars and of the Hessian. If of the first kind on the given curve it is also of the first kind on the Hessian with the same tangent cone.

Doubly infinite systems of lines in hyperspace are studied by Todd [79] by means of their Grassmannian coordinates, interpreted as point coordinates of a variety in another space. The systems considered are those represented by non-ruled surfaces with rational or elliptic prime sections. It is shown that the line systems generate three dimensional loci which are closely related to each other; they can be derived as projections of known loci.

James [32] defines a system of characteristics in terms of which it is possible to express the orders of varieties generated by lines which satisfy given incidence and other conditions. From this system can be obtained the characteristics of the intersection of two such systems and so on.

Curved n-dimensional varieties containing n systems of straight lines are derived by Barrau.[8] Let x_1, x_2 be the homogeneous coordinates of a point on one line S_1; y_1, y_2, those of another. Then $\xi_{ik}=x_i y_k$ determine a point on an H_2 in S_3. (Identity among ξ_{ik}.) Consider three lines x, y, z; then $\xi_{ikl}\equiv x_i y_k z_l$ define a sextic three dimensional variety in S_7, etc.

3. Geometry of the sphere. The line elements of S_2 can be mapped on the minimal lines of S_3 and the points of S_3 on the minimal lines of another S_3 (Beck [9]). The points of the projective planes can be mapped by a (1, 1) correspondence on the isotropic lines of S_3. Two distinct minimal lines determine a sphere in a conformal space. Hence the lines of S_3 are mapped on the ∞^4 spheres of a conformal space.

A linear congruence of lines of S_3 is mapped on the points of a plane, then this map is interpreted in terms of congruences of spheres by means of the polar theory of the circle by Haenzel.[31] The method is novel; no new results are mentioned. By the use of this and various other methods, Matsumura [42] gives a number of further details on the geometry of circles and of spheres.

4. Surfaces and varieties; irregularity. If a surface of genera 1, with non hyper-elliptic plane sections and canonical curves of genus $p>4$ is not the complete section of a system of quadric primals through it, then the complete section is either (a) a normal three dimensional variety V_{p-2} generated by ∞^1 planes, the surface being cut by a cubic variety through $p-4$ of its planes, or (b), if $p=6$, a conical V_4 which projects a surface of Veronese from a point, the surface itself being cut on V_4 by a cubic variety through one of the quadric cones which project the conics of a surface of Veronese.

The only general type of surface having canonical curves of genus 6 for primal sections is the quadric section of $V_3{}^5$ which represent the lines common to three linear complexes of S_4. It may have a double point in two different ways, each imposing one condition (Du Val [17, 18]).

Fubini [23] showed that the only surfaces of S_3 with collinear plane sections are either cones or quadrics, certain cubics and certain quartics. A surface representing pairs of non ordered points of a curve $p>0$ is irregular. If the curve belongs to an irrational involution, so does the surface. The image of the involution is irregular. If a curve L belongs to a cyclic i_p of order p, without fixed points, and the image of i_p is L', for which $\pi'-1=p(\pi-1)$, then i_p is determined by a cyclic Cremona transformation. The image in S_r is a collineation with axes $\epsilon, \epsilon^2, \ldots \epsilon^p=1$. (Godeaux [27]). Dubreil [15] extended Noether's theorem to surfaces and to varieties. The necessary and sufficient condition that every form F vanish simply on C and at the points M satisfying the condition $F=\Sigma a_i F_i$ is that the curve C_r meets (F_i, F_k) outside of C only in first species.

Generalizations of the theorem of Noether extend to linear systems of higher dimensions of curves having a certain number of common base points. The system which serves to characterize the set of base points is called the ideal (Gambier,[24, 25] Legaut,[33, 34] Dubreil,[12–16] Macaulay,[37] Sperner [73]).

The papers of Dubreil extend the results to derive theorems concerning the base for space curves.

Babbage [5] describes a rational surface of order 8 in S_5, mapped on a plane by $|C_7|: P^3 8P^2$. It has a double line corresponding to C_3, $p=1$ through all nine base points. Four types are discussed. A fifth is added,[6] mapped by $|C_9|: 8P^3P^2$.

Roth[65] derives a series of formulas for the regularity of a surface in S_n by means of the Riemann-Roth theorem $\rho = p_a + n - \pi + 1 - i + \delta$. The effective dimension ρ' is given by $\rho' = p_a + n - \pi + 1 - i$. If F is not mappable on a ruled surface and $\pi > n$, then $|C|$ is non-special and virtually effective. If $2\pi = n + 3$ and $2r > n - 1$, F is either regular or referable. If $|C|$ is virtually effective and $\pi > \pi_r(n)$, then F is regular. Torroja[80] shows that every algebraic biuniform correspondence between two spaces or algebraic varieties of different dimensionality has necessarily exceptional elements. It is not possible to establish such a correspondence without exception between the points of a space or variety and groups of points of another space or variety unless the sums of the dimensions in the two cases are equal.

Baker[7] considers V_3^3 of S_4 with 10 nodes as section of Perazzo's $V_4^{(3)}$ of S_5 and points out its connection with the del Pezzo surface, mapped by $|C_3| : 4P'$. Godeaux[26.1] finds 63 pairs of conics and 63 pencils of rational space quartics on Cremona's quartic surface with a tacnode and an adjacent infinitesimal double line. It can be mapped on a double plane with a non singular C_4 for curve of branch-points. The same author[26.2] also shows that the Humbert quartic is invariant under a collineation of period 7.

Further criteria for the regularity of surfaces have been determined by Roth.[65] In S_3 all surfaces of order < 8, with ordinary singularities, are regular or referable to a ruled surface. In S_4, all non-singular ones of order < 10 are regular or referable; with one exception, also those of order 10; in S_5, $n < 12$; $n = 12$ with one exception; in S_6, $n < 14$; $n = 14$ with one possible exception. The exception in each case is that of an elliptic or hyperelliptic surface, normal in the space considered.

The surfaces whose prime sections contain a g^1_3 have been determined by Bronowski.[11.1] The five existing types of rational surfaces were known before.[11] The present paper generalizes to higher genera. There is one type of genus $p = 3i - 1$ and two for all other genera.

5. Involutions on surfaces. Snyder[72.1] points out the rôle played by the general V_3^3 of S_4 in the study of possible irrational involutions of S_3; 104 references are given.

Sauer[67] discusses the topological problem of mapping one plane on another by means of triangles. Consider $x'_i = \phi_i(x)$, each ϕ_i being quadratic; it is multiple if $|\phi|$ have less than three fixed base points. In the system are pencils with composite conics, forming triangles; 10 cases are considered. The three lines through a point are tangent to a curve of class 3. When $p = 1$, these are expressed in terms of elliptic functions.

Godeaux [28] continues his studies of cyclic involutions having no fixed points. Let I_p be a cyclic involution generated by a Cremona transformation of period $p=p_a+1$ having no fixed points on the given regular surface F. Then Φ, image of I_p, is regular with arithmetic genus π_a. $p_a+1=p(\pi_a+1)$, $p^{(1)}-1=p(n^{(1)}-1)$. The canonical system $|C|$ of F contains $p-1$ isolated curves, invariant under T. The order of I_p is odd and the bigenus of Φ is at least $(p-1)/2$. The linear genus of $F>10$. If $p_a>1$, $P_2>10$. The author continues [29] with irregular surfaces. If F' has $p_g-p_a=q'$, has no irrational pencil of curves and contains an involution I_p, then the image F has irregularity q and the varieties V, V' of Picard have the properties: V contains I_p, the image of which is abelian and belongs to V' or V contains pI_p, the images of which on V' do not meet by pairs. Godeaux [30] also considers an F_8 obtained from a Steiner surface $F(y')$ by a quadratic transformation without base points $y_i=\phi_i(x)$. It has eight triple points and three double quartic curves $p=1$ through them. The same author [30.1] considers the fixed fundamental points of an I_2 belonging to an algebraic surface, generalizing an earlier note.[26.3]

An alternate proof of the theorems of Enriques (Bologna Rend., 1908), that a surface of genera 0 and bigenus 1 represents an involution of order 2 having no fixed points and belonging to a surface of genus 1, is given by Godeaux.[30.2] Conversely, if an algebraic surface of $p_g=p_a=1$ possesses an involution of order 2 without fixed points, the image surface of the involution has genera $p_g=p_a=0$, $P_2>0$ (Godeaux).[30.3] A quartic surface $p_a=P_4=1$ with four ordinary uniplanar double points and three conical points represents an involution of order 4 on a double plane of genera 1, with branch curve consisting of three conics mutually bitangent. Seven planes touch the quartic surface along irreducible conics (Godeaux [30.4]). If an I_2 of S_3 has a finite number of fixed points, this must be 8 (Godeaux [30.5]). Godeaux [30.1] gives further details of an I_2 on an algebraic surface, with fixed points that are also fundamental points. Rational correspondences between two irregular surfaces are studied by the same author.[29] These various results are collected and elaborated in *Les surfaces algébriques non rationnelles de genre arithmétique et géométrique nuls. Exposés de Géométrie IV*. Hermann, Paris, 1934.

6. Characteristics of surfaces. The characteristics of a surface V_2^n in S_4 are all expressible in terms of four of them (Severi). For ruled surfaces various modifications must be made on account of multiply secant lines, inflexional tangents and tangents elsewhere secants (Roth [61]).

Loci which have two systems of generating spaces have been discussed by Richmond and Bath.[57] Systems of S_p and S_q in S_{p+q+1} which intersect, generate the loci in question. For all values there exist in S_{p+q+1}, $\infty^p S_p$ and $\infty^q S_q$ such that any S_p intersects every S_q.

If V_3^3 of S_4 contains no plane, and consequently has less than six double points, it contains only complete intersections (Fano). The primals of any given order cut from V_3^3 a complete linear system (Archbold [4]). Baker [7] supplies a number of formulas concerning chordal curves and trisecant curves for rational surfaces of S_4, with the plane map of some of them.

Osculating and cor-osculating spaces of primals of S_h are discussed by del Re.[55,56] The dual of an osculating S_k in S_h, $h<k$ is a cor-osculator. It is not defined at a point of a line lying on the surface or variety.

7. Generation of curves, surfaces and varieties. A family of surfaces of S_6 is obtained from pairs of points of a curve of $p=3$, each point on the surface F mapping two pairs of points of the curve C (Humbert). The projective model is a surface of order 12 in S_6, with 28 conical double points, and 63 S_5 touching it along a curve of order 6 and genus 1. In S_6 consider the V_3^4, which projects a surface of Veronese. F is cut from this cone by a cubic primal touching the cone in 28 points. It is invariant under a Cremona transformation with four fixed points (Godeaux [26]). A rational surface F of order 7 in S_5 is mapped on a plane by the $|C_4|:9P'$, the base points being also the base of a pencil of cubic curves. The spatial sections f of F are of order 7, genus 3, in S_4, hence normal (del Re [53]). The theorem that if two non-singular plane curves are in (1, 1) correspondence they are projectively equivalent is generalized by Severi [71] to apply to varieties in S_r without multiple points.

Bronowski [11] maps a rational normal surface of S_4 having two skew double lines and obtains its characteristic properties.

An alternate proof of the theorem of Welchman concerning two normal C_4 with six intersections is given by Edge,[19] who derives other properties of the configuration employed.

Further incidence theorems are given by Perazzo.[50] In S_5, the ∞^2 lines incident to three planes meet ∞^1 planes, projective along collinear nets of S_3. The ∞^1 planes dual hereto can be defined as planes incident to four lines.

Incidence is studied by Eiesland.[20] The locus of ∞^{n-2} lines incident to n given S_{n-2} in S_n is a ruled V_{n-1}^{n-1} whose equation may be written as a determinant; it may be generated projectively by $n-1$ linear ∞^{n-2} systems of S_{n-1}. Intersection with quadrics give a double S_{n-2}

on V_{n-1}^{n-1}. If n=5, 10 double lines appear at once, intersection of five S_3. These lines do not intersect. There are 20 other double lines, and five double space C_3. Only $V_4{}^4$ has triple points.

Roth [62] derives criteria for surfaces analogous to those of Enriques-Severi for curves, including the condition that a complete intersection becomes composite. The method is that of envelopes, analogues to that of Chisini.

Steed [74] generalizes the analogue of 27 lines on a cubic surface to configurations in hyperspace. There is an independent proof of Schubert's theorem that F_5 of S_4 has 2875 lines.

On a primal of order $2n-3$ of S_n exists a finite number of lines. The methods of enumerative geometry are employed.

Roth [60] discusses the Jacobian of a system of quadratic forms in S_4, incidentally treating the Cremona transformation defined by four general quadric polarities, with special interpretation of the cases in which the system includes one, two, three or four composite quadrics.

The formulas for simultaneous incidence conditions are expressed as a sum of simple conditions and applied to lines and to planes by Todd,[76] who contrasts this method with the earlier proofs involving degenerate forms (Schubert, Palatini, Pieri, Severi).

BIBLIOGRAPHY

1. Aprile, G. Cat. Acc. Gio. Atti (5) 6, No. 14 (1913). 14 p.
2. ———. Cat. Acc. Gio. Atti (5) 8, No. 28 (1915). 13 p.
3. ———. Giorn. di Mat. 70: 196-219 (1933).
4. Archbold, J. W. Camb. Ph. Soc. Proc. 29: 484 (1933).
5. Babbage, D. W. Camb. Ph. Soc. Proc. 29: 95-102 (1933).
6. ———. Camb. Ph. Soc. Proc. 29: 405-406 (1933).
7. Baker, H. F. Camb. Ph. Soc. Proc. 28: 62-82 (1932).
8. Barrau, J. A. Amst. K. Ak. Wet. Proc. 35: 61-69 (1932).
9. Beck, H. Monatsh. f. Math. 39: 359-370 (1932).
10. Bennati, G. Cat. Acc. Gio. Atti (5) 17, No. 6 (1929). 20 p.
11. Bronowski, J. Camb. Ph. Soc. Proc. 29: 478-483 (1933).

11.1 ———. Camb. Ph. Soc. Proc. 30: 170-177 (1934).

12. Dubreil, P. Paris, Thèse, 1930.
13. ———. Variétés Algébriques.
14. ———. C. R. 196: 1270-1272 (1933).
15. ———. C. R. 196: 1637-1639 (1933).
16. ———. Soc. M. F. Bull. 61: 258-286 (1933).
17. DuVal, P. Rom. Acc. L. Rend. (6) 15: 276-279 (1931).
18. ———. Rom. Acc. L. Rend. (6) 15: 345-347 (1931).
19. Edge, W. L. Camb. Ph. Soc. Proc. 30: 19-26 (1934).
20. Eiesland, J. Pal. Circ. Mat. 54: 335-365 (1930).
21. Fano, G. Rom. Acc. L. Rend. (6) 11: 227-232 (1930).
22. ———. Rom. Acc. L. Rend. (6) 11: 329-335 (1930).

23. Fubini, G. Rom. Acc. L. Rend. (6) 1: 469-473 (1925).
24. Gambier, B. Ann. Éc. Norm. (3) 41: 147-264 (1924).
25. ———. Ann. Éc. Norm. (3) 42: 217-291 (1925).
25.1 Gherardelli, G. Rom. Acc. L. Rend. (6) 19: 140-144 (1934).
26. Godeaux, L. Soc. M. F. Bull. 52: 484-503 (1924).
26.1 ———. Bull. Ac. Romaine 9: 145-147 (1925).
26.2 ———. Bull. Ac. Romaine 9: 147-150 (1925).
26.3 ———. Mathesis 40: 353-360 (1926).
27. ———. Brux. Bull. (5) 19: 674-680 (1933).
28. ———. Brux. Bull. (5) 19: 986-991 (1933).
29. ———. Brux. Bull. (5) 19: 992-997 (1933).
30. ———. Tôhoku M. J. 37: 122-126 (1933).
30.1 ———. Brux. Bull. (5) 19: 1237-1243 (1933).
30.2 ———. Liège R. S. Bull. 1933: 154-157.
30.3 ———. Brux. Bull. (5) 20: 8-11 (1934).
30.4 ———. Liège R. S. Bull. 1934: 1-7.
30.5 ———. Liège R. S. Bull. 1933: 186-189.
31. Haenzel, G. Deutsche Math.-Ver. 42: 75-84 (1931).
32. James, C. G. F. London M. S. Proc. (2) 24: 359-380 (1925).
32.1 ———. Camb. Ph. Soc. Proc. 21: 664-684 (1923).
33. Legaut. Paris, Thèse, 1925.
34. ———. Toulouse Ann. (3) 16: 29-132 (1924).
35. Lob, H. Camb. Ph. Soc. Proc. 29: 45-51 (1933).
36. Longhi, A. Lomb. Ist Rend. (2) 66: 771-785 (1933).
37. Macaulay, F. S. London M. S. Proc. (2) 26: 531-555 (1927).
38. Marletta, G. Cat. Acc. Gio. Atti (4) 8, No. 5 (1894). 7 p.
39. ———. Cat. Acc. Gio. Atti (5) 3, No. 1 (1909). 6 p.
40. ———. Cat. Acc. Gio. Atti (5) 6, No. 1 (1913). 26 p.
41. ———. Note e Eserc. Cat. 5, No. 3 (1928-9).
41.1 ———. Cat. Acc. Gio. Atti (5) 18, No. 2 (1931-2). 10 p.
41.2 ———. Cat. Acc. Gio Atti (5) 8, No. 11 (1914). 7 p.
42. Matsumura, S. Mem. Univ. Jap. 5: 169-206, 247-269, 271-285, 301-316, 347-367 (1933).
43. Mentre, P. C. R. 193: 1315-1316 (1931).
44. Miglio, Maria. Cat. Acc. Gio. Atti (5) 15, No. 8 (1925). 9 p.
45. ———. Cat. Acc. Gio. Atti (5) 16, No. 15 (1926). 12 p.
46. Morin, U. Pad. Sem. Rend. 3: 82-94 (1931).
47. Panello, Alma. Pavia, Diss., 1925.
48. ———. Giorn. di Mat. 67: 168-171 (1929).
49. Parisi, Lucia. Cat. Acc. Gio. Atti (5) 17, No. 9 (1929). 13 p.
50. Perazzo, U. Torino Mem. (2) 54: 149-182 (1904).
51. del Pezzo, P. Nap. R. A. Rend. (4) 2: 91-97 (1932).
51.1 Privitera, J. Cat. Acc. Gio. Atti (5) 18, No. 8 (1931-2). 8 p.
52. del Re, Maria. Nap. R. A. Rend. (3) 31: 43-49 (1925).
53. ———. Nap. R. A. Rend. (3) 31: 108-113 (1925).
54. ———. Nap. R. A. Rend. (3) 32: 131-136 (1926).
55. ———. Nap. R. A. Rend. (4) 2: 7-11 (1932).
56. ———. Nap. R. A. Rend. (4) 2: 53-57 (1932).
57. Richmond, H. W., and Bath, F. Camb. Ph. Soc. Proc. 22: 319-324 (1924).
58. Room, T. G. Camb. Ph. Soc. Proc. 29: 331-346 (1933).

59. Rosenblatt, A. Ann. Soc. Pol. 3: 19-50 (1925).
60. Roth, L. Camb. Ph. Soc. Proc. 25: 268-271 (1928-9).
61. ———. Camb. Ph. Soc. Proc. 25: 390-406 (1928-9).
62. ———. Camb. Ph. Soc. Proc. 29: 88-94 (1933).
63. ———. Camb. Ph. Soc. Proc. 29: 178-183 (1933).
64. ———. Camb. Ph. Soc. Proc. 29: 184-194 (1933).
65. ———. Camb. Ph. Soc. Proc. 30: 4-14 (1934).
66. Sapienza, L. Cat. Acc. Gio. Atti (5) 17, No. 7 (1930). 15 p.
67. Sauer, R. Math. Ann. 106: 722-754 (1931).
68. Segre, B. Rom. Acc. L. Rend. (5) 33: 218-222 (1924).
69. ———. Rom. Acc. L. Rend. (6) 2: 476-479 (1925).
70. Sgroi, M. Cat. Acc. Gio. Atti (5) 15, No. 5 (1925). 26 p.
71. Severi, F. Math. Ann. 109: 4-6 (1933).
72. v. Slooten, Bertha. Utrecht, Diss., 1927. 64 p.
72.1 Snyder, V. Am. M. S. Bull. 35: 607-642 (1929).
73. Sperner, E. Hamb. Univ. M. Sem. 7: 149-157 (1929).
74. Steed, D. V. Cal. Univ. Pub. 1: 425-443 (1924).
75. Todd, J. A. Camb. Ph. Soc. Proc. 25: 42-47 (1928-9).
76. ———. Camb Ph. Soc. Proc. 25: 384-389 (1928-9).
77. ———. Camb. Ph. Soc. Proc. 25: 436-444 (1928-9).
78. ———. London M. S. Proc. (2) 30: 513-550 (1930).
79. ———. London M. S. Proc. (2) 33: 328-352 (1931).
80. Torroja, A. Bologna Cong. Atti 4: 51-60 (1930) [1928].
81. Weiss, E. A. Berlin M. G. Ber. 32: 17-30 (1933).
82. ———. Monatsh. f. Math. 39: 345-358 (1931).
83. Weitzenböck, R. Amst. K. Ak. Wet. Proc. 29: 235-244 (1933).
84. Welchman, W. G. Camb. Ph. Soc. Proc. 28: 275-284 (1932).
85. ———. Camb. Ph. Soc. Proc. 28: 416-420 (1932).
86. ———. Camb. Ph. Soc. Proc. 29: 103-115 (1933).
87. ———. Camb. Ph. Soc. Proc. 29: 235-244 (1933).

CHAPTER IV

CREMONA TRANSFORMATIONS

VIRGIL SNYDER

1. Plane quadratic inversions. Patterson[170] examines the writings of Quetelet, Dandelin and others and concludes that the original discovery of the principle of circular and spherical inversion is due to Steiner, that it was known before 1826 but not published.

Fladt[80] considers the type to which a plane quadratic Cremona transformation belongs, by means of the rank of the determinant $|A_{11}B_{22}C_{23}|$, wherein $\rho x'_1 = \Sigma A_{ik}x_i x_k$, etc. When the rank is zero, $x'_i = 1/x_i$ results.

Berdiales[14] makes a general study of inversion. Contact transformations that are also birational can be expressed in terms of linear elements x, y, p. Let $x' = x'(x, y, p)$. In S_3 they are inherent in a linear complex. An example is given by the harmonic division of the bisecants of a space cubic curve. Planes go into cubic surfaces on which the given curve is asymptotic. Osculating rays go into osculating rays (Fano[76]).

Given a triangle, its inscribed and three escribed circles. Relation of the centers to the vertices is expressed by the vanishing of a certain determinant of order 4 (Herbert[113]). A continuation from[112] determines an inversion from this figure.

Dieudonné[58] obtains the generation of the plane quadratic correlation by means of a pencil of conics.

From the formula connecting a distance PQ and its inverse $P'Q'$ under circular inversion, a number of invariant relations is obtained.

Hadamard[110] gives a detailed account of recent progress in the geometry of inversion in the plane.

Excellent summaries of the field which include practically everything up to the time they were published are given by Emch[71] and by Berzolari.[15] Mehmke[150] gives further properties of transformation by reciprocal radii. Further properties of general inversion are given by Peters[172] and by Dean.[54] A generalization of ordinary inversion is proposed by Shaub.[200] For the relations between inversion and mechanisms for constructing circles and circular curves, *see* Lojciansky,[139]

Rademacher and Toeplitz,[179] Zinder,[239] Delaunay,[56] Holst,[118] Hilbert and Cohn Vossen,[114] Julia,[139] Caratheodory,[35] Kowalewsky,[135, 136] Ginzel,[90] Fubini and Cech,[85] Bompiani,[21] Lane,[138] Weatherburn,[229] Delens,[57] Nakajima [167] and Muhlbach.[166]

Applications of quadratic transformations to the resolution of singularities of surfaces have been made by Sander,[186] Brauner,[23] Dumas [60-64] Hodge,[116] Dohogne,[59] and Paul [171]; *see* in particular Hudson,[121] p. 353-381.

For applications to linear systems of curves, the texts of Enriques and Chisini,[75] p. 192-199 and of Coolidge,[49] p. 401-409 should be consulted; also the following memoirs: Hodge,[115] Segre [190, 191] and Todd.[220] Zelenka [238] constructs the conic determined by five pairs of conjugate points by means of plane quadratic transformations.

Particular quadratic transformations in S_3 have been considered by Gallucci,[86] who gives a particular metric interpretation; by Godeaux,[95] giving more detail to the case of the base conic consisting of two intersecting lines and by Hudson and Wren.[123] The last corrects a statement by Aschieri, repeated by Doehlemann, that the general (2, 2) Cremona transformation in S_3 can not have an invariant pair of homologous points. It is shown that two such pairs exist.

Other cases are given by Godeaux,[97] Strazzeri [214] and Palozzi.[169]

General theorems concerning the Jacobian of a web of surfaces, particularly of a homaloidal web have been given by Villa,[225] who extends them to hyperspace.

2. Types. Given the two fundamental equations of condition for a plane Cremona transformation, the most important question is whether they represent a geometric transformation or are simply arithmetic conditions. Barber [11] employs the method of groups, as developed by Coble. If the number of F points of the Cremona transformation is $\leqq 8$, the group is finite. If this number ≥ 9, it is infinite. Any two consecutive involution conditions generate the group.

Montesano [162] gives the details of a complete series of composition of a plane Cremona transformation in terms of its quadratic components, in the fewest possible number of steps. Further details are given in [162] for the corresponding reduction in terms of Jonquières components.

The problem of types of Cremona transformations with points of only one multiplicity (symmetric forms) has been completely solved, and the arrangement of those with two sets only into definite types has been accomplished. The triadic types have been discussed by Montesano in a number of earlier essays and recently by Farnum.[78] Those having a group of simple base points in each plane are completely determined, and a considerable number of others added. The problem of reduction is considered from another point of view by Torrance,[221] who supplements the results obtained by Montesano.

de Martino [149] constructs series of Cremona tables having no two fundamental points of the same order. The simplest is of order 23. The problem of the determination and location of the isolated invariant points is considered by Brigaglia.[24] The formula given by Marletta [147] does not agree with that of Doehlemann [Math. Ann. 39: 567-600 (1891)] but does agree with that of Zeuthen [C. R. 78: 1553-1556 (1871)]. The author shows how Doehlemann's result should be modified to be correct. For (1, 1) correspondences, $\rho_i=0$, $\theta_i=0$.

A complete enumeration of all types of plane Cremona transformations with not more than nine base points is given by Taylor.[216] Seven types are found.

Scott [189] presents a new method of expressing the relation between x and y near (0, 0) by means of a finite number of terms without the use of infinitesimals nor expansion in infinite series. Extensive use is made of algebraic processes.

3. Reducibility of curves under Cremona transformations. Any plane algebraic curve can be transformed into another in which all the singularities are ordinary, that is, with distinct tangents at each, by means of Cremona transformations. But by this procedure alone, real curves can not always be transformed into real curves. The necessary modifications are supplied in order to make this possible by Carnevali.[36]

The condition that one given plane algebraic curve can be transformed into another given such curve by Cremona transformations is discussed by Keller.[33] The method is to reduce each to its simplest form by Jonquières transformations, as outlined by Castelnuovo, assuming the presence of nested singularities. If $p>1$, the problem is solved. For $p=1$, $|C_{3n}|: 9P^n$ is still undecided. The concept of group of Cremona transformations is that of Coble.

Williams [231] determines a transformation by means of a pencil of curves, when k base points are taken as consecutive on a given curve. The method includes that of mapping certain rational surfaces upon a plane.

4. Systems of cubic curves. The isogonal transformations of a plane cubic into itself by means of central and non-central involutions, considered as Cremona transformations of the whole plane, are discussed by Audinarayanan.[6] Elliptic functions are employed.

Further study of plane curves of genus 1 is made by Machytka.[140] Central and non-central involutions are employed to establish the results:

1) There are $4g^1_2$ and corresponding products which transform the pairs of g^1_2 into pairs of a different given g^1_2.

2) The general E and g^1_2 have four pairs in common.

3) On c_3 exist $\infty^1 G_8$ containing $3E$, $4g^1_2$ and the identity.

5. Finite groups. Finite groups of plane Cremona transformations leave systems of cubic curves through three to eight points invariant, and also the system of sextics through eight double points. There is an essential difference between those with three to six base points on the one hand and those with seven or eight on the other. For $n=7$, if the points are arbitrary, the Geiser involution is the only case possible. Similarly, for $n=8$ the Bertini involution is unique. When the base points have particular positions larger groups result (Bydzovsky [32]).

The properties of G_{16} are further developed by Machytka.[141] All three of the groups G_8, G_{16}, G_{32} of plane quadratic transformations are generated in a different way by the same author.[142] Given five points $\alpha, \beta, \gamma_1, \gamma_2, \gamma_3$ in the plane. They determine a G_4, defined by quadratic I_2 with α, β, γ_i as vertices, and the other two γ_k as a pair of conjugates. The cubics of a pencil determined by these five points and by the four fixed points of $\gamma_1, \gamma_2, \gamma_3$ with α, β a pair of conjugates. The fixed points have a common first tangential, variable with the curve in the pencil. The operations of G_4 generate on each C_3 the three irrational involutions. By adjoining one Cremona transformation G_8 is generated; another generates G_{16} and finally with three, G_{32} results.

Linear systems of plane cubics invariant under a quadratic involution of the first class are given by Machytka.[143] Given a central involution, as the center describes a line a pencil $|C_3|$ results, each curve of the pencil invariant under an abelian G_8. The same author [144] applies the same methods to determine the degenerate Bertini involutions in the plane. If the eight base points P_i are such that the ninth base point of a pencil of cubics determined by them coincides with one of the points P_i, the involution reduces to the Geiser on the other seven. The pencil $|C_6| : 7P_i^2B^2C^2$, B and C arbitrary, contains one composite C_6. For ten positions of C all curves of the pencil are composite.

Marletta [146] considers a (1, 2) correspondence between the points of two planes; it is determined by eight pairs. A (2, 2) correspondence is determined by nine pairs if quadratic and by 17 if cubic.

Meyer employs a system of confocal conics to establish a quadratic Cremona transformation of lines. Associated lines are conjugate as to a certain conic of the system.[151]

The generation of certain Cremona groups by means of projective transformations and one Cremona transformation is discussed by Plamitzer.[175] The same procedure is generalized to S_3.

Theorems of closure (Poncelet, Steiner, Emch) are generalized by means of mapping circles of S_3 on points of a hyperbolic space. If a polygon $X_1, X_2, \ldots$ is circumscribed to a circle M and has its vertex X_i lying on a conic K_1, then an infinite number of such polygons exist, one for every point of the first conic K_i (Toda [218]).

That certain anharmonic ratios are preserved by Cremona transformations is extended by Sibirani [209] to apply to configurations on $C_n: O^{n-1}$ and to $C_n: O^{n-2}(2n-2)O^2$. Deaux [55] derives various properties of conics from the direct circular transformation expressed in terms of the complex numbers z, z'. Only elementary methods are employed.

Another derivation of G_{16} (for the plane) and of G_{32} (for space) is given by Moody.[165] The former is expressed in terms of cubic Cremona involutorial transformation, determined as follows: let (i, k, l, m, n) be any five points in the plane, no three collinear. Let C_i be the cubic curve through i and touching the lines $i\ p$ at p, etc. Let T_i be the cubic involution with i as vertex and leaving C_i invariant. Then $T_i, \ldots T_n$ define an abelian group of order 16, with $T_iT_kT_lT_mT_n=1$. The process can be extended to S_3 but not further, without modification.

Ciani [38] describes various geometric applications of the theory of groups of substitutions. Only linear transformations are considered; each is interpreted as a collineation.

6. Plane involutions. The Jonquières perspective plane involutorial case in which all the simple F points are consecutive to the $n-1$ fold point has been derived and the linear system of curves left invariant by it all determined by Dahy [50] in a simple well-written presentation.

An elementary property of a plane involution was given by Godeaux [96] and an alternate proof given by Deaux.[55] If (AA'), (BD), (CX) and (AA') $(BX)(CE)$ are pairs of elements of a form of the first species, conjugate in two involutions, then X is a fixed element in (AA') (DE).

Ayres [8] generalizes a method given by White [Plane Cubic Curves] to generate plane curves of order n by pairs of involutions on a line.

Curves which are transformed into themselves by plane involutorial transformations are determined by Emch.[72] Pairs of conjugate points are mapped on a surface F_{n+1} of order $n+1$, which is intersected by other surfaces also invariant and the curve of intersection determined.

An involution in the complex plane can be defined by a series of quadratic equations in the complex variable z. Let $f_i=0$ be such an equation, and λ_i a real number. Then $\Sigma\lambda_i f_i=0$ defines a (1, 2) between points (λ) and points z. The two points z define an involution in the complex plane (Grace [109]).

The plane involutional transformations of order 5 are expressed in symmetric form; all the fundamental points are of the same order 2. They are therefore particular cases of the Geiser transformation $c_1 \sim c_8 : 7P^3$. They are of two kinds according to the association of conjugate F points. In one the locus of invariant points is a cubic curve, while in the other it is a straight line. Both are discussed by Bydzovsky,[28, 29] who applies it to the $C_6 : 10P^2$ left invariant.

Further properties have been determined by Sisk.[203]

Jonzon [103] discusses groups of birational transformations which leave elliptic and hyperelliptic curves invariant.

Plane involutions, order $n>2$. The involutions of order greater than 2 are associated with Cremona transformations only when cyclic. Those mentioned here are not of that kind. de Vries [226] considers I_3 of S_2 of the first class, that is, such that an arbitrary line contains one and only one pair of associated points. Two pencils $|c_2|$, $|c'_2|$ of conics with one common base point furnish an example. As P describes a line, $P'P'' \equiv p$ envelopes a curve. There are five types, excluding those in which the points of each group are always collinear. This is followed [227] by a similar study of I_4 of the first class in the plane. The curve of coincidences is a curve of order 5 with the vertex of the pencil as triple point, hence eight tangents to C_5 can be drawn from P. They envelope a curve of class 8. Two pencils of conics and a system of $|C_3| : 5P$ furnish examples.

Sharpe [197] discusses planar I_t by means of $(1, t)$ correspondence defined by systems of curves. It is a summary only.

Pomey [177] constructs two I_3 in S_2 by means of a net of conics. Hollcroft [117] derives the defining systems of curves to generate the (3, 3) planar correspondences.

7. Quadratic space transformations. Inversion (spherical) as a Cremona transformation is discussed by Franceschi.[84] Quadratic (general) inversion is applied by Vycichlo [228] to the solution of problems in descriptive geometry, problems of contact and the construction of space curves by conjugate projections. Besides the tangents at particular points, the osculating conic must be known.

Meyer [152] applies quadratic inversion to problems in closure, generalizing from S_2 to S_3. Given a quadric surface as a point locus and a space cubic curve. Consider a tetrahedron inscribed in the surface, circumscribed to the curve (and dually). There are 12 linear equations of condition, giving one general solution. If more than one exists, then there are ∞^2 solutions.

A quadratic transformation in S_3 is determined by two F points and four pairs of conjugate points, or by four fixed points, the F cone and the principal plane (Mikan [158]). The same author considers a (2, 2) Cremona correspondence in S_3 and its inverse. A bundle of lines is transformed into ∞^2 conics, ∞^1 of which meet their associated lines. The point of intersection generates a space quintic curve of genus 2. A complex $|C_5|$ is defined by ∞^3 vertices. The complex has seven base points. The $|C_5|$ through a point generate a cubic surface. The curves are isologues for ∞^3 Cremona transformations.[156] The (2, 2) Cremona correspondence of S_3 has six fixed points, hence imposes 18 conditions. ∞^8 have the same fixed points. One can be constructed by 12 pairs of points.[155]

Cubic transformations. The (3, 3) Cremona transformations defined by three bilinear equations are classified into 24 varieties by Pobanz.[176]

Vaidyanathanaswami [224] gives an elaborate analysis of cubic transformations associated with a desmic system and applies the results to plane geometry.

8. Hyperspace transformations. Given two pencils (A), (B) in the plane (x) and (A'), (B') in (x') in projective correspondence. If they have only one variable intersection, the process determines a plane Cremona transformation. In the plane every Cremona transformation can be generated in this way. It may also be applied to higher dimensions (Glagoleff [91]).

The G_{32} of correlations and collineations on the Kummer surface (Klein's group) is studied in connection with the general quadratic line complex by Bouligand.[22]

The Cremona group isomorphic with that of the 27 lines on a cubic surface (Black [18]) is a direct generalization of Coble's $G_{n'}$. Coble's P_6^2 of the $G_{6'}$ of S_4 is generated by adjoining the inversion $x'_i=1/x_i$. The $G_{6,2}$ is obtained, isomorphic with the group of the 27 lines of the cubic surface. The order for each transformation for $n=6$ and $n=7$ is found.

Reduction of algebraic surfaces. Geppert [89] gives a basis of classification of algebraic surfaces according to genera and plurigenera into three categories, (a) $P_{24}=0$, (b) $P_{24}=1$, (c) $P_{24}>1$. The first includes those reducible to rational $(p=0)$ and irrational ruled surfaces; (b) those having pluricanonical curves of order zero; and (c) those with proper canonical and pluricanonical curves. If $p^{(1)}>1$, each can be reduced to a definite normal form in hyperspace. Godeaux [108] enumerates unsolved problems in algebraic geometry and suggests that the most important are:

A. Determine the conditions for rationality of an algebraic variety of three dimensions.

B. Classify the involutions belonging to three space.

C. Determine whether the general cubic hypersurface of S_4 is rational. *See also* Snyder.[205, 210]

Godeaux[93] continues his study of surfaces of genera zero and of bigenus 1. The images of sections of F_4 and of F_3 are mapped on S_9. It is shown that the Jacobian of a general web of quadrics is birationally equivalent to a double quadric having a curve of branch points of order 8.

Hudson[122] mentions that many of her results[121] concerning incidence relations for Cremona space transformations had been anticipated by Montesano.[161]

Conte[45] gives an extension of del Pezzo's Cremona transformation in which planes are transformed into quadric cones, to cones of order n having an $(n-1)$-fold base line with $n-1$ fixed tangent planes and $n-1$ isolated points. The feature is the existence of homaloidal systems with variable multiple points.

Conte[46] gives the detailed description of a quartic surface obtained by transforming a general quadric by the (2, 2) Cremona transformation in which the image of a plane in each system is a quadric cone with a variable vertex. The same author[47] gives similar details concerning the transformation of a general cubic surface by the same transformation, and adds the effect of the presence of $1-4$ nodes on the given surface.

9. Special transformations in S_3. Albanese[2] extends his researches of plane algebraic curves to surfaces. By means of a certain system of curves on the surface F, it can be transformed into a surface of S_r with only multiple curves and simple multiple points. Then by a finite series of quadratic transformations, a surface still in hyperspace is obtained, having no singular points.

Formulas in S_3 analogous to those already known for the plane were found by Hudson,[120] including the theorem of Tummarello[223] that the genera of irrational fundamental curves are the same in S_3 and S'_3. The 31 formulas given include F curves of the first and second species but do not include condition of contact. The latter can be removed by Levi's theorem [Atti di Torino 33: 56-76 (1897)].

Various special Cremona transformations of S_3 are concerned with restricted systems; thus, Dohogne[59] gives a synthetic treatment of a transformation that transforms the congruence of bisecants of a space c_3 into a similar congruence for c'_3. $\Phi_{2n+1}: c_3{}^n \sim c'_{2n+1}: c'_3{}^n$.

10. Fundamental elements. Piazzolla-Beloch[179] treats those transformations that have only one fundamental curve of the first kind. Let $f_m{}^{(i)}$ on $|\phi_\mu|$ be the only multiple curve. If $i>\mu/4$, there are only a few possible forms, but if $i \leqq \mu/4$, and f is not a straight line, f must lie on a cone, with vertex at a fundamental point and each generator meets the curve once. If T and T^{-1} each have only one fundamental curve, T can always be reduced to the monoidal type.

Godeaux [94] makes an extensive study of fundamental curves of the second kind. The tangent planes to the surfaces $|\phi_m|$, images of the planes in the other space, may be all variable, while in the other space they may be all variable, part of them fixed, all fixed along the image curve. A new proof is given that if one fundamental curve of the second kind is of multiplicity r, its image is of order kr; if one in the second space is of multiplicity s, and its image of order $k's$, then $k=k'$ This theorem was first proved by Montesano.

The essay of Stuyvaert [215] is first concerned with quadratic and reciprocal transformations in S_n, and then sketches some (1, 1) transcendental transformations.

The stereographic generation of the Jonquières S_2 transformation is treated by Tinto.[217] A ruled surface of order n and having an $n-1$ fold line is given. Two points on it are taken as centers of projection. A point in a given transversal plane is projected upon the surface R_n from one of them and this image point on the surface is then projected back upon the plane from the other.

A further study of the behavior of a homaloidal web of surfaces along a fundamental curve of the second kind is given by Torrey.[222]

Sharpe [196] points out errors in the paper of Godeaux on the Jonquières space transformation.

The cubic line complex associated with the involutorial [3, 3] transformation is mapped on the points of S_3 by Horsfall.[119] Two quadratic and the general linear complex are included as composite forms.

Burniat [26] considers certain fundamental points of space Cremona transformations. Let $|\phi_m|$ of S_3 have a fixed cone at a point O, of multiplicity p, and ∞^2 have O to multiplicity $p+1$, with variable tangent cones. Then T must be monoidal and each cone must have a generator of multiplicity $p-1$. Equations of T are derived but the discussion is synthetic.

Coble [42] discusses the behavior of the rational plane sextic curve and its related Cayley symmetroid under regular Cremona transformations. Under Cremona transformations C_6 has only a finite number of types; with each is associated a symmetroid.

Let I_2 be an involution of pairs of points P_1, P_2 in S_3 and V_3 the map of the involution. Let Γ be a fundamental curve of I_2 such that corresponding to any point P the image is the whole curve. Hence Γ comprises ∞^2 pairs of points P_1P_2 and its image is a surface Φ on V_3. But every point of Γ has itself among its images, hence is a coincidence. The image is therefore on the locus of branch points on V_3.

If $\pi \sim \Psi$, the system $|\pi+\Psi|$ of order $n+1$ is invariant under I_2. The individual surfaces $\Sigma\pi_i\Psi_i$ belong to this system and are themselves invariant. If I_2 is rational, V_3 is rational. Since Γ is rational, the image surface Φ is rational.

The multiplier λ of Montesano is such that the multiplicity s of C on $|\Psi|$ and its order ν are connected by the relation $s=\lambda v$. The curve on Φ as curve of branch points is of order s (Godeaux [106]).

The theory of isolated fundamental points is continued in.[105] Consider F points with variable cones of a pencil for images. $F_n: O^{pq}$, $q \geq 1$, tangent cones $|\Gamma|$ composite, q cones G of a pencil $|G|$. $|\Gamma|$ is of dimensionality 2 or 3. Points in the domain of the first order are transformed into rational curves Γ'_q, points on Γ' are images of generators of the same cone G. Γ'_q is multiple, $p-1$ fold on $|\Gamma'|$ and part of the components of $|\Gamma|$ may be fixed. The transformation may be monoidal.

11. Reduction of surfaces by Cremona transformations. The classification of algebraic surfaces from the standpoint of birational transformations (not Cremona transformations) is proposed by Enriques.[74] Let π be the genus of a plane section, p_g, p_a, P_i have the usual meanings. Ruled surfaces can be transformed into cylinders $f(x, y)=0$ of $P_4=0$, $P_6=0$. Hyperelliptic surfaces $\begin{matrix} 1 & 0 & g & h \\ 0 & 1/\delta & h & g' \end{matrix}$, $p_a=-1$, $p_g=P_4=1$, δ is the divisor, ∞^3 surfaces existing for each value 1, 2, . . . of δ.

If $p_a=P_4=1$, the surface is regular with no proper canonical curve $z^2=f_6(x, y)$, and an F_4 and F_6 of S_4 are examples. If $P_4+P_6>1$, $p^{(1)}=1$, and ∞^1 families exist; if $p^{(1)}>1$, only a finite number of families exists. Multiple planes with various curves of branch points are considered, also integrals of exact differentials and existence criteria.

The system of invariants under a general birational transformation of an algebraic curve, and the smaller set under linear transformations are both known. Are there invariants for a Cremona transformation? Coble [43] points out that, in addition to the canonical C_{2p-2} of S_{p-1} for all birational transformations, for a Cremona transformation there is also a particular rational surface ϕ on which lies the non-hyperelliptic canonical curve. *See also*,[44] p. 23-64.

12. Tables of characteristics. In case a Cremona transformation of S_3 is regular, that is, has no contact conditions along base curves and no fixed tangent cones at isolated fundamental points, it is possible to construct two tables of incidence, analogous to those of Cayley for

Cremona transformations in S_2. This is particularly explicit, provided the given transformation has no fundamental curves of the second kind.

Let $O_1, \ldots, O_{h-1}$ be isolated base points, and $O_h, \ldots, O_p$, the system of base curves of a homaloidal system. Let a surface π be of order n, have each O_k to multiplicity x_k and O_i to multiplicity x_i, $k<h$, $i \geq h$. Then we may write

$$\pi: nx_1 \ldots x_{h-1};\, x_n \ldots x_p.$$

These numbers define the characteristic group for the system of elements $O_1 \ldots O_{h-1}O_h \ldots O_p$. Similarly for each fundamental surface of the first kind (image of a base curve) or of the second kind (image of an isolated F point). This is table A. In the same way, a table B for a curve, image of a line and of each principal curve can be constructed. In (x'), $A'=B$ and $B'=A$, if horizontal and vertical rows are interchanged (Montesano [161]). If only one fundamental curve exists in each space, the two genera are equal.

Similarly, if all the fundamental curves in one space are rational, those of the other are also rational. The sum of the number of base points and base curves is the same in both spaces.

It was then proved by Tummarello [223] that the genus of each fundamental curve in (x) is equal to that of a certain fundamental curve in (x'). The determinants A, B have the property that the reciprocal of A is B, and the reciprocal of B is A, hence all four determinants which define incidence of curves or of surfaces in each space with its fundamental elements are known when any of them in either space are given.

These properties were all found independently by Hudson [120] by a partly different method,[121] p. 206.

13. Series of composition. The fact of the incidence tables giving so much information concerning a regular Cremona transformation of S_3 suggests something analogous to series of composition. Various cases of this are known. Montesano [163] constructs such a system, different from those already known. Start with the (3, 3) Cremona transformation whose base elements are four skew lines r_i and their two transversals u, v. The latter are fundamental lines of the second kind. The inverse is of the same form r'_i, u', v'. A line meeting u, v is transformed into a line meeting u', v', hence the point ranges on such lines are projective. Moreover, the points (r, u) and (r, v) are transformed into the points (r', u') and (r', v') and the projectivity can never be degenerate. Now consider the product T_1 and T_2, such that if T_1 is defined by (u_1, v_1) (u_2, v_2), then T_2 is defined by (u_2, v_2) (u_3, v_3). The fundamental lines r_i of the first kind may all be different or partly the same, hence the order of the product may have different values. The properties of the lines (u_1, v_1), (u_3, v_3) are as before, except that, given a point on u_1 and one on u_3, there may be two lines passing through these points and be images one of the other. Similarly for any number of such component T_i. Conversely, given a T determined by two skew fundamental lines u, v of the second kind, and all the other base lines consisting of lines of the congruence u, v of various multiplicities. If u, v enter symmetrically, then the transformation is expressible as the product of T_i as defined above. In one sense this is analogous to a generalized monoidal transformation. In the regular monoidal transformation a bundle of lines is transformed into a bundle. Here a linear line congruence is transformed into a similar congruence (with distinct directrices). But such a congruence can be transformed into a bundle.

A rational three dimensional variety, representable on S_3 in two different manners, establishes a Cremona transformation between two linear spaces S_3, S'_3. By considering sections and their plane representations, most of the Cremona transformations in which the homaloids of one or both systems are cubics are obtained (Semple [193]).

14. Cremona transformations in hyperspace. The general theory rapidly becomes complicated as the number of dimensions increases. The types involving quadrics in one system have been completely enumerated. Semple [194] gives the detailed study of the (2, 3) case in which the base is an elliptic quintic curve of S_4. In the inverse, the homaloids are Segre 10-nodal cubics, the intersection of two of which is the projection of the quartic Veronese surface of S_5. The base of this system is a quintic surface. The highest order of the inverse, when one system is composed of quadrics, is 7.

A particular (4, 4) transformation of S_4 associated with certain projectively generated loci is given by Todd.[219] Given 5 points A_i in S_4, vertices of projective systems of $|S_3|$. A symmetric (4, 4) Cremona transformation is generated by them. It can also be expressed by four bilinear equations and includes the reciprocal case $x'_i=1/x_i$.

In Cremona transformations of S_r defined by systems of bilinear equations, the base is expressed by the vanishing of several determinants of a matrix. Given a matrix of p columns, q rows, each element a linear function of the point coordinates in S_n. The locus defined by the vanishing of all the determinants of order p has the freedom $pq(n+1)-p^2-q^2+1$ (Room [183]).

When each of the bilinear equations $\Sigma a_{ik}x_ix'_k=0$ has every $a_{ik}=a_{ki}$, the transformation is an involution. The case $n=4$ is considered by Alderton.[3] The description of $|M_3^4|$, conjugate of the $|S_3|$ of S_4, was added by Wong.[235]

Given five planes π_i in S_4; relate the systems (A_i) to them so that sets of corresponding S_3 intersect in these planes. The homaloids then contain the planes. The $M_4^{(4)}$ through all the planes π_i form a homaloidal system and they are all determinantal. That of Veneroni [Lomb. Rend. (2) 34: 640-644 (1901)] is an example. This form may also be defined as the direct generalization of cubic surfaces in S_3 through four skew lines, the transversals u, v then being fundamental lines of the second kind. The $V_4^{(4)}$ through 5 planes form a homaloidal system, the transversals of the planes forming a system of the second kind. It can be generalized to the V_r determined by $r+1$ S_{r-2} as base (Snyder [210]).

For the cubic Cremona transformation a more extensive study of the intersections of cubic hypersurface in S_4 must first be made. A first step in this direction has been taken by Semple.[192] When composite, the residual intersections and the curves of intersection of the components are also discussed. In terms of these a homaloidal system can be built up. Only the main categories of forms are considered.

Archbold [4] derives the involutions of S_3 by means of projections of the involution in S_4 formed by linear systems of quadrics. Two methods are possible: (a) obtain $|H_2|$ of grade 2 and freedom 4, which requires postulation and equivalence, not completely developed for S_4; (b) get the intersection of two quadrics of the system; it is always rational. This can be mapped on a plane; then get the residual intersection $|c|$; then on $|c|$ find systems of grade 2.

The use of Cremona transformations of S_r, $r>3$ to transform surfaces with singular curves and singular points into surfaces having neither has been made by Babbage,[10] who discusses 16 surfaces in S_3 in detail. The Segre-Zeuthen invariant and the Severi invariant, as well as the genera of the surfaces, play the most important part, $I=\mu_2-2\mu_1+3\mu_0-4+\Sigma s$, $\omega-1=\mu_2+v_2-6\mu_1+9\mu_0+\Sigma s(2s-3)$; $\mu_0=$order, $\mu_1=$rank, $\mu_2=$class of the given surface; v_2 is the number of pinch points on multiple curves, s the number of isolated multiple points. A number of properties which contribute to the study of the cubic variety in S_4 are given in.[212] Thus:

Given V_3 of S_4 and a line l not lying on it. Let M be one of the points in which l meets V. A point P of S_4 determines a plane Pl, meeting V in a cubic curve through M. Let M' be the first tangential of M as to this curve. The line $M'P$ meets the polar quadric of M' as to V in M''. The harmonic conjugate of P as to $M'M''$ is P'.

The transformation $P\sim P'$ is involutorial and leaves V invariant. The locus of M' is the cubic surface γ, intersection of V with the tangent space at M.

If l meets V in M, N, P, the three involutions I_M, I_N, I_P generate an infinite discontinuous birational group, leaving V invariant. Each I transforms S_3 into a V_8.

Given two skew lines l_1, l_2 both on V, and a plane π not on V. Given a point y. The $S_3 = \pi$, y meets l_i in P_i. The line P_1P_2 meets V in K. The line yK meets the polar quadric of K as to V in K'. The harmonic conjugate of yKK' is y'. This is also involutorial.

Given a line l on V and a rational surface R_n not on V, having $n-1$ points on l. A point (y) determines a plane y, meeting R_n in one point P. The plane meets V in a conic C_2. Let p be the polar of P as to C_2. The conjugate of y in the harmonic homology P, p is y'. It generates an involution leaving V invariant.

Williams [232] defines a transformation of the lines of S_3 by means of two quadratic reguli. A line t meets two lines of each and these have another transversal. The system is involutorial. The same problem was considered simultaneously and from nearly the same point of view by Clarkson.[40] Mapped on the points of a quadric hypersurface of S_5, each defines a non-linear transformation under which Ω remains invariant.

The monoidal Cremona transformations of S_4 are discussed by Franceschi [82, 83] with numerous analytic details.

Wong [236] generalizes the quartic transformation between two planes by means of the bisecants of a space cubic curve to T_{r^2} between S_r, S'_r, by means of a ruled manifold of $r-1$ dimensions in S_{r+1}.

15. Invariants of surfaces and of varieties. The general theory of transforming a given algebraic curve into another having no multiple points is given by Albanese.[1] Given $f_n(x) = 0$, genus p. Cut it by $|\phi|$, $\lambda > n-3$, in $g_m{}^r$, $r = \lambda n - \dfrac{(n-1)(n-2)}{2}$, $m = \lambda n$. In $g_m{}^r$ is contained $g_{m_1}{}^{r-1}$, $m_1 \leq m-2$, and this contains $g_{m_2}{}^{r-2}, \ldots, g_{m_s}{}^{r-s}$ for every s for which $r-s > 2$. The proof is claimed to be simpler than that of Severi [Ven. Atti. 79:933-943 (1920)]. By projecting the transformed non-singular curve upon a plane, one with only ordinary multiple points (with distinct tangents) is obtained.

Fano [77] continues his studies of algebraic varieties of three dimensions that are regular, having all genera zero and invariant under a continuous group of birational transformations. He shows that such varieties are either rational or can be mapped upon a cone projecting a surface on which $P_3 = 0$, $P_2 > 1$. An involution of S_3 is rational or irrational according as it admits, or not, a continuous group of birational transformations, having a finite number of continuous parameters.

By means of two planes in S_5, Edwards[68] derives the periodic (2, 2) correspondences between two planes. The V_3 of joining lines of corresponding points is cut by S_4. The section is a surface invariant under two birational correspondences. The conditions for periodicity are then derived.

16. Involutorial transformations in S_r, $r \geq 3$. The enumeration of involutorial Cremona transformations given by Sharpe and Snyder[198.1] excluded monoidal types and also those having contact along a fundamental curve. Godeaux[100] considers further various I discussed by Sharpe and Snyder; $|F_4|$ with a single base curve may be defined by congruences of elliptic $|C_4|$, intersections of homologous quadrics of two homologous nets, each with eight base-points. Babbage[9] develops an extension of the same process to higher spaces, such that by projection contact conditions in S_3 can be secured. Given a complete system $|\Phi|$ of hypersurfaces in S_r such that r members of the system meet in two variable points, and a rational variety in S_n.

Two spaces S_{n-r-1}, S'_{n-r-1} can then be constructed such that $|S_{n-r}|$ through S_{n-r-1} meet the rational variety in one point, and $|S_{n-r}|$ through S'_{n-r-1} meet it in two points. If these are cut by S_3, S'_3 respectively, a (1, 2) correspondence is defined. Sections by S_{n-r} contain prime sections of $|\Phi|$. All I_2 defined by cubic surfaces of grade 2 are found. These include ten new ones having contact conditions.

Schoonmaker[188] discusses a non-linear point-plane duality on points of a cubic surface. By using a pencil $|F_3|$, an involution transforming the planes of space into surfaces of order 11 is obtained. Every conic of a linear congruence is transformed into itself. Two similar involutorial transformations, one of order 9 and one of order 11 are developed by the same methods by Snyder and Schoonmaker.[212]

Following a suggestion proposed by Montesano in 1888, Purcell[178] defines certain involutorial Cremona transformations between two spaces S, S' by means of two linear line congruences G in S, G' in S' connected by a birational relation θ, and a reciprocity ϕ between S and S'. Through a point P of S passes a line g of G, to which corresponds a line g' of S' defined by θ. The line g' is cut by π, image of P in ϕ, in P', conjugate of P. Some particular cases of this transformation, in which only a finite number of invariant points appear have been developed by the same methods by Godeaux.[107]

A series of involutorial transformations of space can be generated by a pencil of surfaces $|F_n|: d^{n-2}$, the points on the line being made projective with the surfaces of the pencil. A point P in space determines the surface of the pencil passing through it and this in turn a point D on the rational carrier d. The line PO meets the surface determined by P in P', its conjugate in the transformation. The (1, 1) correspondence can be replaced by (k, 1). The simplest case is that of pairs of planes and points of a line. The case $k=1$ gives a transformation of order 3. This is always rational (Dye [66]), although Montesano strongly was of the opinion that the process included irrational involutions (Snyder [207]). A similar process is applied to obtain those involutorial transformations in S_3 in which the conjugates of planes are $F_n : d^{n-1}$. The equations and the maps of all such involutions are obtained by Dye.[66] The I_4 defined by webs of quartic surfaces with a basis double conic are derived by Dye and Sharpe,[67] thus supplementing and completing those derived by Aroldi.

Sharpe [197] derives the equations and a map on S'_3 of the involutorial transformation obtained and described by Montesano but only for the case in which the planes through the $n-2$ fold line are paired in an involution. That in which each plane is invariant is not discussed.

Lairesse [137] discusses an I_3 determined by polarity as to a net of quadrics through a space cubic curve and a system containing a line. All are included in the general bilinear system, but various properties are more easily determined by the particular cases.

The tetrahedral I_{19} is treated from a slightly different stand point by Godeaux.[100] In a later paper by the same author,[101] three other cases already derived by Sharpe and Snyder are studied by the same method. The I defined by the web of quadrics through six points has been studied further by Bolus [19, 20] when the points approach coincidence in various ways. A systematic discussion of all the cases is made by Hyden.[124]

17. Involutions defined by a (k, 1) correspondence. The projectivity between the quadrics of a pencil and a pair of generators on another as first considered synthetically by Montesano [Rom. Acc. L. Rend 4: 171-178 (1888)] with the equations and numerous properties added by Davis [51, 52, 53] is replaced by a (k, 1) correspondence by Snyder [207] and the system mapped on hyperspace. The same idea is applied to $|F_n| : d^{n-2}$, in which d is a straight line and the residual is noncomposite, by Carroll.[30] For $n=2$, this gives a generalization of a type synthetically discussed by Montesano [Giorn. di Mat. 3: 52-64 (1893)]. The congruence of bisecants of the elliptic quartic curve, base of the pencil of quadrics, is left invariant. For $n>2$ various fundamental straight lines of the second kind appear, each one illustrating Montesano's theorem [Rom. Acc. L. Rend (5), 18: 421-427 (1921)]. Every plane through d is invariant; in each the plane involution is of the Rufini type in which every conic of a pencil remains invariant and a number of simple base points are collinear. The case of the residual base curve being composite is considered by Rusk.[185] A wealth of new configurations of fundamental lines of the second kind is shown.

When the $(n-2)$-fold base line of the pencil is a curve, the number of forms is restricted. This was considered by Black,[16] restricting the residual curve to be noncomposite. A fundamental curve may be both of the first and of the second kind. In a second paper[17] the cases of composite residual curves are all included.

When all the surfaces of the pencil are ruled, the generators through the associated point on the directrix (necessarily rectilinear except in a few cases) are all parasitic and the surface generated by them is a factor of every homaloid. All the surfaces of the proper homaloidal web have the same tangent planes along the residual base curve as those of the surface removed. The transformations furnish examples of the largest contact conditions compatible with a homaloidal web (Snyder.[209, 210])

Involutions in hyperspace. The study of cyclic involutions is extended to three dimensional varieties by Godeaux.[99] The image of a coincidence is a four-fold point on the image variety of the involution. The hyperplane sections of the tangent hypercones is a surface of Veronese for I_2. For the cyclic I_3 either the image of a coincidence is a four-fold point, sections of the tangent hypercone being a rational normal ruled surface of order 4, or it may be a nine-fold point, the associated surface being the image of the system of all plane cubic curves.

James[127] corrects a statement in a previous paper[126] concerning involutions on a normal rational quartic curve. Any I on C_4 can be obtained by one procedure, by means of $\infty^3|S_2|$ each meeting C_4 in three points; *see also* Segre [Rom. Acc. L. Rend. (5), 18: 22-30 (1921)].

18. Involutions on V_3^3 of S_4. The involutions of the general cubic variety of S_4 are important in the study of the rationality or nonrationality of the variety. Four new involutorial transformations were found by Snyder and Lehr[211] which generate infinite discontinuous Cremona groups belonging to the variety, but which do not admit continuous transformations.

Snyder[205] maps V_3 of S_4 doubly on the points of S_3 and discusses various I of S_3 belonging multiply to a linear line complex. Various forms of infinite discontinuous groups are generated which leave V_3 invariant. The general Cremona transformations of S_4 are discussed by Archbold.[5] Isolated points $\sim$ primals; points may go into surfaces or curves and curves $\sim$ primals, surfaces or curves. This is illustrated by (3, 3) transformations defined by bilinear forms.

The invariant loci of the Cremona transformations defined by $x_i = 1/y_i$, wherein y_i are linear functions of x_i, can be expressed in terms of $\pm 2^h$ fixed points arranged in $\binom{h}{2}$ pyramids of 2^{h-1} vertices, each inscribed in the other. Multiply perspective cases were considered by del Re.[181] Inversion as to systems of quadrics in $[h]$ is defined by polarity, with special reference to $h=3$ by del Pezzo.[173]

Let x_i be the non-homogeneous coordinates of one space, y_i, of another joined by $y_i = (a_i x_i + b_i)/(c_i x_i + d_i)$. If $c_i \neq 0$, this can be reduced to inversion by linear transformations; if $c_i = 0$, to identity. It is a generalization of the T_{tet} of Hudson. For $n=3$, interesting conical quadratic types result (del Re [182]).

19. Polarities in S_4. Given four quadric primals in S_4. The polar S_3 of a point meet in a point. An $S_3 \sim \Omega_3^{(4)} : V_2^{10}$. A point of the base line $C_1 \sim C_4$ but if C_1 meets V_2^{10} in r points, r lines come off as components of image C_4. Ω is mapped on S_3 by means of surfaces through intersections of S_3 and V_2^{10}. The base curve is γ_{10} of genus 11, residual being C_6, $p=3$ (Roth [184]).

Wong [233, 237] discusses further properties of Cremona transformations defined by polarities. The first is concerned with the image of a plane in S_r, with more details for $r=4$. The second defines a point-line transformation in S_4 by means of three quadric polarities. Points of a line $\sim$ lines of a cubic ruled surface; points of a plane $\sim$ lines of a V_3^3 whose sections by arbitrary S_3 is a general cubic surface.

By projecting the image of a plane in a [4, 4] of S_4 defined by polarities upon S_3, a sextic surface with a double curve of order 7 is obtained (Wong [234].)

20. Cremona transformations connected with line complexes in S_r. Given an $r-1$ complex Γ of ∞^{r-1} lines in S_r, one line through every point. It may be generated by secants of linear spaces or of rational varieties. A Cremona transformation of S_r can be defined by two primals Σ, Σ', points P, P' belonging to same line of Γ. Hyperspace transformations may be interpreted by lines. Any transformation that leaves the quadric $V_2^{(4)}$ of S_5 invariant can be interpreted as a transformation of lines of S_3. Clarkson [39] considers four pencils of lines in four different planes. A line t meets a line of each pencil, and these four skew lines have a second transversal t'. The involutorial transformation (t, t') is then interpreted in terms of points. Two of the pencils can be replaced by a quadric regulus, or all four by two such reguli.

The same process [41] is applied to a ruled cubic and a pencil of lines and to a ruled quartic surface having a proper double cubic curve and no rectilinear directrix. The latter has a linear involutorial transformation in S_5, the transversal t' being the conjugate of the given one as to the linear complex to which the given ruled quartic belongs. Another variation is given.[41] Given two plane Cremona transformations H_a in the plane a, H_β in β. Let $A \sim A'$ by H_a, $B \sim B'$ by H . A line t meets a, β in A and B. Then t' is defined as the line $A'B'$.

A line to line Cremona transformation in S_3 is defined by two pencils of quadrics by Snyder and Clarkson.[213] Given a fixed plane π. An arbitrary line p is bisecant of a single C_4 of the system and meets π in P; through P passes a second bisecant p' of the same C_4. The involution of $p \sim p'$ has the property that the lines of each conjugate pair intersect.

Haenzel [111] considers transformations of linear line congruences by means of polar S_3. Only collineations and correlations are included. A more extended use of polarity is made by Jolles.[129] Minute study is made of the differences arising according as directrices are real or imaginary. The characteristic numbers arising from such transformations of lines are determined. Three characteristics appear as analogous to class. If all three classes are each equal to one, the lines meet two fixed arbitrary planes. The procedure is interpreted in terms of points on $V_2^{(4)}$ of S_5.

Beck [12] considers involutorial transformations in conformal space. Lines are mapped on points, with a metric interpretation of points on a quadric hyperplane of S_4. It is applied to various cases treated projectivily by de Vries, Clarkson, etc.

21. Cyclic involutions of order $n>2$. In a series of notes Godeaux [101, 102, 108] considers involutions belonging to algebraic surfaces. Every such involution is generated by a birational transformation that leaves the surface invariant. An invariant point is said to be perfect or imperfect, according as the projectivity of associated directions through it is the identity or not. This difference furnishes the principal basis of classification, when the point is transformed into an exceptional curve. Various particular cases are studied in detail, $n=11$.

Certain algebraic space curves can be generated by means of involutions. Given two cubics C_3, C'_3 in S_3, with $[C_3, C'_3]=0$. Let G_1, G_2 be their congruences of bisecants. If a (1, 1) correspondence between G_1, G_2 be given, then a curve is defined as the locus of the point of intersection of ∞^1 conjugate bisecants. Given two projective bundles of lines, a C_3 is generated in this way. This can be generalized to a rational congruence defined by a line and a rational C_n meeting the line in $n-1$ points. If the Cremona transformation is a projectivity, the order of the curve is n_1+n_2-2.

Browne [25] discusses the algebraic problem when the product of two involutions in S_n is an invariant. No geometric applications are given.

22. Double planes. A surface which can be written non-homogeneously in the form $z^2=f(x, y)$ is called a double plane. The curve $f(x, y)=0$ in the plane $z=0$ is its curve of branch points.

Campedelli [34] considers those in which the curve of branch points is of order 10 and finds 28 types of surfaces birationally distinct. Among them are:

1) If p of C_{10} is 36, $p_g=p_a=6$, $P_2=15$, canonical curves are hyperelliptic and represented by double conics.

2) $c_{10}:P_9$. Surface is ruled, $p_a=-4$, $p_g=a$.

3) $c_{10}:P_8$. Rational.

4) $c_{10}:P_7$ (or P_6). $P_2=5$, $p^{(1)}=1$.

5) $c_{10}:P_4P_6$. $p=2$, $p^{(1)}=1$, $P_2=3$. Similarly for $c_{10}:P_7P_3$.

6) $c_{10}:P_7P_3$. Reducible by quadratic transformations $p_a=P_2=1$.

The same author [33] discusses double surfaces having a curve of branch points of order 8. If C_8 is non-singular, $p_g=p_a=p=3$, the canonical curves are hyperelliptic and are represented by lines on the double plane, $P_2=6$.

$C_8:P_4$, $p=2$, $P_2=3$. $C_8:2P_4$, $p=1$, $P_2=1$. These types were found by Enriques [Mem. Soc. It. (3) 10:1-106 (1896)]. C_8 is reducible to a particular C_6 and two lines, $p_a=p_g=0$, $P_2=1$. This case, found by Enriques, has been studied extensively by Godeaux [Brux. Bull. (5) 13:114-133 (1927)]. Several properties have been found by Johnson,[84] who provides alternate proofs and various corrections to the results of Godeaux.

Canonical system compounded with an involution. On an algebraic surface Φ, if the canonical system $|K^*|$ is not invariant under Γ, then (a) within $|K^*|$ are two linear systems $|K_1^*|$, $|K_2^*|$ compounded with Γ, of dimensions r_1, r_2 such that $r_1+r_2+2=p_g$; (b) every special complete linear system, invariant in Γ is contained totally or partially in $|K_1^*|$ or $|K_2^*|$. If Γ has D fixed points, then (c) if π_g is the geometric genus of the image F of T on Φ, either $|K^*|$ is invariant in Γ or $|K_1^*|$ has the dimension p_g+1, $|K_2^*|$ has the dimension $p_g-\pi_3-1$. (d) If $\pi_g=0$, then $|K^*|$ is compounded with Γ (Maroni [148]).

23. Montesano's multiplier. The theorem of Montesano [Rom. Acc. L. Rend. (5) 27:396-400 (1918); 30:447-451 (1921)], that the multiplicities of fundamental curves of the second kind in S_3 of orders m, m' are respectively, $\lambda m'$, λm, in which λ is a positive integer, has been generalized by Marletta [145] to apply to extensive categories of Cremona transformations in S_n.

Given a Cremona transformation T between the points of S_n, S'_n, both contained in the same S_{2n}, and also two spaces O_{n-1}, O'_{n-1} of $n-1$

dimensions. A point P in S_n determines a primal Σ_n through O_{n-1}. Its image P' determines Σ'_n through O'_{n-1}. Let the locus of the intersection Σ_n, Σ'_n be a primal Ω.

If a point F in S_n has the property that Σ_n determined by F and O_{n-1} meets Ω in a variety ψ which is λ fold on Ω, then the variety ϕ' obtained by projecting ψ upon S'_n from O'_{n-1} has the property that each of its points is the image of F. If ϕ' is of dimensionality k, F is said to be a fundamental point of the species k.

Emch [79] defines an involutorial Cremona transformation of S_r by means of a series of quadratic systems $a_r\lambda_r^2+b_r\lambda_r+c_r=0$ of primes.

24. Generalization of types of involutions from S_2 to S_3. Space involutorial transformations represent generalizations of the four types of plane involutions in two ways. If I in space leaves a rational surface F invariant, then the plane map of F has this property and the map is invariant under one or another of the four types. Examples of this are given by Godeaux [100] and by Snyder.[205] If in I each plane of a pencil is invariant, the space involutions of this category may be classified according to the nature of the transformation in each plane. Examples of the Geiser type are given by Carroll [37] and Rusk,[185] those of homology by Snyder,[204] the Jonquières type by Purcell [178] and those of the Bertini type by Snyder and Lehr [211] and by Godeaux.[104]

The last case has been more extensively developed by Sharpe and Dye.[198] Given a cubic surface F and two fixed points O_1, O_2 on it. A conic is uniquely fixed by a point P and touching F at O_1 and O_2. It meets F in P', conjugate of P in I. Each plane through O_1O_2 is transformed into itself. If F is mapped on a plane, the plane map of I is an involution of the Bertini type. Let O_3 be the third point of intersection of the line O_1O_2 with F and O_P be the tangential of O_3 in the section of F by the plane O_1O_2P. The line PP' passes through O_P (Snyder and Lehr [211]). When the cubic surface describes a pencil, the transformation is defined for S_3 (Black [16]). The points O_1, O_2 may be both fixed or one or both describe rational curves projective with the pencil. The process is then carried into S_4 by means of nodal V_3.

BIBLIOGRAPHY

1. Albanese, G. Rom. Acc. L. Rend. (5) 33: 13-14 (1924).
2. ———. Pal. Circ. Mat. 48: 321-332 (1924).
3. Alderton, Nina. Univ. Cal. Pub. 1: 345-358 (1923).
4. Archbold, J. W. London M. S. Proc. (2) 34: 340-359 (1931).
5. ———. Camb. Ph. Soc. Proc. 27: 502-510 (1931).
6. Audinarayanan, S. Ind. M. S. 17: 197-207 (1927-8).
7. ———. Ind. M. S. 18: 215-225 (1930).

8. Ayers, F. Am. M. S. Bull. 34: 143-144 (1928).
9. Babbage, D. W. Camb. Ph. Soc. Proc. 27: 404-420 (1931).
10. ———. Camb. Ph. Soc. Proc. 29: 319-330 (1933).
11. Barber, S. F. Am. J. 56: 109-121 (1934).
12. Beck, H. Amst. K. Ak. Wet. Proc. 32: 490-494 (1929).
13. Bennati, G. Cat. Acc. Gio. Atti (4) 17, No. 6 (1929). 10 p.
14. Berdiales, B. Pub. C. M. Buen. Aires (1924). 39 p.
15. Berzolari, L. Encyklopädie, III C 11, p. 1781-2218.
16. Black, A. H. Am. M. S. Trans. 34: 795-810 (1931).
17. ———. Am. M. S. Bull. 40: 417-420 (1934).
18. Black, H. L. Ann. of Math. (2) 28: 433-450 (1927).
19. Bolus, F. Liège R. S. Mém. (3) 15 No. 2 (1930).
20. ———. Liège R. S. Mém. (3) 16 No. 3 (1931).
21. Bompiani, E. Geom. Diff. (Appendix), p. 720-721.
22. Bouligand, G. Enseign. Math. 28: 204-207 (1930).
23. Brauner, K. Hamb. M. G. Mitt. 6: 1-55 (1928).
24. Brigaglia, A. Pal. Circ. Mat. 53: 310-312 (1929).
25. Browne, E. T. Ann. of Math. (2) 29: 483-492 (1928).
26. Burniat, P. Brux. Bull. (5) 18: 223-233 (1931).
27. ———. Brux. Bull. (5) 19: 163-178 (1932-3).
28. Bydzovsky, B. Prag Ber. No. 11 (1929). 26 p.
29. ———. Rozpravy 38: No. 2 (1929). 17 p.
30. ———. C. R. Cong. math. pays slav. p. 314-317 (1930).
31. ———. Univ. Masaryk Pub. 65: 214-224 (1930).
32. ———. Bol. Cong. 4: 43-44 (1930).
33. Campedelli, L. Rom. Acc. L. Rend. (6) 15: 358-362 (1931).
34. ———. Rom. Acc. L. Rend. (6) 15: 203-208 (1931).
35. Carathéodory, C. Conformal Representation (1932), p. 109-111.
36. Carnevali, E. Lomb. Ist. Rend. (2) 66: 434-438 (1933).
37. Carroll, Evelyn T. Am. J. 54: 707-717 (1931).
38. Ciani, E. Giorn. di Mat. 68: 46-68 (1930).
39. Clarkson, J. M. Am. M. S. Bull. 38: 533-540 (1932).
40. ———. Am. M. S. Bull. 39: 149-154 (1933).
41. ———. Am. M. S. Bull. 39: 434-440 (1933).
42. Coble, A. B. Toronto Cong. Proc. 1: 924 (1928) [1924].
43. ———. Am. M. S. Bull. 34: 142-143 (1928).
44. ———. Algebraic Geometry (1929).
45. Conte, L. Giorn. di Mat. 68: 121-129 (1930).
46. ———. Giorn. di Mat. 69: 113-120 (1931).
47. ———. Giorn. di Mat. 70: 83-87 (1932).
48. Coolidge, J. L. Wash. Nat. Ac. Sci. Proc. 14: 435-438 (1928).
49. ———. Treatise on Alg. Curves. 1931.
50. Dahy, E. Brux. Bull. (5) 13: 42-55 (1927).
51. Davis, H. A. Am. J. 52: 53-71 (1930).
52. ———. Am. J. 53: 72-80 (1931).
53. ———, and Davis, Tyr. Tôhoku M. J. 33: 254-259 (1931).
54. Dean, M. W. Am. J. 52: 586-600 (1930).
55. Deaux, R. Mathesis 46: 264-282 (1931).
56. Delaunay, N. Toronto Cong. Proc. 1: 911-913 (1928) [1924].
57. Delens, P. C. Méthodes de Géométrie Diff. 1927.

58. Dieudonné, J. Mathesis 44: 277-282 (1930).
59. Dohogne, N. Liège R. S. Mém. (3) 14, No. 7 (1928). 20 p.
60. Dumas, G. Bol. Congr. 4: 419-42 (1930-1) [1928].
61. ———. Acta Soc. Helv. 1928: 129-135.
62. ———. Enseign. Math. 27: 324 (1929).
63. ———. Comm. Helv. Math. 1: 120-126 (1929).
64. ———. Comm. Helv. Math. 4: 230-247 (1932).
65. Durairajan, N. Ind. M. S. 18: 106-109 (1929).
66. Dye, L. A. Am. M. S. Trans. 32: 251-261 (1930).
67. Dye, L. A., and Sharpe, F. R. Am. J. 54: 499-504 (1931).
68. Edwards, B. A. London M. S. J. 2: 72-79 (1927).
69. Emch, A. Tôhoku M. J. 24: 68-87 (1924).
70. ———. Am. M. S. Bull. 30: 392 (1924).
71. ———. National Research Council, Bull. 63: 1-55.
72. ———. Tôhoku M. J. 25: 63-76 (1925).
73. ———. Tôhoku M. J. 37: 100-109 (1933-4).
74. Enriques, F. Soc. M. F. Bull. 52: 602-609 (1924).
75. Enriques, F., and Chisini, O. Lezioni, vol. 3 (1924).
76. Fano, G. Bol. Congr. 4: 35-42 (1930) [1928].
77. ———. Rom. Acc. L. Rend. (6) 15: 3-5 (1931).
78. Farnum, Fay. Am. J. 50: 357-370 (1928).
79. Finkoff, S. Ann. di Pisa (2) 2: 59-88 (1932).
80. Fladt, K. J. f. Math. 170: 64-68 (1933).
81. Franceschi, O. Ras. di Mat. 7: 73-78, 109-113 (1927).
82. ———. Giorn. di Mat. 65: 80-98 (1927).
83. ———. Giorn. di Mat. 66: 131-139 (1928).
85. Fubini, G., and Cech, E. Geom. Diff. I (1926); II (1928).
86. Gallucci, G. Note e Eserc. Cat. 6: 19-27 (1931).
87. Gennaro, A. Ann. di Pisa 15 (1927). 50 p.
88. ———. Ras. de Mat. 1: 75-79 (1929).
89. Geppert, H. Deutsche Math.-Ver. 41: 18-39 (1931).
90. Ginzel, I. Sächs. Ber. 84: 51-58 (1932).
91. Glagoleff. C. R. 196: 666-667 (1933).
92. Godeaux, L. Lomb. Ist. Rend. (2) 43: 116-119 (1910).
93. ———. Brux. Bull. (5) 13: 114-133 (1927).
94. ———. Brux. Bull. (5) 15: 314-351 (1929).
95. ———. Mathesis 41: 147-155 (1927).
96. ———. Mathesis 42: 228 (1928).
97. ———. Anais de Soc. Porto 15: 157-168 (1928).
98. ———. Brux. Bull. (5) 16: 762-777 (1930).
99. ———. Brux. Bull. (5) 17: 29-39 (1931).
100. ———. Brux. Bull. (5) 17: 516-526 (1931).
101. ———. Brux. Bull. (5) 16: 450-467 (1930).
102. ———. Brux. Bull. (5) 17: 1131-1150 (1931).
103. ———. Brux. Bull. (5) 17: 1356-1364 (1931).
104. ———. Brux. Bull. (5) 17: 991-1000 (1931).
105. ———. Brux. Bull. (5) 19: 163-178 (1933).
106. ———. Brux. Bull. (5) 19: 503-507 (1933).
107. ———. Am. J. 56: 214-218 (1934).
108. ———. Problèmes non résolues (1933).

109. Grace, J. H. Camb. Ph. Soc. Proc. 24: 210-217 (1928).
110. Hadamard, J. Nouv. Ann. (6) 2: 257-270, 289-320 (1927).
111. Haenzel, G. J. f. Math. 166: 167-181 (1931).
112. Herbiet, V. Mathesis 38: 8-13 (1924).
113. ———. Mathesis 43: 178-179 (1929).
114. Hilbert, D., and Cohn-Vossen, S. Anschauliche Geom. (1932), p. 25ff.
115. Hodge, W. V. D. London M. S. Proc. (2) 29: 145-167 (1929).
116. Hodge, W. V. D. London M. S. Proc. (2) 30: 133-143 (1929).
117. Hollcroft, T. R. Am. J. 49: 553-567 (1927).
118. Holst, C. P. Nieuw Tidschr. 15: 183-190 (1928).
119. Horsfall, I. O. Am. M. S. Bull. 38: 872-878 (1932).
120. Hudson, Hilda P. London M. S. Proc. (2) 26: 453-469 (1927).
121. ———. Cremona Trans. (1927).
122. ———. London M. S. J. 3: 3 (1928).
123. ———, and Wren, T. L. London M. S. Proc. (2) 24: xxiii (1926).
124. Hyden, J. A. Cornell, Diss., 1934.
125. James, C. G. F. Camb. Ph. Soc. Proc. 21: 664-684 (1923).
126. ———. Camb. Ph. Soc. Proc. 22: 24-25 (1924).
127. ———. Camb. Ph. Soc. Proc. 22: 201-216 (1924).
128. Johnson, Roberta F. Am. J. 56: 199-213 1934).
129. Jolles, S. Math. Zschr. 27: 427-452 (1927).
130. Jonzon, B. Upsala, Diss., 1930. 65 p.
131. Julia, G. Principes Géométriques, I (1930).
132. Kahler, E. Math. Zschr. 30: 188-204 (1928).
133. Keller, O. H. J. f. Math. 169: 193-218 (1933).
134. Klima, J. Prag. Ber. No. 3 (1929). 22 p.
135. Kowalewski, G. Natürliche Geom. (1931).
136. ———. Sächs. Ber. 83: 298-304 (1931).
137. Lairesse, J. Liège R. S. Mém. (3) 14-15 (1929-30). 15 p.
138. Lane, E. P. Am. J. 48: 204-214 (1927).
139. Lojcianski, L. G. Lening Ann. Ist. Polyt. 30: 143-150 (1930).
140. Machytka, B. Časopis 53: 136-144, 272-281 (1924).
141. ———. Spisy Praha 27 (1925). 52 p.
142. ———. Rozpravy 34, No. 17 (1925). 28 p.
143. ———. Časopis 58: 60-69 (1929).
144. ———. Časopis 58: 219-225 (1929).
145. Marletta, G. Pal. Circ. Mat. 49: 252-262 (1925).
146. ———. Cat. Acc. Gio. Atti (5) 16, No. 5 *bis* (1928-9). 8 p.
147. ———. Cat. Acc. Gio. Atti (5) 18, No. 2 *bis* (1930). 6 p.
148. Maroni, A. Ven. Ist. Atti 89: 559-566 (1930).
149. de Martino, P. Giorn. di Mat. 68: 130-148 (1930).
150. Mehmke, R. Hamb. M. G. Mitt. 7: 78-89 (1931).
151. Meyer, F. W. Hamb. M. G. Mitt. 6: 352-374 (1928-29).
152. ———. Hamb. M. G. Mitt. 6: 433-443 (1928-29).
153. Miglio, Maria. Cat. Acc. Gio. Atti (5) 18, No. 19 (1932). 10 p.
154. Mikan, M. Rozpravy 37, No. 2 (1928). 8 p.
155. ———. Rozpravy 37, No. 41 (1928). 10 p.
156. ———. Rozpravy 37, No. 42 (1928). 8 p.
157. ———. Rozpravy 39, No. 21 (1929). 18 p.
158. ———. Časopis 58: 70-73 (1929).

159. ———. Rozpravy 40: No. 1 (1930). 6 p.
160. Mlodjowsky, B. K. Rec. Math. 31:341-344 (1924).
161. Montesano, D. Nap. R. A. Atti (2) 18: No. 2 (1926). 25 p.
162. ———. Nap. R. A. Rend. 34: 42-50, 87-97, 123-136 (1928).
163. ———. Nap. R. A. Atti (2) 18 (1926). 44 p.
164. ———. Rom. Acc. L. Rend. (6) 11: 872-879, 1069-1077 (1930).
165. Moody, Ethel I. Am. J. 53: 460-474 (1931).
166. Müllbach, R. Heidelb. Sitzber. No. 11 (1928). 10 p.
167. Nakajima, S. Tôhoku M. J. 31: 21, 36, 210 (1928).
168. Oaricescu, O. Roum. Ac. Bull. 30: 94-98 (1927).
169. Palozzi, G. Rom. Acc. L. Rend. (6) 14: 14-20 (1931).
170. Patterson, B. C. Isis 19: 154-180 (1933).
171. Paul, M. Mathesis 46: 215-218 (1932).
172. Peters, J. W. Am. J. 51: 599-611 (1929).
173. del Pezzo, P. Nap. R. A. Rend. (4) 2: 91-97 (1932).
174. Piazzolla-Beloch, Margarete. Pad. Sem. Rend. 1: 184-195 (1930).
175. Plamitzer, A. Prac. mat. fiz. 35: 1-55 (1929).
176. Pobanz, J. F. Cal. Univ. Pub. 1, No. 19: 401-423 (1924).
177. Pomey, L. C. R. 196: 1451-1453 (1933).
178. Purcell, E. J. Am. J. 55: 381-389 (1933).
179. Rademacher, H., and Toeplitz, O. Zahlen und Figuren (1930).
180. Ramumurti, B. Ind. M. S. 19, N. and Q.: 177-181 (1932).
181. del Re, Maria. Nap. R. A. Rend. (3) 28: 203-211 (1922).
182. ———. Nap. R. A. Rend. (3), 35: 208-217 (1929).
183. Room, T. G. London M. S. Proc. (2) 36: 1-28 (1933).
184. Roth, L. London M. S. Proc. (2) 30: 118-126 (1929).
185. Rusk, Evelyn C. Am. J. 56: 96-108 (1934).
186. Sander, K. Halle, Diss., 1927.
187. Schaake, G. Amst. K. Ak. Wet. Versl. 33: 804-815 (1924).
188. Schoonmaker, Hazel E. Am. J. 51: 439-464 (1929).
189. Scott, Charlotte A. Camb. Ph. Soc. Proc. 23: 206-232 (1923-5).
190. Segre, B. Ann. di Mat. (4) 7: 71-102 (1929).
191. ———. Bol. Congr. 4: 129-131 (1930) [1928].
192. Semple, J. G. London R. S. Trans. 228A: 331-376 (1929).
193. ———. Camb. Ph. Soc. Proc. 25: 145-167 (1928-9).
194. ———. London M. S. Proc. (2) 32: 369-400 (1931).
195. Severi, F. Trattato I, p. 309ff; Vorles., p. 57-65.
196. Sharpe, F. R. Am. M. S. Bull. 30: 393 (1924).
197. ———. Ann. of Math. (2) 31: 637-640 (1930).
198. Sharpe, F. R., and Dye, L. A. Am. M. S. Trans. 36: 292-305 (1934).
198.1 Sharpe, F. R., and Snyder, V. Am. M. S. Trans. 20: 185-202 (1919); 21: 52-78 (1920).
199. Shaub, H. C. Am. J. 49: 367-382 (1927).
200. ———. Am. M. S. Bull. 37: 25 (1931).
201. ———. Am. M. S. Bull. 38: 33 (1932).
202. Sibirani, F. Tôhoku M. J. 24: 243-258 (1925).
203. Sisk, A. Cornell, Diss., 1932.
204. Snyder, V. Am. M. S. Bull. 30: 101-124 (1924).
205. ———. Am. M. S. Bull. 34: 405 (1928).
206. ———. Am. M. S. Bull. 35: 600-630 (1929).

207. ———. Am. M. S. Bull. 36: 89-93 (1930).
208. ———. Bol. Congr. 4: 13-21 (1930) [1928].
209. ———. Zurich Cong. Int. 2: 160 (1932).
210. ———. Am. M. S. Trans. 35: 341-347 (1933).
211. Snyder, V., and Clarkson, J. M. Am. M. S. Bull. 40: 441-448 (1934).
212. Snyder, V., and Lehr, Marguerite. Am. J. 53: 186-195 (1931).
213. Snyder, V., and Schoonmaker, Hazel. Am. J. 54: 299-304 (1931).
214. Strazzeri, V. Pal. Circ. Mat. 54: 314-330 (1929).
215. Stuyvaert, M. Brux. Bull. (5) 10: 597-603 (1925).
216. Taylor, Mildred. Am. J. 54: 123-128 (1931).
217. Tinto, F. J. Edinb. M. S. Proc. 37: 48-50 (1918-9).
218. Toda, K. Jour. Hirosh Univ. A2: 117-126 (1931).
219. Todd, J. A. Camb. Ph. Soc. Proc. 26: 323-333 (1930).
220. ———. London R. S. Proc. A131: 286-306 (1931).
221. Torrance, C. C. Am. J. 53: 911-919 (1931).
222. Torrey, Marian. Am. J. 54: 305-313 (1931).
223. Tummarello, A. Giorn. di Mat. 68: 92-107 (1930).
224. Vaidyanathanaswami, R. Ind. M. S. Supp. 17 (1927). 92 p.
225. Villa, M. Lomb. Ist. Mem. (3) 13: 177-192 (1931).
226. de Vries, J. Arch. Néer (3) 13: 58-67 (1930-31).
227. ———. Arch. Néer (3) 13: 68-71 (1930-31).
228. Vycichlo, F. Časopis 61: 291-300 (1931).
229. Weatherburn, C. E. Differential Geom. 1927, p. 162-164.
230. White, H. S. Am. M. S. Bull. 35: 16-18 (1929).
231. Williams, A. R. Am. M. S. Bull. 35: 605 (1929).
232. ———. Am. M. S. Bull. 38: 554-559 (1931).
233. Wong, B. C. Ann. of Math. 27: 330-336 (1926).
234. ———. Ann. of Math. 28: 251-262 (1927).
235. ———. Am. J. 49: 383-388 (1927).
236. ———. Am. M. S. Bull. 35: 829-831 (1929).
237. ———. Ann. of Math. (2) 30: 547-551 (1929).
238. Zelenka, A. Časopis 56: 86-95 (1927).
239. Zindler, K. Wien. Ber. 140: 399-406 (1931).

CHAPTER V

MULTIPLE CORRESPONDENCES

VIRGIL SNYDER

Multiple correspondences in the plane and space have had in view the (a) method of generation, (b) the curves of branch points and (c) neutral elements.

1. (1, n) correspondences. de Vries[83] generates an I_4 in S_2 by means of two pencils of conics and by systems of cubics. The latter as given is incorrect. The curve of coincidences is $C_5: P^3$, class 14. The tangents from P envelope a curve of class 8. The same author[82] gives an I_3 of S_2, such that an arbitrary line contains one pair of associated points. It can be generated by two pencils of conics with a common base-point. As P describes a line, the line joining its associates envelopes a curve of class n. There are five types, other than those having triads of collinear points.

In a net of conics, one curve of the system exists having a given line of the plane for principal axis. Two principal axes determine a Cremona involution, the line being the element. These include a particular Geiser involution and a Jonquières of order 4. An interesting metrical case is furnished by conics touching the sides of a triangle. The cases mentioned are now produced as the product of three quadratic transformations, which generate an abelian group of order 8 (Rao[59]).

The (2, 1) correspondences between two complex planes is defined by $fw+f'=0$, f and f' being quadratic in z. By it circles are transformed into bicircular quartic curves. It can be expressed in terms of homographies and one circular inversion (Purushottam[57]).

van Dieren[15] studies properties of plane null systems. In an (n, m) system, $u_n=0$, $f(x, u)=0$, $a_x^m u_a^n=0$. An $(n, 1)$ system has n^2+n+1 singular lines.

Ditmarsch[16] maps the line elements (P, l) of a plane a upon the lines of space which touch a monoid. Let $M_n: O^{n-1}$ be a monoid with vertex at O, and OP meet it in R. Then r, the line of intersection of the tangent plane p to M at R and the plane (O, l) is the image of the element (P, l). The representation is (1, 1) with singular elements.

Sharpe[69] gives the various systems of curves for generating I_t in S_2. Particular I_3 in S_2 generated by means of a net of conics are given by Pomey.[56]

Another I_2 defined by the complex variable is given by Grace.[34] Let $f_i=0$ be a quadratic equation in z. If a_i is real $\Sigma a_i f_i(z)=0$ determines a (1, 2) correspondence between points a_i and points z_i.

The algebraic expression of the condition that the product of two involutions shall be an involution is given as an invariant by Browne.[6]

Algebraic space curves can be generated by means of a $(1, \gamma)$ correspondence between the congruences of bisecants of two space cubic curves having no point in common. This is also applied to congruences meeting a line c_i and a rational curve of order n which meets it in $n-1$ points. The curve considered is the locus of points of intersection of those ∞^1 lines of the congruence that meet their associates (Hulin [37]). Godeaux [29, 31, 32] continues the study of cyclic involutions belonging to algebraic surfaces. The differences between perfect and imperfect invariant points are made clearer by transforming them into curves (exceptional curves of the generating transformation). Particular attention is paid to transformation of period 11.

The idea is extended [30] to cyclic involutions on three dimensional varieties.

The image of a coincidence is a four-fold point on the image of the involution. The prime sections of the tangent primal cones is a surface of Veronese for I_2. For I_3 either a rational ruled surface of order 4 (normal) results, or it may have a nine-fold point, with prime sections of the tangent cone being the surface, image of plane cubic curves.

2. (2, n) correspondences. Periodic (2, 2) correspondences between two planes are considered by Edwards [18] by means of hyperspace. Given two planes in S_5 in (2, 2) correspondence. Join corresponding points, generating a V_3. Cut this by an S_4. The surface of section is invariant under two birational transformations. Conditions for periodicity are given.

A type of symmetric (2, 2) correspondence is given by Gherardelli.[27] The (3, 3) correspondences between two planes are derived and classified by Holicroft [35] by means of their generating systems of curves.

Suppose an (l, l') point correspondence in the plane has a curve of fixed points. It is required to determine the number of its isolated fixed points, heretofore known only for the case $l=l'=1$. If of order n, the number x of such points is $x=n+l+l'-\mu-\delta+\gamma$, defined in terms of characteristics of hyperspace (Marletta [26]).

Further comments on the $(2, n)$ correspondence were made by Vycichlo.[84]

3. Double planes. Double planes with curves of branch points of orders 8 and 10 have been discussed by Campedelli.[10, 11] If C_8 has P_4, then $p=2$ and $P_2=3$. If C_8 has $2P_4$, $p=1$, $P_2=1$. These types had already been found by Enriques (1896). Various criteria for reducibility are discussed.

If C_{10} is nonsingular, $p_g=g_a=6$, $P_2=5$. Canonical curves are hyperelliptic, represented by double conics. In all, 28 types are discussed.

Johnson [41] gives the conditions that the Enriques' F_6, having the edges of a tetrahedron for double lines, can be reduced to a double plane with a composite double curve of order 8 consisting of two lines, a conic and a quartic or two lines and two cubics. In each case the surface belongs to three other involutions, making a non-cyclic S_4.

4. Methods of generation. Algebraic surfaces may have their canonical systems compounded with an involution. In the case of a curve having this property, it must be hyperelliptic. A surface may have its canonical system compounded with an I_n, $n>2$. An F_5 of S_3 with a tacnode is an example. Two other examples are given by Segre [68]; an F_5 of S_3 with two oscnodes with a double curve passing through both, $p_a=p_g=4$, $p^{(1)}=9$; I_4 and another F_5 of S_3 with two oscnodes, compounded with I_3.

Other systems compounded with involutions are noticed by Maroni.[49] If the canonical system $|K^*|$ is not invariant under Γ, then (a) with $|K^*|$ are two linear systems $|K_1^*|$, $|K_2^*|$ compounded with Γ, of dimensions r_1, r_2 and $r_1+r_2+1=p_g-1$; (b) every special complete linear system invariant in Γ is contained totally or partially in $|K_1^*|$ or $|K_2^*|$. A number of further properties are obtained.

Properties of curves and surfaces. The condition that a composite algebraic curve can be considered as the limit of an irreducible curve was considered by Albanese.[1] Each component has a Riemann surface. When they have been constructed for all the components and various cuts obliterated, the result is shown to be a Riemann surface. This is illustrated by pencils of cubics and by systems of $|C_6|:8P^2$.

Baker [3] gives an elaborate enumeration of characteristic conditions which satisfy the system of conics in S_3 which satisfy eight conditions.

Carnot's theorem concerning products of distances cut from a triangle by an algebraic plane curve has been extended to algebraic curves in space by Ichida.[38] The products are determined for the planes of a trilateral for curves that are complete intersections.

Cayley's problem of the in-and-circumscribed triangle of a plane C_4 has been studied by the methods of correspondences by Edge.[17] For a nonsingular C_4 the number is 288. The same method is then extended to singular C_4 and to higher algebraic curves with ordinary singularities.

Deaux [13] derives C_4, $p=0$ from C_4, $p=1$ of S_3. The pedal of a point with respect to a system of generators of a hyperboloid of one sheet and the line of striction is a rational C_4, obtained by means of two elliptic C_4. The methods of elementary projective geometry are employed.

Topology is employed to obtain the forms of real algebraic curves by Brusotti.[7] The curve consists of mutually skew circuits Σ and perhaps also of infinite branches. Conversely, every such system Σ admits an algebraic model. The method is that of small variations.

The moduli of a system of plane algebraic curves with given genus make a single irreducible entity. For $p<11$, it is rational. For $p<10$ the variable curve in a continuous system Σ, locus of linear systems having arbitrary base groups has particular moduli. If the moduli are general, the multiple points have restricted positions (Segre [68]).

Ichida [39] gives an elementary proof of Grace's theorem [Camb. Trans. 16 (1898)] on double sixes. The quintic symmetroid is defined as the locus of a point (x) in S_3 for which the primal $\Sigma x_i F_i=0$ in S_4 is a cone, $F_i=0$ being independent primals of order 2 in S_4. This can be written as a determinant $|ahg|=0$, first column linear, second quadratic, third cubic. [Continued from Milne [52]; *see* [53].]

The projective generation of curves and surfaces in S_4 is considered by White.[91] Given two linear forms $\Sigma a_i \beta u_i=0$, $\Sigma u'_i=0$ in S_4, they generate $V_3{}^3$ with six double points. The F_6 common to two $V_3{}^3$ of system has a unique trisecant from every point of S_4. These lines project upon S_3 into a complex of order 8.

Pencils of linear complexes generated by projective systems of linear congruences are discussed by Deaux.[14] Milne [54] considers the contact conics of a quintic. Given such a C_2, 495 C_3 pass through the points of contact and touch C_5 in ten residual points at which it is touched by two further five-tangent conics. The method is to construct the canonical C_8, $p=5$, of S_4 and obtain systems of tangent quadric primals, then project into S_2.

The Steiner surface has an analogue in S_n. Let $x_i=u_i{}^n$, $i=1, \ldots, 4$, each x_i being a linear function of three homogeneous variables. The equations define parametrically a rational surface of order n^2, having three lines of multiplicity n. If n is even, multiple lines are concurrent in a point of multiplicity $3(n-1)$ on the surface. If n is odd, the multiple lines are not concurrent, but coplanar, forming a triangle. If $n>2$, other double curves also exist.[92]

A linear element in S_2 is determined by the line element u and the point x, satisfying the condition $\Sigma u_i x_i=0$. Put $p_{ik}=x_i u_k$. Then $p_{11}+p_{22}+p_{33}=0$. In the matrix $||p_{ik}||$ all quadratic determinants vanish. These nine coordinates fix a linear element (P, l) in S_2. The p_{ik} are mapped on the points of S_8 (Schaake [62]).

5. Point groups on curves and surfaces. Comessatti [12.1] has given an interesting interpretation of Severi's series of groups of points on an algebraic surface, based on the notion of the characteristic series of a surface on a four dimensional variety V_4. The analogy of this to the characteristic series of a curve C on a surface Φ, which is known to be the difference between the canonical series of C and the series cut on C by the canonical system of Φ, might lead one to suppose the characteristic series of F on V_4 could be expressed by linear combinations of known series. It was shown by Segre [69] that it can not.

Hollcroft [36] discusses the involution defined by a web of surfaces $\sum_{i=1}^{4} \lambda_i f_i = 0$. $J(f)$ has $p_a = \frac{1}{3}(2n-3)(4n-5)(4n-7)$. Consider the map $y_i = f_i(x)$. Line in $y \sim C_n^2$, $p = (n-1)(n^2-n-1)$. This is well known for $n=2$.

The surface of branch points L of the involution and the Steinerian S of the web are reciprocal surfaces.

The M_3^6 of S_5, the image of the general cubic line complex of S_3, is irrational. If the complex contains three planes (or three bundles) it is birationally equivalent to the general cubic primal of S_4, and possibly is still irrational. Yet all the genera of M_3^6 are zero. It can be mapped on an involution of S_3. Both M_3^6 of S_5 and M_3^4 of S_4 are particular cases of M_3^{2p-2} of S_{p+1} (Fano [20]).

6. Double fives. Five lines of S_3 constitute a quintuple if each group of four among them have a single transversal. These five transversals also form a quintuple. The aggregate of ten such lines is called a double five.

The condition that a simplex can be inscribed in and circumscribed about a quadric three dimensional variety in S_4 is expressed by a symmetric determinant of order 6. A proper choice of signs of certain quadratic radicals leads to a consistent result (Segre [66]).

The same author [67] showed that this leads to a double five in S_3, proved in another manner by Weitzenböck.[88] The general theory of double fours and of double fives is given in the dissertation of Maurer.[57]

Gambier [24] applies the same idea to a system of circles and of spheres, including a metric generalization of the Morley-Peterson property to S_n; this was also applied to orthogonal systems by methods of pure geometry.[26] Another application [25] is made to the exceptional case $n=4$ in the problem of tangent cycles discussed by Tzitzeica [79] and Barbilian.[4] A new proof of double fives by parametric representation is given by Weiss.[87] Ciani [12] shows that the whole idea of double fives is due to him. [*Cf.* Ann. di Mat. (3) 8:1-21 (1902)]. An alternative proof for the characteristic property of a double five is given by Schaake.[62.1]

Fano [21] proves that an involution in S_3 is rational or irrational according as, considered as an algebraic three dimensional variety M_3, there is a finite continuous group of birational transformations under which M_3 is invariant.

7. Reducibility. Burniat [8] discusses a (1, 16) transformation in connection with a net of desmic plane quartic curves. Purushottam [57] and Stolzenberg [74] consider multiple correspondences between curves, defined by quadratic relations. Ragonesi [58] proves that the quadratic (m_1, m_2) transformations between two planes can be of indices (1, 1), (1, 2), (2, 2), (1, 3) and (1, 4) only. Veneroni [80] studies the symmetric (2, 2) correspondences under the restructive assumption that a general line of the plane contains one and only one pair of homologous points. The focal surface of a congruence of conics in S_3, defined by the incidence of homologous elements in a quadratic correspondence, is determined by Souris.[71–73]

Krames [43] employs quadratic transformations in connection with kinematic systems of circles. *See also* Federhofer [22] and Mayer.[51]

An application to non-euclidean geometry is made of (1, 2) transformations in S_2 and S_3 by Ota.[55]

Warnock [86] considers some of the groups of line configurations by determining transformations that leave the quadratic hypersurface in S_5 invariant. It is not exhaustive.

8. Application to lines, spheres, etc. The cubic line complex determined by the generators of the net of quadrics through eight points is discussed by Godeaux.[33] Particular nets may have more than eight singular points. Complexes defined in this way may have 8, 9, 10 or 12 principal points, but cannot have just 11.

Study [75] proposes a simplified and more logical basis for the Lie line-sphere transformation by replacing the two bilinear equations proposed by Lie by four such equations, denoting the planes (of one system) on Ω_4^2 in S_5, namely:

$$\begin{aligned} 0+z_4\xi_1+z_1\xi_2-z_3\xi_3&=0,\\ -z_4\xi_0+0-z_2\xi_2+z_0\xi_3&=0,\\ -z_1\xi_0+z_2\xi_1+0-z_4\xi_3&=0,\\ z_3\xi_0-z_0\xi_1+z_4\xi_2+0&=0, \end{aligned}$$

and a corresponding one for the other system of planes.

Bennati [5] discusses two remarkable ∞^r complexes of S_4, one of lines and one of normal C_r.

Kol [47] defines congruences of conics in S_4 by means of enumerative geometry. James [40] associates complexes of conics in S_4 with the Weddle surface. A conic of the system, one through each point, meets a directrix C_6, $p=2$ in four points. Those conics of the system which touch a given S_3 have a Weddle surface for locus of points of contact. A Cremona transformation is determined. Through P passes one C_2 of the system and its four points on C_6 determine a pencil of conics. The polar lines of P as to this pencil meet in P'. It is (4, 4) : a plane is transformed into a surface of order 6.

Elesland [19] defends the method of Lie as opposed to that of Study, maintaining that mistakes would have been avoided by a more careful use of the analysis involved.

A further study of complexes of spheres is made by Frank.[6] All the irreducible invariants of three quadric surfaces are directly interpreted as invariants of the concomitant linear complexes involved (Williamson [93]).

9. Systems of curves. (*a*) *Conics.* Calapso [9] gives a synthetic summary of a problem concerning the osculating linear complexes of conics to the asymptotic lines of a surface, usually treated by the methods of differential geometry. A metric application to obtain focal properties of a conic is discussed by Jonescu.[14]

Thomsen [77] raises the question: what is the largest number of singly infinite systems of conics possible on an algebraic surface? On a cubic it is 27, in the pencils of plane sections through the lines. Results are obtained largely by tensor analysis from the condition that each curve shall meet a line twice (de Vries [81.1]).

(*b*) *Cubics.* Godeaux [28] considers the congruence of space C_3 having five given bisecants.

(*c*) *Quartics.* (*a*) $p=1$. A congruence of elliptic space quartics is determined by five points and two skew bisecants. It can be generated by two pencils of quadrics and further curves of the system obtained by quadratic transformation (de Vries [82]).

The doctoral dissertation of Tol [78] treats various systems of elliptic space quartic curves.

de Vries [81] discusses two congruences of rational C_4 of S_3.

Kol [44, 45] derives various characteristics of systems defined by five points. The points are the vertices of a simplex, with regard to which the transformation $x'_i=1/x_i$ is employed. Lines are transformed into normal quartics through the vertices.

(*b*) $p=0$. A congruence of rational C_4 is generated by the residual intersection of cubic surfaces of a pencil and quadrics of a pencil, two fixed lines being common to both. This system was investigated by de Vries.[81] This author generalizes to congruences of rational curves of odd order, intersection of cones of two pencils, then mapped on the points of a plane through their vertices. For $n=5$, it is the congruence of Reye.

Schaake [62, 63] discusses ∞^2 systems through seven points and ∞^6 through five points, without further conditions. A study of the number of rational quintics through ten given points is given by the same author.[64]

Quadratic congruences of planes in S_4 were considered by Welchman [89] in connection with fifth incidence theorems. Given two curves C_4, C''_4, the planes that meet the first in three points, the second in one, also meet a third. These planes form a quadratic congruence. The C_4 may be replaced by two C_2; let two C_1 also be given, meeting both. The planes meeting all will meet another conic. The configuration is symmetric; there are three other systems of planes meeting three conics and two lines (Welchman [90]).

An extension of the theorem of the fifth associated line to spaces of more dimensions is made by Room.[60]

A transformation of the ∞^6 $|C_4|$, $p=0$ through five points can be made from the lines of S_4 by $x'_i=1/x_i$. These $|C_4|$ project into S_3 from a fixed point T. A curve C_4 through T of $S_3 \sim$ line of S_4 meeting a fixed C_4 (Schaake [64]).

An alternate discussion was given by Salmon-Rogers in Geometry of Three Dimensions, vol. I, p. 365.

BIBLIOGRAPHY

1. Albanese, G. Pal. Circ. Mat. 52: 105-150 (1928).
2. Aprile, G. Giorn. di Mat. 70: 196-219 (1932).
3. Baker, W. E. London M. S. Proc. (2) 36: 96-116 (1933).
4. Barbilian, D. Buch. Bull., p. 12-16 (1929).
5. Bennati, G. Cat. Acc. Gio. Atti (5) 17, No. 6 (1930). 6 p.
6. Browne, E. T. Ann. of Math. (2) 29: 483-492 (1928).
7. Brusotti, L. Bol. Congr. 4: 139-145 (1930) [1928].
8. Burniat, B. Liège R. S. Mém. (3) 16, No. 7 (1931).
9. Calapso, R. Brux. Bull. (5) 18: 68-70 (1931).
10. Campedelli, L. Rom. Acc. L. Rend. (6) 15: 203-208 (1931).
11. ———. Rom. Acc. L. Rend. (6) 15: 358-362 (1931).
12. Ciani, E. Rom. Acc. L. Rend. (6) 17: 215-217 (1933).

12.1 Comessatti, A. Rom. Acc. L. Rend. (6) 16: 555-560 (1932).

13. Deaux, R. Boll. Congr. 4: 177-184 (1930) [1928].
14. ———. Bol. Congr. 4: 185-192 (1930) [1928].
15. v. Dieren, Catherine. Utrecht, Diss., 1928, 78 p.
16. v. Ditmarsch, A. J. Utrecht, Diss., 1928, 31 p.
17. Edge, W. L. London M. S. Proc. (2) 36: 142-172 (1933).
18. Edwards, B. A. London M. S. J. 2: 72-79 (1927).

19. Eiesland, J. Tôhoku M. J. 32: 59-80 (1929-30).
20. Fano, G. Bol. Congr. 4: 115-122 (1930) [1928].
21. ———. Rom. Acc. L. Rend. (6) 15: 3-5 (1932).
22. Federhofer, K. Graphische Kinematik (1932).
23. Frank, P. Hamb. M. G. Mitt. 7: 119-132 (1932).
24. Gambier, B. C. R. 190: 157-159 (1930).
25. ———. Bull. Sci. M. 55: 75-96 (1931).
26. ———. J. de Math. (9) 11: 377-387 (1931).
27. Gherardelli, G. Boll. d. U. M. It. 6: 184-187 (1927).
28. Godeaux, L. Amst. K. Ak. Wet. Proc. 31: 1001-1004 (1928).
29. ———. Brux. Bull. (5) 16: 450-467 (1930).
30. ———. Brux. Bull. (5) 17: 29-39 (1931).
31. ———. Brux. Bull. (5) 17: 1131-1150 (1931).
32. ———. Brux. Bull. (5) 17: 1356-1364 (1931).
33. ———. Brux. Bull. (5) 18: 570-575 (1932).
34. Grace, J. H. Camb. Ph. Soc. Proc. 24: 210-217 (1928).
35. Hollcroft, T. R. Am. J. 49: 553-564 (1927).
36. ———. Am. M. S. Trans. 35: 855-868 (1933).
37. Hulin, G. Liège S. R. Mém. (3) 14-15 (1928-30), 14 p.
38. Ichida, A. Tôhoku M. J. 35 II: 276-279 (1932).
39. ———. Jap. M. J. 9: 63-65 (1932).
40. James, C. G. F. Camb. Ph. Soc. Proc. 22: 201-216 (1924).
41. Johnson, Roberta F. Am. J. 56: 199-213 (1934).
42. Jonescu, D. V. Mathesis 46: 398-400 (1932).
43. Krames, J. L. Zyklographie, p. 257-263 (1929).
44. v. Kol, J. W. A. Amst. K. Ak. Wet. Proc. 31: 620-625 (1928).
45. ———. Amst. K. Ak. Wet. Proc. 31: 334-339 (1928).
46. ———. Amst. K. Ak. Wet. Proc. 32: 978-984 (1929).
47. ———. Nieuw Arch. (2) 16 II: 31-38 (1929).
48. Marletta, G. Cat. Acc. Gio. Atti (5) 16, No. 2 *bis* (1928), 6 p.
49. Maroni, A. Ven. Ist. Atti 89: 559-565 (1930).
50. Maurer, A. Bonn, Diss., 1929.
51. Mayer, Joanna I. Tôhoku M. J. 36: 1-12 (1932).
52. Milne, W. P. London M. S. Proc. (2) 33: 165-176 (1931).
53. ———. London M. S. Proc. (2) 36: 207-216 (1933).
54. ———. London M. S. Proc. (2) 36: 217-234 (1933).
55. Ota, T. Tôhoku Sc. R. 26: 187 (1926).
56. Pomey, L. C. R. 196: 1451-1453 (1933).
57. Purushottam, S. Ind. M. S. 17: 102-105 (1928).
58. Ragonesi, R. Cat. Acc. Gio. Atti (5) 18, No. 23 (1931). 17 p.
59. Rao, A. N. Ind. M. S. 18: 68-87 (1929).
60. Room, T. G. London M. S. Proc. (2) 31: 455-486 (1930).
61. ———. London M. S. Proc. (2) 36: 1-28 (1933).
62. Schaake, G. Bol. Congr. 4: 45-50 (1930) [1928].
62.1 ———. Amst. K. Ak. Wet. Proc. 31: 715-717 (1928).
63. ———. Amst. K. Ak. Wet. Proc. 32: : 487-489 (1929).
64. ———. Amst. K. Ak. Wet. Proc. 33: 332-336 (1930).
65. Segre, B. Bol. Congr. 4: 129-132 (1930) [1928].
66. ———. Rom. Acc. L. Rend. (6) 2: 539-572 (1925).
67. ———. Rom. Acc. L. Mem. (6) 2: 578-592 (1927).

68. ———. Rom. Acc. L. Rend. (6) 16: 316-320 (1932).
69. ———. Rom. Acc. L. Rend. (6) 17: 917-918 (1933).
70. Sharpe, F. R. Am. J. 50: 627-635 (1928).
71. Souris, R. Brux. Bull. (5) 17: 1365-1377 (1931).
72. ———. Brux. Bull. (5) 18: 71-82 (1932).
73. ———. Brux. Bull. (5) 18: 165-169 (1932).
74. Stolzenberg, R. Säch. Ber. 82: 57-67 (1930).
75. Study, E. Berlin Monatsber. 1926: 360-380.
76. Telling, Miss H. G. Camb. Ph. Soc. Proc. 29: 470-477 (1933).
77. Thompsen, G. Math. Ann. 108: 260-295 (1932-3).
78. v. Tol, M. G. Groningen, Diss., 1932. 132 p.
79. Tzitzeica, G. Buch. Bull. 1: 16-21 (1929).
80. Veneroni, E. Amst. K. Ak. Wet. Versl. (4) 28: 335-339 (1925).
81. de Vries, J. Amst. K. Ak. Wet. Proc. 31 II: 1071-1075 (1928).
81.1 ———. Amst. K. Ak. Wet. Proc. 32: 100-104 (1929).
82. ———. Amst. K. Ak. Wet. Proc. 32: 251-253 (1929).
83. ———. Arch. Néerl. (3) 13: 58-67 (1930).
84. ———. Arch. Néerl. (3) 13: 68-71 (1930).
85. Vycichlo, F. Časopis 62: 20-23 (1932).
86. Warnock, W. G. Tôhoku M. J. 36: 303-319 (1931).
87. Weiss, E. A. Amst. K. Ak. Wet. Proc. 35 II: 969-973 (1932).
88. Weitzenböck, R. Amst. K. Ak. Wet. Proc. 31 I: 133-137 (1928).
89. Welchman, W. G. Camb. Ph. Soc. Proc. 28: 275-284 (1931-2).
90. ———. Camb. Ph. Soc. Proc. 28: 416-420 (1931-2).
91. White, F. P. Camb. Ph. Soc. Proc. 21: 216-227 (1922-3).
92. Williams, A. R. Am. M. S. Bull. 39: 621-626 (1933).
93. Williamson, J. Edinb. M. S. Proc. 42: 29-40 (1923-4).

CHAPTER VI

SURFACES AND VARIETIES

AMOS H. BLACK

1. Cubics in S_3. An alternate proof of the property of a double six of lines on a cubic surface is given by Kubota[42] by methods of descriptive geometry. A totally different one was given by Todd.[84] Those double sixes whose cubic surface and cubic envelope are apolar are determined by Bishara[7] in a new way. The relation between a double six and its Schur quadric, in regard to which the pairs of double six are polar lines, is proved by elementary methods by Jolliffe.[41] The arrangement of the 27 lines in threes in the form of a determinant from which three double sixes can be at once determined, and from them the Schur quadrics, is given by Bishara,[8] suggesting some alterations in Dixon's earlier discussion. Desmic quartics are generated by the quadrics.

Piazzolla-Beloch[49] shows that C_5, $p=1$ on cubic surfaces have concatenated circuits. No quadric can have such curves but on surfaces of order >2 various curves of order 5 exist with this property; *see also.*[50] Timms[83] projects an F_2 of S_9 into S_3, giving all forms of cubic surfaces with a node. A systematic and comprehensive classification is included. In a somewhat similar manner, Roth[58] projects the apparent double F of a cubic manifold in S_5 upon S_4. This surface now has a triple curve and has improper nodes. These are the surfaces of S_5 which represent the congruence of bisecants of a C of S_3.

Various metric properties of the four-nodal cubic surface are described by Saddler.[68] Ursell[88] discusses the Cayley problem of seven lines; in symbolic notation the determinant of order 35 is expressed and reduced. The seven lines are mutually skew and the postulation of each is five. Given the seven lines, if three are selected, two elliptic C_6 can be found, having these three as quadrisecants and the other four as trisecants (Richmond[52]). The theorem of E. Toeplitz that, "if three quadrics are polars of three points as to a cubic surface, then their (2, 2, 2) invariant vanishes," is generalized to space of any odd number of dimensions by Anderson.[2] The method is that of algebraic symbolism.

2. Quartics and ruled surfaces in S_3. The ruled quartic surface with a non-composite cubic for double curve always belongs to a linear complex, which may be special or non-special, and has a finite number of torsal generators (Deweck [17]). The bisecant curves, that is, curves meeting every generator of a ruled surface twice, are further studied by Welchman,[90] extending the formula of Segre for such curves. The same author [89] generalizes the concept of incidence scrolls. In S_3, lines meeting three fixed mutually skew lines generate a quadric surface. In S_4, lines meeting five mutually skew planes generate a ruled surface $p=1$. Not all scrolls can be generated in this way.

Schaake [69] extends the formula for the order of the surface of trisecants of an algebraic curve of order n in S_3 to the case in which the curve has v_i points of multiplicity i and h apparent double points. The order of the ruled surface is $(n-2)\left\{h-\frac{n(n-1)}{6}\right\}+\frac{1}{6}\Sigma v_i i(i-1)(i-2)$.

Room [53] determines the quadrics with regard to which a rational scroll in S_3 is its own polar reciprocal. That ∞^6 correlations in S_3 transform a generator into a generator is not true (*see* Waerden).[91]

It is shown by Mathews [46] that the bitangents to a desmic surface at an arbitrary point on it are in two triples; those of one belong to a cubic cone and those of the other to its Hessian. Emch [23] uses the Geiser S_3 transformation on eight points to obtain some further theorems concerning the Weddle surface. Danielsson [15] considers a line $x_1y_1=x_2y_2=x_3y_3$ through the triple point of a Steiner surface. Any plane through such a line and a double line is always a tangent plane, with point of contact on the double line. The other tangent plane at that point describes a pencil as the point describes the line. Axes are conjugate in the quadratic involution of the bundle of lines. They coincide in four positions. At points where the lines meet F_4, tangent cones from them have plane C_4: 3 cusps for residual intersection.

By means of a quadratic transformation in the plane of projection of a Steiner surface from the triple point, a (1, 1) transformation leaving the surface invariant is obtained, expressed in terms of the projective lemniscate on it (Ciani [13]). The triple points of an algebraic surface are studied by means of quadratic Cremona transformations by Paul.[48] Danielsson [16] discusses double fours and double fives on a quartic surface having a double line or double conic. A plane C_4 has groups of eight double tangents which have the same property for $\infty^1 C_9$. There are four points through which two double tangents can be drawn. Wiman [95] expresses the tangent to a C_4, $p=1$ of S_3 by means of elliptic functions and considers the transformation $u'=u+c$. The locus of the intersection of osculating planes at corresponding points is an elliptic ruled surface of order eight. The tangent developable is included in the family, $c=0$.

It had been touched upon by Sturm (Verwandtschaften IV, p. 208), la Gournarie (Recherches, 1867) and Cayley [Camb. Trans. 11.2: 277-289 (1869); Edin. Trans. 25: 1-110 (1868)].

In 1892 Wiman, in his doctoral dissertation at Lund, gave a comprehensive study of sextic ruled surfaces. Various additions and alternate discussions have been made from time to time. He [96, 97] now comes back to the same subject and reconsiders the whole problem from the later standpoint. With the exception of a few tacnodal forms, supplied by Snyder (Am. J., 1901-1905), the earlier list was found to be correct. Although Snyder had no criterion for completeness, his list is the only one containing all the forms thus far found. With the new memoir of Wiman, the problem is thus completely solved. The most systematic treatment, arranging the results both as point loci and as envelopes, is that of Edge.[22]

The theory of isolated singular points of algebraic surfaces including an interpretation of the Zeuthen-Segre invariant $I=\delta-\sigma-4p$ for points of multiplicity higher than two, including uniplanar, tacnodal, etc., points was extended by Babbage.[3]

Turnbull [87] derives the condition that eight points in S_3 are associated without the use of a coordinate system. The result is expressed in algebraic vector symmetric form. A similar extension to surfaces having triple curves was made by Roth.[57] A three-dimensional variety has a double surface, triple curve, etc. Relations analogous to the characteristic of a surface in S_3 are derived. The double surface of a Del Pezzo variety must be ruled.

3. Quintic surfaces in S_3. The basic study of rational quintic surfaces was made by Montesano in a series of papers (1890-1915) by means of birational transformations of other known surfaces. Williams [92] obtains for certain of these surfaces the same results and some new ones by means of a different set of transformations. That having two distinct double lines is given in [93] and with the double lines approaching coincidence in.[94] In the latter case if the lines are coplanar, the surface is not rational but is rational when the lines are skew. Hubbs [35] considers all the rational quintic surfaces having no multiple curves. He confirms results obtained synthetically by Montesano and derives the equation of each type.

The 28 bitangents of a plane C_4 of genus three can be obtained from a net of quadric surfaces through eight points and the plane map of a cubic surface. The same method, applied to one dimension higher, gives the system of tri-tangent planes of a quintic symmetroid (Roth [56]).

It was shown by Fano [25] that the only surfaces in S_3 having projective plane sections are of one of four types: cones, quadrics, ruled cubics and the developable on the C_3 of S_3.

4. Sextic surfaces in S_3. Two papers by Souris [79, 80] describe sextic surfaces as focal surfaces of congruences of conics. Given a net of quadric surfaces with eight base points and a bundle of planes in quadratic correspondence with it, $\Sigma a_i\phi_i=0$, $\Sigma \frac{x_i}{a_i}=0$. Corresponding elements intersect in a conic, generating a congruence of order 2. Let the a_i be point coordinates in a plane (a). When two conics through P coincide, S_1 in (a) touches C_2 in (a). The locus is the focal surface. It is of order 6, has 45 conical double points and is the envelope of four systems of cubic surfaces of index two.

5. Configurations, surfaces and varieties in S_n. The general problem of the maximum number of lines that lie on a non-ruled algebraic surface is considered by Hollcroft.[32] Three categories are considered: independent, concurrent and coplanar. Dependent lines are illustrated but not completely classified.

Fano [26] finds that a configuration F_5 in S_5, discussed earlier by Baker,[5] is invariant under more collineations than indicated by Baker.

Welchman [90] extends Segre's formula to apply to curves meeting each generator of a scroll in hyperspace in two points. Todd [85] maps the system of $\infty^4|F_4|$ of S_3 through six mutually skew lines on the prime sections of a V_3^4 of S_4. It has 36 nodes instead of 20 and is determinantal. It defines a Cremona transformation of S_3 in which planes are transformed into surfaces of order 19 having the given lines five fold and other base curves. If the lines belong to a linear complex, the transformation is involutorial.

A system of lines in S_3 is said to be connected if it is possible to go from any line to the next and eventually back to the one started from. Points of intersection are not to be regarded as double points. If only consecutive lines meet and the last meets the first, a closed polygon with $p=0$ results. A necessary and sufficient condition that an n-lateral have $p=0$ is that it is simply connected. If $p>0$, the system is multiply connected (Piazzolla-Beloch [51]).

Involutions of points on an algebraic surface may be rational or irrational, except that irrational involutions do not exist on rational surfaces. It was shown by Enriques [24] that if on a regular surface, corresponding to an arbitrary n, n points can be chosen such that they all belong to the same group of $I_n{}^r$, $r\geq 2$, the surface is rational. If the given surface is irregular, it can be transformed into a ruled surface.

Maroni [45] generalizes the formula of Torrelli in various directions.

The number of apparent double points of surfaces and of varieties can be expressed when certain conditions are known. Those having but one apparent double point are determined by Edge.[21] Del Pezzo's quintic surface of S_5 is the only ruled example in S_5. Comessatti [14] considers the canonical series of an algebraic surface F. Let V be its variety of pairs of ordered points, Ω its surface of coincidences; the surface Ω is transformed into F. The characteristic series of Ω within V is the canonical series of Severi.

The surfaces of sectional genus 4 are classified into seven families of normal surfaces of hyperspace, by methods similar to those used by Castelnuovo for lower genera (Roth [61]). Lob [43] obtained the Clifford-Morley chain theorem by means of projection from higher spaces. By regarding the various terms of $|C_3|$ of a plane as point coordinates in S_9, loci representing various linear systems of cubics are discussed, with a sketch of its extension to C_n (Yerushalmy [100]).

By methods of differential geometry, Mentre [47] maps a quadratic line complex upon a non-special linear complex. The non-ordered pairs of points of a plane are mapped on the points of S_4 by Schaake.[70] Babbage [4] derives the rational double plane by means of a Veronese cone, that is, any cone in S_6 whose sections by $|S_5|$ are Veronese surfaces. The Geiser and Bertini types are derived and discussed.

The triple planes with branch curves of order not greater than 12 are discussed by Du Val.[19] If n is the number of sheets of a multiple plane with genus, by a prime section, π, then $2\pi+2n-2$ is the order of the curve of branch points. Let $\pi=0$. Then C_4 to be branch-curve of a triple plane, must have three cusps. The surface is rational; it is the projection of a ruled cubic form from a general line. If $\pi=1$, C_6 has six cusps on C_2, surface rational, or C_6 has nine cusps, surface elliptic. If $\pi=2$, C_8 is envelope of $\infty^1\{C_8\}$ index three, with 12 cusps. If $\pi=3$, C_{10} has 18 cusps, or 21 cusps. If $\pi=4$, the Grassmannian of projection must be from a restricted point.

The forms in S_r, $r>3$, may or may not contain planes. V_4 in S_r may have three, four, five or at most six. If V_4 contains six planes, and consequently 39 nodes, it contains a regulus with a generator in each plane (Roth [59]).

Du Val [20] constructs loci whose plane sections are elliptic. First, Segre's V_4^6, representing the pairs of points of two planes $\sim$ the congruence of lines meeting two planes in S_5. This can be mapped on S_4, by its trace of each S_1 of the congruence. The prime sections are represented by quadrics through two lines, traces of the directrix planes on S_4. Now let the planes approach coincidence. The problem was first studied by Scorza [72] and later by Timms.[83]

Particular systems of n-dimensional varieties containing n-systems of straight lines are developed by Barrau.[6] If x_1, x_2 are homogeneous point coordinates on a line, y_1, y_2 those of another, then $\xi_{ik}=x_iy_k$ determines a point on a quadric surface in S_3. In case of three, $\xi_{ikl}=x_iy_kz_l$ defines a sextic three dimensional variety in S_7. This is generalized to S_n.

The theory of correspondence as applied to curves is extensively developed but that of surfaces is still almost unknown, except for those invariant under continuous birational transformations, and certain involutions. The chief difficulty is not knowing valence on a surface. Albanese [1] considers two distinct algebraic surfaces, F, F' with general moduli, in (α, β) correspondence, and shows by transcendental methods that the valence is necessarily zero. On a surface of general moduli every correspondence is a valence correspondence.

The order of the double curve on the surface of trisecants of a given curve as given by Cayley (*Collected Mathematical Papers,* V, 187) and by Salmon-Rogers (*Analytic Geometry of Three Dimensions,* II, 97) is incorrect. Gambier [27] discusses types of singularities on ruled surfaces by means of plane birational transformations. Tacnodal and oscnodal conics may appear, which were not included in the classification of Edge [22]; *see also* Snyder.[78]

The harmonic ruled surfaces were found by Chiellini.[12] The hyperbolic paraboloid is the only algebraic surface. An additional note is given by Hilton.[31]

Relations among certain systems of curves enabled Segre [74] to determine a condition for regularity of a linear system of forms and of surfaces having their canonical system compounded with an involution. The varieties of Jacobi have geometric genus one and canonical variety of order two.

Severi [75] gives a comprehensive résumé of the foundations of algebraic geometry.

Sisam [77] shows that if through each point of a variety there pass just six right lines which lie on the variety, then the variety is a cubic hypersurface in four dimensions. The same author [76] gives the condition that a non-composite, non-conical three dimensional variety of order 5 be ruled. Roth [60] in studying ruled forms of four dimensions, distinguishes between those generated by a congruence, thus having ∞^2 lines, and those having ∞^3 lines, generated by ∞^1 planes. This is followed by a general study of possible types and a full set of references.

By a precisely similar method to Steiner's well known generation of the cubic surface, Roth [62] obtains from a polarity in S_n a rational form $V_{n-1}{}^n$, the points of which correspond to those of a given prime. The discussion is restricted to the four-dimensional case. Properties similar to those of Steiner are found; *see also*.[63] The same author [67] develops formulas concerned with the singularities of a primal in S_4 whose space section is a surface containing a double curve of general type; such a primal, therefore, has a double surface on which there is a triple curve containing a number of quadruple points. It is thus a particular case of the primal to be considered in developing the theory of reciprocal polars in four dimensions. In fact, some of the equations given are the first steps toward such a theory.

Schur's sextic is generalized to a higher space by Roth,[66] who shows that if the points of S_n are placed in (1, 1) correspondence with the quadrics of a given Σ_n in S_{n-1}, then to the cones of Σ_n correspond the points of a rational $V_{n-1}{}^n$. Room [54] represents the (self-dual) double-six on the Grassmannian manifold. There exist simplexes which at the same time are inscribed in and circumscribed about the manifolds in S_N whose points represent S_m's of S_n, where $N+1=\binom{n+1}{m+1}$. That in S_n there is one and only one normal rational curve osculating $r+4$ S_r's, where $n=r^2+4r+1$, was proved by Grace.[29] Room [55] proves the same theorem by a different method, with extensions.

A figure in S_9, consisting of 16 points and 16 S_4's, which can be derived from Kummer's configuration and which projects into it from a suitable chosen S_5 is discussed by Todd.[86] It is shown that all 16_6 configurations in S_3 may be derived by projection from the same figure in S_9 by a suitable choice of the vertex of projection.

It is shown by Hanumanta-Rao [30] that the planes of S_5 can be represented by the locus V_9 in space of 19 dimensions. This locus is a double locus on a certain other manifold V_{14}, and the tangent spaces of V_9 generate a W_{18}. Segre has a memoir [Ann. di Mat. 27: 75-123 (1918)] on the subject of these loci, in which he arrives at his results by a series of short steps.

In Segre's representation of the lines of S_3 by the points of a quadric in S_5, the chords of a rational C_3 are represented by the points of a Veronese surface $V_2{}^4$ lying on the quadric (where the $V_2{}^4$ is the map in S_5 of a plane by means of its conics). Brown [11] generalizes Segre's results. The S_k's of S_n may be considered as points of a (Grassmannian) variety. $V_{(n-k)(k+1)}$ in S_x, $x=\binom{n+1}{k+1}-1$. Those S_k's of S_n which are chordal ($k+1$ secant) to a rational normal C_n in S_n are represented by the points of a Veronese variety V_{k+1} lying in the Grassmannian variety (where the Veronese variety V_{k+1} is the map in S_x of an S_{k+1} by means of its $(n-k)-ic$ primals).

The number of $(q+1)$ – secant S_{q-1}'s of the non-developable varieties V_k^n which are rational loci each of $\infty^1 (k-1)$ – spaces is shown by Wong[98] to be $\binom{n-qk}{q+1}$ in a space of $qk+q+k+1$ dimensions. The number of apparent s-fold points on a $V^m{}_{(s-1)t}$ in S_{st+1} is finite. Wong[99] finds this number and also determines its upper and lower limits.

Roth[64] shows that a plane form is the locus generated by ∞^1 planes, and in four-dimensions is a dual construct of a scroll. An arbitrary space section is a scroll of general type in ordinary space, the enumerative properties of which have been exhaustively treated by James.[37, 38] Further a paper by James[36] on its dual form is complementary to the above paper. Together they give a complete account of scrolls and place forms in four dimensions. The last named paper depends on some formulas for the intersection of line systems in higher space found by James.[38]

Some formulas for multiple tangents and multiple secants of surfaces in higher space are found by Roth[65] by means of the "functional method." The same author[65.1] supplements the above work except that in plane of "functional method," formulas for multiple tangents to a surface in n dimensions $(4 \leqq n \leqq 6)$ may be deduced from the formulas for multiple secants by means of elementary correspondence processes.

The invariant postulation of a manifold of stated nature, but undefined position, on a hypersurface is discussed by Hollcroft.[33] The same author[34] uses the above method in finding the properties of hypersurfaces of order n in n dimensions. Properties of a general hypersurface of order four in S_4 are discussed in detail.

From differential geometry Sun[81] establishes certain properties of the curve of intersection of a surface and its tangent plane. The curve of intersection expressed parametrically is

$$y = -2bx^2/3 - b_u x^3/6 - \ldots$$
$$x = -2a'y^2/3 - a_u y^3/6 - \ldots$$

It is the osculating nodal cubic. Takami[82] shows that if P', P'' be the two residual points of intersection of F_5 and a principal tangent at P on F_5, then the locus L of P such that $P = \frac{1}{2}(P' + P'')$ is a curve of order 200 which touches the parabolic curve of F_5 in 1200 points.

Segre[73] shows that there are projective transformations which preserve lines of curvature. By the transformation $x'_i = 1/x_i$ certain surfaces $\sim$ duals. Godeaux[28] maps the images of sections of genus zero by a system of F_4 and of F_3 on S_9. The Jacobian of $\infty^3|H_2|$ is birationally equivalent to a double quadric having a curve of branch points of order eight.

A complete enumeration of the F_8 with an elliptic pencil of conics is given by Longo.[44]

The surfaces whose prime sections are hyperelliptic curves have been classified by Castlenuovo [Pal. Circ. Mat. 4: 73-88 (1890)].

Those of genus p are rational except ruled surfaces and two sheeted surfaces. There are exceptions for $p=1$. Such surfaces contain pencils of conics. Two plane representations $A \equiv |C_{p+3}| : P^{p+1}P^2$ and $(B_k) \equiv |C_{p+2+k}| : P^{p+k}kP^2$. Every (B_k), $k \geq 2$ is a special case of (B_{k-2}) and (B_1) is included in (A) (du Val[18]). Bronowski[9] shows that the theorem of Palatini [Torino Atti 41: 634 (1905-6)] must be modified.

Campedelli[11.1] shows that the theorem of Enriques that "if an algebraic surface has any linear system of curves of genus π, grade n, $n > 2\pi - 2$, it can be mapped on a ruled surface" is true without restriction.

Enriques[24.1] showed that elliptic surfaces $p_g=0$, $p_a=1$, with canonical curves of virtual order zero are distributed into seven families, representable on double, triple or quadruple planes. The surfaces of hyperspace with a triply infinite system of rational normal curves have been found by Bompiani.[8.1] They include two ∞^1 systems of rational normal curves C^v, $C^{v'}$ with $v+v'=\mu$, belonging to spaces of dimensions $(v+1)(v'+1)-1$ and having the order $2vv'$. They represent plane curves of order μ with one multiple point of order v, and one of order v'. By means of the notation developed by Turnbull (Determinants, Matrices and Invariants), Saddler[68.1] considers various forms of the equation of a cubic surface. The envelope of a threefold cutting four lines in four points which lie in a plane is a cubic manifold. Some of the generalizations to higher space lead only to particular cases.

Consider the system of quadric surfaces expressed in plane coordinates, and having two essential linear parameters. They all touch eight associated planes. They intersect in pairs in C_4, $p=1$ all having the same cross-ratio, equal to that of the tetrahedral complex defined by the system of quadrics. (Danielson.[16.1]) Edge[22.1] maps the ∞^2 quadric surfaces of a net on the points of a plane. The cones $\sim$ points of C_4, in (1, 1) correspondence with a C_6 of S_3. The contact C_3 of C_4 associated with a tritangent plane of C_6 is an in-and-circumscribed triangle of C_4. There are eight such triangles of C_4 belonging to the system of contact cubics. The quadrics of the system having ∞^1 of index 2 envelope an octadic surface with the base points of the quadrics for nodes. de Vries[88.1] maps a monoid upon a plane by a central projection from its vertex; special discussions of orders 3, 4, 5. Roth[67.1] classifies the surfaces of sectional genus 5 and order 7 into 13 types. Those of lower order were known before.[57.1]

BIBLIOGRAPHY

1. Albanese, G. Boll. U. M. It. 11: 131-138 (1932).
2. Anderson, J. Edinb. M. S. Proc. (2) 3: 289-295 (1933).
3. Babbage, D. W. Camb. Ph. Soc. Proc. 29: 212-230 (1933).
4. ———. London M. S. J. 8: 130-133 (1933).
5. Baker, H. F. London M. S. J. 6: 176-185 (1931).
6. Barrau, J. A. Amst. K. Ak. Wet. Proc. 35: 61-69 (1932).
7. Bishara, S. London M. S. J. 8: 178 (1933).
8. ———. London M. S. Proc. (2) 35: 241-248 (1933).

8.1 Bompiani, E. Rom. Acc. L. Rend. (6) 19: 89-93 (1934).

9. Bronowski, J. Camb. Ph. Soc. Proc. 29: 69 (1933).
10. ———. London M. S. J. 8: 308-312 (1933).
11. Brown, L. M. London M. S. J. 5: 168-176 (1930).

11.1 Campedelli, L. Rom. Acc. L. Rend. (6) 18: 377-381 (1933).

12. Chiellini, A. Boll. U. M. It. 11: 201-205 (1932).
13. Ciani, E. Lomb. Inst. Rend. (2) 63: 755-760 (1930).
14. Comessatti, A. Rom. Acc. L. Rend. (6) 16: 555-560 (1932).
15. Danielsson, O. Math. Ann. 102: 790-795 (1930).
16. ———. VII Skan. Math. Congress, 31-37 (1930).

16.1 ———. Math. Ann. 109: 521-524 (1933-4).

17. Deweck, M. Mathesis 43: 239-245 (1929).
18. Du Val, P. London M. S. J. 8: 306-307 (1933).
19. ———. London M. S. J. 8: 199-206 (1933).
20. ———. London M. S. J. 8: 206-210 (1933).
21. Edge, W. L. Camb. Ph. Soc. Proc. 28: 285-299 (1931).
22. ———. Ruled Surfaces, p. 324 (1931).

22.1 ———. London M. S. J. 9: 50-54 (1933).

23. Emch, A. Am. M. S. Bull. 30: 226 (1924).
24. Enriques, F. Rom. Acc. L. Rend. (6) 17: 109-111 (1933).

24.1 ———. Rom. Acc. L. Rend. (6) 17: 195-199 (1933).

25. Fano, G. Rom. Acc. L. Rend. (6) 1: 473-477 (1925).
26. ———. Lomb. Ist. Rend. (2) 65: 93-96 (1932).
27. Gambier, B. C. R. 195: 745-747 (1932).
28. Godeaux, L. Brux. Bull. (5) 13: 114-133 (1927).
29. Grace, J. H. Camb. Ph. Soc. Proc. 25: 421 (1929).
30. Hanumanta Rao, C. V. Camb. Ph. Soc. Proc. 26: 72-81 (1930).
31. Hilton, H. Boll. U. M. It. 12: 138-139 (1933).
32. Hollcroft, T. R. Mess. of Math. 81-88 (1928).
33. ———. Am. M. S. Bull. 36: 421-426 (1930).
34. ———. Tôhoku M. J. 35: 1-7 (1932).
35. Hubbs, H. N. Am. M. S. Bull. 38: 295-302 (1931).
36. James, C. G. F. London M. S. Proc. (2) 27: 513-540 (1928).
37. ———. Camb. Ph. Soc. Trans. 23: 201 (1925).
38. ———. London M. S. Proc. (2) 24: 359-380 (1925).
39. ———. London M. S. Proc. (2) 27: 513-540 (1928).
40. ———. Camb. Ph. Soc. Proc. 25: 390 (1929).
41. Joliffe, A. E. Mess. of Math. 55: 136-139 (1925).
42. Kubota, T. Tôhoku Sc. R. (1) 14: 391-398 (1925).
43. Lob. H. Camb. Ph. Soc. Proc. 29: 45-51 (1933).

44. Longo, Concettina. Cat. Acc. Gio. Atti. (5) 15, No. 2 bis (1925-26).
45. Maroni, A. Ann. di Mat. (4) 10: 125-144 (1931).
46. Mathews, R. M. Am. M. S. Bull. 34: 646-648 (1928).
47. Mentre, P. C. R. 193: 1315-1316 (1931).
48. Paul, M. Mathesis 46: 215-218 (1932).
49. Piazzolla-Beloch, M. Pal. Circ. Mat. 54: 83-88 (1928).
50. ———. C. R. 187: 749 (1928).
51. ———. Rom. Acc. L. Rend. (6) 12: 380-386 (1930).
52. Richmond, H. W. London M. S. J. 7: 113-117 (1931).
53. Room, T. G. London M. S. J. 7: 112 (1931).
54. ———. London M. S. J. 7: 4-6 (1931).
55. ———. London M. S. Proc. (2) 33: 367-383 (1931).
56. Roth, L. London M. S. Proc. (2) 30: 425-432 (1930).
57. ———. Camb. Ph. Soc. Proc. 28: 300-310 (1932).
57.1 ———. Camb. Ph. Soc. Proc. 29: 88-98 (1933).
58. ———. Camb. Ph. Soc. Proc. 29: 178-183 (1933).
59. ———. London M. S. Proc. (2) 35: 249-256 (1933).
60. ———. London M. S. Proc. (2) 35: 380-406 (1933).
61. ———. Camb. Ph. Soc. Proc. 29: 184-194 (1933).
62. ———. London M. S. Proc. (2) 30: 118-126 (1930).
63. ———. London M. S. Proc. (2) 30: 305-318 (1930).
64. ———. London M. S. Proc. (2) 33: 115-144 (1932).
65. ———. Camb. Ph. Soc. Proc. 25: 390 (1929).
65.1 ———. London M. S. Proc (2) 32: 154-162 (1931).
66. ———. London M. S. J. 5: 291-295 (1930).
67. ———. London M. S. Proc. (2) 35: 540-550 (1933).
67.1 ———. Camb. Ph. Soc. Proc. 30: 123-133 (1933-4).
68. Saddler, W. Edinb. M. S. Proc. (2) 1: 204-208 (1927-29).
68.1 ———. Edinb. M. S. Proc. (2) 4: 1-11 (1934).
69. Schaake, G. Amst. K. Ak. Wet. Versl. 33: 160-168 (1924).
70. ———. Amst. K. Ak. Wet. Proc. 34: 1116-1123 (1931).
71. ———. Amst. K. Ak. Wet. Proc. 35: 156-157 (1932).
72. Scorza, G. Ann. di Mat. (3) 15: 217 (1908).
73. Segre, B. Boll. U. M. It. 11: 3-8 (1932).
74. ———. Boll. U. M. It. 11: 265-267 (1932).
75. Severi, F. Abh. Hamb. Sem. 9: 335-364 (1933).
76. Sisam, C. H. Mess. of Math. 58: 89-96 (1928).
77. ———. Am. J. 52: 607-610 (1930).
78. Snyder, V. Am. M. S. Bull. 37: 791-793 (1931).
79. Souris, R. Brux. Bull. (5) 17: 1365-1377 (1931).
80. ———. Brux. Bull. (5) 18: 71-82 (1932).
81. Sun, D. Tôhoku M. J. 38: 245-251 (1933-34).
82. Takami, M. Tôhoku M. J. 38: 279-287 (1933-34).
83. Timms, G. London R. S. Proc. 119: 213-248 (1928).
84. Todd, J. A. Camb. Ph. Soc. Proc. 26: 323-333 (1930).
85. ———. Camb. Ph. Soc. Proc. 29: 52-68 (1933).
86. ———. London M. S. J. 8: 126-129 (1933).
87. Turnbull, H. W. Camb. Ph. Soc. Proc. 22: 481-487 (1925).
88. Ursell, H. D. Camb. Ph. Soc. Proc. 25: 31-38 (1928-29).
88.1 de Vries, J. Amst. K. Ak. Wet. Proc. 37: 47-50 (1933-4).

89. Welchman, W. G. Camb. Ph. Soc. Proc. 29: 235-244 (1933).
90. ———. Camb. Ph. Soc. Proc. 29: 382-388 (1933).
91. Waerden, v. d. B. L. Zent. blatt. f. Math. 3: 130 (1930).
92. Williams, A. R. Am. M. S. Bull. 34: 761-770 (1928).
93. ———. Am. M. S. Bull. 34: 631-639 (1928).
94. ———. Am. M. S. Bull. 34: 436 (1928) (Abstract).
95. Wiman, A. Math. Ann. 108: 420-433 (1932-33).
96. ———. Acta. Math. 57: 341-387 (1931).
97. ———. Acta. Math. 59: 1-62 (1932).
98. Wong, B. C. Am. M. S. Bull. 39: 392-394 (1933).
99. ———. Am. M. S. Bull. 36: 102-106 (1930).
100. Yerushalmy, J. Am. J. 54: 129-144 (1932).

NOTES: 1. The arrangement is alphabetical, by author. **2.** The numbers such as 8284-0189-6 are the Standard Book Numbers, intended to identify every book published in the U.S.A. The "8284" part of the number refers to Chelsea Publishing Company, the "0189" is Chelsea's internal identification number, and "6" is a so-called "check" digit. **3.** Additional titles are in preparation and will be announced from time to time. **4.** The prices are subject to change without notice.

THE THEORY OF SETS AND FUNCTIONS

By P. S. ALEKSANDROV

Translated from the Russian by PROF. B. D. SECKLER.

—Approx. 400 pp. 6x9. 8284-0189-6. **In prep.**

TOPOLOGIE

By P. ALEXANDROFF and H. HOPF

A standard work.

Partial Contents: PART ONE (Chaps. I and II): Foundations of Set-Theoretic Topology. PART TWO (Chaps. III-VII): The Topology of Complexes. PART THREE (Chaps. VIII-X): Topological Invariance and Related Topics. PART FOUR (Chaps. XI-XIV): Linking in Euclidean Space. Continuous Mappings of Polyhedra. (XI. Linking Theory. The Alexander Duality Theorem. . . . XIII. Homotopy and Extension Theorems for Mappings. XIV. Fixed Points.) APPENDIXES.

—1935-65. xiii + 636 pp. 6x9. 8284-0197-7. **$9.50**

NIEDERE ZAHLENTHEORIE

By P. BACHMAN

TWO VOLUMES IN ONE.

—2 vols. in 1. 1902/10-68. 902 pp. 5⅜x8. 8284-0217-5. **$15.00**

ZAHLENTHEORIE

By P. BACHMANN

A standard work.

—6 vols. in 5. Vol. I (Elements of Number Theory), xii + 284 pp. Vol. II (Analytical Number Theory), xviii + 494 pp. Vol. III (Kreisteilung), xii + 299 + 2 pp. Vol. IV, pt 1 (The Arithmetic of Quadratic Forms), xvi + 688 pp. Vol. IV, pt 2 (The Arithmetic of Quadratic Forms), xxii + 537 pp. Vol. V (General Arithmetic of Number Fields), xxii + 548 pp. 1892/1894/1872/1898/1923/1905-69. 5⅜x8. Vols. I and II will be issued in one volume. Five vol. set. **In prep.**

STRING FIGURES, and other monographs

By BALL, CAJORI, CARSLAW, and PETERSEN

FOUR VOLUMES IN ONE:

String Figures, *by W. W. Rouse Ball;*

The Elements of Non-Euclidean Plane Geometry, *by H. S. Carslaw;*

A History of the Logarithmic Slide Rule, *by F. Cajori;*

Methods and Theories for the Solution of Problems of Geometrical Construction, *by J. Petersen*

—528 pp. 5⅜x8. 8284-0130-6. Four vols. in one. **$6.00**

THEORIE DES OPERATIONS LINEAIRES

By S. BANACH

—1933-63. xii + 250 pp. 5⅜x8. 8284-0110-1. **$4.95**

DIFFERENTIAL EQUATIONS

By H. BATEMAN

CHAPTER HEADINGS: I. Differential Equations and their Solutions. II. Integrating Factors. III. Transformations. IV. Geometrical Applications. V. Diff. Eqs. with Particular Solutions of a Specified Type. VI. Partial Diff. Eqs. VII. Total Diff. Eqs. VIII. Partial Diff. Eqs. of the Second Order. IX. Integration in Series. X. The Solution of Linear Diff. Eqs. by Means of Definite Integrals. XI. The Mechanical Integration of Diff. Eqs.

—1917-67. xi + 306 pp. 5⅜x8. 8284-0190-X. **$4.95**

ERGODENTHEORIE, u.a.

By H. BEHNKE, P. THULLEN, and E. HOPF

TWO VOLUMES IN ONE.

THEORIE DER FUNKTIONEN MEHRERER KOMPLEXER VERAENDERLICHEN, by *H. Behnke* and *P. Thullen.*

ERGODENTHEORIE, by *E. Hopf.*

—1934/1937-68. 210 pp. 5½x8½. 8284-0068-7. Two vols. in one. **Prob. $4.95**

L'APPROXIMATION

By S. BERNSTEIN and CH. de LA VALLÉE POUSSIN

TWO VOLUMES IN ONE:

Leçons sur les Propriétés Extrémales et la Meilleure Approximation des Fonctions Analytiques d'une Variable Réelle, *by Bernstein.*

Leçons sur l'approximation des Fonctions d'une Variable Réelle, *by Vallée Poussin.*

—1925/1919-69. 363 pp. 6x9. 8284-0198-5. Two vols in one. **In prep.**

CONFORMAL MAPPING

By L. BIEBERBACH

Translated from the fourth German edition by F. STEINHARDT.

Partial Contents: I. Foundations; Linear Functions (Analytic functions and conformal mapping, Integral linear functions, $w=1/z$, Linear functions, Groups of linear functions). II. Rational Functions. III. General Considerations (Relation between c. m. of boundary and of interior of region, Schwarz's principle). IV. Further Study of Mappings ($w=z+1/z$, Exponential function, Trigonometric functions, Elliptic integral of first kind). V. Mappings of Given Regions (Illustrations, Vitali's theorem, Proof of Riemann's theorem, Actual constructions, Potential-theoretic considerations, Distortion theorems, Uniformization, Mapping of multiply-connected plane regions onto canonical regions).

"Presented in very attractive and readable form."—*Math. Gazette.*

—1952. vi + 234 pp. 4½x6½. 8284-0090-3. Cloth **$3.25**
8284-0176-4. Paper **$1.50**

BASIC GEOMETRY

By G. D. BIRKHOFF and R. BEATLEY

"is in accord with the present approach to plane geometry. It offers a sound mathematical development . . . and at the same time enables the student to move rapidly into the heart of geometry."—*The Mathematics Teacher.*

"should be required reading for every teacher of Geometry."—*Mathematical Gazette.*

—3rd ed. 1959. 294 pp. 5⅜x8. 8284-0120-9. **$6.00**

VORLESUNGEN UEBER DIFFERENTIALGEOMETRIE, Vols. I, II

By W. BLASCHKE

TWO VOLUMES IN ONE.

Partial Contents: VOL. I. 1. Theory of Curves. 2. Extremal Curves. 3. Strips. 4. Theory of Surfaces. 5. Invariant Derivatives on a Surface. 6. Geometry on a Surface. 7. On the Theory of Surfaces in the Large. 8. Extremal Surfaces. 9. Line Geometry.

VOL. II. *Affine Differential Geometry.* 1. Plane Curves in the Small. 2. Plane Curves in the Large. 3. Space Curves. 5. Theory of Surfaces. 6. Extremal Surfaces. 7. Special Surfaces.

—3rd ed. (Vol. I); 2nd ed. (Vol. II). 1930/23-67. 589 pp. 6x9. 8284-0202-7. Two vols. in one. **$12.00**

KREIS UND KUGEL

By W. BLASCHKE

Isoperimetric properties of the circle and sphere, the (Brunn-Minkowski) theory of convex bodies, and differential-geometric properties (in the large) of convex bodies. A standard work.

—x + 169 pp. 5½x8½. 8284-0059-8. Cloth **$3.50**
8284-0115-2. Paper **$1.50**

INTEGRALGEOMETRIE

By W. BLASCHKE and E. KÄHLER

THREE VOLUMES IN ONE.

VORLESUNGEN UEBER INTEGRALGEOMETRIE, Vols. I and II, by *W. Blaschke.*

EINFUEHRUNG IN DIE THEORIE DER SYSTEME VON DIFFERENTIALGLEICHUNGEN, by *E. Kähler.*

—1936/37/34-49. 222 pp. 5½x8½. 8284-0064-4. Three vols. in one. **$6.00**

FUNDAMENTAL EXISTENCE THEOREMS, *by G. A. BLISS.* See EVANS

VORLESUNGEN UEBER FOURIERSCHE INTEGRALE

By S. BOCHNER

—1932-48. 237 pp. 5½x8½. 8284-0042-3.

ALMOST PERIODIC FUNCTIONS

By H. BOHR

Translated by H. COHN. From the famous series *Ergebnisse der Mathematik und ihrer Grenzgebiete,* a beautiful exposition of the theory of Almost Periodic Functions written by the creator of that theory.

—1951-66. 120 pp. 6x9. Lithotyped. 8284-0027-X. **$2.95**

WISSENSCHAFTLICHE ABHANDLUNGEN

By L. BOLTZMANN

—1909-68. 1,976 pp. 5⅜x8. 8284-0215-9.
Three vol. set. **$39.50**

LECTURES ON THE CALCULUS OF VARIATIONS

By O. BOLZA

A standard text by a major contributor to the theory. Suitable for individual study by anyone with a good background in the Calculus and the elements of Real Variables.

—2nd (corr.) ed. 1961. 280 pp. 5⅜x8. 8284-0145-4.
Cloth **$4.50**
8284-0152-7. Paper **$1.19**

VORLESUNGEN UEBER VARIATIONSRECHNUNG

By O. BOLZA

A standard text and reference work, by one of the major contributors to the theory.

—1909-63. ix + 715 pp. 5⅜x8. 8284-0160-8. **$9.50**

THEORIE DER KONVEXEN KOERPER

By T. BONNESEN and W. FENCHEL

"Remarkable monograph."

—*J. D. Tamarkin, Bulletin of the A. M. S.*

—1934. 171 pp. 5½x8½. 8284-0054-7. **$4.50**

THE CALCULUS OF FINITE DIFFERENCES

By G. BOOLE

A standard work on the subject of finite differences and difference equations by one of the seminal minds in the field of finite mathematics.

Numerous exercises with answers.

—4th ed. 1958. xii + 336 pp. 5⅜x8. 8284-0121-7.
Cloth **$4.95**
8284-0148-9. Paper **$1.39**

A TREATISE ON DIFFERENTIAL EQUATIONS

By G. BOOLE

Including the Supplementary Volume.

—5th ed. 1959. xxiv + 735 pp. 5⅜x8. 8284-0128-4. **$6.95**

HISTORY OF SLIDE RULE, *By F. CAJORI.* See BALL

INTRODUCTORY TREATISE ON LIE'S THEORY OF FINITE CONTINUOUS TRANSFORMATION GROUPS

By J. E. CAMPBELL

Partial Contents: CHAP. I. Definitions and Simple Examples of Groups. II. Elementary Illustrations of Principle of Extended Point Transformations. III. Generation of Group from Its Infinitesimal Transformations. V. Structure Constants. VI. Complete Systems of Differential Equations. VII. Diff. Eqs. Admitting Known Transf. Groups. VIII. Invariant Theory of Groups. IX. Primitive and Stationary Groups. XI. Isomorphism. XIII. Construction of Groups from Their Structure Constants . . . XIV. Pfaff's Equation . . . XV. Complete Systems of Homogeneous Functions. XVI. Contact Transformations. XVII. Geometry of C. T.'s. XVIII. Infinitesimal C. T's. XX. Differential Invariants. XXI.-XXIV. Groups of Line, of Plane, of Space. XXV. Certain Linear Groups.

—1903-66. xx + 416 pp. 5⅜x8. 8284-0183-7. **$6.50**

THEORY OF FUNCTIONS

By C. CARATHÉODORY

Translated by F. STEINHARDT.

Partial Contents: **Part One.** Chap. I. Algebra of Complex Numbers. II. Geometry of Complex Numbers. III. Euclidean, Spherical, and Non-Euclidean Geometry. **Part Two.** Theorems from Point Set Theory and Topology. Chap. I. Sequences and Continuous Complex Functions. II. Curves and Regions. III. Line Integrals. **Part Three.** Analytic Functions. Chap. I. Foundations. II. The Maximum-modulus principle. III. Poisson Integral and Harmonic Functions. IV. Meromorphic Functions. **Part Four.** Generation of Analytic Functions by Limiting Processes. Chap. I. Uniform Convergence. II. Normal Families of Meromorphic Functions. III. Power Series. IV. Partial Fraction Decomposition and the Calculus of Residues. **Part Five.** Special Functions. Chap. I. The Exponential Function and the Trigonometric Functions. II. Logarithmic Function. III. Bernoulli Numbers and the Gamma Function.

Vol. II.: **Part Six.** Foundations of Geometric Function Theory. Chap. I. Bounded Functions. II. Conformal Mapping. III. The Mapping of the Boundary. **Part Seven.** The Triangle Function and Picard's Theorem. Chap. I. Functions of Several Complex Variables. II. Conformal Mapping of Circular-Arc Triangles. III. The Schwarz Triangle Functions and the Modular Function. IV. Essential Singularities and Picard's Theorems.

"A book by a master . . . Carathéodory himself regarded [it] as his finest achievement . . . written from a catholic point of view."—*Bulletin of A.M.S.*

—Vol. I. 2nd ed. 1958. 310 pp. 6x9. 8284-0097-0. **$6.50**

—Vol. II. 2nd ed. 1960. 220 pp. 6x9. 8284-0106-3. **$5.50**

ALGEBRAIC THEORY OF MEASURE AND INTEGRATION

By C. CARATHÉODORY

Translated from the German by FRED E. J. LINTON. By generalizing the concept of point function to that of a function over a Boolean ring ("soma" function), Prof. Carathéodory gives an algebraic treatment of measure and integration.

—1963. 378 pp. 6x9. 8284-0161-6. **$8.00**

VORLESUNGEN UBER REELLE FUNKTIONEN

By C. CARATHÉODORY

This great classic is at once a book for the beginner, a reference work for the advanced scholar and a source of inspiration for the research worker.

—3rd ed. (c.r. of 2nd). 1968. 728 pp. 5⅜x8. 8284-0038-5. **$12.00**

NON-EUCLIDEAN GEOMETRY, *by H. S. CARSLAW*. See BALL

COLLECTED PAPERS (OEUVRES)

By P. L. CHEBYSHEV

One of Russia's greatest mathematicians, Chebyshev (Tchebycheff) did work of the highest importance in the Theory of Probability, Number Theory, and other subjects. The present work contains his post-doctoral papers (sixty in number) and miscellaneous writings. The language is French, in which most of his work was originally published; those papers originally published in Russian are here presented in French translation.

—1962. Repr. of 1st ed. 1,480 pp. 5½x8¼. 8284-0157-8.
Two vol. set. **$27.50**

TEXTBOOK OF ALGEBRA

By G. CHRYSTAL

In addition to the standard topics, Chrystal's *Algebra* contains many topics not often found in an Algebra book: inequalities, the elements of substitution theory, and so forth. Especially extensive is Chrystal's treatment of infinite series, infinite products, and (finite and infinite) continued fractions.

OVER 2,400 EXERCISES (with solutions).

—7th ed. 1964. 2 vols. xxiv + 584 pp.; xxiv + 626 pp. 5⅜x8.
8284-0084-9. Cloth. Each vol. **$4.95**
8284-0181-0. Paper. Each vol. **$2.35**

MATHEMATICAL PAPERS

By W. K. CLIFFORD

One of the world's major mathematicians, Clifford's papers cover only a 15-year span, for he died at age 34. [Included in this volume is Clifford's English translation of an important paper of Riemann.]

—1882-67. 70 + 658 pp. 5⅜x8. 8284-0210-8. **$15.00**

MODERN PURE SOLID GEOMETRY

By N. A. COURT

In this second edition of this well-known book on synthetic solid geometry, the author has supplemented several of the chapters with an account of recent results.

—2nd ed. 1964. xiv + 353 pp. 5½x8¼. 8284-0147-0. **$7.50**

MODERN PURE GEOMETRY FOR HIGH-SCHOOL MATHEMATICS TEACHERS

By N. A. COURT

—1969. Approx. 100 pp. 6x9. **In prep.**

SPINNING TOPS AND GYROSCOPIC MOTION

By H. CRABTREE

Partial Contents: Introductory Chapter. CHAP. I. Rotation about a Fixed Axis. II. Representation of Angular Velocity. Precession. III. Discussion of the Phenomena Described in the Introductory Chapter. IV. Oscillations. V. Practical Applications. VI-VII. Motion of Top. VIII. Moving Axes. IX. Stability of Rotation. Periods of Oscillation. APPENDICES: I. Precession. II. Swerving of "sliced" golf ball. III. Drifting of Projectiles. IV. The Rising of a Top. V. The Gyro-compass. ANSWERS TO EXAMPLES.

—2nd ed. 1914-67. 203 pp. 6x9. 8284-0204-3. **$4.95**

GESAMMELTE MATHEMATISCHE WERKE

By R. DEDEKIND

—1930/31/32-68. 1,359 pp. 6x9. 8284-0220-5.
Three vol. set.

THE DOCTRINE OF CHANCES

By A. DE MOIVRE

In the year 1716 Abraham de Moivre published his *Doctrine of Chances,* in which the subject of Mathematical Probability took several long strides forward. A few years later came his *Treatise of Annuities.* When the third (and final) edition of the *Doctrine* was published in 1756 it appeared in one volume together with a revised edition of the work on Annuities. It is this latter two-volumes-in-one that is here presented in an exact photographic reprint.

—3rd ed. 1756-1967. xi + 348 pp. 6x9. 8284-0200-0. **$7.95**

THEORIE GENERALE DES SURFACES

By G. DARBOUX

One of the great works of the mathematical literature.

An unabridged reprint of the latest edition of *Leçons sur la Théorie générale des surfaces et les applications géométriques du Calcul infinitésimal.*

—Vol. I (2nd ed.) xii+630 pp. Vol. II (2nd ed.) xvii+584 pp. Vol. III (1st ed.) xvi+518 pp. Vol. IV. (1st ed.) xvi+537 pp.
[216] **In prep.**

COLLECTED PAPERS

By L. E. DICKSON

—1969. 4 vols. Approx. 3,400 pp. 6½x9¼. **In prep.**

HISTORY OF THE THEORY OF NUMBERS

By L. E. DICKSON

"A monumental work . . . Dickson always has in mind the needs of the investigator . . . The author has [often] expressed in a nut-shell the main results of a long and involved paper *in a much clearer way than the writer of the article did himself.* The ability to reduce complicated mathematical arguments to simple and elementary terms is highly developed in Dickson."—*Bulletin of A. M. S.*

—Vol. I (Divisibility and Primality) xii + 486 pp. Vol. II (Diophantine Analysis) xxv + 803 pp. Vol. III (Quadratic and Higher Forms) v + 313 pp. 5⅜x8. 8284-0086-5.
Three vol. set. **$22.50**

STUDIES IN THE THEORY OF NUMBERS

By L. E. DICKSON

—1930-62. viii + 230 pp. 5⅜x8. 8284-0151-9. **$4.50**

ALGEBRAIC NUMBERS

By L. E. DICKSON, et al.

TWO VOLUMES IN ONE.

Both volumes of the *Report of the Committee on Algebraic Numbers* are here included, the authors being L. E. Dickson, R. Fueter, H. H. Mitchell, H. S. Vandiver, and G. E. Wahlen.

Partial Contents: CHAP. I. Algebraic Numbers. II. Cyclotomy. III. Hensel's p-adic Numbers. IV. Fields of Functions. I′. The Class Number in the Algebraic Number Field. II′. Irregular Cyclotomic Fields and Fermat's Last Theorem.

—1923/28-67. ii + 211 pp. 5⅜x9. 8284-0211-6.
Two vols. in one. **$4.95**

INTRODUCTION TO THE THEORY OF ALGEBRAIC EQUATIONS, *by L. E. DICKSON. See* SIERPINSKI

VORLESUNGEN UEBER ZAHLENTHEORIE

By P. G. L. DIRICHLET and R. DEDEKIND

The fourth (last) edition of this great work contains, in its final form, the epoch-making "Eleventh Supplement," in which Dedekind outlines his theory of algebraic numbers.

"Gauss' *Disquisitiones Arithmeticae* has been called a 'book of seven seals.' It is hard reading, even for experts, but the treasures it contains (and partly conceals) in its concise, synthetic demonstrations are now available to all who wish to share them, largely the result of the labors of Gauss' friend and disciple, Peter Gustav Lejeune Dirichlet (1805-1859), who first broke the seven seals . . . [He] summarized his personal studies and his recasting of the *Disquisitiones* in his *Zahlentheorie.* The successive editions (1863, 1871, 1879, 1893) of this text . . . made the classical arithmetic of Gauss accessible to all without undue labor."—*E. T. Bell,* in *Men of Mathematics* and *Development of Mathematics.*

—Repr. of 4th ed. 1893-1968. xv + 657 pp. 5⅜x8.
8284-0213-2. **$13.50**

WERKE

By P. G. L. DIRICHLET

The mathematical works of P. G. Lejeune Dirichlet, edited by L. Kronecker.

—1889/97-1969. 1,086 pp. 6½x9¼. 8284-0225-6.
Two vol. set. **In prep.**

THE INTEGRAL CALCULUS

By J. W. EDWARDS

A leisurely, immensely detailed, textbook of over 1,900 pages, rich in illustrative examples and manipulative techniques and containing much interesting material that must of necessity be omitted from less comprehensive works.

There are forty large chapters in all. The earlier cover a leisurely and a more-than-usually-detailed treatment of all the elementary standard topics. Later chapters include: Jacobian Elliptic Functions, Weierstrassian Elliptic Functions, Evaluation of Definite Integrals, Harmonic Analysis, Calculus of Variations, etc. Every chapter contains many exercises (with solutions).

—2 vols. 1921/22-55. 1,922 pp. 5x8.
8284-0102-0; 8284-0105-5. Each volume **$9.50**

TRANSFORMATIONS OF SURFACES

By L. P. EISENHART

Many of the advances in the differential geometry of surfaces in the present century have had to do with transformations of surfaces of a given type into surfaces of the same type. The present book studies two types of transformation to which many, if not all, such transformations can be reduced.

—2nd (Corr.) ed. 1962. ix + 379 pp. 5⅜x8.
8284-0167-5. **$6.50**

MATHEMATISCHE ABHANDLUNGEN

By G. EISENSTEIN

—Repr. of edition of 1847. iv+336 pp. 6½x9¼. **In prep.**

INTRODUCTION TO THE ALGEBRA OF QUANTICS

By E. B. ELLIOTT

—Repr. of 2nd ed. 1913-64. xvi + 416 pp. 5⅜x8.
8284-0184-5. **$6.00**

LOGARITHMIC POTENTIAL, and Other Monographs

By G. C. EVANS and G. A. BLISS

TWO VOLUMES IN ONE.

THE LOGARITHMIC POTENTIAL, by *G. C. Evans.*

FUNDAMENTAL EXISTENCE THEOREMS, by *G. A. Bliss.*

—1927/1934-68. viii + 150 + iv + 107 pp. 5⅜x8. **In prep.**

ASYMPTOTIC SERIES

By W. B. FORD

TWO VOLUMES IN ONE: *Studies on Divergent Series and Summability* and *The Asymptotic Developments of Functions Defined by MacLaurin Series.*

—1916/1936-60. x + 341 pp. 6x9. 8285-0143-8.
Two vols. in one. **$6.50**

AUTOMORPHIC FUNCTIONS

By L. R. FORD

"Comprehensive . . . remarkably clear and explicit."—*Bulletin of the A. M. S.*

—2nd ed. (Corr. repr.) 1929-51. x + 333 pp. 5⅜x8.
8284-0085-7. **$6.50**

THE CALCULUS OF EXTENSION

By H. G. FORDER

—1941-60. xvi + 490 pp. 5⅜x8. 8284-0135-7. **$6.50**

GRUNDGESETZE DER ARITHMETIK

By G. FREGE

TWO VOLUMES IN ONE.

—1893/1903-69. xxvi + 253 + xvi + 266 pp. 5⅜x8.
Two vols. in one. **In prep.**

CURVE TRACING

By P. FROST

This much-quoted and charming treatise gives a very readable treatment of a topic that can only be touched upon briefly in courses on Analytic Geometry. Teachers will find it invaluable as supplementary reading for their more interested students and for reference. The Calculus is not used.

Partial Contents: Introductory Theorems. II. Forms of Certain Curves Near the Origin. Cusps. Tangents to Curves. Curvature. III. Curves at Great Distance from the Origin. IV. Simple Tangents. Direction and amount of Curvature. Multiple Points. Curvature at Multiple Points. VI. Asymptotes. VIII. Curvilinear Asymptotes. IX. The Analytical Triangle. X. Singular Points. XI. Systematic Tracing of Curves. XII. The Inverse Process. CLASSIFIED LIST OF THE CURVES DISCUSSED. FOLD-OUT PLATES.

Hundreds of examples are discussed in the text and illustrated in the fold-out plates.

—5th (unaltered) ed. 1960. 210 pp. + 17 fold-out plates. 5⅜x8. 8284-0140-3. **$4.50**

LECTURES ON ANALYTICAL MECHANICS

By F. R. GANTMACHER

Translated from the Russian by PROF. B. D. SECKLER, with additions and revisions by Prof. Gantmacher.

Partial Contents: CHAP. I. Differential Equations of Motion of a System of Particles. II. Equations of Motion in a Potential Field. III. Variational Principles and Integral-Invariants. IV. Canonical Transformations and the Hamilton-Jacobi Equation. V. Stable Equilibrium and Stability of Motion of a System (Lagrange's Theorem on stable equilibrium, Tests for unstable E., Theorems of Lyapunov and Chetayev, Asymptotically stable E., Stability of linear systems, Stability on basis of linear approximation, . . .). VI. Small Oscillations. VII. Systems with Cyclic Coordinates. BIBLIOGRAPHY.

—Approx. 300 pp. 6x9. 8284-0175-6. **In prep.**

THE THEORY OF MATRICES

By F. R. GANTMACHER

This treatise, by one of Russia's leading mathematicians gives, in easily accessible form, a coherent account of matrix theory with a view to applications in mathematics, theoretical physics, statistics, electrical engineering, etc. The individual chapters have been kept as far as possible independent of each other, so that the reader acquainted with the contents of Chapter I can proceed immediately to the chapters that especially interest him. Much of the material has been available until now only in the periodical literature.

Partial Contents. VOL. ONE. I. Matrices and Matrix Operations. II. The Algorithm of Gauss and Applications. III. Linear Operators in an n-Dimensional Vector Space. IV. Characteristic Polynomial and Minimal Polynomial of a Matrix (Generalized Bézout Theorem, Method of Faddeev for Simultaneous Computation of Coefficients of Characteristic Polynomial and Adjoint Matrix, . . .). V. Functions of Matrices (Various Forms of the Definition, Components, Application to Integration of System of Linear Differential Eqns, Stability of Motion, . . .). VI. Equivalent Transformations of Polynomial Matrices; Analytic Theory of Elementary Divisors. VII. The Structure of a Linear Operator in an n-Dimensional Space (Minimal Polynomial, Congruence, Factor Space, Jordan Form, Krylov's Method of Transforming Secular Eqn, . . .). VIII. Matrix Equations (Matrix Polynomial Eqns, Roots and Logarithm of Matrices, . . .). IX. Linear Operators in a Unitary Space. X. Quadratic and Hermitian Forms.

VOL. TWO. XI. Complex Symmetric, Skew-symmetric, and Orthogonal Matrices. XII. Singular Pencils of Matrices. XIII. Matrices with Non-Negative Elements (Gen'l and Spectral Properties, Reducible M's, Primitive and Imprimitive M's, Stochastic M's, Totally Non-Negative M's, . . .). XIV. Applications of the Theory of Matrices to the Investigation of Systems of Linear Differential Equations. XV. The Problem of Routh-Hurwitz and Related Questions (Routh's Algorithm, Lyapunov's Theorem, Infinite Hankel M's, Supplements to Routh-Hurwitz Theorem, Stability Criterion of Liénard and Chipart, Hurwitz Polynomials, Stieltjes' Theorem, Domain of Stability, Markov Parameters, Problem of Moments, Markov and Chebyshev Theorems, Generalized Routh-Hurwitz Problem, . . .). BIBLIOGRAPHY.

—Vol. I. 1960. x + 374 pp. 6x9. 8284-0131-4. **$7.50**
—Vol. II. 1960. x + 277 pp. 6x9. 8284-0133-0. **$6.50**

UNTERSUCHUNGEN UEBER HOEHERE ARITHMETIK

By C. F. GAUSS

In this volume are included all of Gauss's number-theoretic works: his masterpiece, *Disquisitiones Arithmeticae*, published when Gauss was only 25 years old; several papers published during the ensuing 31 years; and papers taken from material found in Gauss's handwriting after his death.

These papers (pages 457-695 of the present book) include a fourth, fifth, and sixth proof of the Quadratic Reciprocity Law, researches on biquadratic residues, quadratic forms, and other topics.

—1889-65. xv + 695 pp. 6x9. 8284-0191-8. **$8.75**

THEORY OF PROBABILITY

By B. V. GNEDENKO

This textbook, by Russia's leading probabilist, is suitable for senior undergraduate and first-year graduate courses. It covers, in highly readable form, a wide range of topics and, by carefully selected exercises and examples, keeps the reader throughout in close touch with problems in science and engineering.

The translation has been made from the fourth Russian edition by Prof. B. D. Seckler. Earlier editions have won wide and enthusiastic acceptance as a text at many leading colleges and universities.

"extremely well written . . . suitable for individual study . . . Gnedenko's book is a milestone in the writing on probability theory."—*Science*.

Partial Contents: I. The Concept of Probability (Various approaches to the definition. Space of Elementary Events. Classical Definition. Geometrical Probability. Relative Frequency. Axiomatic construction . . .). II. Sequences of Independent Trials. III Markov Chains IV. Random Variables and Distribution Functions (Continuous and discrete distributions. Multidimensional d. functions. Functions of random variables. Stieltjes integral). V. Numerical Characteristics of Random Variables (Mathematical expectation. Variance... Moments). VI. Law of Large Numbers (Mass phenomena. Tchebychev's form of law. Strong law of large numbers...). VII. Characteristic Functions (Properties. Inversion formula and uniqueness theorem. Helly's theorems. Limit theorems. Char. functs. for multidimensional random variables...). VIII. Classical Limit Theorem (Liapunov's theorem. Local limit theorem). IX. Theory of Infinitely Divisible Distribution Laws. X. Theory of Stochastic Processes (Generalized Markov equation. Continuous S. processes. Purely discontinuous S. processes. Kolmogorov-Feller equations. Homogeneous S. processes with independent increments. Stationary S. process. Stochastic integral. Spectral theorem of S. processes. Birkhoff-Khinchine ergodic theorem). XI. Elements of Queueing Theory (General characterization of the problems. Birth-and-death processes. Single-server queueing systems. Flows. Elements of the theory of stand-by systems). XII. Elements of Statistics (Problems. Variational series. Glivenko's Theorem and Kolmogorov's criterion. Two-sample problem. Critical region . . . Confidence limits). TABLES. BIBLIOGRAPHY. ANSWERS TO THE EXERCISES.

—4th ed. 1968. 527 pp. 6x9. 8284-0132-2. **$9.50**

A SHORT HISTORY OF GREEK MATHEMATICS

By J. GOW

A standard work on the history of Greek mathematics, with special emphasis on the Alexandrian school of mathematics.

—1884-68. xii + 325 pp. 5⅜x8. 8284-0218-3. **$6.50**

COMMUTATIVE NORMED RINGS

By I. M. GELFAND, D. A. RAIKOV, and G. E. SHILOV

Partial Contents: CHAPS. I AND II. General Theory of Commutative Normed Rings. III. Ring of Absolutely Integrable Functions and their Discrete Analogues. IV. Harmonic Analysis on Commutative Locally Compact Groups. V. Ring of Functions of Bounded Variation on a Line. VI. Regular Rings. VII. Rings with Uniform Convergence. VIII. Normed Rings with an Involution and their Representations. IX. Decomposition of Normed Ring into Direct Sum of Ideals. HISTORICO-BIBLIOGRAPHICAL NOTES. BIBLIOGRAPHY.

—1964. 306 pp. 6x9. 8284-0170-5. **$6.50**

THE ALGEBRA OF INVARIANTS

By J. H. GRACE and A. YOUNG

An introductory account.

Partial Contents: I. Introduction. II. The Fundamental Theorem. III. Transvectants. V. Elementary Complete Systems. VI. Gordan's Theorem. VII. The Quintic. VIII. Simultaneous Systems. IX. Hilbert's Theorem. XI. Apolarity. XII. Ternary Forms. XV. Types of Covariants. XVI. General Theorems on Quantics. APPENDICES.

—1903-65. vii + 384 pp. 5⅜x8. 8284-0180-2. **$4.95**

DIE AUSDEHNUNGSLEHRE

By H. G. GRASSMANN

—2nd ed. 1862-1969. 388 pp. 6x9.

GYROSCOPIC THEORY

By G. GREENHILL

This work is intended to serve as a collection in one place of the various methods of the theoretical explanation of the motion of a spinning body, and as a reference for mathematical formulas required in practical work.

—1914-67. vi + 277 pp. + Fold-out Plates. 6½x10¾. 8284-0205-1. **$9.50**

LES INTEGRALES DE STIELTJES et leurs Applications aux Problèmes de la Physique Mathématique

By N. GUNTHER

The present work is a reprint of Vol. I of the publications of the V. A. Steklov Institute of Mathematics, in Moscow. The text is in French.

—1932-49. 498 pp. 5⅜x8. 8284-0063-6. **$6.95**

ONDES: Leçons sur la Propagation des Ondes et les Equations de l'Hydrodynamique

By J. HADAMARD

"[Hadamard's] unusual analytic proficiency enables him to connect in a wonderful manner the physical problem of propagation of waves and the mathematical problem of Cauchy concerning the characteristics of partial differential equations of the second order."—*Bulletin of the A. M. S.*

—Repr. 1949. viii + 375 pp. 5½x8½. 8284-0058-X. **$6.00**

REELLE FUNKTIONEN. Punktfunktionen

By H. HAHN

—1932-48. xi + 415 pp. 5½x8½. 8284-0052-0. **$6.00**

ALGEBRAIC LOGIC

By P. R. HALMOS

"Algebraic Logic is a modern approach to some of the problems of mathematical logic, and the theory of polyadic Boolean algebras, with which this volume is mostly concerned, is intended to be an efficient way of treating algebraic logic in a unified manner.

"[The material] is accessible to a general mathematical audience; no vast knowledge of algebra or logic is required . . . Except for a slight Boolean foundation, the volume is essentially self-contained."—*From the Preface.*

—1962. 271 pp. 6x9. 8284-0154-3.

LECTURES ON ERGODIC THEORY

By P. R. HALMOS

CONTENTS: Introduction. Recurrence. Mean Convergence. Pointwise Convergence. Ergodicity. Mixing. Measure Algebras. Discrete Spectrum. Automorphisms of Compact Groups. Generalized Proper Values. Weak Topology. Weak Approximation. Uniform Topology. Uniform Approximation. Category. Invariant Measures. Generalized Ergodic Theorems. Unsolved Problems.

"Written in the pleasant, relaxed, and clear style usually associated with the author. The material is organized very well and painlessly presented."

—*Bulletin of the A.M.S.*

—1956-60. viii + 101 pp. 5⅜x8. 8284-0142-X. **$3.25**

INTRODUCTION TO HILBERT SPACE AND THE THEORY OF SPECTRAL MULTIPLICITY

By P. R. HALMOS

A clear, readable introductory treatment of Hilbert Space.

—1957. 2nd ed. (C.r. of 1st ed.) 120 pp. 6x9. 8284-0082-2. **$3.50**

ELEMENTS OF QUATERNIONS

By W. R. HAMILTON

Sir William Rowan Hamilton's last major work, and the second of his two treatises on quaternions.

—2nd ed. 1899/1901-68. 1,185 pp. 6x9. 8284-0219-1. Two vol. set. **$29.50**

RAMANUJAN:
Twelve Lectures on His Life and Works

By G. H. HARDY

The book is somewhat more than an account of the mathematical work and personality of Ramanujan; it is one of the very few full-length books of "shop talk" by an important mathematician.

—1940-59. viii + 236 pp. 6x9. 8284-0136-5. **$4.95**

GRUNDZUEGE DER MENGENLEHRE

By F. HAUSDORFF

The original, 1914 edition of this famous work contains many topics that had to be omitted from later editions, notably, the theories of content, measure, and discussion of the Lebesgue integral. Also, general topological spaces, Euclidean spaces, special methods applicable in the Euclidean plane, the algebra of sets, partially ordered sets, etc.

—1914-49. 484 pp. 5⅜x8. 8284-0061-X. **$7.50**

SET THEORY

By F. HAUSDORFF

Hausdorff's classic text-book is an inspiration and a delight. The translation is from the Third (latest) German edition.

"We wish to state without qualification that this is an indispensable book for all those interested in the theory of sets and the allied branches of real variable theory."—*Bulletin of A. M. S.*

—2nd ed. 1962. 352 pp. 6x9. 8284-0119-5. **$7.50**

VORLESUNGEN UEBER DIE THEORIE DER ALGEBRAISCHEN ZAHLEN

By E. HECKE

"An elegant and comprehensive account of the modern theory of algebraic numbers."

—*Bulletin of the A. M. S.*

—1923-69. 264 pp. 5½x8½. 8284-0046-6.

INTEGRALGLEICHUNGEN UND GLEICHUNGEN MIT UNENDLICHVIELEN UNBEKANNTEN

By E. HELLINGER and O. TOEPLITZ

"Indispensable to anybody who desires to penetrate deeply into this subject."—*Bulletin of A.M.S.*

—1928-53. 286 pp. 5⅜x8. 8284-0089-X. **$4.95**

THEORIE DER ALGEBRAISCHE FUNKTIONEN EINER VARIABELN

By K. HENSEL and G. LANDSBERG

Partial Contents: PART ONE (Chaps. 1-8): Algebraic Functions on a Riemann Surface. PART TWO (Chaps. 9-13): The Field of Algebraic Functions. PART THREE (Chaps. 14-22): Algebraic Divisors and the Riemann-Roch Theorem. PART FOUR (Chaps. 23-27): Algebraic Curves. PART FIVE (Chaps. 28-31): The Classes of Algebraic Curves. PART SIX (Chaps. 32-37): Algebraic Relations among Abelian Integrals. APPENDIX: Historical Development. Geometrical Methods. Arithmetical Methods.

—1902-65. xvi + 707 pp. 6x9. 8284-0179-9. **$9.50**

Grundzüge Einer Allgemeinen Theorie der LINEAREN INTEGRALGLEICHUNGEN

By D. HILBERT

—1912-53. 306 pp. 5½x8¼. 8284-0091-1. **$4.95**

SQUARING THE CIRCLE, and other Monographs

By HOBSON, HUDSON, SINGH, and KEMPE

FOUR VOLUMES IN ONE.

SQUARING THE CIRCLE, by *Hobson.* A fascinating account of one of the three famous problems of antiquity, its significance, its history, the mathematical work it inspired in modern times, and its eventual solution in the closing years of the last century.

RULER AND COMPASSES, by *Hudson.* "An analytical and geometrical investigation of how far Euclidean constructions can take us. It is as thoroughgoing as it is constructive."—*Sci. Monthly.*

THE THEORY AND CONSTRUCTION OF NON-DIFFERENTIABLE FUNCTIONS, by *Singh.* I. Functions Defined by Series. II. Functions Defined Geometrically. III. Functions Defined Arithmetically. IV. Properties of Non-Differentiable Functions.

HOW TO DRAW A STRAIGHT LINE, by *Kempe.* An intriguing monograph on linkages. Describes, among other things, a linkage that will trisect any angle.

"Intriguing, meaty."—*Scientific American.*

—388 pp. 4½x7½. 8284-0095-4. Four vols. in one. **$4.95**

SPHERICAL AND ELLIPSOIDAL HARMONICS

By E. W. HOBSON

"A comprehensive treatise . . . and the standard reference in its field."—*Bulletin of the A. M. S.*

—1931-65. xi + 500 pp. 5⅜x8. 8284-0104-7. **$7.50**

ELASTOKINETIK: Die Methoden zur Angenäherten Lösung von Eigenwertproblemen in der Elastokinetik

By K. HOHENEMSER

—(Erg. der Math.) 1932-49. 89 pp. 5½x8½. 8284-0055-5. **$2.75**

ERGODENTHEORIE, *by E. HOPF.* See BEHNKE

RULER AND COMPASSES, *by H. P. HUDSON.* See HOBSON

PHYSIKALISCH-MATHEMATISCHE MONOGRAPHIEN

By W. v. IGNATOWSKY, et al.

THREE VOLUMES IN ONE.

Of the many well-known monographs in the series published by the Steklov Institute of the Academy of Sciences of the U.S.S.R., only a few were originally published in a language other than Russian. Two of the French-language works have been reprinted and are listed elsewhere in this catalogue: Gunther's book on *Stieltjes Integrals* and Lappo-Danilevski's three-volume work on *Systems of Differential Equations.*

CONTENTS: 1. *Untersuchungen einiger Integrale mit Besselschen Funktionen und ihre Anwendung auf Beugungserscheinungen,* by Ignatowsky. 2. *Kreisscheibenkondensator,* by Ignatowsky. 3. *Table of a Special Function,* by Bursian and Fock.

—1932-66. 16 + 232 pp. 6¼x9¼. 8284-0201-9. Three vols. in one. **$5.50**

GESAMMELTE WERKE

By C. G. J. JACOBI

The collected mathematical works of the celebrated mathematician. [Volume 8 of the collected works is Jacobi's *Vorlesungen über Dynamik*, also available as a separate volume (see below).]

—1881/1891-1968. 4,032 pp. 6½x9¼. Eight vol. set.

VORLESUNGEN UEBER DYNAMIK

By C. G. J. JACOBI

—1884-67. viii+300 pp. 6½x9¼.

A TREATISE ON THE LINE COMPLEX

By C. M. JESSOP

"The best introduction to the subject."—*Virgil Snyder.*

—1903-68. xv + 364 pp. 5⅜x8. 8284-0223-X. **$7.95**

THE CALCULUS OF FINITE DIFFERENCES

By CHARLES JORDAN

"... destined to remain the classic treatment of the subject . . . for many years to come."—*Harry C. Carver, Founder and formerly Editor of the* ANNALS OF MATHEMATICAL STATISTICS.

—3rd ed. 1965. xxi + 655 pp. 5⅜x8. 8284-0033-4. **$7.50**

THEORIE DER ORTHOGONALREIHEN

By S. KACZMARZ and H. STEINHAUS

The theory of general orthogonal functions. *Monografje Matematyczne*, Vol. VI.

—1935-51. viii + 294 pp. 6x9. 8284-0083-0. **$4.95**

KAHLER, "System von . . . ," see Blaschke

DIFFERENTIALGLEICHUNGEN: LOESUNGSMETHODEN UND LOESUNGEN

By E. KAMKE

Everything possible that can be of use when one has a given differential equation to solve, or when one wishes to investigate that solution thoroughly.

PART A: General Methods of Solution and the Properties of the Solutions.

PART B: Boundary and Characteristic Value Problems.

PART C: Dictionary of some 1600 Equations in Lexicographical Order, with solution, techniques for solving, and references.

"A reference work of outstanding importance which should be in every mathematical library."
—*Mathematical Gazette.*

—3rd ed. 1944. 692 pp. 6x9. 8284-0044-X. **$12.50**

DIFFERENTIALGLEICHUNGEN, *by KAEHLER*.
See BLASCHKE

HOW TO DRAW A STRAIGHT LINE, *by KEMPE*.
See HOBSON

ASYMPTOTISCHE GESETZE DER WAHRSCHEINLICHKEITSRECHNUNG

By A. A. KHINTCHINE

—1933-48. 82 pp. 5½x8½. 8284-0036-9. Paper **$2.00**

VORLESUNGEN UEBER CONTINUIERLICHE GRUPPEN MIT GEOMETRISCHEN UND ANDEREN ANWENDUNGEN

By S. LIE

An expository work by the originator of the Lie theory of continuous groups.

—1893-69. xii + 810 pp. 6x9. **In prep.**

THE THEORY OF MATRICES

By C. C. MacDUFFEE

"No mathematical library can afford to be without this book."—*Bulletin of the A. M. S.*

—2nd ed. 1932-47. 116 pp. 6x9. 8284-0028-8. **$3.25**

COMBINATORY ANALYSIS, Vols. I and II

By P. A. MACMAHON

TWO VOLUMES IN ONE.

A broad and extensive treatise on an important branch of mathematics.

—1915/16-60. 660 pp. 5⅜x8. 8284-0137-3.
Two vols. in one. **$8.50**

INTRODUCTION TO COMBINATORY ANALYSIS, *by P. A. MACMAHON. See* KLEIN

FORMULAS AND THEOREMS FOR THE FUNCTIONS OF MATHEMATICAL PHYSICS

By W. MAGNUS and F. OBERHETTINGER

Gathered into a compact, handy and well-arranged reference work are thousands of results on the many important functions needed by the physicist, engineer and applied mathematician.

Translated by J. WERMER.

—1954. 182 pp. 6x9. 8284-0051-2.

THEORY OF NUMBERS

By G. B. MATHEWS

CHAPTER HEADINGS: I. Elementary Theory of Congruences. II. Quadratic Congruences. III. Binary Quadratic Forms; Analytical Theory. IV. Binary Quadratic Forms; Geometrical Theory. V. Generic Characters of Binary Quadratics. VI. Composition of Forms. VII. Cyclotomy. VIII. Determination of Number of Improperly Primitive Classes for a Given Determinant. IX. Applications of the Theory of Quadratic Forms. X. The Distribution of Primes.

—2nd ed. 1892-62. xii + 323 pp. 5⅜x8. 8284-0156-X. **$4.95**

KURVENTHEORIE

By K. MENGER

Partial Contents: CHAP. I. The Old Concept of a Curve. II. The New Concept of a Curve . . . IV. The Sum of Curves. V. Properties under Deformation, etc. VI. Fundamental Theorem of Curve-Theoretical Concept of Order . . . VIII. Regular Curves. IX. Rational Curves . . . XII. The Universal Curve.

—2nd ed. 1932-67. vi + 376 pp. 6½x9¼. 8284-0172-1.
$12.00